The Histology of Fishes

Editors

Frank Kirschbaum

Faculty of Life Sciences, Unit of Biology and Ecology of Fishes
Humboldt University
Berlin, Germany

Krzysztof Formicki

Department of Hydrobiology, Ichthyology and Biotechnology of Reproduction
West Pomeranian University of Technology in Szczecin
Szczecin, Poland

CRC Press
Taylor & Francis Group
Boca Raton London New York

CRC Press is an imprint of the
Taylor & Francis Group, an **informa** business

A SCIENCE PUBLISHERS BOOK

Cover credit:
- Top left image: Fig. 8 from Chapter 3. Reproduced by permission of the author, François J. Meunier
- Top right image: Fig. 4c from Chapter 5. Reproduced by permission of the author, Frank Kirschbaum
- Bottom left image: Fig. 9 from Chapter 10. Reproduced by permission of the authors, Mari Carmen Uribe and Harry J. Grier
- Bottom right image: Fig. 23 (Part of Inner Ear) from Chapter 16. Reproduced by permission of the author, Tanja Schulz-Mirbach

CRC Press
Taylor & Francis Group
6000 Broken Sound Parkway NW, Suite 300
Boca Raton, FL 33487-2742

First issued in paperback 2021

© 2020 by Taylor & Francis Group, LLC
CRC Press is an imprint of Taylor & Francis Group, an Informa business

No claim to original U.S. Government works

Version Date: 20190430

ISBN 13: 978-1-03-208823-5 (pbk)
ISBN 13: 978-1-4987-8447-4 (hbk)

Library of Congress Cataloging-in-Publication Data
Names: Kirschbaum, Frank, editor.
Title: The histology of fishes / editors, Frank Kirschbaum, Faculty of Life Sciences, Unit of Biology and Ecology of Fishes, Humboldt University, Berlin, Germany, Krzysztof Formicki, Department of Hydrobiology, Ichthyology and Biotechnology of Reproduction, West Pomeranian University of Technology in Szczecin, Szczecin, Poland.
Description: Boca Raton, FL : CRC Press, [2019]	"A science publishers book."	Includes bibliographical references and index.
Identifiers: LCCN 2019007827	ISBN 9781498784474 (hardback)
Subjects: LCSH: Fishes--Histology.
Classification: LCC QL639 .H57 2019	DDC 597--dc23
LC record available at https://lccn.loc.gov/2019007827

Visit the Taylor & Francis Web site at
http://www.taylorandfrancis.com

and the CRC Press Web site at
http://www.crcpress.com

Preface

The science of fine structure, development, and function of tissues and organs is histology [Greek histós 'tissue', 'fabric' and logos 'word', 'science']. The beginning of histology started with the invention of the first compound microscope at the end of the 16th century. Histology as an independent discipline of science separated in the early 20th century from descriptive anatomy. Since then, huge progress has occurred due to the development of new microscopes (transmission and scanning electronmicroscopy, confocal microscopes) and research and staining techniques (histochemistry, immunohistochemistry, fluorescence techniques) (see Chapter 1, Introduction to Histological Techniques) with a concomitant increase of our understanding of the fine structure and function of tissues and organs.

It was our conviction that this complexitity of new techniques and results in histology could only be demonstrated in a multi-author book of many specialists in the field sharing their knowledge and experience. Thus, a book with 18 chapters comprising the work of 28 authors and co-authors from 10 countries arose. We must admit that we sometimes underestimated the task and time in order to motivate all our collegues to meet the deadlines. An additional intention of this monograph was the integration of functional aspects and a comparative view on the diversity of histological features of fishes (actually more than 30.000 species described) with their amazing ecological specializations. To match with these expectations we also encouraged the authors to compile very detailed lists of references and to produce high quality figures. We therefore believe that this monograph with its eighteen chapters represents a useful compilation of the actual state-of-the-art of histology helpful for scientists and students in many disciplines. Though this monograph treats normal histology it will be useful as a guideline for fish pathology and fish histopathology.

In Chapter 1—**Introduction to Histological Techniques**—a description of the history of histology over five centuries is presented starting with the invention of the first compound microscopes at the end of the 16th century. The detailed description of the various histological techniques (fixation, staining, fluorescence techniques) represents a useful introduction for people at the start of their histological experience. The chapter is completed by the description of freezing, histochemical and immunohistochemical techniques. In Chapter 2—**Integument**—differences and common features of the integument of various taxa of fish are described comprising jawless, cartilagenous and teleost fishes. Particular intention is paid to the latest terminology of the various cell types of the epidermis. Specialisations of the integument of weakly electric mormyrid and gymnotiform fishes are also mentioned. In Chapter 3—**Fish Skeletal Tissues**—the amazing diversity, dynamics, and evolution of osteichthyan skeletal tissues is described. Fish skeletal tissues comprise mineralized bony tissues, cartilage, which can mineralize, teeth and some specialized unmineralized conjunctive tissues, which all serve various mechanical and physiological functions. It is shown that fishes consist of two main groups, those with bony tissues with embedded cells–cellular bone—and those which possess bone devoid of enclosed cell–acellular bone. The amazing characteristics of alive bony tissues is demonstrated which show a continuous more or less remodelling assuming various physiological functions, in particular exchanges between bone and the internal environment thus participating in various homeostatic activities and hormonal control. It is fascinating to learn how the improvement of techniques of sectioning undecalcified material and the recent great development of 3D computering has greatly increased our understanding of the structure and function of skeletal elements. The chapter also describes a bone anomaly termed hyperostosis which represents an abnormal swelling of bone. In Chapter 4—**Muscular System**—the three basic types of muscle, striated muscle, smooth muscle, and cardiac muscle are described both at the light and transmission electronmicroscopical level. The specialized extrinsic eye muscles are also treated. A detailed description of the complex biochemical interactions underlying the process of muscle contraction is presented. The anatomical organisation of the muscle fibers of the axial muscle in myomeres and their functional implication is described. The different muscle fiber types—slow-red fibers, intermediate or pink fibers, white or fast fibers—and their function are shown based on biochemical and histochemical characteristics. A comparison of the heart of non-osteichthyan fishes (agnatha, elasmobranchs) terminates this chapter. In Chapter 5—**Structure and Function of Electric Organs**—an overview is presented on the eight groups of electric fishes which have evolved independently based on convergent evolution. Both the anatomy and the fine structure of the electric organs are described as well as functional aspects. Differences and similarities of the organs due to the convergent development are mentioned. Ontogenetic data are presented including larval electric organs which are found both in mormyrid and gymnotiform fishes. In Chapter 6—**Digestive System**—the anatomy and histology of the gastrointestinal tract, comprised of four main parts, is discussed. The morphological and physiological adaptations of different fish species to digestion of food as well as gas exchange through the digestive tract are presented. Histology of different parts of the gastrointestinal tract is also described, indicating types of epithelial cells and functions they perform. The change of the morphology, histology, and function of the gastrointestinal tract during ontogeny is also discussed. The glands of the gastrointestinal tract are described in Chapter 7—**Glands of the Digestive Tract**—they comprise the liver and the pancreas which both are characterized by exocrine and endocrine functions. Important functions of the liver comprise synthesis and secretion of lipoproteins and proteins and neutralisation of digestion products which are described in the chapter. The digestive enzymes produced in the exocrine part of the pancreas as well as the peptide hormones delivered by the endocrine part are subject of the second part of the chapter. One element of the gastrointestinal tract which emerges during ontogeny from the esophagus and which is not treated in Chapter 6, is the swim bladder. This organ is treated in Chapter 8—**Swim Bladder**—the different functions of the swim bladder—hydrostatic organ, respiration, transport of sound waves, sound production—are described. The anatomy and histology of the swim bladder are presented as well as the process of filling the swim bladder with gas. In Chapter 9—**Kidney**—the anatomy of the kidney and the anatomy and function of the nephrons, the functional elements of the kidney, are described. But in this chapter only the kidney of teleost fishes is treated. Particular attention is paid to anatomical and functional differences of the nephrons of freshwater and marine teleosts due to the adaptation to different environmental conditions. The anatomy of fish gonads varies considerably in relation to the phylogenetic position of the taxon. In Chapter 10—**Ovaries and Eggs**—the description is restricted to teleosts. First, anatomical and histological features of the ovary are treated with mention of the adaptations that occur in viviparous teleosts. The complicated processes that occur during folliculogenesis and oogenesis are described. The six stages of oogenesis—oogonial proliferation, chromatin-nucleolus stage, previtellogenesis, vitellogenesis, maturation, ovulation—are treated in detail and are accompanied by fascinating colour photomicrographs. The ontogenetic development of the ovary is described and the cyclic ovarian changes regulated by environmental factors. One feature of eggs, their envelopes, were not treated in detail in the preceding chapter. This information is presented in Chapter 11—**Egg Envelopes**—mainly based on scanning electron microscopy, the amazing diversity of the egg envelopes and their relation to the ecological function, though often not very well understood, are presented. Similarly diverse are the micropyle structures which can be grouped into four categories. The development of the egg envelopes is also described and data on the surprising physical resistance of egg envelopes are presented. Chapter 12 describes—**Testis Structure, Spermatogenesis, and Spermatozoa in Teleost Fishes**—most fishes exhibit external fertilisation, only a smaller part practises insemination. Both aspects are treated in this chapter. Three types of testis structure are described: the anastomosing tubular testis type, the unrestricted spermatogonial testis type, and the restricted spermatogonial testis type. In the course of spermatogenesis spermatozoa differentiate into spermatocytes, spermatids and finally spermatozoa. The diversity of the ultrastructure of spermatozoa is virtually astonishing visualized in the many excellent drawings and ultrastructural pictures. This diversity is seen both in spermatozoa which fertilize the eggs in water (aquasperm) and in those which fertilize eggs internally (introsperm). Many of the fine structural details are not well understood concerning their functions. A surprizing type of spermatozoa are the aflagellate spermatozoa found in mormyroidei and the biflagellate aquasperm found in many families of the Euteleostomi. At the end of the chapter another astonishing fact is shown: In species of the family Hemitripteridae and in some of the family Cottidae two sperm types exist: so called euspermatozoa and paraspermatozoa. Additional investigations will probably reveal even more fascinating aspects of sperm ultrastructure and function. In Chapter 13—**Cardiovascular System and Blood**—the histological and structural characteristics of the fish heart are reviewed. The external and internal views of the heart reveal the presence of six distinct components: sinus venosus, atrium, atrioventricular segment, ventricle, conus arteriosus and bulbus arteriosus. The different parts of the heart are analyzed following the direction of blood, from the caudal to the cranial end. Because teleosts constitute the most evolved fish group and have experienced the widest radiation in vertebrate evolution, they constitute the backbone of this chapter. However, data from other fish groups—hagfishes, elasmobranches, Chondrostean and Holosteans—are also presented. Vascular histology, with special emphasis on the distribution of collagen and elastin, and structural details of the blood cells across several fish groups, are also treated. In Chapter 14—**Immune System of Fish**—it is shown that the immune system of fish has two components, the non-specific and the specific immunity. Both systems act together. The non-specific immunity is a very affective system acting in a non-specific way against pathogens and it is comprised of transferrins and lectins, lytic enzymes (chitinase, lyzosyme) and phagocytotic activity and it is a characteristic feature of fish. But there also exists in addition a great diversity of lymphoid and lymphomyeloid organs comprizing the thymus, kidney and spleen, the Leydig organ, the epigonal organ, the pericardial tissue, the orbital and the preorbital tissues. However, the specific immunity of fish is weaker than in higher vertebrates. These two kinds of immune system in their common action probably represent one aspect of the evolutionary success of fish. In Chapter 15—**Gills: Respiration and Ionic-Osmoregulation**—the remarkable ability of fish to uptake the O_2 needs for aerobic metabolism and to keep ionic and osmotic levels independent of the diverse characteristics of the habitat they inhabit are described. The adaptations of marine and freshwater elasmobranchs are treated, as well as adaptations of euryhyaline and marine teleosts, but emphasis is put on freshwater teleosts. In some fish the gills as the main organ for respiration do not fulfil this task sufficiently and therefore they have to emerge from the water and take O_2 from atmospheric air using air-breathing organs. The remarkable gill plasticity is described in euryhyaline teleosts and in carp and goldfish after exposure to hypoxia and changes in temperature leading to gill remodelling and intense reversible interlamellar cell proliferation thus reducing or increasing the respiratory surface area. In Chapter 16—**Sensory Organs**—the different categories of the sensory organs are described: they comprise chemoreceptors (olfactory and gustatory systems), photoreceptors (visual system), and mechanoreceptors (auditory and mechanosensory lateral line systems), which are found in all species, and electroreceptors, which are found only in a small subset of taxa. Sensory systems allow fishes to perceive and interpret their complex sensory environments comprising a remarkably diverse range of marine, estuarine, and freshwater habitats, so it is not surprising that their sensory systems are structurally and functionally diverse which is demonstrated in an impressive way in this chapter based on a range of methods for the visualization of cells and tissues like light and electron microscopy, µCT, etc.

The central nervous system, in particular the brain, integrates sensory information and transforms them into appropriate behaviours. Due to the morphological plasticity of the brain the ecological and sensory specialisations of fish can be visualised in brain morphology. This is exemplified in Chapter 17—**Morphology and Ecomorphology of the Fish Brain: The Rhombencephalon of Actinopterygians**—first, an overview on the various fish taxa of high rank is presented which show the surprising variation in external brain morphology as an reflection of the species biology and the modification of the primary sensory and motor centers in response to environmental needs. In the main part of the chapter the morphological and histological features of the fish brain with special reference to the brainstem in relation to ecomorpholgy are provided. In a comparative way five actinopterygian species ranging from non-specialized to extremely specialized species belonging to major taxa were chosen as model objects for the brainstem description: (1) The senegal bichir (*Polypterus senegalus*), (2) the Siberian sturgeon (*Acipenser baerii*), (3) the spotted gar (*Lepisosteus oculatus*), (4) the European eel (*Anguilla anguilla*), (5) the common carp (*Cyprinus carpio*). Briefly mentioned are the brain of cartilaginous fish (Chondrichthyes), coelacanths, and lungfishes for additional comparative purposes. In Chapter 18—**Endocrine System**—the description is restricted to the endocrine system of teleosts. They possess, apart from the classical endocrine organs of higher vertebrates, additional organs like the corpuscles of Stannius, the urophysis, the ultimobranchial body, and the pseudobranch. It is shown that the endocrine system, together with the central nervous system, is organized in a hierarchical manner. The highest hierarchic level is the hypothalamus triggering the pituitary by releasing (or inhibiting) hormones to stimulate the classical endocrine glands: thyroid gland, ovary and testis, adrenal gland, or the liver being a source of hormones (insulin like growth factor-I) as well as a target organ. The classical endocrine glands secrete their hormones into the blood circulation to affect their target organs. At each hierarchic level like the hypothalamus, the pituitary, the endocrine glands and the target organs further endogenous and exogenous factors can influence their regulation.

The 18 chapters of the monograph reflect the amazing diversity and complexity of the fine structure of tissues and organs of fishes and thereby convey an impression of the biological complexity at the morphological, physiological, and functional level of the so-called lower vertebrates. We would like to thank all contributors for their very valuable knowledge, their expertise, their effort and their contribution to the book. They are the real backbone of the book. In particular we would like to thank J.F. Webb who did a big job in harmonising as guest editor the different subchapters of Chapter 16, Sensory Organs. We are very sad to convey the death of Professor Dr. H.J. Grier who passed away during the preparation of the book and took part in the completion of Chapter 10, Ovaries and Eggs.

Last not least we would like to thank the publisher for the very constructive cooperation and patience over a period of years and his generosity concerning the number of color pictures which is so essential for the quality of a book on histology.

The editors dedicate this book to their respective wives, Yvonne (Frank Kirschbaum) and Marzena (Krzysztof Formicki).

May 2019

F. Kirschbaum
K. Formicki

Contents

Chapter **1**

Introduction to Histological Techniques

Anna Pecio* and Rafal P. Piprek

Early Steps in Histology

Although today's definition of histology sounds: "the scientific study of fine detail of biological cells and tissues of organisms performed by examining the structure under a light microscope or electron microscope", we should be aware that the current state is the legacy of five centuries of history and evolution of histology, which goes hand-in-hand with that of the invention of the microscope and the improvement of its resolution.

First compound microscopes were invented probably around 1595 by the Dutch Zacharias Jensen. It was many years before the birth of two known persons: the Englishman Robert Hook (1635–1703) and Antony van Leeuwenhoek (1632–1723), who were making important discoveries with microscopes giving the major milestones in the history of histology in the 17th century. The first cellular observations were documented by Robert Hooke in 1665 in the illustrated book "Micrographia". This volume described Hooke's own experiments made with a microscope: flies, feathers and snowflakes, and he correctly identified fossils as remnants of once-living things. Hook was the first who published the record of the word "cell", while discussing the structure of cork. "Micrographia" as a popular publication has inspired Antony van Leeuwenhoek, who grinded lenses to construct simple microscopes. Leeuwenhoek's skill enabled him to build microscopes that magnified over 200 times. His curiosity to observe almost anything that could be placed under his lenses gave him the possibility to be the first who had seen living cells and tissues. All his observations were documented in letters, written since 1673 to the newly-formed Royal Society of London. Most of his descriptions of microorganisms are even nowadays recognizable.

Another excellent researcher was the Italian physician, anatomist, botanist, histologist and biologist Marcello Malpighi (1628–1694), who constructed one of the first microscopes for studying tiny biological entities. During his 40 years of research, he developed several methods to study living tissue and structures of plant and animals and initiated the science of microscopic anatomy, giving the foundations of major areas of research in botany, embryology, and human anatomy for future generations of biologists. As a tribute to his great discoveries, many microscopic anatomical structures were named after him: the basal layer, renal corpuscles, as well as insect excretory organs.

In the mid-18th century, Ernst Abbe working in Germany in the Zeiss Factory improved microscopes with spherical and chromatic aberration. The researchers, who studied until this time with the help of rudimentary microscopes, received new opportunities to study cellular details. At that time, new techniques using a variety of chemicals to preserve tissues in a life-like state and to prepare them for later staining had developed.

Department of Comparative Anatomy, Institute of Zoology and Biomedical Research, Jagiellonian University, 9 Gronostajowa
St., 30-387 Krakow, Poland.
Email: rafal.piprek@uj.edu.pl
* Corresponding author: anna.pecio@uj.edu.pl

Further great milestones in the development of histology as an academic discipline occurred in the 19th century, when Marie François Xavier Bichat (1771–1802), a French anatomist and pathologist, known as the father of histology, introduced in 1801 the concept of "tissue" in anatomy. He distinguished 21 types of elementary tissues of which the organs of the human body are composed (Nafziger 2002). The term "histology" appeared nearly twenty years after that, in 1819, when a German anatomist and physiologist, Karl Mayer (1787–1865), published a book: "*Ueber Histologie und eine neue Eintheilung der Gewebe des menschlichen Körpers*" (in English: On histology and a new classification of tissues of the human body).

In 1906, the Nobel Prize in Physiology or Medicine was awarded to histologists: to the Italian pathologist Camillo Golgi (1843–1926) and the Spanish neuroscientist Santiago Ramon y Cajal (1852–1934), who gave the first descriptions of brain structure. Cajal won the prize for his original pioneering investigations of microscopical structures and Golgi for a revolutionary method of staining individual nerve and cell structures, which is termed "black reaction". This method uses a weak solution of silver nitrate and is particularly valuable in tracing the processes and the most delicate ramifications of cells.

Histological Techniques

The examination of the details of cells and tissues using microscopes requires a proper preparation of the tissues dissected from the organisms. This special preparation of tissue is called "histological techniques" or histoprocessing. A process of tissue preparation allowing the examination of normal tissue structure or cell details on the slides under the microscope takes place in the following steps: fixation, dehydration, clearing, infiltration, embedding, microtomical cutting, staining, mounting.

Fixation is the process of preservation of tissues in its normal condition; it means to preserve the cell volume and fix all chemical components: proteins, carbohydrates, fats, etc. The material may be fixed using physical or chemical methods. The simplest physical method is heating specimens immersed in boiling saline or by cooking in a microwave oven, which causes the coagulation of proteins and melting of lipids (Bernhard 1974). The opposite method, by freezing, requires specimens to be immersed in isopentane cooled to its freezing points (−170°C), in liquid nitrogen (−196°C) or solid carbon dioxide (−75°C). Fixation by freezing is a rapid method especially useful in histochemistry and immunochemistry because of preserving proteins (Pearse 1980, Bald 1983).

Chemical methods are very popular and used in laboratories for most histological and histochemical purposes. They were used also by early histologists, who used potassium dichromate, alcohol and the mercuric chloride to harden tissues. All nowadays used chemicals have additional properties such as: preventing shrinkage or swelling, rapid rate of penetration, which allows the fixing of a specimen of 3–5 mm within 24 hours and preserves tissue from degradation. The appropriate rate of penetration allows the maintaining of the structure of the cell and of sub-cellular components such as cell organelles (e.g., nucleus, endoplasmic reticulum, mitochondria, etc.).

The chemical fixatives are administered in two ways: (1) through perfusion (fixatives are infused in the animals' body through diffusion in a very short time) or (2) through immersion of the prepared tissue in fixative solution. Fixatives may be simple, e.g.: aldehydes (formaldehyde, glutaraldehyde), acetone, alcohols (ethanol or methanol), acids (glacid acetic acid, picric acid, trichloroacetic acid) and many others such as mercuric chloride, chromium trioxide, osmium tetroxide, etc., and compound fluids (e.g.: Bouin's fluid, Carnoy's fluid, Zenker's fluid).

The main action of aldehyde fixatives is to cross-link amino groups in proteins through the formation of methylene bridges ($-CH_2-$), in the case of formaldehyde, or by C5H10 cross-links in the case of glutaraldehyde. Buffered aqueous 2–5% solution of formaldehyde at pH 7.2–7.4 is used in routine practice for most histological and histochemical examination in light microscopy. This solution is an appropriate fixative as well for immersion or perfusion fixation (Gerrits and Horobin 1996). Formaldehyde reacts with several parts of protein molecules and preserves most lipids and carbohydrates because it does not react with them (Kiernan 1990). However, formaldehyde fixation leads to degradation of mRNA, miRNA and DNA in tissues. For nucleic acids, the recommended fixative is acetic acid, but it does not fix proteins though.

Glutaraldehyde is the most widely used fixative for electron microscopy, usually as a 2.5% solution in phosphate buffered to pH 7.2–7.4. As single fixative, it is useful only when small pieces (0.5–1.0 mm) are processed. In larger specimens, glutaraldehyde causes a tightly cross-linked proteinaceous matrix that is difficult to penetrate (Gerrits and Horrobin 1996). This process, while preserving the structural integrity of the cells and tissue, can damage the biological functionality of proteins, particularly enzymes, and can also denature them.

Aldehyde fixation does not add electron density to tissues, so post fixation in osmium tetroxide is usually practiced. Additionally, osmium tetroxide is used as secondary fixative, increasing electron density and is also effective as stain.

Methanol, ethanol, and acetone may be used alone for fixing films and smears of cell or unfixed cryostat sections, but they are not suitable for blocks of tissue because they cause considerable shrinkage and hardening.

Trichloroacetic and picric acids are widely useful by biochemists for the demonstration of proteins because they precipitate them and cause coagulation forming salts with the basic group of proteins. Acetic acid does not fix the proteins, but it coagulates nucleic acids and is included in fixative mixtures to preserve chromosomes.

All above mentioned liquids are used as individual fixatives, but histologist and biochemists more often employ a mixture of different agents in order to offset undesirable effects of individual substances and to obtain more than one of the chemical fixation.

For general histology and for the preservation of nucleic acids and macromolecular carbohydrates, the best result is given by non-aqueous, Carnoy's fluid. The others, Bouin's and Zenker's fluids, are aqueous fixatives excellent for preserving morphological features, but are not compatible with histochemical techniques.

The choice of fixative depends first of all on the structural or chemical components of the tissue that are to be demonstrated. However, some fixatives (e.g., Zenker's fluid, osmium tetroxide) are unsuitable for glycol methacrylate embedding procedures (GMA), since they contain substances which interfere with the polymerization reaction (Gerrits and Horobin 1996).

Dehydration is the processing of removing all water from tissues. This process is followed by clearing and infiltration, which allows for the penetration by a medium, in which the tissue is finally embedded (paraffin wax, celloidin, plastic media). Paraffin and celloidin infiltration developed at the end of the 19th century (Klebs 1869, Duval 1897). Celloidin is also a common embedding medium, which allows the assessment of excellent morphologic details, but is difficult to remove and puts significant restrictions on success with immunostaining.

Nowadays, the most frequently used medium for light microscopy is paraffin wax. Dehydration in alcohol is followed by a hydrophobic clearing agent (such as xylene) to remove the alcohol, and finally molten paraffin wax. The most commonly used paraffin wax is usually a mixture of straight chain or n-alkanes with a carbon chain. Paraffin wax can be purchased with melting points at different temperatures, the most common for histological use being about 56°C–58°C, but often the temperature is increased to 60°C to decrease viscosity and improve infiltration. Today's companies offer a wide assortment of paraffin waxes: e.g., Paraplast, Paramat, in which added plastic polymers increased the paraffin elasticity and improved tissue penetration or Carbowax Polyethylene Glycol, which is an excellent embedding medium for histochemistry.

Embedding in paraffin allows immunostaining to be performed, but preservation of cellular details is relatively poor. The sections embedded in paraffin media can be cut typically at 5 μm, but it does not provide a sufficiently hard matrix for cutting very thin sections. To enhance cutting of thin sections of the tissues, plastic embedding media (methacrylates, polyester resins, epoxy resins) were introduced in the mid-20th century (Litwin 1985). These media facilitate obtaining semithin sections (0.5–2.0 μm) for light microscopical examination, but also allow ultrathin sectioning of biological specimens for electron microscopy. The epoxy resins are definitely the most popular embedding media for electron microscopy at present, whereas the glycol methacrylate has remained in use for light microscopy because of its large number of advantages, when compared to embedding media such as paraffin or celloidin (Litwin 1985, Gerrits and Horobin 1996, Titford 2009). Glycol methacrylate is polar, which means it is permeable to aqueous solutions of stains and reagents and provides the possibility to perform histochemical reactions without removing it (Leduc and Bernhard 1962). Shrinkage of tissues after is very limited and Glycol methacrylate is therefore suitable for quantitative investigations as well for embedding of heterogeneous tissues, including uncalcified bone (Chappard et al. 1983, Hanstede and Gerrits 1983). Glycol methacrylate embedding can be carried out in the cold or at ambient temperature, and thus allows to demonstrate enzymatic activity on sections in contrast to paraffin embedding performed at 60°C (Higuchi et al. 1979).

Now, the commercially available plastic media such as Technovit 7100 and HistoResin are widely used and less toxic (Gerrits and Smid 1983).

Staining is a technique used in microscopy to enhance contrast and highlight colors of various components of the tissue for viewing in the microscopic image. Antony van Leeuwenhoek was the first, who in 1673 used dye extracted from the saffron crocus bulb to study his specimens. In the 18th century, dyes used in histological staining were the naturally occurring dyes of plants (madder, saffron, indigo), animals (shellfish purple, cochineal), and minerals. Researchers found numerous ways to use dyes to stain tissues because of their extremely universal character. For example, carmine is one of the first dyes used by early botanists (e.g., John Hill who published his work in 1770) and it is still used for staining chromosomes and nuclei (Clark and Kasten 1983, Bracegirdle 1986). The stained compounds—cochineal (= carminic acid) is obtained from the dried Mexican insect *Coccus cacti*. The dried and grinded bodies of females treated with alum and calcium salts result in carmine—a bright red color (Dapson 2007). In nowadays' histology, carmine with potassium salt is used to stain glycogen in popular Best's carmine method (published in 1906) or used for counterstaining of fat in frozen sections by modified Grenatcher's and Orth methods (Clark 1981). Another dye, still useful in laboratories today because of its versatility, is haematoxylin. It is known since 1863, when it was used by Wilhelm van Waldeyer, a German anatomist. Haemtoxylin is obtained from the logwood tree, *Haematoxylon campechianum*, and is a weak dye applied in different solutions, based on its oxidized form, haematein. Haematein with an oxidizer as a mordant can be used to identify a variety of cellular acidic structures, staining them blue-purple (Titford 2005).

The development of histological staining was accelerated by the invention of the first synthetic dye in 1856 by William Perkin. These aniline dyes became more and more popular in the industry and many of them proved to be useful in histological staining techniques (Titford 2009).

A dye molecule contains two elements: a chromogen and an auxochrome. The chromogen is a colored part and consists of a specific arrangement of atoms—a chromophore, which is responsible for the absorption of visible light. The auxochrome, responsible for attaching the chromogen to the substrate, is a part of the ionizable substituent, or a substituent that reacts to form a covalent bond or coordinate with metal (mordant) ion. When auxochromes are amines, they are classified as basic (cationic) dyes, whereas when they are derived from carboxylic, sulphonic acids or from phenolic hydroxyl groups, they are acid (anionic) dyes.

Staining of sections, after removing the paraffin wax, is performed in large excess of staining solution at room temperature. When a solution of a dye is allowed to act slowly until the desired effect is obtained the staining is progressive, whereas in regressive staining the tissue is overstained, and is followed by the process of differentiation, which is a controlled removal of the dye in the staining process in water or alcohol. The success of staining depends on the affinities of the dyes to some components of tissues.

Histological staining methods are usually selected according to the type of microscope (LM, TEM, fluorescent microscopes) and of the biological tissue to be observed. There are many different histological staining techniques suitable for examination of particular components of tissues or types of cells. The most popular staining techniques are described below.

Haematoxylin and eosin (H&E stain) method is commonly used more than any other technique in general histology. Haematoxylin is a basic dye, which binds and stains nucleic acids (DNA and RNA), thus it stains the nuclei of cell. Eosin, an acidic dye, stains the cytoplasm pink. There are many haematoxylin methods used in laboratories today. For example: the alum-based haematoxylin introduced by Ehrlich (1886) or Mayer (1903), Weigert's iron haematoxylin introduced in 1904 used in trichrome methods, or Harris's method (1900) resulting in a shorter staining time (Titford 2009).

There is a multitude of histological methods for connective tissues, which are based on the use of the mixtures of anionic dyes that stain in different colors collagen and cell cytoplasm. The first reported triple stains were made by Gibbes in 1880 (Bracegirdle 1986). The most popular trichrome stains in present time are: Mallory's, Masson's trichrome and Van Gieson methods. The trichrome stain is a complex method, in which tissues of different fine structure and consistency are stained by dyes with small molecules staining tissues with very fine structures, and larger molecules dyes staining more porous tissues. In different trichrome methods, phosphotungistic acid is used as this compound enhances the dye's affinity to collagen (Putchtler and Isler 1958). Many methods such as Verhoeff's or orcien stain are used to demonstrate elastic fibers in connective tissues, whereas reticulin fibers, also called argyrophilic tissue because of its affinity to silver salts, are identified by various silver stains.

Another method based on a silver reaction is the Bielschowsky silver method, a classic stain for the identification of neurofibrils in nerve cells (Sheehan and Hrapchak 1980). Staining of nervous tissue was developed in the early 1900s in Spain by Santiago Ramón y Cajal. His gold chloride sublimate method enables to identify astrocytes (Sheehan and Hrapchak 1980). Another stain for glial cells, known as Holtzer's method, was introduced in 1921. Now, for the depiction of neurons, Cajal or Golgi stain can be used, as well as cresyl violet and Nissl stains, which show the tigroid bodies in the perikaryon of neurons. Phosphotungstic Acid-Haematoxylin (PTAH) can be used to stain myelin in Schwann cells in the peripheral nervous system and oligodendrocytes in the central nervous system (Titford 2009).

For blood smears, the mixture of methylene blue and its oxidation products with eosin dissolved in methanol and glycerol, known as Romanowsky stains, was introduced in 1891 (Power 1982). In 1901–1902 Lieshman, Wright and Giemsa introduced changes in this formula to improve staining. Today, these methods are used in laboratories in many variations and not only for the smears, but also for paraffin embedded sections and for the analysis of bone marrow biopsies (Horobin and Kiernan 2002, Jay 2002).

Very popular in some laboratories are stainings with toluidine blue or methylene blue. Both dyes can be used for staining frozen or hydrated paraffin, as well as semithin epoxy sections useful for preliminary study for transmission electron microscopy.

There are also many simple staining methods providing the information concerning the chemical nature of the tissue. For example, for staining mucins and mucosubstances, the most popular dyes are alcian blue and safranin stains. For staining neutral lipids, fatty acids and phospholipids oil fast red or Sudan stains are used. Some stains as alizarin red are used to identify calcium in tissue sections, and azan stain can be used to differentiate osteoid from mineralized bone.

During the past 40 years, great changes in histological techniques have occurred, since many old and popular stains were replaced by more recent staining methods in fluorescence microscopy, histochemistry and immunohistochemistry.

Fluorescence Techniques

Until the early 20th century, most of the histological studies were conducted using light microscopical and traditional histological staining. However, between 1911 and 1913, two German physicists—Otto Heimstädt and Heinrich Lehmann-constructed the first fluorescence microscope based on the previous UV-microscope. The first fluorescence microscope was employed to observe autofluorescence in bacteria, plant, and animal tissues, as well as substances such as albumin, elastin, and keratin. Soon after, in 1914, the Austrian zoologist Stanislaus von Prowazek used fluorescence microscope to study dye binding in fixed tissues. Thereafter, in 1941, the American immunologist Albert Coons developed a technique for labeling antibodies with fluorescent dyes, thus giving birth to immunofluorescence.

Fluorescence is the emission of light by a substance that has absorbed light. The emitted light has a longer wavelength, and therefore lower energy, than the absorbed radiation. Many biomolecules, collagen, cellulose, cytochromes, chlorophyll, and biological structures such as mitochondria and lysosomes naturally emit light when they absorb light, which is called autofluorescence (Goldys 2009).

Commonly used fluorescence dyes (fluorochromes) include: green fluorescein (as FITC fluorescein isothiocyanate), red rhodamine, Texas red, and cyanine dyes. They are usually used conjugated with antibodies or molecular probes. Rhodamine conjugated with phalloidin (toxic heptapeptide) produced by the death cap mushroom (*Amanita phalloides*) binding filamentous actin (F-actin) is used to identify cytoskeleton. Moreover, there is a list of fluorescence dyes that specifically mark given organelles, e.g.: DAPI, Hoechst, ethidium bromide, propidium iodide, acridine orange are specific for the nucleus; MitoTracker and MitoFluor dyes are used for mitochondria; ER tracker, fluorescent ceramide, fluorescent sphingomyelin identify Golgi/ER; Lysotracker mark lysosomes (Albani 2007). Additionally, acridine orange stains DNA green, and RNA orange; this divergence in color is an effect of different dye concentration binding DNA and RNA.

The majority of fluorescent dyes can be used on fixed samples (cells or tissues), frozen or paraffin-embedded. For instance, DAPI and Hoechst are the most popular fluorescent dyes used to stain nuclei in living or fixed cells or tissues and thus is used as a counterstain. However, other fluorescent dyes require procedures on living cells, e.g., MitoTracker, which accumulates in the inner mitochondrial membrane due to the proton gradient. Fluorescent dyes can be used to distinguish between living and dead cells; e.g., acridine orange stains living cells green, whereas propidium iodide stains DNA of dead cells red. In fluorescent techniques, samples are fixed similarly to light microscopical techniques, usually in 4% formaldehyde; samples can be frozen or eventually embedded in paraffin. After incubation with the dye solution and rinsing, samples are mounted in unstable aqueous medium (e.g., glycerol), thus long-term storage of fluorescence slides is difficult.

Reporter fluorescent molecules are one of the most advanced tools enabling structure visualization (Chale and Kain 2006). GFP (green fluorescence protein) is the most popular reporter protein used nowadays. Similar fluorescent proteins naturally occur in many marine organisms; however, GFP traditionally refers to the protein first isolated from the jellyfish *Aequorea victoria* in 1962 (Shimomura et al. 1962). In 2008, the scientists Roger Y. Tsien, Osamu Shimomura, and Martin Chalfie were awarded the Nobel Prize in Chemistry for their discovery of this protein. Transgenic organisms carrying *gfp* gene inserted into the genome express GFP in appropriate cells depending on the promoter located upstream. Cells or organisms can be genetically modified so that every protein of interest carries a fluorescent reporter molecule (tag). The location of a protein can then be traced via the fluorescent signal in a living organism (Crivat and Taraska 2012).

Due to many advantages in the 21st century, fluorescence microscopy revolutionized cell biology, enabling observation of live cells and highly specific multiple labeling of individual organelles and molecules. Nowadays, super-resolution fluorescence microscopes such as photo-activated localization microscopes (PALM) allow the tracking of even single molecules within a cell due to a resolution of ~ 10–55 nm (Galbraith and Galbraith 2011). In a typical fluorescence microscope, an image is disrupted due to the fact that light is registered from different layers located above and under the focal plane, which impairs the quality of image and resolution ability. Nowadays, many studies are carried out using a confocal microscope, which is a type of fluorescence microscope equipped in a special pinhole. A sample in this microscope is horizontally scanned with a narrow beam of laser light, due to which a confocal microscope registers light exclusively from the focal plane, and eliminates out-of-focus light. This enables achieving high resolution (0.1 μm) contrast and sharp focus images.

Fluorescence microscopy also has few disadvantages. For instance, the fluorescence is not permanent. As the samples are viewed photobleaching occurs, the fluorescence fades and thus using antifade reagents is needed (Widengren et al. 2007, Hinkeldey et al. 2008, Ji et al. 2008).

Electron Microscopical Techniques

The invention of the electron microscope by Ernst Ruska and Max Knoll in 1931 in Berlin was one of the most major steps in histology and allowed to overcome the barrier to higher magnification and resolution. Using early electron microscopes, a resolution of 2–10 nm was obtained, whereas present electron microscopes achieve 50 pm resolution and magnifications of up to about 10,000,000 x, compared to light microscopy with diffraction enabling to achieve only about 200 nm resolution and useful magnifications below 2000 x (Erni et al. 2009). Electron microscopes use an electron lens system that is analogous to glass lenses of an optical light microscope.

There are two main types of electron microscopes: **transmission electron microscope (TEM)** and **scanning electron microscope (SEM)**. In TEM, electrons are transmitted through the specimen that is partially transparent to electrons and partially scatters them. The electron beam passing through the specimen is viewed. In SEM, an electron beam is focused on a sample surface coated with metal which reflects electrons, thus the sample surface is horizontally scanned and an image of special surface with large depth, almost 3D is obtained (Gordon 2014).

High resolution and magnification achieved in electron microscopes requires extremely thorough sample preparation. Artifacts invisible in the light microscope, such as swelling, shrinking or cracks in cell membranes can be identified in the electron microscope, thereafter proper fixation in a solution of proper osmolarity is essential for electron microscopical techniques. The process of specimen preparation for electron microscopy consists in sample fixation, contrasting and embedding. The best method of fixation is perfusion in which a fixative is injected into the bloodstream and thus reaches and effectively penetrates the tissue. More time is required for immersion fixation in which a sample is immersed in the solution of a fixative which defunds inside the sample. Immersion fixation also requires small sample size to avoid tissue damage. The diversity of fixatives in TEM is limited compared to light microscopical techniques. Glutaraldehyde (0.5–6%) is the most commonly used fixative in electron microscopy (Prento 1995). It is usually used in combination with formaldehyde or rarely with tannic acid. Usually, Karnovsky's fixative is used, i.e., glutaraldehyde and formaldehyde in phosphate or cacodylate buffer. pH (7.2–7.4) and osmolarity (270–400 mOsm for mammals) are two factors crucial for proper fixation. Too low osmolarity leads to cell swallowing; in contrast, too high osmolarity leads to shrinking, both effects disrupt cell and tissue structure. Therefore, fixating compounds are dissolved in buffers (phosphate or cacodylate buffer). Afterwards, the buffer is used to rinse the fixative from the samples, and osmium tetroxide (0.5–2.5%) is then used as secondary fixative (Ito and Karnovsky 1968, Wilgglesworth 1988). There are two reasons of using osmium tetroxide for electron microscopy: first, this compound binds lipids (exactly unsaturated fatty acids) and thus stabilizes cell membranes, and second, osmium as heavy metal reflects electrons and provides contrasting of samples visible under the microscope. Then samples are dehydrated and embedded. Since extremely thin sections are required for transmission electron microscopy, samples must be embedded in substances with high hardness. Polymerizing plastics as epoxy or acrylic resins meet these conditions (Finck 1960, Armbruster et al. 1982). The most commonly used epoxy resins are: Epon 812 or Polybed 812. They are non-polar and polymerize at higher temperature (60–80°C) within 24–48 hours. For some purposes, such as histochemical and immunohistochemical studies, polar resins, such as LRWhite, LRGold or Lowicryl K4M appear appropriate. These resins polymerize by heating or UV irradiation.

Samples are cut using glass or diamond knifes and ultramicrotomes. Thickness of samples depends on the destination (LM or TEM). Such samples can be cut for semithin (0.5–1 μm) or ultrathin sections (50–70 nm). Semithin sections are placed on microscope slides, stained with methylene blue and Azur II and viewed under a light microscope. Ultrathin sections are placed on copper (sometimes nickel for immunohistochemistry) grids. Contrasting with osmium tetroxide gives insufficient results and for better visualization of organelles, uranyl acetate and lead citrate are used for additional contrasting (Thaete 1979). Then, samples are viewed in TEM. Other histological staining techniques and even immunolocalization can also be used for both semithin (on slides) and ultrathin (on grids) sections. For these procedures, chemical removal of the resin is sometimes required.

For **scanning electron microscopy (SEM)**, samples are usually fixed as for TEM, postfixed in osmium tetroxide, dehydrated, and dried using liquid carbon dioxide at a critical point of this gas, i.e., at 31°C and -7^6 hPA. Dried samples are coated with carbon and then with gold or platinum using the method of vacuum sputtering (Ito and Karnovsky 1968). Sputter-coated samples are viewed in SEM.

Freeze-fracture technique is sometimes used for cytological studies, especially when cell membranes and cell junctions are studied. The samples are frozen and then fractured in a high vacuum and at -100°C (Severs 2007). A replica is made from the surface of fractured sample, which is a subject of further research in the electron microscope.

Freezing Techniques

The freezing technique involves hardening of material by freezing it (Crang and Klomparens 1988, Kellenberger et al. 1992). Thus, this method is simple and of short duration, which enables cutting samples even without previous fixation and embedding. In addition, the great advantage of this method is the preservation of all substances, including lipids, in the tissues that are rinsed during the sample preparation for paraffin or celloidin embedding. Moreover, low temperature prevents the proteins from being kept in a non-denatured form and thus allows for the maintenance of enzyme activity and can be used to test enzyme activity. It does not alter the antigenicity of proteins and therefore this technique is widely used in histochemistry and immunohistochemistry. Before freezing, the tissue may be gently fixed (e.g., 4% formaldehyde) or frozen as unfixed tissue. There is approximately 70% of water in the tissue; thus, during freezing, crystals of water form, which can destroy the cell structure. Cryoprotection is used to avoid crystallization of water and to protect cell structure, for example 30% sucrose, 30% glycerol, 10% DMSO is used before freezing. Then the tissue is frozen usually by immersion (plunge freezing) in liquid gas, for example in liquid nitrogen (temperature –196°C). Another method is jet freezing in which a stream of liquid gas is streamed directly on a sample. In the third method, a sample is frozen under high pressure of gas (McDonald 2009). Tissues can be also frozen in liquid carbon dioxide (–78.5°C), which requires a longer process. Frozen materials are cut at lower temperature (–15–25°C) using a cryostat for 4–10 µm sections. Cryoultramicrotomes, working at temperatures of –120–160°C, are used to obtain thinner sections (50–100 nm) that are destined for electron microscopy.

Histochemistry

Histochemistry is the field of science revealing localization of chemical compounds and enzymes within cells (cytochemistry) or tissues (Pearse 1980). It is a combination of histological techniques and chemical reactions. Histochemical methods are characterized by high specificity since they are aimed at the detection of a concrete substance or group of substances, whereas histological methods are intended to visualize various structures in the tissue. The main idea in histochemistry is that in order to localize a substance in a tissue, a substrate is introduced into the sample and a colored product appears. The product should be visible and insoluble in the reaction medium in order not to diffuse and persist in the exact location of the studied substance. In the end, the image can be analyzed in the light, fluorescent or even electron microscope.

The procedure begins with the fixation of the tissue, which depends on the studied substance, e.g., formaldehyde is used for protein fixation, osmium tetroxide for lipid, and Carnoy's fixative for nucleic acids. Sometimes unfixed samples are studied or fixation is carried out after the reaction. Next, samples can be embedded in paraffin in case of localization of proteins, carbohydrates and nucleic acids, but not when lipids are aimed to be detected, because alcohol used in this process rinses lipids. In all cases, the freezing method is the most suitable for histochemistry because it does not change the chemical composition, does not rinse lipids and does not denature proteins and thus does not deactivate enzymes.

Examples of Histochemical Reactions

(i) Detection of proteins—reaction with ninhydrin and Schiff's reagent. Ninhydrin causes formation of aldehyde groups in aminoacid molecules by oxidative deamination whereas aldehydes are shown by reaction with Schiff's reagent. Schiff's reagent is commonly used in histochemistry. It is a water solution of fuchsin that is colorless due to the presence of sodium bisulfate; in reaction with aldehydes, a chromogenic (colorful) product is produced (Hardonk and Van Duijn 1964). Purple places in the tissue indicate the presence of the studied compound. However, immunohistochemistry is the most popular technique since it allows the detection of small amount of a concrete protein.

(ii) Detection of neutral polysaccharides such as glycogen—PAS (periodic acid-Schiff) reaction. First, carbohydrates are oxidized by (0.5–1%) periodic acid, and as a result glycol groups in carbohydrates are transformed into aldehyde groups. In a second step, Schiff's reagent is used to detect polysaccharides (purple) (Kiernan 1990).

(iii) Detection of acidic polysaccharides such as glycosaminoglycans and mucopolysaccharides, present mainly in mucus and extracellular matrix, especially in cartilages. Here, alcian blue is used as a cationic dye that, via electrostatic impact, binds polyanions such as acidic polysaccharides (Kiernan 1990). The specificity of reaction depends here on the pH of the solution in which the reaction takes place. In pH 1, very acidic glycosaminoglycans are stained, whereas in pH 2.5 hialuronic acid, sialomucins and slightly acidic glycosaminoglycans are stained.

(iv) Detection of lipids, triglycerides and lipoproteins—here dyes slightly soluble in water and alcohols with high affinity to lipids, such as Sudan III, Sudan IV, Oil Red O, are used (Pfüller et al. 1977). This staining method is based on physical processes rather than on a chemical reaction, since the lipophylic dye defunds from a water or alcohol solution tissues containing fat due to its higher affinity to lipids. Additionally, osmium tetroxide can be used to specifically stain lipids in tissues because it binds unsaturated fatty acids present in molecules of membrane phospholipids.

(v) Detection of nucleic acids—Feulgen reaction. First, DNA is hydrolized in 1M hydrochloric acid due to which aldehyde groups are formed. Next, Schiff's reaction is used to visualize nucleic acids (Chieco and Derenzini 1999). Moreover, fluorescent dyes such as DAPI and Hoechst are used to detect nucleic acids. DAPI (4'6-diamidino-2-phenylindole) intercalates in the minor groove of DNA in sites of high A–T concentration. DAPI also binds RNA but does not reveal high fluorescence in this case.

(vi) Detection of calcium by alizarin and alizarin red S—in order to detect calcium deposits or stain of bone tissue, these two anthraquinone dyes are used. First, the tissue (usually skeletal tissue) should be fixed in neutral solution, for example in 80% ethanol. Acidic fixatives rinse calcium and thus disrupt its detection. Alizarin dyes chelate metal atoms; calcium combines with two molecules of the dye (Kiernan 1990). This procedure is especially useful in histology for staining bone tissue and hydroxyapatite deposition in red.

(vii) Histoenzymatic reaction—Histochemical detection of enzymes does not consist in localizing them as specific substances, but on the visualization of their activity (Kiernan 1990). A substrate is introduced in the sample, and then a colorful product is produced due to enzyme action. An enzyme must be active, thus a sample must be properly prepared, for example freezing without fixation or gentile fixation is suitable; however, usually embedding in paraffin completely inactivates enzymes. Reactions are carried out in a specific incubation medium containing substrate and additional substances (coenzymes, ions) at optimal pH and temperature. The most popular histoenzymatic reactions are detection of dehydrogenases, oxygenases and peroxidases. Here, as a result of substrate oxidation or reduction, a chromogenic product appears. The activity of dehydrogenases is detected by using the properties of tetrazole salts which are colorless and soluble in oxidized form; however, via reduction, they are turned into insoluble and colorful compounds. A typical incubation medium contains the substrate, e.g., succinate or lactate, coenzyme (NAD, NADP) and tetrazole salt (NBT—nitro blue tetrazolium or TNBT—tetranitro blue tetrazolium); as a result, a purple blue product appears. For the detection of the activity of oxygenases and peroxidases, benzydine and diaminobenzidine (DAB) are used as substrates which are colorless and water soluble in a reduced form; however, in the presence of H_2O_2 and enzyme activity, they turn into colorful and insoluble products.

Immunohistochemistry

Immunohistochemistry (IHC), consisting in detection of specific antigens by antibodies in tissues or cells (immunocytochemistry, ICC), is characterized by the highest specificity. A given antibody binds a specific antigen (usually a protein), and exactly an epitope, i.e., a small fragment usually composed of several amino acids. Interestingly, the first use of antibodies for immunolabeling was performed in 1934 (Marrack 1934), the first fluorescent detection using antibodies was reported in 1941 (Coons et al. 1941), and the first enzymatic labeling was described in 1966 (Avrameas et al. 1966).

The antibody is an immunoglobulin produced by lymphocytes B. Two types of antibodies are used. Monoclonal antibodies are synthesized by one clone of cells, deriving from one mother cell (Liu 2014). These antibodies are obtained by animal immunization (usually mouse), followed by isolation of lymphocytes B from the spleen and fusion of these cells with cells of myeloma. Such hybrids have the ability to produce antibodies and prevent cell divisions. Since monoclonal antibodies are produced by a single clone of cells, they bind exactly at the same epitope. Therefore, a high specificity is achieved. However, polyclonal antibodies are obtained usually via rabbit immunization (Hanly et al. 1995). After injection of the antigen into the organism, blood is taken, and antibodies are isolated from the serum. Thus, a mixture of many antibodies binding any epitope of a single antigen is obtained. In this case, the specificity is lower because of the possibility of binding other antigens (cross reaction) is increased; however, signal strength here is higher since a single antigen in bound by several antibodies, thus the sensitivity is higher.

Antibodies are labeled in various manners to enable their detection and detection of an antigen. Rarely, a direct method of immunohistochemistry is used. Here, an antibody binding an antigen is labeled with a marker. However, usually an indirect method is used, where two antibodies are used: first, the primary antibody binds an antigen, and second, the secondary antibody binds the primary antibody and is labeled (conjugated with a marker molecule). The indirect method provides a stronger signal detection since many secondary antibodies (labeled) can bind one primary antibody. Antibodies can be conjugated (labeled) with fluorochromes, enzymes or colloid gold. In the immunofluorescence (IF) method, antibodies are labeled with fluorescent compounds (fluorochromes such as FITC, TRITC, Cy5) and samples are destined for fluorescent (or confocal) microscopy. When samples are destined to bright light microscopy, secondary antibodies conjugated with enzymes are used, usually horseradish peroxidase (HRP) or alkaline phosphatase (AP, ALP). An antigen is immunolocalized owing to a colorful product of enzyme activity, similar to a histoenzymatic reaction (see above). DAB (diaminobenzidine) is used as a chromogen for peroxidase, and Fast Red or Fast Blue for phosphatase. When samples are destined to electron microscopy, secondary antibodies are labeled with colloid gold, which reflects electrons. Several compounds such as biotin, avidine and streptavidine can be used to increase the number of bound secondary antibodies or enzymes and thus to intensify the signal.

Sample preparation for immunolocalization begins with tissue fixation which should be gentle in order not to disrupt antigenicity of antigens. Usually, the material is fixed in 4% formaldehyde and freezing techniques are the most suitable. The reaction with antibodies can also be carried out on paraffin sections.

Fixation and embedding often affect protein conformation, epitopes become hidden and as a result, the ability of antibody binding is lost. Therefore, after removal of the embedding substance, epitope-retrieval is performed (Shi et al. 2011). Slides are usually boiled in buffer at pH 6 or pH 9 (HIER, heat-induced epitope retrieval). Eventually, incubation with photolytic enzymes, such as proteinase K or pronase, is used to expose binding sites for antibodies. In immunotechniques, sections are subjected to intensive rinsing and even boiling, which can damage the samples. Therefore, using highly adhesive slides, coated with poly-L-lysine or silane, are required.

Afterwards, optionally endogenous enzymes are blocked when histoenzymatic reactions are used for antigen detection. Endogenous peroxidase is blocked with hydrogen peroxide and phosphatase with levamisole. Always before incubation with antibodies, unspecific binding must be blocked. Antibodies can unspecifically bind different molecules via electrostatic or hydrophobic interactions. In order to increase the reaction specificity and simultaneously the reliability of studies, solution of serum (e.g., FBS or goat serum) or eventually albumin (BSA) is applied onto the section, and samples become saturated with protein solution. Protein block is followed by incubation with primary, and then secondary antibodies. Double immunolabeling is possible due to the application of two primary antibodies (differing in species origin) against two different proteins. Also, counterstaining is performed to visualize the background. For bright light microscopy, the background is stained with hematoxylin and sections are dehydrated and mounted in nonpolar resin. For fluorescent microscopy, DAPI or Hoechst are used as counterstain and sections are mounted in an aqueous medium such as glycerin.

Immunogold Staining

Immunogold staining (IGS) is a combination of electron microscopical and immunohistochemical techniques, which allows subcellular localization of a protein of interest (D'Amico and Skarmoutsou 2008). Colloid gold or nanogold, conjugated to a secondary antibody, is used in this technique. The samples are gently fixed in PFA with or without glutaraldehyde (strong fixation and osmium tetroxide postfixation should be omitted to avoid disruption of immunoreactivity), dehydrated, embedded in hydrophilic polar resin, and cut on formvar-carbon coated nickel grids. The grids with ultrathin sections are applied onto drops of the following solutions: blocking solution, primary antibodies, secondary antibodies, contrast staining with uranyl acetate and lead citrate. Labeled samples can be viewed in transmission or scanning electron microscope.

Indeed, there are two types of immunolabeling with respect to the sequence of labeling and embedding: pre-embedding and post-embedding. Method described above is the post-embedding method where the tissue after fixation is first embedded in paraffin, resin or frozen, then cut and immunolabeling is performed on sections (in a following sequence: fixation, embedding, cutting, immunolabeling, imaging). In the pre-embedding technique, immunolabeling is done prior to embedding and sectioning (in a sequence: fixation, immunolabeling, embedding in paraffin, resin or freezing, cutting, imaging). The pre-embedding version of IHC is also known as **whole mount immunohistochemistry** (WM-IHC or WIHC).

References Cited

Albani, J.R. 2007. Principles and Applications of Fluorescence Spectroscopy. Wiley-Blackwell.

Armbruster, B.L., E. Carlemalm, R. Chiovetti, R.M. Garavito, J.A. Hobot, E. Kellenberger and W. Villiger. 1982. Specimen preparation for electron microscopy using low temperature embedding resins. J. Microsc. 126: 77–85.

Avrameas, S. and J. Uriel. 1966. Method of antigen and antibody labelling with enzymes and its immunodiffusion application. C.R. Acad. Sci. Hebd. Seances Acad. Sci. D 262: 2543–2545.

Bald, W.B. 1983. Optimizing the cooling block for the quick freeze method. J. Microsc. 131: 11–23.

Bernhard, G.R. 1974. Microwave irradiation as a generator of heat for histological fixation. Stain Technol. 49: 215–224.

Bichat, X. 1801. Anatomie générale appliquée à la physiologie et à la médecine. Brosson, Gabon, Paris.

Bracegirdle, B. 1986. A History of Microtechnique. Lincolnwood, Il. Science Heritage LTD.

Chale, M. and S.R. Kain. 2006. Green Fluorescent Protein: Properties, Applications and Protocols, Second Edition, John Wiley and Sons.

Chappard, D., C. Alexandre, M. Camps, J.P. Montheard and G. Riffat. 1983. Embedding iliac bone biopsies at low temperature using glycol and methyl methacrylates. Stain Technol. 58: 299–308.

Chieco, P. and M. Derenzini. 1999. The Feulgen reaction 75 years on. Histochem. Cell Biol. 111(5): 345–358.

Clark, G. and F.H. Kasten. 1983. History of Staining. Baltimore. Williams and Wilkins.

Coons, A.H., H.J. Creech and R.N. Jones. 1941. Immunological properties of an antibody containing a fluorescent group. Proc. Soc. Exp. Biol. Med. 47: 200–202.

Crang, R.F.E. and K.L. Klomparens. 1988. USA. pp. xii–233. *In*: R.F.E. Crang and K.L. Klomperens (eds.). Artifacts in Biological Electron Microscopy. Plenum Press, New York.

Crivat, G. and J.W. Taraska. 2012. Imaging proteins inside cells with fluorescent tags. Trends Biotechnol. 30(1): 8–16.

D'Amico, F. and E. Skarmoutsou. 2008. Quantifying immunogold labelling in transmission electron microscopy. J. Microsc. 230(Pt 1): 9–15.

Dapson, R.W. 2007. The History, chemistry and modes of action of carmine and related dyes. Biotech. Histochem. 82: 13–15.

Duval, M. 1879. Technique de l'emploi du colloidion humide pur la practique des coupes microscopiques. J. Anat. Physiol. XV: 185.

Erni, R., M.D. Rossell, C. Kisielowski and U. Dahmen. 2009. Atomic-resolution imaging with a sub-50-pm electron probe. Phys. Rev. Lett. 102(9): 096101.

Finck, H. 1960. Epoxy resins in electron microscopy. J. Biophys. Biochem. Cytol. 7: 27–30.

Galbraith, C.G. and J.A. Galbraith. 2011. Super-resolution microscopy at a glance. J. Cell Sci. 124: 1607–1611.

Gerrits, P.O. and L. Smid. 1983. A new loss toxic polymerization system for embedding of soft tissues in glycol methacrylate and subsequent preparing of serial sections. J. Microsc. 132: 81–85.

Gerrits, P.O. and R.W. Horobin. 1996. Glycol methacrylate embedding for light microscopy: basic principles and trouble-shooting. J. Histotechnol. 19: 297–311.

Goldys, E.M. 2009. Fluorescence Applications in Biotechnology and Life Sciences, Wiley-Blackwell.

Gordon, R.E. 2014. Electron microscopy: a brief history and review of current clinical application. Methods Mol. Biol. 1180: 119–135.

Hanly, W.C., J.E. Artwohl and B.T. Bennett. 1995. Review of polyclonal antibody production procedures in mammals and poultry. ILAR J. 37(3): 93–118.

Hanstede, J.G. and P.O. Gerrits. 1983. The effects of embedding in water soluble plastics on the final dimensions of liver sections. J. Microsc. 131: 79–86.

Hardonk, M.J. and P. Van Duijn. 1964. The mechanism of the Schiff reaction as studied with histochemical model systems. J. Histochem. Cytochem. 12: 748–751.

Higuchi, S., M. Suga, A.M. Danneberg and B.H. Schofield. 1979. Histochemical demonstration of enzyme activities in plastic and paraffin embedded tissue sections. Stain Technol. 54: 5–12.

Hinkeldey, B., A. Schmitt and G. Jung. 2008. Comparative photostability studies of bodipy and fluorescein dyes by using fluorescence correlation spectroscopy. Chem. Phys. Chem. 9: 2019–2027.

Horobin, R.W. and J.A. Kiernan. 2002. Conn's Biological Stains. Oxford. Bios. Scientific Publishers Ltd.

Ito, S. and M.J. Karnowsky. 1968. Formaldehyde glutaraldehyde fixatives containing trinitro compounds. J. Cell. Biol. 36: 168.

Jay, V. 2001. Paul Ehrlich. Arch. Pathol. Lab. Med. 125: 724–725.

Ji, N., J.C. Magee and E. Betzig. 2008. High-speed, low-photodamage nonlinear imaging using passive pulse splitters. Nat. Methods 5(2): 197–202.

Kellenberger, E., R. Johansen, M. Maeder, B. Bohrmann, E. Stauffer and E. Villiger. 1992. Artefacts and morphological changes during chemical fixation. J. Microsc. 168: 181–201.

Kiernan, J.A. 1990. Histological & Histochemical Methods: Theory and Practice. Pergamon Press.

Klebs, E. 1869. Die Einschmelzungs-Methode ein Beitrag zur mikroskopischen Technik. Arch. Mikrosk. Anat. 5: 164–166.

Leduc, E.H. and W. Bernhard. 1962. Water soluble embedding media for ultrastructural cytochemistry. Symp. Int. Soc. Cell Biol. 1: 21–28.

Lendrum, A.C., D.S. Fraser, W. Slidders and R. Henderson. 1962. Studies on the character and staining of fibrin. J. Clin. Pathol. 15: 401–413.

Litwin, J. 1985. Light microscopic histochemistry on plastic sections. Prog. Histochem. Cytochem. 16: 1–84.

Liu, J.K.H. 2014. The history of monoclonal antibody development—Progress, remaining challenges and future innovations. Ann. Med. Surg. (Lond.) 3(4): 113–116.

Marrack, J. 1934. Nature of antibodies. Nature. 133: 292.

Mayer, A.F.J.K. 1819. Über Histologie und eine neue Eintheilung der Gewebe des menschlichen Körpers. Bonn, Adolph Marcus.

Nafziger, G.F. 2002. Historical Dictionary of the Napoleonic Era. Scarecrow Press.

Pearse, A.G.E. 1980. Histochemistry, Theoretical and Applied. 4th edition. Vol. 1. Preparative and Optical Technology. Vol. 2. Analytical Technique. Edinburgh: Churchill-Livingstone.

Pfüller, U., H. Franz and A. Preiss. 1977. Sudan Black B: Chemical structure and histochemistry of the blue main components. Histochemistry 54(3): 237–250.

Power, K.T. 1982. The Romanowsky stains: A review. A. J. Med. Technol. 48: 519–523.

Prento, P. 1995. Glutaraldehyde for electron microscopy: a practical investigation of commercial glutaraldehydes and glutaraldehyde-storage conditions. Histochem. J. 27(11): 906–913.

Severs, N.J. 2007. Freeze-fracture electron microscopy. Nat. Protoc. 2(3): 547–576.

Sheenan, D.C. and B.B. Hrapchak. 1980. Theory and Practice of Histotechnology. St. Luis, MO C.V. Mosby Company.

Shi, S.R., J. Shi and C.R. Taylor. 2011. Antigen retrieval immunohistochemistry. J. Histochem. Cytochem. 59(1): 13–32.

Shimomura, O., F.H. Johnson and Y. Saiga. 1962. Extraction, purification and properties of aequorin, a bioluminescent protein from the luminous hydromedusan, Aequorea. J. Cell. Comp. Physiol. 59(3): 223–239.

Stevens, A. 1990. The haematoxylins and eosin. pp. 126–138. *In*: J.D. Bancroft and A. Stevens (eds.). Theory and Practice of Histological Techniques. London, Churchill Livingstone.

Thaete, L.G. 1979. Lead and uranium stain artefacts in electron microscopy: a technique for minimizing their occurrence. J. Microsc. 115(2): 195–201.

Titford, M. 2005. The long history of hematoxylin. Biotech. Histochem. 80: 73–78.

Titford, M. 2009. Progress in the development of microscopical techniques for diagnostic pathology. J. Histotechnol. 32: 9–19.

Widengren, J., A. Chmyrov, C. Eggeling, P.A. Lofdahl and C.A.M. Seidel. 2007. Strategies to improve photostabilities in ultrasensitive fluorescence spectroscopy. J. Phys. Chem. A 111(3): 429–440.

Wilgglesworth, V.B. 1988. Histological staining of lipids for the light and electron microscope. Biol. Rev. Camb. Philos. Soc. 63(3): 417–431.

Chapter 2

Integument

Frank Kirschbaum[1],* and *Shaun P. Collin*[2]

INTRODUCTION

"The body of the fish is enclosed in the *integumentum commune*, or briefly the *integument*, *tegment* or *outer skin*, a sheath consisting essentially of two layers, the *epidermis* and the *corium = dermis* or *cutis*. A *hypodermis*, a *subcutaneous connective tissue*, is almost never formed in fishes, and is never comparable to the corresponding layer of higher vertebrates which often contains considerable amount of *fat* tissue. The corium and the hypodermis are of *mesodermal* origin, whereas the epidermis is derived from the *ectoderm*" (Harder 1975). The integument has protective functions including reducing injuries to internal organs, inhibiting the penetration of bacteria, viruses, and toxic substances (Genten et al. 2009) and mediating respiration, osmoregulation, and excretion. The integument is also important for communication and camouflage due to pigmentation patterns and the capacity for colour change. The thickness of the integument varies considerably with respect to both regional differences across the body and between species. In general, pelagic species have a thinner epidermis (in general, ca. 20–50 μm thick, rarely 100 μm) than benthic species (Pfeiffer 1968), e.g., the epidermis is about 300–455 μm thick in *Petromyzon* sp., Dipnoi, Polypteriformes, *Misgurnus* sp., *Malapterurus* sp., *Anguilla* sp., and *Lota* sp. The coelacanth *Latimeria chalumnae* possesses a similarly thick epidermis (Pfeiffer 1968), although this ancient sarcopterygian fish is more benthopelagic (Balon 2003). Fish integuments can attain a maximal thickness of about 10 mm as seen in some sharks (Selachii) and pufferfishes (Tetraodontiformes, e.g., *Orthagoriscus* sp.) (Bertin 1958).

In addition to the general functions describe above, the integument can also develop special epidermal structures for more specialised functions. Some examples will now be described: (1) Keratinized epidermal cells are found in so-called breeding tubercles: The keratinized cells develop on a pad of hypertrophied epithelial cells (Harder 1975). However, the keratin-like substance is not identical to the keratin found in the epidermis of mammals (Burgess 1956, Aisa 1959). In general, the breeding tubercles develop during the breeding season and are predominantly found in males. They aid in establishing a tight contact during the spawning act. The breeding tubercles occur in at least 15 families of fishes within four orders: Salmoniformes, Gonorhynchiformes, Cypriniformes, and Perciformes (Harder 1975). (2) Species of the family Heterocongridae (Anguilliformes) possess thick mucous cells at the end of their tail. These fish burrow into the sand with their rear end and the mucous cements the sand of the tube walls together

---

[1] Humboldt University, Faculty of Life Sciences, Unit of Biology and Ecology of Fishes, Philippstr. 13, Haus 16, D-10115 Berlin, Germany.
[2] School of Life Sciences, La Trobe University, Bundoora 3086, Victoria, Australia.
  Email: s.collin@latrobe.edu.au
* Corresponding author: frank.kirschbaum@staff.hu-berlin.de

(Casimir and Fricke 1971). (3) In the Scaridae and some Labridae (Perciformes), upon nightfall, the mucous gland of the gill cover secretes mucous, which protects the fish during sleep (Winn 1955). (4) Species of the cichlid genus *Symphysodon* (Perciformes) secrete an opaque substance in the neck region and along the anal and dorsal fins that forms a coat on the epidermis, which is eaten by their fry for several weeks (Hildemann 1959). (5) Luminescent organs, or photophores (Bertin 1958) are also epidermal derivatives (see Herring (1982), Anctil et al. (1984), and Best and Bone (1976) for reviews of their function and physiology). Many species of deep-sea sharks also possess ventral photogenic organs or photophores that produce bioluminescence (Claes et al. 2014). The skin of some species of catsharks is also biofluorescent (Gruber et al. 2016). (6) Fish larvae often possess a mucous gland on the head, which serves as attachment organ (Ilg 1952, Britz et al. 2000).

The epidermis is composed of a stratified squamous epithelium; the number of cell layers in fish varies from two in larvae to more than ten in adults (Whitear 1986a). The basic epidermal cell is quite small (a few micrometers) and contains many tonofilaments and desmosomes, thereby providing strong adhesion between adjacent cells. These "filament-containing" cells (Henrikson and Matoltsy 1968) have also been called Malpighian cells. The superficial epithelial cells differ to some extent from the cells below. They contain, in addition to the tonofilaments, actin filaments and in teleosts they develop at the surface microridges or micropapillae (reviewed by Whitear 1986a). The superficial epithelial cells secrete surface secretory vesicles. The cells of the basal epidermal layer are cuboidal or columnar and show mitotic activity, although mitosis is also detected in other epidermal cell layers (Linna et al. 1975, Bullock et al. 1978, Spitzer et al. 1979, 1982).

Single-celled glands are embedded in between the basic epidermal cells. Two different cell types were distinguished in older literature: goblet or mucoid cells and serous cells (Harder 1975). In recent literature, a different terminology has been adopted (reviewed by Whitear 1986a). In this chapter, we will follow this revised terminology: **Goblet cells** are exocrine unicellular glands producing membrane-bound globules. These cells generally contain mucoid substances and remain intact until they are discharged. However, there are also serous goblet cells, which produce protein (Blackstock and Pickering 1980, Sato and Kurotaki 1964, Whitear 1977). **Sacciform cells** are characterized by a large vacuole, which contains a secretion. The supply of the secretion is performed intracellularly by small vacuoles or channels (reviewed by Whitear 1986a). **Club cells** are large cells and are found in the middle layer of the epidermis. There are two types of club cells. In eels, the club cells have a secretory vacuole and a peculiar peripheral cytoplasm. In ostariophysans, the vacuole is missing. The cytoplasm of the club cells contains coiled filaments associated with desmosomes, but no tonofilaments. The term **Granular cells** should be applied, following the suggestion of Whitear (1986a), to cells in the epidermis of lampreys. These cells are characterized by electron-dense membrane-bound granules and possess filaments that are thicker than tonofilaments (Lethbridge and Potter 1982). **Thread cells** are typical for hagfishes. They can be quite abundant in the epidermis (Spitzer et al. 1979). The thread material is comprised of a unique protein; most of the slime produced by hagfishes originates from the thread cells (after Whitear 1986a). **Venom cells** are found in venomous fish, e.g., in catfishes and stingrays. They are comprised of multicellular holocrine glands (Halstead and Halstead 1978). **Jonocytes** are found in gill epithelia and in the oropharyngeal epithelium of teleosts. They are involved in osmoregulation (see Chapter "Gills: Respiration and Ionic-Osmoregulation"). In lampreys and selachians, cells of apparently similar function have been detected (after Whitear 1986a) and may be considered jonocytes. Many sensory skin organs like taste buds, neuromasts, and electroreceptors are comprised of **Epithelial Sensory Cells** and **Supporting Cells**. For details, see Chapter "Sensory Organs". **Merkel Cells** are special epidermal cells associated with nerve endings. Their mechanoreceptive function now seems well established (Iggo and Muir 1969, Hartschuh et al. 1986).

The basement membrane is the boundary between the epidermis and the dermis and is also the product of both these regions (Whitear 1986b). The dermis consists of two layers: the stratum spongiosum (laxum) and the stratum compactum. The stratum spongiosum is situated below the basement membrane and contains collagen and reticulin fibres, fibroblasts, nerves, capillaries, and pigment cells (Genthen et al. 2009). The stratum compactum is composed of densely-packed bundles of collagen fibres running parallel to the skin surface. The dermis is capable of mineralization and, in most fishes, bony scales develop in the dermis. The scales are embedded in scale pockets and lost scales can regenerate (Kirschbaum 1975). While the basement membrane marks the outer boundary of the dermis, there also exists an inner boundary termed the dermal endothelium (Whitear et al. 1980). It consists of a single layer of cells connected by desmosomes, which represents a complete sheet penetrated by only blood vessels, nerves or insertions of myosepta.

Integument of Non-Teleost Fishes

In this section, we will restrict the description of the integument to agnathan (jawless) fishes and sharks. The integument of hagfishes is shown in Fig. 1A. The epidermis is composed of a superficial and a basal layer of epithelial cells, with numerous thread cells and mucous goblet cells situated between these two layers. The underlying dermis is represented by a thick layer of collagen fibres. In the river lamprey *Lampetra fluviatilis*, a similar organisation of the integument is seen (Fig. 1B). The unicellular glands are comprised of club cells and mucous goblet cells. Figure 1C shows a cross section of the integument of the lamprey *Geotria australis* using transmission electron microscopy. Very prominent are the many sacciform cells filled with mucous globules. A tangential section through one such sacciform cell is shown in Fig. 1D. A cross section through a superficial epithelial cell of the cornea of the shorthead lamprey *Mordacia mordax* shows the glycocalyses extending from the microplicae, which stabilises the mucins of the tear film.

The epidermis covers the whole body, including the eye. However, the outer cell layer of the cornea is composed of specialized epithelial cells, which are characterized by a variety of microprojections increasing the cell surface area (Collin and Collin 2000, 2006). They are thought to improve the movement of oxygen, nutrients, and metabolic products across the outer cell membranes. The smooth optical surface of the cornea is maintained by a tear film, which adheres to these microprojections. As an example of these specialized epithelial cells, which are found in different vertebrate taxa, the cornea of three lampreys are shown in Fig. 1: Fig. 1F shows the ultrastructure of the corneal surface epithelial cells of the shorthead lamprey *Mordacia mordax* in cross section with deep microholes filled with mucus-secreting granules. Surface microplicae and microholes in the cornea of the lamprey *Geotria australis* are shown in the scanning electron micrograph presented in Fig. 1G. Note the rows of microholes along the cell borders. In Fig. 1H, high power scanning electron microscopy reveals the microplicae and microholes in corneal surface cells of the river lamprey *Lampetra fluviatilis*.

The integument of sharks is comprised of a simple epidermis and a thick dermis (Bertin 1958). In the epidermis, superficial epithelial cells and a basal layer of epithelial cells can be distinguished (Fig. 2A), within which lie mucous cells. An essential part of the integument are placoid scales (Figs. 2A and B and Fig. 3). They are arranged in the integument in a regular pattern (Figs. 2B and 3). They originate in the dermis and sit on a platform (Figs. 2A and B) and protrude into the epidermis. The outer layer of the placoid scale is composed of dentine, apparently a product of the epidermis (Bertin 1958). The central part of the scale is enamel (Fig. 2A), produced by dermal cells which extend into the placoid scale. For comparison, see also the chapter "Fish Skeletal Tissues".

Integument of Teleost Fishes

The basic structure of the teleost integument has already been described in the introduction. Here, we show some examples of the integument of various teleost taxa and modifications found in weakly electric teleosts. The epidermis in fish embryos and larvae is composed of a double layered epithelium (Fig. 2C), which develops into a stratified epidermis with two types of single-celled glands: goblet cells and club cells (Fig. 2D). In the zebrafish *Danio rerio* (Fig. 2D), these club cells produce an alarm substance, which is released after injury of the epidermis. In teleosts with large scales, as shown in Figs. 2E and F, the epidermis covers both the upper and the lower parts of the scale. In between the epidermis and the scale, there lies a thin layer of dermal cells, which are responsible for the growth of the scale. The European eel, *Anguilla anguilla*, is an example of a teleost with a thick epidermis and a high number of unicellular glands, goblet cells and club cells (Figs. 2G and H). The small scales in this species do not penetrate the epidermis. Some teleosts have evolved a complex ctenoid scale anatomy (Fig. 4), which serves several protective functions. The pattern of microridges over the skin epithelial surface (see Introduction) appears

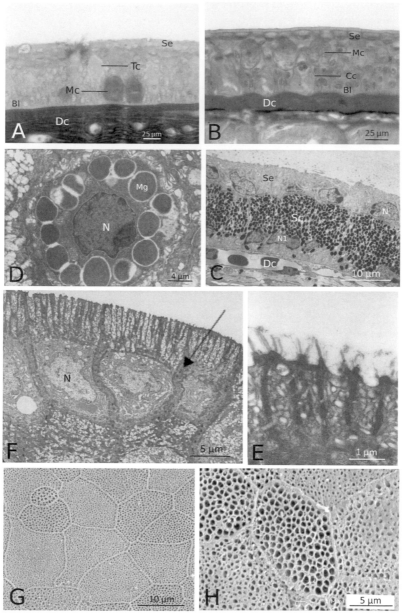

Fig. 1. Integument of the hagfish *Myxine glutinosa* (**A**), lampreys (**B–E**), and corneal epithelium of lampreys (**F–H**). **A.** Cross section through the integument of *M. glutinosa* showing the epidermis with the superficial epithelial cells (Se), the basal layer of epithelial cells (Bl), thread cells (Tc), and mucous goblet cells (Mc) on top of the underlying dermal collagen (Dc). **B.** Cross section through the integument of the river lamprey *Lampetra fluviatilis* depicting the epidermis with the superficial epithelial cells (Se), the basal layer of epithelial cells (Bl), club cells (Cc), and mucous goblet cells (Mc) on top of the underlying dermal collagen (Dc). **C.** Cross section through the integument of the lamprey *Geotria australis* showing the superficial epithelial cells (Se), their nuclei (N), sacciform cells (Sc) and their nuclei (N1) and the underlying dermal collagen layer (D), TEM. **D.** Tangential section through the epithelium of *Geotria australis* depicting a club cell, its nucleus (N) and the surrounding mucous granules, TEM. **E.** Cross section through a superficial epithelial cell of the cornea of the shorthead lamprey *Mordacia mordax* showing the epithelial protrusions (microplicae) and glycocalyses, which stabilise the mucins of the tear film, TEM. **F.** Corneal surface epithelial cells of *M. mordax* showing deep microholes filled with mucus-secreting granules (arrow), TEM. **G.** Microplicae and microholes in the cornea of *G. australis*. Note the rows of microholes along the cell borders, SEM. **H.** The corneal surface of a hexagonal cell of *L. fluviatilis* showing numerous small microplicae but the predominant feature is the presence of many large microholes, through which mucous granules can be expelled, SEM. N, nucleus.

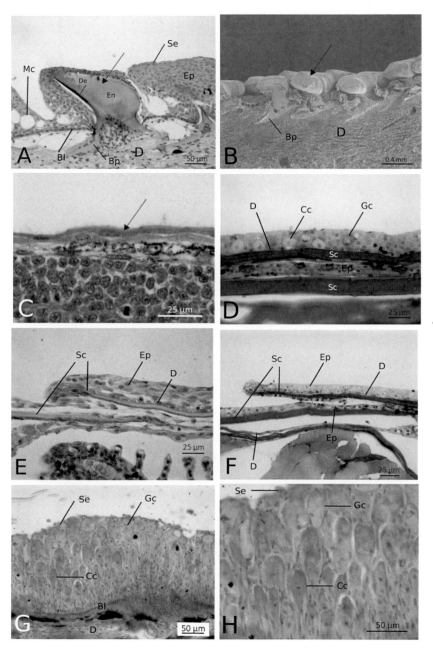

Fig. 2. Integument of chondrichthyan (**A, B**) and teleost (**C–H**) fishes. **A.** Cross section through the skin of a young dogfish *Acanthias* sp. depicting a placoid scale (arrow) integrated in the epidermis (Ep) and embedded in a root-like manner (Bp) in the dermis (D). De, dentine; En, enamel; Mc, mucous cell. **B.** Cross section through the skin of the wobbegong shark *Orectolobus ornatus* showing placoid scales (arrow) embedded in the dermis (D), Bp, basal plate. **C.** Thin, double-layered epidermis (arrow) of a 49 hour old embryo of the zebrafish *Danio rerio*. **D.** Cross section through the integument of an adult *D. rerio* showing goblet cells (Gc), club cells (Cc), a thin layer of dermis (D) covering the surface of a scale (Sc) and additional epidermal cells (Ep) above the second underlying scale (compare to **D**). **E** and **F.** Cross sections of the integument showing epidermis (Ep) and dermal tissues (Dt) enveloping the large scales (Sc) in the congo tetra *Phenacogrammus interruptus* (**E**) and *D. rerio* (**F**). **G, H.** Cross section of the integument of the European eel *Anguilla anguilla* showing the superficial (Se), the basal epithelial cell layer (Bl), many goblet (Gc) and club cells (Cc), and a high power micrograph (**H**) of the same section. D, dermis.

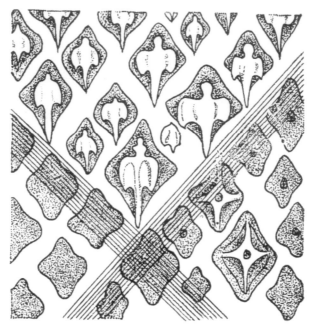

Fig. 3. Arrangement of placoid scales in the integument of a young catshark *Scyliorhinus* sp. The placoid scales of the lower part were ground down to a different degree to show the anatomy of the placoid scale: the basal plates are seen on the left and the tubes of the pulp on the right. Some collagen fibres of the dermis are also indicated. After: P. GRASSE, "Traité de zoologie, Tome 13 Volume 2—Agnathes poissons Anatomie ethologie systematique", © Armand Colin, Paris, 1958.

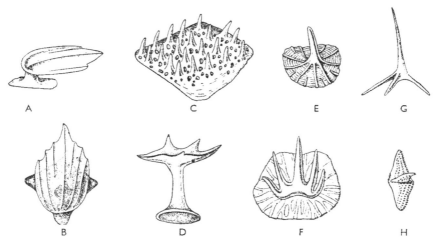

Fig. 4. Examples of modified ctenoid scales in perciform teleosts. **A** and **B.** *Centriscus* sp. (Centriscidae). **C.** Triggerfish *Balistes* sp. (Balistidae). **D.** The Louvar *Luvarus imperalis* (Luvaridae). **E.** The batfish *Ogcocephalus* sp. (Ogcocephalidae). **F.** The filefish *Monacanthus* sp. (Monacanthidae). **G.** The burrfish *Chilomycterus* sp. (Diodontidae). **H.** The searobin *Peristedion* sp. (Peristediidae). After: P. GRASSE, "Traité de zoologie, Tome 13 Volume 2—Agnathes poissons Anatomie ethologie systematique", © Armand Colin, Paris, 1958.

to be fingerprint-like and may be species-specific. The complex pattern of epidermal microridges in the sandlance *Limnichthyes fasciatus is* shown in Fig. 5A.

Modifications of the general structure of the integument are seen in the two groups of weakly electric fish, the South American knifefishes (Fig. 5B) and the African mormyrid fishes (Figs. 5C and D), where the epidermis is composed of three distinct layers in both taxa (Genthen et al. 2009), although Szabo (1974)

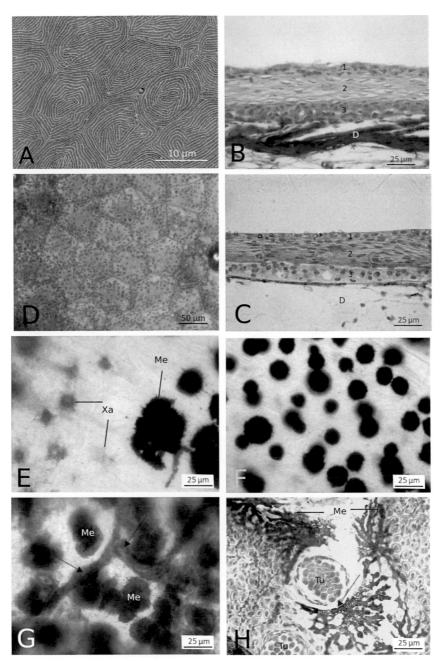

Fig. 5. Integument and pigment cells of teleost fishes. **A.** Fingerprint pattern of microridges of the skin epithelial surface of the sandlance *Limnichthyes fasciatus*, SEM. **B–C.** Cross section through the integument of weakly electric teleosts, the glass knifefish *Eigenmannia virescens* (**B**) and the mormyrid fish *Pollimyrus isidori* (**C**) showing the three layers (1–3) of the epidermis and the underlying dermis (D). **D.** Tangential section through the second epidermal cell layer of *P. isidori*, note the regular hexagonal pattern. **E–G.** Pigment cells in the anal fin of the zebrafish *Danio rerio*. **E.** Melanophores (Me) and (Xa) are seen. **F.** Melanophores (black pigment) and xanthophores (yellow pigment) with pigment granules concentrated in the centre of the pigment cells. **G.** Melanophores (Me) surrounded by iridophores (arrows). **H.** Melanophores (Me) in the integument of *E. virescens* surrounding tuberous electroreceptors (Tu). Note the two nuclei (arrow) of the melanophore typical for this pigment cell type.

distinguishes four layers (two different layers in the basal laycr). Tangential sections through the middle layer in mormyrid fishes reveal a regular pentagonal/hexagonal pattern (Fig. 1D). The epidermis of these fishes has a high electrical resistivity, where the current produced by the electric organ enters the fish via the pores of the electroreceptors.

Pigment Cells

The pigment pattern of fishes is due to the presence of specialized cells, chromatophores, which populate different parts of the integument. They contain pigment-containing organelles, which can be translocated within the cell, producing different colour patterns. The movement of the organelles is regulated via nerves and/or hormones. Within the pigment cells, filamentous components and microtubules have been revealed (Schliwa 1982). However, the mechanism of the movement of the pigment cell organelles is still not well understood (reviewed by Schliwa 1986). The mechanism of the rapid colour change is called 'physiological colour change' but there is also a change in the number of pigment cells or the regional amount of pigment within the cell. This latter process is slower and is termed 'morphological colour change'. The pigment cells originate apparently from the neural keel, which is the equivalent of the neural crest (Armstrong 1980). The undifferentiated and colourless precursor cells migrate into different parts of the body and specifically into the integument. Four different types of chromatophores are distinguished based on the characteristics of the pigments: melanophores containing black or brown pigments (Figs. 5E–H), erythrophores with red or yellowish pigments, xanthophores with primarily yellowish pigments (Figs. 5E and F), and leucophores or iridophores with colourless pigments (primarily guanine) (Fig. 5G). There is growing support for the notion that all pigment cells originate from a pluripotent stem cell (Bagnara et al. 1979). Additional support comes from the observation that there are mosaic pigment cells with more than one pigment organelle (reviewed by Schliwa 1986).

Acknowledgements

We thank Prof. R. Krahe for his offer to use his microscope-digital-camera device for taking the photomicrographs. Prof. Scholz kindly offered the use of histological preparations of the "Zoologische Lehrsammlung der Humboldt-Universität zu Berlin". We also thank Michael Archer and H. Barry Collin for histological assistance.

References Cited

Aisa, E. 1959. I tubercoli nuziali in Rutilus rubilio Bp. var. rubella (trasimenicus) Bp. del Lago Trasimeno. Boll. Zool. 26: 601–609.

Anctil, M., L. Descarries and K.C. Watkins. 1984. Distribution of noradrenaline and (^{3}H) serotonin in photophores of *Porichthys notatus*. An electron-microscopic radio-autographic study. Cell Tissue Res. 235: 129–136.

Armstrong, P.B. 1980. Time-lapse cinemicrographic studies of cell motility during morphogenesis of the embryonic yolk sac of *Fundulus heteroclitus* (Pisces: Teleostei). J. Morphol. 165: 13–29.

Bagnara, J.T., J. Matsumoto, W. Ferris, S.K. Frost, W.A. Turner, T.T. Chen and J.D. Taylor. 1979. Common origin of pigment cells. Science 203: 410–515.

Balon, E.K. 2003. Coelacanthiformes (Coelacanths). pp. 189–196. *In*: M. Hutchins, D.A. Thoney, P.V. Loiselle and N. Schlager (eds.). Grzimeks Animal Life Encyclopedia, 2nd edition. Vol. 4, Fishes I. Farmington Hills, MJ: Gale Group.

Bertin, L. 1958. Peau et pigmentation. Glandes cutanees et organs lumineux. Ecailles et sclerifications dermiques. Denticules cutanes et dents. pp. 433–531. *In*: P.P. Grasse (ed.). Traite de Zoologie, Vol. XIII, Prem. Fasc., Masson, Paris.

Best, A.C.G. and Q. Bone. 1976. On the integument and photophores of the alepocephalid fishes *Xenodermichthys* and *Photostylus*. J. Marine Biol. Assoc. UK 56: 227–236.

Blackstock, N. and A.D. Pickering. 1980. Acidophilic granular cells in the epidermis of the brown trout, *Salmo trutta* L. Cell Tissue Res. 210: 359–369.

Britz, R., F. Kirschbaum and A. Heyd. 2000. Observations on the structure of larval attachment organs in three species of gymnotiforms (Teleostei: Ostariophysii). Acta Zoologica. 81: 57–68.

Bullock, A.M., R. Marks and R.J. Roberts. 1978. The cell kinetics of teleost fish epidermis: mitotic activity of the normal epidermis at varyıng temperatures in plaice, *Pleuronectes platessa*. J. Zool. 184: 423–428.

Burgess, B.H.O. 1956. Absence of keratin in teleost epidermis. Nature (London) 178: 93–94.

Casimir, M.J. and H.W. Fricke. 1971. Zur Funktion, Morphologie und Histochemie der Schwanzdrüse bei Röhrenaalen (Pisces, Apodes, Heterocongridae). Marine Biol. 9: 339–346.

Claes, J.M., D.-E. Nilsson, N. Straube, S.P. Collin and J. Mallefet. 2014. Isoluminance counterillumination drove bioluminescenent shark radiation. Scientific Rep. 4: 4328. doi: 10.1038/srep04328.

Collin, H.B. and S.P. Collin. 2000. The corneal surface of aquatic vertebrates: microstructures with optical and nutritional functions? Phil. Trans. R. Soc. Lond. B 355: 1171–1178.

Collin, S.P. and H.B. Collin. 2006. The corneal epithelial surface in the eyes of vertebrates: environmental and evolutionary influences on structure and function. J. Morph. 267: 273–291.

Genten, F., E. Terwinghe and A. Danguy. 2009. Atlas of Fish Histology. Science Publishers, Enfield, Jersey, Plymouth.

Gruber, D.F., E.R. Loew, D.D. Deheyn, D. Akkaynak, J.P. Gaffney, W.L. Smith, M.P. Davis, J.H. Stern, V.A. Pieribone and J.S. Sparks. 2016. Bioflurescence in catsharks (Scyliorhinidae): Fundamental description and relevance for elasmobranch visual ecology. Scientific Reports 6: 24751. DOI: 10.1038/srep24751.

Halstead, B.W. and L.G. Halstead. 1978. Poisonous and venomous marine animals of the world. Rev. Edition. Darwin Press, Princeton.

Harder, W. 1975. Anatomy of Fishes. E. Schweizerbartsche Verlagsbuchhandlung. Nägele und Obermiller, Stuttgart.

Hartschuh, W., E. Weihe and M. Reinecke. 1986. The Merkel cell. pp. 605–620. *In*: J. Bereiter-Hahn, A.G. Matoltsy and K.S. Richards (eds.). Biology of the Integument 2 Vertebrates. Springer-Verlag, Berlin, Heidelberg.

Henrikson, R.C. and A.G. Matoltsy. 1968. The fine structure of telcost epidermis. I. Introduction and filament containing cells. J. Ulrastruct. Res. 21: 194–212.

Herring, P.J. 1982. Aspects of the bioluminescence of fishes. Oceanogr. Mar. Biol. Ann. Rev. 20: 415–417.

Hildemann, W.H. 1959. A cichlid fish, Symphysodon discus, with unique nurture habits. Am. Nat. 93: 27–34.

Iggo, A. and A.R. Muir. 1969. The structure and function of a slowly adapting touch corpuscule in hairy skin. J. Physiol. 200: 763–796.

Ilg, L. 1952. Über larvale Haftorgane bei Teleosteern—Zoologische Jahrbücher, Abteilung für Anatomie und Ontogenie der Tiere. 72: 577–600.

Kirschbaum, F. 1975. Untersuchungen über das Farbmuster der Zebrabarbe *Brachydanio rerio* (Cyprinidae, Teleostei). Wilh. Roux`s Arch. 177: 129–152.

Lethbridge, R.C. and I.C. Potter. 1982. The skin. pp. 377–448. *In*: M.W. Hardisty and I.C. Potter (eds.). The Biology of Lampreys, Vol. III. Academic Press, London, New York.

Linna, T.J., J. Finstad and R.A. Good. 1975. Cell proliferation in epithelial and lympho-haematopoietic tissues of cyclostomes. Am. Zool. 15: 29–38.

Pfeiffer, W. 1968. Über die Epidermis von *Latimeria chalumnae* J.L.B. Smith 1939 (Crossopterygii, Pisces). Z. Morph. Tiere. 63: 419–427.

Sato, M. and M. Kurotaki. 1964. Electron microscopy and histochemistry of coloured bodies found in the epidermal cells of the goby, *Chaenogobius castanea*. Annot. Zool. Jpn. 37: 215–220.

Schliwa, M. 1982. Chromatophores: Their use in understanding microtubule-dependent intracellular transport. Methods Cell Biol. 25: 285–312.

Schliwa, M. 1986. Epidermis. pp. 65–77. *In*: J. Bereiter-Hahn, A.G. Matoltsy and K.S. Richards (eds.). Biology of the Integument 2 Vertebrates. Springer-Verlag, Berlin, Heidelberg.

Spitzer, R.H., S.W. Downing and E.A. Koch. 1979. Metabolic-morphologic events in the integument of the pacific hagfish (*Eptatretus stoutii*). Cell Tissue Res. 197: 235–255.

Spitzer, R.H., E.A. Koch, R.B. Reid and S.W. Downing. 1982. Metabolic-morphologic characteristics of the integument of teleost fish with mature lymphocystis nodules. Cell Tissue Res. 222: 339–357.

Szabo, T. 1974. Anatomy of the specialized lateral line organs of electroreception. pp. 13–58. *In*: A. Fessard (ed.). Electroreceptors and other specialized receptors in lower vertebrates. Springer Verlag, Berlin, Heidelberg, New York.

Whitear, M. 1977. A functional comparison between the epidermis of fish and of amphibians. Sympos. Zool. Soc. London 39: 291–313.

Whitear, M., A.K. Mittal and E.B. Lane. 1980. Endothelial layers in fish skin. J. Fish Biol. 17: 43–65.

Whitear, M. 1986a. Epidermis. pp. 8–38. *In*: J. Bereiter-Hahn, A.G. Matoltsy and K.S. Richards (eds.). Biology of the Integument 2 Vertebrates. Springer-Verlag, Berlin, Heidelberg.

Whitear, M. 1986b. Dermis. pp. 39–64. *In*: J. Bereiter-Hahn, A.G. Matoltsy and K.S. Richards (eds.). Biology of the Integument 2 Vertebrates. Springer-Verlag, Berlin, Heidelberg.

Winn, H.E. 1955. Formation of a mucous envelope at night by parrot fish. Zoologica (New York) 40: 145–148.

Chapter 3

Fish Skeletal Tissues

François J. Meunier

INTRODUCTION

Fish skeletal tissues are comprised of mineralized bony tissues, cartilages (which can mineralize), teeth, (Francillon-Vieillot et al. 1990, Ricqlès et al. 1991), and some specialized unmineralized conjunctive tissues (Meunier and Géraudie 1980, Meunier 1987). The bony skeleton is comprised of different bones that have various morphologies and body localizations: vertebrae, head skeleton, jaws, fin endoskeleton, myorhabdos, fin rays, scales. Fish skeletal tissues and especially mineralized tissues serve various mechanical and physiological functions as does the tetrapod skeleton (Doherty et al. 2014 for a review).

Nowadays, living fishes constitute two main taxa: the Chondrichthyes (cartilaginous fishes) and the Osteichthyes (bony fishes). The first comprehensive studies that describe the skeletal tissues of the Osteichthyes date from the mid nineteenth century (Williamson 1849, 1851, Kölliker 1859, Hertwig 1876–82, and others). Owing to these various works, it rapidly appeared that there are two main groups of fishes: fishes that have bony tissues with embedded cells (the osteocytes) called "cellular bone" and fishes that are completely devoid of enclosed cells, i.e., "acellular bone" (= osteocytic bone and anosteocytic bone of Weiss and Watabe 1979).

Cartilages, bones, teeth and conjunctive tissues have been, for a long time, considered as inert tissues in Teleostei. In fact, this is not true: the skeleton is formed of well alive tissues having various physiological functions (Francillon et al. 1990, Ricqlès et al. 1991, Witten et al. 2004a, Doherty et al. 2014). Thus, bony tissues show a more or less developed remodelling, one of the consequences of which is to insure exchanges between bone and the internal environment of fishes in order to participate in various homeostasic activities under hormonal control (Lopez 1973). Moreover, skeleton grows, which leads to changes of bone morphology and proportion (modelling of bone) without loss of any action for mechanical functions and of the transmission of thrust by the vertebral axis during swimming. Therefore, the fish skeleton is a morpho-functional complex that has been neglected too much during a long time by the ichthyophysiologists, in contrast to other anatomical systems playing other functions like feeding, blood circulation, homeostasy, and reproduction. Moreover, the infraclass of Teleostei is known to regroup more than 30,000 species (Froese and Pauly 2019), the biology and environmental constraints of which are very diverse, but the histophysiological characteristics of their skeleton have been studied, at the best, on some hundred species only.

UMR 7208 (CNRS-IRD-Sorbonne Universités-MNHN), BOREA, Département Adaptations du Vivant, Muséum national d'Histoire naturelle, C.P. 026, 43 rue Cuvier, 75231 Paris cedex 05, France.
Email: francois.meunier@mnhn.fr

Fish bones and teeth have mineral components and so the skeleton of osteichthyes can fossilize relatively frequently that allows paleomorphological and paleohistological studies. Therefore, from a technical point of view on the one hand and skeleton history on the other hand, there is no discontinuity between paleohistology (distant past) and histology (present time) (Meunier 2011); the recent past (archeozoological material) is also in the scope of histo-morphological bone techniques even if this kind of fossil material shows some fragility contrary to old fossil material.

The mineral component of skeletal tissues has been, for a long time, a true technical difficulty and did not allow the deciphering at the histological level of the cellular and extracellular structures of bony tissues and associated tissues. The first step was to remove the mineral component and to subsequently use classical paraffin microscopy. Afterwards, studying bone after removing the mineral, an essential characteristic of bone, appears as an aberration. Even if techniques for sectioning undecalcified material (adapted from petrographic techniques) are relatively old (mid XIXth century; Agassiz 1833–44, Williamson 1849, and see Meunier and Herbin 2014), they have not been successfully employed for about a century. This fact can explain the relative recurrent lack of interest of the ichthyologist's community for an accurate study of skeletal elements in the Osteichthyes. Thus, all along the past five to six decades, sophisticated improvements to manage the mineralized tissues in association with the recent great development of 3D computering (Sanchez et al. 2012, Qu et al. 2013, Qu 2015, and others) had allowed important progress of the knowledge of osteichthyan mineralized skeleton (bones, cartilages and teeth), and the biology of these specific tissues. We shall present the essential knowledge acquired and open new directions for future studies.

Remarks. The otoliths, that are mineralized (CO_3Ca_2) structures of the inner ear, are out of the scope of the present chapter because they do not belong to the skeleton. On the contrary, some specific bony tissues have been described out of the normal skeleton as the rocker bones localized on the swim bladder of the Carapidae (Parmentier et al. 2008) or the mineralized plates that surround the lungs of the extinct and extant Actinistia (Brito et al. 2010, Cupello et al. 2017).

Histological Structure of Fish Bone

Generalities and Definitions

The tissues that constitute the skeleton belong to the conjunctive tissue group. To understand and to facilitate comparisons between the various species that we want to describe in this chapter, it is necessary to give some fundamental definitions about conjunctive tissues and their derivates.[1]

Conjunctive Tissues[2]

Conjunctive tissues have a mesodermal origin and they are composed of specific cells, the fibrocytes, embedded in an organic fibrillary matrix (collagenous fibres), surrounded by mucopolysaccharidic substances (the proteoglycans) and a lot of other organic complex molecules. A conjunctive tissue plays a fundamental function as a frame for the organs in a general way. Certain conjunctive tissues are specialized as true organs that mineralize to form a skeleton that assumes the frame of the fishes: bones, and cartilages.

Cartilaginous Tissue. This is a specialized conjunctive tissue enriched with proteoglycans, comparable to chondroitine sulfates. These substances give the cartilage its peculiar consistence and colour. The collagenous fibres of the cartilage are of collagen type II. The fibrocytes differentiate in chondroblasts that are organized in a chondrogenic "membrane" the perichondre, that overlays the cartilaginous element

---

[1] These definitions are suitable for all vertebrates (see Francillon-Vieillot et al. 1990, Ricqlès et al. 1991).

[2] The conjunctive tissues differ completely from epithelial tissues which, in general, are of ectodermal origin and are constituted of joined cells the basal part of which lies on a basal membrane. Epithelia build up covering membranes or they line organ cavities; many constitute glandular tissues (see, e.g., Chapter Endocrine Systems).

(Fig. 1A). The chondroblasts surround themselves progressively with a cartilaginous matrix and thus break away from the perichondre. They become chondrocytes that are incorporated into the cartilage. Chondrocytes are localized in regular ovoid or spherical lacunae, and they are deprived of radiating cytoplasmic processes (Figs. 1A, B). There exists a variety of cartilage types according to their organic composition (Benjamin 1988, 1989, 1990, Benjamin et al. 1992).

The chondrocytes keep their potential of multiplication contrary to the osteocytes (see below). During multiplication, chondrocytes frequently form isogenic groups, i.e., namely a series of cells all coming

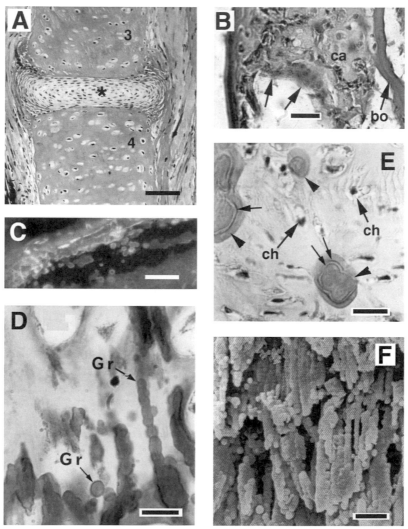

Fig. 1. A. *Protopterus annectens* (Dipnoi). Third (3) and fourth (4) distal segment of the pelvic fin (trichrome staining). Both segments are constituted of cartilaginous tissue, overlaid by the perichondre and separated by a ligament joint (asterisk). The chondrocytes are more or less circular without cellular extensions (Scale bar = 10 μm) (From Meunier and Laurin 2012). **B.** *Salmo trutta fario* (Salmonidae). Cross section of a ceratohyal of a small trout showing chondroclasts (arrows) that destroy cartilaginous tissue (APS-Groat-PIC) (bo, bone; ca, cartilage) (Scale bar = 50 μm). **C.** *Cyprinus carpio* (Cyprinidae). Cross section of the hemal cartilage. Vital labelling (tetracycline) allows localization of active mineralizing cartilage: mineralized spherules progressively fuse with each other (Scale bar = 20 μm). **D.** *Cyprinus carpio* (Cyprinidae). Cross section of the hemal cartilage showing calcified cartilage (Azan staining) (Gr, mineralized granules) (Scale bar = 10 μm). **E.** *Cyprinus carpio* (Cyprinidae). Detail of calcified cartilage showing liesegang lines (arrows) in the mineralized granules (arrow-heads) (ch, chondrocytes) (Scale bar = 20 μm). **F.** *Cyprinus carpio* (Cyprinidae). MEB after destruction of the unmineralized matrix. Spherical mineralization of the cartilage is obvious (Scale bar = 15 μm) (From Zylberberg and Meunier 2008).

from the same initial chondrocyte. The cartilage can mineralize (Figs. 1C, D): this property, that is exceptional for Mammals, is more frequent in teleosts (Meunier 1979, Zylberberg and Meunier 2008) and more or less general in the Chondrichthyes (Kemp and Westrin 1979). Cartilage growth results from a peripheral thickening owing to the perichondre activity and/or with the division of the chondrocytes (interstitial growth). Clastic cells, the chondroclasts, can destroy the cartilaginous tissue (Fig. 1B) (Meunier 1979, Zylberberg and Meunier 2008). The extracellular matrix shows various staining intensities with lines, some of which are concentric and parallel (Fig. 1E), others forming spherical shape (Figs. 1D, F). These lines are Liesegang lines that characterize an active mineralizing process named spheritic mineralization (Ørvig 1951). The rings of Liesegang indicate that the individual granules increase through periodic centrifugal deposit of mineral. The mineralizing front grows owing to calcified granules that progressively fused with each other.

Bony Tissue. This is a conjunctive tissue enriched with collagenous fibres of type I (Zylberberg 2004). Normally, it mineralizes with hydroxyapatite crystals (Francillon-Vieillot et al. 1990, Ricqlès et al. 1991). The cells that synthesize bone substance are the osteoblasts, which lie at the surface of bone and form a periost (Fig. 2A). When the osteoblasts are imbedded in the bone matrix, they become osteocytes

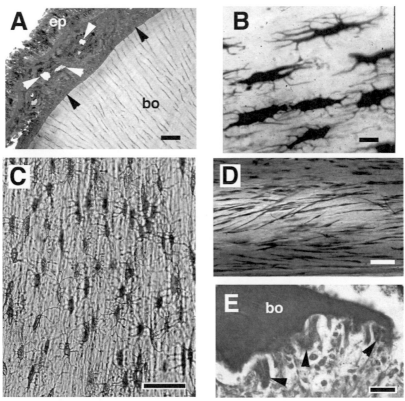

Fig. 2. A. *Sparus aurata* (Sparidae) (Semi-thin section; Toluidine blue staining). Cross section of a dorsal fin ray showing the periostic membrane (black arrow-heads) at the surface of bone, and the numerous osteoblastic canalicles in the bony tissue (bo). The white arrow-heads point to vascular canals in the dermis (ep, epidermis) (Scale bar = 25 μm) (From Sire and Meunier 2017). **B.** *Cyprinus carpio* (Cyprinidae; dorsal fin ray; Von Ebner staining). Detail of star-shape osteocytes with their cytoplasmic canalicles (Scale bar = 10 μm). **C.** *Barbus barbus* (Cyprinidae) (groung section). Axial section of a dorsal spiny ray showing the star shape osteocytes with their osteocytic canalicles (Scale bar = 50 μm) (From Meunier and Herbin 2014). **D.** *Katsuwonus pelamis* (Scombridae) (Dorsal spiny ray; ground section; polarized light). Osteocytes are elongated and the few osteocytic canalicles are localised at the cell extremities (Scale bar = 100 μm). **E.** *Eigenmannia virescens* (Ramphichthyidae; Gymnotiformes). Cross section of a 5-day-old regenerating tail showing multinucleated osteoclasts (arrow-heads) on the surface of the bony stump (bo) (APS-Groath-PIC) (Scale bar = 25 μm) (From Kirschbaum and Meunier 1981).

that lie in an osteocytic lacuna (Figs. 2A–D). The osteocytes are star shaped and more or less flattened, especially in pseudo-lamellar and lamellar bone, but globular in woven bone; they show at their margin thin ramified cytoplasmic processes running in micro-tunnels of the bony matrix, the osteocytic canalicles (Figs. 2B, C). These osteocytic canalicles form a more or less developed network in the collagenous matrix, which participates in bone nutrition (Cao et al. 2011). Yet, many bony fishes, notably many teleostean taxa belonging to the Acanthopterygii, do not have osteocytes in their bony tissues (see below). The mineral component of bone is a calcium phosphate that crystallizes as hydroxyapatite (see below) on the fibrillary organic matrix. The mineralization of fish bone generally represents about 60–65% of the dry weight (Table 1).

Bone is destroyed by specialized clastic cells, the osteoclasts, which are generally multinucleated cells (Fig. 2E) (Kirschbaum and Meunier 1981, Sire et al. 1990). Many teleosts have mononucleated osteoclasts (Witten 1997, Witten et al. 2001, Witten and Huysseune 2009) besides multinucleated ones. That is the reason why, for a long time, scientists claimed that teleostean fishes do not have osteoclasts (Blanc 1953). Because of its mineralized condition, bone growth implies only a peripheral secretion of new matrix owing to the activity of the periostic cells, the osteoblasts. Effectively, osteocytes cannot divide because they are locked in the stiff mineralized matrix. There is no possibility of internal growing, contrary to cartilage. But internal bone can renew if it is destroyed before: this represents bone remodelling (Francillon-Vieilot et al. 1990, Ricqlès et al. 1991).

Chondroid Bone. Chondroid bone is a peculiar mineralized skeletal tissue with mixed characters of bone and cartilage (Beresford 1981, Benjamin 1989, Meunier and Huysseune 1992). Its localization in the skeleton and its functions are also specific (Haines and Mohuidin 1968). Frequent in various compartments of the splanchnocrane (Benjamin 1989, 1990), the chondroid bone assumes topographic links between bony and cartilaginous elements, especially in articulation zones (Huysseune 1986, 1989, 2000, Huysseune and Verraes 1986, Huysseune and Sire 1990).

Table 1. The mineral rate of bony tissues (ash % of dry bone) in several teleostean fishes. Min-Max, minimal and maximal values; +, cellular bone; –, acellular bone. [(1) Sbaihi et al. 2007; (2) Bareille et al. 2015; (3) Meunier et al. 2008; (4) Kacem et al. 2004; (5) Kacem et al. 2000; (6) Casadevall et al. 1990].

Species	Family	Bone	Bony type	Mineralization rate (%)	
				Mean	Min/Max
Anguilla anguilla[1]	Anguillidae	vertebra	+	49,6	
Anguilla obscura[2]	Anguillidae	vertebra	+		48,1–52
Conger conger[2]	Congridae	vertebra	+		49,5–50,2
Sardina pilchardus[6]	Clupeidae	whole skeleton	–		57,3–61,3
Cyprinus carpio[3]	Cyprinidae	mandible	+	56,5	53,2–60,4
Cyprinus carpio[3]	Cyprinidae	spiny ray	+	59,4	56,5–61,4
Cyprinus carpio[3]	Cyprinidae	rib	+	58,3	53,8–63,8
Cyprinus carpio[3]	Cyprinidae	abdominal vertebra	+	50,2	44,7–57,8
Cyprinus carpio[3]	Cyprinidae	caudal vertebra	+	55,3	52,7–57,3
Oncorhynchus mykis[4]	Salmonidae	vertebra	+		49,6–52,7
Salmo salar[5]	Salmonidae	vertebra	+		48,4–52,3
Esox lucius[3]	Esocidae	mandible	–	65,2	
Esox lucius[3]	Esocidae	opercule	–	63,8	
Micromesistius potassou[6]	Gadidae	whole skeleton	–		66,8–70,8
Coryphaenoides zaniophorus[3]	Macrouridae	mandible	–	65,2	
Coryphaenoides zaniophorus[3]	Macrouridae	dorsal spiny ray	–	63,1	

Teeth Tissues

In fishes, teeth originate in the dental lamina, an epithelial-mesenchymal formation (Huysseune 2006, Soukup et al. 2008, Smith et al. 2009, Jernvall and Thesleff 2012, and many others). A tooth is formed of a cone of dentine around a pulp cavity (Figs. 3A, B). It is covered by a layer of hypermineralized substance (Figs. 3A, B), either enamel or enameloid. This hypermineralized layer is frequently thicker at the top of the teeth, especially in predatory fishes (Suga et al. 1983).

Dentine. Dentine is the main component of teeth. It is a mesodermal formation whose structure is very close to that of bony tissue: specific cells (the odontoblasts) and a mineralized collagenous fibrillary matrix (type I collagen). The growth of dentine is centripetal. Thus, the odontoblasts that synthesize dentine stay at the surface of the pulp cavity and they send polarized cytoplasmic processes inside the dentine (Lison 1954, Peyer 1968, Schmidt 1971). This dentine penetrated by odontoblastic canalicles, more or less at right angle, is named orthodentine (Fig. 3C). The mineralization rate of dentine is higher than that of bone by about 10–15%. In Teleostei, teeth are fixed directly on the jaw owing to a bone attachment; in certain fishes, teeth are fixed with a mineralized ligament and in some others, by an unmineralized ligament (for example in numerous Siluriformes) (see Fink 1981).

In teeth of certain fishes, dentine can be crossed by vascular canals with odontoblasts on their walls. If these odontoblasts send cytoplasmic processes in the dentine tissue, it is named osteodentine (Fig. 3D): for example, the teeth of the pike *Esox lucius* (Lison 1954, Peyer 1968, Schmidt 1971, Texereau et al. 2018). In other species like the Gadidae and the Esocidae, there are no odontoblastic processes starting from walls of the vascular canals (Fig. 3E); this dentine is named vasodentine (Lison 1954, Peyer 1968, Schmidt 1971).

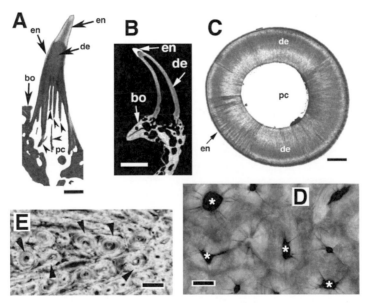

Fig. 3. A. *Lepisosteus platostomus* (Lepisosteide) (Ground section; transmitted light). Axial section of a fang showing the cone of dentine (de) with an apical cape of enamel (en). In the pulp cavity (pc), dentinous pleats (arrow-heads) characterize a plicidentine organization (bo, bone) (Scale bar = 500 µm) (From Meunier and Brito 2017). **B.** *Amia calva* (Amiidae) (Ground section; microradiography). Axial section of a caniniform tooth showing the cone of dentine (de) with an apical cape of enameloid (en) (bo, bony jaw) (Scale bar = 1 mm) (From Germain and Meunier 2018). **C.** *Lophius piscatorius* (Lophiidae) (Ground section; transmitted light). Transversal section of a caniniform tooth showing the orthodentine (de) around the pulp cavity (pc) and with an external layer of enameloid (en). The pulp cavity is regularly circular (Scale bar = 250 µm) (From Meunier 2015). **D.** *Lepidopus caudatus* (Trichiuridae) (Ground section; transmitted light). Cross section of a tooth showing denteons (arrow-heads) with odontoblastic canalicles (= osteodentine) (Scale bar = 10 µm) (From Texereau et al. 2018). **E.** *Anarhichas denticulatus* (Anarhichadidae) (Ground section; transmitted light). Cross section of a tooth showing denteons (arrow-heads) without odontoblastic canalicles (= vasodentine); compare with Figure **D** (Scale bar = 100 µm) (From Meunier and Germain 2018).

In many teleostean fishes, the pulpar cavity of the teeth is perfectly circular (Fig. 3C). But recently, several studies have shown in certain piscivorous actinopterygian fishes that the dentine walls of the pulp cavity are pleated (Meunier et al. 2013, 2015a, 2018a, Meunier 2015, Germain et al. 2016, Meunier and Brito 2017). This three-dimensional organization of dentine tissue is named plicidentine (Peyer 1968, Schultze 1969, 1970, Schmidt 1971, and many others). Contrary to the teeth of numerous fossil Sarcopterygii (Bystrow 1939, Schultze 1969, 1970), these plies are simply not ramified in teleostean teeth; this teleostean plicidentine is of the simplexodonte type (Meunier et al. 2015a, b). Such a simplexodonte plicidentine exists also in the teeth of Polypteridae and Amiidae-two basal taxa of the Actinopterygii (Germain and Meunier 2018), and in the Devonian †*Cheirolepis* (Meunier et al. 2018c). This peculiar spatial organization of the dentinous tooth tissue is very well revealed with non destructive X-ray tomography (Meunier et al. 2015b, Germain et al. 2016, Germain and Meunier 2018).

Enamel. Enamel is only associated with teeth and their derivates. It is not a true skeletal tissue. It is an ectodermal production (linked with dentine: epidermo-dermal interaction) of the basal lamina of the epidermis (tegumentary odontods) or of the buccal epithelium (teeth) (Lison 1954, Ørvig 1967, Peyer 1968, Schmidt 1971). Enamel is very rich in mineral (95% of dry weight), that is deposited as hydroxyapatite crystals on an organic matrix constituted of specific proteins, the amelogenines and the enamelines, without collagenous fibres. But in numerous cases, particularly in teleosts, the authors considered that the external hypermineralized layer of teeth is a mixed production with a mesodermic participation (Lison 1954, Peyer 1968, Schmidt 1971); this peculiar epidermo-dermal production is named enameloid. In polypterids, lepisosteids and coelacanthids, the hypermineralized layer of the teeth is true enamel (Castanet et al. 1975, Ischiyama et al. 1999, Sasagawa and Ishiyama 2005a, b, Sasagawa et al. 2009, 2012, 2013, 2014).

Variability of Components of Bony Tissues

Cellular Variability

The osteocytes of the Teleostei are typically star-shaped cells with cytoplasmic processes (Figs. 2B, C), that enter more or less deeply in bony tissues owing to microscopic tunnels named canalicles. The number and the shape of these more or less long processes vary according to the taxa, as already described by Stephan as early as in 1900. For example, osteocytes in Salmonids and Thunas are weakly elongated and their canalicles, poorly ramified, are localized only in the extremities of the cells (Fig. 2D) (Stephan 1900). Moreover, in a number of teleostean species, bone is wholly deprived of osteocytes (Figs. 4A–C); in this case, bone is named "acellular bone" (Kölliker 1859, Moss 1961, 1965) or "anosteocytic bone" (Weiss and Watabe 1979).[3] In acellular bone, it seems that the osteoblasts withdraw before the front of bone synthesis instead of merging progressively in the bone matrix to become osteocytes (Moss 1961, 1965). Thus, in certain cases in acellular bone, osteoblasts maintain more or less long and branched cytoplasmic processes (Figs. 4A, D) in bony tissue (Meunier and François 1992, Ricqlès et al. 1991, Sire and Meunier 2017). This last type of bone deprived of osteocytes but with incorporated osteoblastic processes is called "tubular acellular bone" (Hughes et al. 1994).

Specific canalicles, the so-called canalicles of Williamson (Williamson 1849, Goodrich 1913, and others), are developed in the extinct and extant Holostei that regroup taxa with cellular bone (Ørvig 1951, Schultze, 1966, Meunier 2011, and others). These canalicules of Williamson are considered as playing a trophic function (Sire and Meunier 1994). The actual hypothesis is that the cells and canalicles of Williamson could have been secondarily acquired (Sire and Meunier 1994).

Acellularisation of bone is also known in fossil taxa: the Heterostracae, and the Osteostracae (Halstead 1963, 1969, 1973), but also in the extant Dipnoi (Géraudie and Meunier 1984). Acellularisation is a heterochronic process that appears in evolved teleostean taxa (Meunier 1987, Davesne et al. 2019). The aspidine of Heterostracan, a fossil group of agnathan fishes that lived in Paleozoic time, is a true acellular bony tissue (Ørvig 1951, Denison 1963, Halstead 1963, 1965, 1973).

---

[3] In the specialized studies on fish bone, it is "cellular/acellular" bone that is currently used.

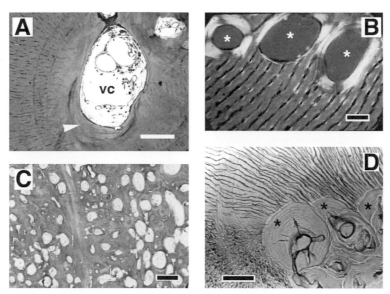

Fig. 4. A. *Sparus aurata* (Sparidae) (semi-thin section; Toluidine blue staining). Cross section of a dorsal fin ray showing osteoblastic canalicles in the left of the figure. Osteocytes are lacking in the bony tissue. At the bottom of the vascular cavity (vc), secondary bone is seen (arrow-head) (Scale bar = 50 μm) (From Sire and Meunier 2017). **B.** *Trachurus trachurus* (Carangidae) (Polirized light). The section shows three secondary osteons (asterisks) and primary vascular bone on the bottom (Scale bar = 100 μm). **C.** *Trachurus trachurus* (Carangidae). Hyperostotic supraoccipital bone. Numerous vascular cavities are seen and osteocytes are lacking as do osteoblastic canalicles (Scale bar = 250 μm). **D.** *Lethrinus nebulosus* (Lethrinidae). Cross section of a dorsal fin ray. Osteocytes are lacking in the bony tissue that is crossed by numerous osteoblastic canalicles. Three secondary osteons (asterisks) are seen on the bottom right (Scale bar = 150 μm).

To answer the evolutionary origin of the osteoblastic canalicles in Acanthopterygii, the authors consider that the formation of the osteoblastic canalicles could be related to the transformation of cellular bone into acellular bone. The osteoblasts are no longer incorporated in the bone matrix but they leave cell processes in the matrix. These osteoblastic processes probably play a nutritive function of bony tissue instead of osteocytes (Sire and Meunier 2017). This is completely different from the evolutionary origin of the canalicles and cells of Williamson that are more subject to discussion (Sire and Meunier 1994).

Fibrillary Matrix Organization

The collagenous fibres of the bony organic matrix belong to type I (Francillon-Vieillot et al. 1990, Ricqlès et al. 1991, Zylberberg 2004), and are deposited by the osteoblasts as thin microfibrils that are packed to form fibres clearly recognizable with the electronic microscope (Zylberberg 2004). The authors have defined three modes of arrangement for the collagenous fibres in bony matrix (Ricqlès et al. 1991): (i) they form an anarchic intermingle network; (ii) they form successive strata in which the fibres have the same direction and are parallel to each other; (iii) they also form successive strata, but in a given stratum the fibres are parallel and their direction in two successive strata differs from an angle near 90° (Giraud-Guilles 1988). These three specific models allow defining respectively: woven fibred bone, pseudo-lamellar bone and lamellar bone (Francillon-Vieillot et al. 1990, Ricqlès et al. 1991). In the lamellar bone category, one can range the isopedine (= elasmodine of Schultze 1996) of elasmoid scales (Meunier and Castanet 1982, Meunier 1987–88). Isopedine is constituted of a series of strata in each of which collagenous fibres are parallel to each other but the direction of which changes from one stratum to the next one with a given angle obviously different from 90° (Meunier and Castanet 1982, Meunier 1987–88; see also paragraph "Bony tissue typology: mineral component"). This very specific model of lamellar network looks like a plywood like structure (Giraud et al. 1978, Meunier 1987).

Mineral Matrix Variability

The mineral component of the bony tissues is generally about 60–65% of the dry weight in Teleostei, as in Tetrapods (Francillon-Vieillot et al. 1990). In fact, relatively recent data has shown a relatively high variability of this parameter in the bony fishes as a whole (Table 1). This mineralization rate can either decrease, or increase, by about 30% relative to this mean value of 60%. Moreover, the mineralization of bone matrix, at the histological level, is clearly heterogeneous as it is revealed by the qualitative (Figs. 5A, B) and quantitative microradiography (Table 2) (Meunier 1984a). The measurement of the mean degree of mineralization of bone may be quantitatively evaluated by exposing an aluminium calibration step-wedge and a plane-parallel calcified tissue section simultaneously to the same beam of X-rays (Boivin and Baud 1984). This last technique allows the measurement of the mineral quantity per volume of bone substance at the histological level: it is the bone mineralization degree (BMD) in g/cm^3.

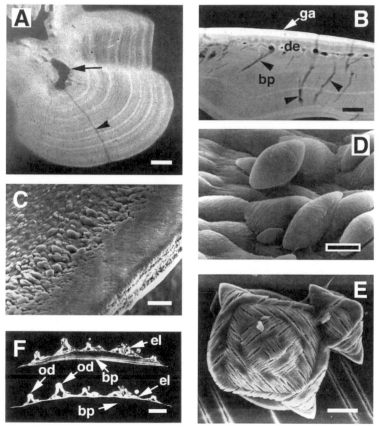

Fig. 5. A. *Cyprinus carpio* (Cyprinidae) (Ground section; X-ray). Cross section of a dorsal spiny ray. The microradiography shows the heterogeneous mineralization of bone. Note the several concentric white lines (= hypermineralization) that were deposited during the slow growth periods. The arrow and the arrow-head respectively point to a vascular cavity and a vascular canal (Scale bar = 250 μm). **B.** *Polypterus senegalus* (Polypteridae) (Ground section; microradiography). Cross section of a scale showing the hypermineralized ganoïne (= enamel layer) (ga), the dentine layer (de) and the basal plate (bp) with several ascending vascular canals (arrow-heads) (Scale bar = 250 μm). **C.** *Astronotus ocellatus* (Cichlidae) (Scale; SEM after destruction of the unmineralized matrix). The mineralizing front shows numerous Mandl's corpuscles (Scale bar = 50 μm). **D.** *Astronotus ocellatus* (Cichlidae) (Scale; SEM after destruction of the unmineralized matrix). Detail of ovoid Mandl's corpuscles (Scale bar = 10 μm). **E.** *Adioryx spinifer* (Holocentridae) (Scale; SEM after destruction of the unmineralized matrix). Detail of a polyedric Mandl's corpuscle (Scale bar = 50 μm). **F.** *Latimeria chalumnae* (Latimeriidae) (Scale; ground section; upper: polarized light; lower: Xray). The section crosses the posterior area of the scale and shows the mineralized external layer overlaying the unmineralized basal plate (Scale bar = 1 mm) bp, basal plate; el, external layer; od, odontode (From Castanet et al. 1975).

Table 2. The degree of mineralization (in g of hydroxyapatite/cm³) of the bony tissues in various bones of several teleostean fishes. Min-Max, minimal and maximal values; +, cellular bone; –, acellular bone). (1) Meunier 1984a; (2) Meunier et al. 2018b.

Species	Family	Bone	Bony type	Fresh (F), Sea (S) Water	Mineralized degree: g/cm³	
					Mean	Min/Max
Anguilla Anguilla[1]	Anguillidae	dentar	+	F	0,98	0,75–1,16
Cyprinus carpio[1]	Cyprinidae	dentar	+	F	0,92	0,80–1,05
Cyprinus carpio[1]	Cyprinidae	spiny ray	+	F	1,00	0,87–1,17
Ictalurus melas[1]	Ictaluridae	spiny ray	+	F	1,10	0,87–1,33
Esox lucius[1]	Esocidae	mandible	–	F	1,13	0,86–1,43
Esox lucius[1]	Esocidae	opercule	–	F	1,28	1,01–1,44
Coryphaenoides coelorhynchus[1]	Macrouridae	dentar	–	S	1,05	0,79–1,36
Coryphaenoides zaniophorus[1]	Macrouridae	dorsal spiny ray	–	S	1,20	0,89–1,46
Nezumia aequalis[1]	Macrouridae	dentar	–	S	1,13	0,88–1,25
Perca fluviatilis[1]	Percidae	dentar	–	F	1,15	0,84–1,36
Drepane africana[1]	Drepanidae	dorsal spiny ray	–	S	1,25	1,05–1,42
Trachurus mediterraneus[1]	Carangidae	cleithrum	–	S	1,24	0,96–1,47
Lepidonothoten squamifrons[2]	Nototheniidae	mandible	–	S	1,17	0,91–1,41
Lepidonothoten squamifrons[2]	Nototheniidae	premaxilla	–	S	1,18	0,90–1,44
Lepidonothoten squamifrons[2]	Nototheniidae	vertebra	–	S	1,05	0,87–1,53
Champsocephalus gunnari[2]	Nototheniidae	mandible	–	S	1,07	0,86–1,36
Champsocephalus gunnari[2]	Nototheniidae	premaxilla	–	S	1,17	0,95–1,56
Champsocephalus gunnari[2]	Nototheniidae	vertebra	–	S	1,21	0,98–1,41
Thunnus alalunga[1]	Scombridae	pectoral ray	+	S	1,12	0,92–1,20
Euthynnus pelamis[1]	Scombridae	dorsal spiny ray	+	S	1,08	0,87–1,37

The variability of the mineral rate has an important biological signification in any studied species. When the demand for mineral elements (essentially calcium and phosphate) is important, for example during vitellogenine production at the time of spawning in salmon and eel (Persson 1997, Persson et al. 1998, Witten et al. 2004b, Sbaihi et al. 2007, 2009, Kacem and Meunier 2009), the skeleton is requested and it can lose a significant part of its stockage; the mineral rate of bone can decrease by up to 40% of dry bone weight in salmon (Kacem et al. 1998, Kacem and Meunier 2003). The variability of the mineralization rate studied with incineration techniques is confirmed with quantitative microradiography (Meunier 1984a, Meunier et al. 2018b). The BMD varies from 1,0 g/cm³ in cellular bone of the teleostean fishes that live in fresh water to 1,5 g/cm³ in the teleostean fishes with acellular bone that live in sea water (Table 2).

It is also necessary to notice in fishes with elasmoid scales, the mode of progression of mineralization in the basal plate, with Mandl's corpuscles (Schönborner et al. 1981, Meunier 1984b, c, 1997, Zylberberg et al. 1992), which progressively grow and then merge with each other to form a rough mineralization front (Figs. 5C–E). In teleostean elasmoid scales, the mineralizing front of the basal plate grows more slowly than the collagenous matrix synthesis. In other species as the coelacanths, the basal plate is constituted of an unmineralized collagenous matrix (Fig. 5F) (Castanet et al. 1975, Meunier et al. 2008).

Other Structural Components of Bony Tissues

The Vascular Network

Bony tissue is an 'alive' tissue that must be "fed". Osteocytes with their expended cytoplasmic processes assume this trophic role at the cellular level (Cao et al. 2011). But when bony tissue is thick, especially in the cortical bone, bony tissue is, in a way, supplied by blood vessels that bring metabolites in the most internal areas of the bone. These vessels take specific "tunnels", the vascular canals that cross the bone matrix (Figs. 5B–D). According to the presence/absence of vascular canals, their abundance and their spatial arrangement, various bony tissue types can be defined: avascular bone, vascular bone with radial or longitudinal canals or plexiforme bone (Francillon-Vieillot et al. 1990, Ricqlès et al. 1991). When the walls of the vascular canals are edged with osteoblasts, these cells can deposit bony tissue that constitutes primary osteons that are not separated from the surrounding bone by a cementing line. During bone remodelling, the osteoclasts can create a tunnel in the mineralized matrix that constitutes a secondary vascular canal. If new bone is deposited by osteoblasts on the walls of this secondary vascular canal, it is separated from the surrounding bony tissue by a cementing line (Fig. 6A) called reversal cementing line. This new cylindrical bony tissue is a secondary osteon (Fig. 6B) (Francillon-Vieillot et al. 1990, Ricqlès et al. 1991).

The Cementing Lines

Fish bony tissue can show, here and there, very thin linear structures, about 1 μm thick, that are chromophilic (Fig. 6A), notably with hematoxyline dye (Castanet 1981). They contain more proteoglycans and less collagenous microfibrils than the surrounding bone; moreover, they are generally weakly hypermineralized (Fig. 5A) compared to the surrounding bony tissue (Castanet 1979). These lines mark temporal disruptions of the osteogenic sequences and they have true function of a cement between two phases of bone matrix deposit. They are called cementing lines.

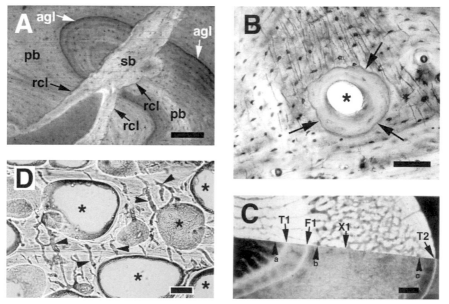

Fig. 6. A. *Cyprinus carpio* (Cyprinidae) (Ground section; Erlich hematoxyline staining). Cross section of a dorsal spiny ray showing primary bone (pb) and secondary bone (sb), that are separated by a reversal cementing line (rcl). In the pimary bone, an arrested growth line (agl) is seen (Scale bar = 100 μm). **B.** *Eusthenopteron foordi* (Tristichopteridae; Devonian). Cross section of a scale; detail of the external layer showing primary cellular bony tissue and a secondary osteon (asterisk) that is limited by a reversal cementing line (arrows) (Scale bar = 100 μm) (From Zylberberg et al. 2010; Fig. 2D). **C.** *Hoplosternum littorale* (Callichthyidae). Cross section of a pectoral spiny ray of a male (upper X-ray; lower vital labelling: T = tetracycline, F = fluoresceine, X = xylenol orange). The fluorochromes are separated by equal intervals of 60 days and they show various speed of growth. Between a and b: avascular slow growing bone; between b and c: spongy fast growing bone that corresponds to the reproductive period (Scale bar = 200 μm) (X-ray) (From Boujard and Meunier 1991). **D.** Unnamed teleostean fish (See Meunier and Herbin 2014). Ground section of a hyperostotic bone. The acellular bony tissue shows primary vascular canals (arrow-heads) and secondary vascular cavities (asterisks), bordered by thin layers of secondary bone (Scale bar = 150 μm).

The disruption of the bone depositing activity is of two types: (i) a stop of the osteogenic processes during growth phases of fish because of bad climatic conditions. These cementing lines are "arrested growth lines" (or A.G.L.) (Fig. 6A) (Castanet 1979, 1981) and they are very useful for the sclerochronological applications (Castanet et al. 1992, 1993, Panfili et al. 2002; see also paragraph "Cyclical bone growth"); (ii) a local new osteogenic phase in an area where bony tissue has been readably destroyed. The cementing lines are called reversal cementing lines (R.C.L.) (Figs. 4A, D, 6A, B). The A.G.L. are observed only in primary bone, resulting in the periostic activity, whereas the R.C.L. delimit an area of secondary bone, and so indicate an erosion-reconstruction process responsible of bone remodelling. The distinction between these two types of cementing lines is essential for the interpretation of the bone dynamic at the histological level (Fig. 6A). When bone remodelling affects a vascular canal, the new layers of bone matrix form a secondary osteon (Figs. 4A, D, 6A, B).

Typology of Bony Tissues

Generally in the same fish, bony tissues differ from one bone to another one, even in several areas of the same bone, and also all along the life of the animal (Ricqlès et al. 1991). Bony tissue characteristics can also vary from one species to another one. Because of these numerous bony tissue descriptions, histologists have constructed typological classifications of bone. Then they have understood that these typologies could have functional significance (Francillon-Vieillot et al. 1990, Ricqlès et al. 1991); therefore, it is now possible to establish a functional classification of bony tissue. They are essentially based on the components of bone (cells, structure of the organic matrix, level of mineralization) and/or the modalities of bone vascularization.

Cellular Component

Among Osteichthyes, numerous taxa lack osteocytes. Therefore, we can distinguish **cellular bony tissues** (e.g., coelacanth, gar-pike, eel, carp, trout, salmon, thuna) and **acellular bony tissues** (e.g., pike, wrasse, tilapia, turbot, gilt head, anglerfish) (Kölliker 1859, Stephan 1900, Moss 1961, 1965, Meunier 1987, Meunier and François 1992, and others). Among Osteichthyes, acellular bone is a particularity of various teleostean groups: essentially the Acanthomorpha, with the exception of the Thunninae (Scombridae) and *Lampris guttatus* (Lampridae), and some other less derivate taxa as the Esocidae, or the Paracanthopterygii (Gadiformes) (Meunier 2011, Davesne et al. 2018, Davesne et al. 2019).

Fibrillary Component

The collagenous fibres show three typical spatial networks: woven fibered bone, pseudo-lamellar bone and lamellar bone. Isopedinc (basal plate) of the elasmoid scales is a very derivate example of lamellar bone tissue, frequently partly or even totally deprived of the mineral component (Fig. 5F) and with a specific mineralizing process: merging of Mandl's corpuscles (Meunier 1987, Zylberberg et al. 1992). The type of fibrillary network arrangement is linked to the speed of their bone matrix deposit (see below).

Mineral Component

The mineral rate of bone is variable but it is not used to differentiate various bony types in fishes. Yet it is possible to describe unmineralized derivate bony tissues, for example the basal plate of elasmoid scales (Meunier 1984c, 1987, 1997, Zylberberg et al. 1992): see, for example, scales of †*Eusthenopteron* (Zylberberg et al. 2010), *Latimeria* (Fig. 5F) (Castanet et al. 1975, Meunier et al. 2008), Ostariophysii (Schönborner et al. 1981), Dipnoi (Meunier and François 1980), Osteoglossiformes (Meunier 1984b) or the fin rays of the Dipnoi (Géraudie and Meunier 1984). When mineralization is lacking and associated with the absence of enclosed cells (the osteocytes), as is the case with the isopedin of scales, then the skeletal tissue is a derived "acellular uncalcified bone" (Meunier 1987, 2011).

Vascular Component

Presence or absence of vascular canals in bone allows the distinguishing of **vascular** and **avascular bony tissues,** respectively. Moreover, according to the importance of the spatial network of primary and secondary vascular canals, it is necessary to differentiate **primary bony tissues** (with no or few remodelling) from **secondary bony tissues** (with important remodelling). The structure of the vascular network in bone is also an attractive information about the metabolic environment in which the osteogenic process occurs. Effectively, a slow osteogenesis (low metabolic level) is frequently characterized by a low vascularization of bone or, even, by a total absence of the vascular canals. On the contrary, a quick osteogenesis (high growth in juveniles and/or high metabolic activity) is characterized by the incorporation of the periostic vascular vessels in bone: for example, bony tissues of the Thunninae (Amprino and Godina 1956). In certain cases, osteogenesis produces spongy bone tissue, i.e., very rich in vascular canals and spaces, as for example in pectoral spiny rays of the Atipa male (*Hoplosternum littorale*; Callichthyidae; Siluriforme) (Fig. 6C) at the time of spawning (Meunier et al. 2002). In this siluroid fish, the instantaneous speed of bone deposit can reach 25 to 50 µm/day (Boujard and Meunier 1991). The primary bony tissues deposited by a periostic osteogenesis, not preceded by the destruction of pre-existing bone, shows a great regularity of the organic matrix: concordance of the bone layers (Fig. 5A) with, eventually, the presence of concentric growth marks (see "Cyclical bone growth"). The secondary bony tissues formed after a previous destruction of pre-existing primary bone (or secondary bone), as in the case of bone remodelling with the successive action of the osteoclasts and then the osteoblasts, show a typical broken aspect (Figs. 4C, D; 6D) with discordances underlined by C.L.R.

When an osteichthyan bone preparation is observed with a photonic microscope, it appears that the histological bony structure is a combination of these various types of bony tissues: vascularised lamellar bone, fibro-lamellar bone (woven fibered bone + lamellar bone) with eventually vascular canals (I and/or II osteons) (Figs. 4B, 6A), remodelled spongy bone (spongiosa with numerous bone trabeculae) and so on. The precise deciphering of the histological organization of bone can allow formulation of interesting hypothesis on biological events that have affected the body structure (Francillon-Vieillot et al. 1990, Ricqlès et al. 1991).

Bony tissue is a sort of black box that registers various biological facts that have affected the animal during its life (Castanet et al. 1993, Meunier 2002). The deciphering of these biological messages is one of the objects of squeletochronological techniques (Castanet et al. 1993, Meunier 2002).

Physiology-Pathology of Fish Bone

Bone Development

The analysis of the skeleton development requires several notions based on precise definitions. They refer to bone localization in the body, or to their mode of formation. First, it is necessary to distinguish between two main concepts: "**exoskeleton**[4]/**endoskeleton**" (Patterson 1977, Francillon-Vieillot et al. 1990). The "exoskeleton" corresponds to the superficial skeleton that has a dermic origin; this is a direct ossification of a fibrillary matrix; it is a **membranous bone**. These bones have generally a peripheral position in the body. Frequently localized in the dermis of skin, they are always covered at least by the epidermis and a part of the dermis, more precisely the loose dermis. The cartilage does not at all participate in the formation of the exoskeleton: for example, the frontal, the nasal, the cleithrum or the lepidotrichia are membranous bones. The endoskeleton results from the transformation of a cartilaginous anlagen. All along its development, this last one is submitted to an enchondral ossification process that implicates the previous destruction of

---

[4] This word is convenient but must be prohibited and replaced by "superficial skeleton". Effectively, the components of the superficial skeleton are always overlaid externally by the epidermis, plus frequently by the loose dermis. So it is not strictly outside the body like in the arthropods or the molluscs.

the cartilage and, then, its replacement by bone. Moreover, cartilaginous anlagen looks like a model on which the periostic bone can be deposited. So there exists strict bone-cartilage relations in the endoskeleton that have generally an internal localization in the body: for example, the mesethmoid, the orbitosphenoid, the scapula, and the hypurals.

Remark: All along the development of the fish organism, the "exoskeleton" and the "endoskeleton" can get more and more closer and, then, fused to give only one complex skeletal element. In that case, the bone has a mixed origin, for example in the skeleton of the head: pterotic, palatin, angulo-articular. Frequently, dermal bones penetrate the body and fuse to endoskeletal bones (Patterson 1977), and then developmental studies are necessary to decipher the true origin of each component of the skeletal structure.

There are two modes of ossification in fishes as in other vertebrates, in accordance with the initial model, either a conjunctive tissue (membranous bone) or a cartilage (enchondral bone).

Membranous Bone

The first step is the differentiation of a skeletogenic mesenchyme. It is constituted of star-shaped cells named scleroblasts that differentiate to become osteoblasts. These osteoblasts secrete collagenous fibrils, proteoglycans and various other complex molecules, all these elements form the pre-osseous matrix (or osteoid tissue). Then this matrix mineralizes owing to the intervention of alkaline phosphatases secreted by the osteoblasts and of the chemical transformation of proteoglycans: this leads to the deposit of hydroxyapatite micro-crystalites on or in collagenous fibres; then the osseous matrix is obtained (Ricqlès et al. 1991, Zylberberg et al. 1992). This process continues by means of a peripheral addition of new bone resulting from the activity of the surrounding periost. The osteoblasts are progressively incorporated in the osseous matrix and they become true osteocytes lying in their lacuna from which starts the canalicles that house the thin cytoplasmic processes. This ossification type is characteristic of dermal bones, for example the frontal, the parietal, and the nasal. The dermal bones are frequently associated with dental tissues (odontodes) (Ørvig 1977): for example scales (Fig. 5F), some cranial bones, so as dermotrichia, in basal living Osteichtyes and in numerous fossil fishes. A variety of membranous ossification can be observed on the internal bones around cartilaginous anlagen: for example the extremities of Meckel's cartilage (Francillon 1974, 1977), and the branchial arches. This leads to the thickening (centrifugal growth) of bones.

Enchondral Ossification

This process is very important because enchondral ossification allows the replacement of a cartilaginous element by bone. In the Osteichthyes, the enchondral ossification processes are frequent even if they are not so spectacular than on the epiphyseal ossification that is responsible for the length growth of the long bones of tetrapods. There are very few examples of description of length increase of bones in Teleostei (Haines 1935, 1937): Meckel's cartilage (Francillon 1974, 1977), branchial arches or the hemal arcualia in the precaudal vertebrae of *Cyprinus carpio* (Meunier 1979, Zylberberg and Meunier 2008). An example of a length increase is known on the axial bones of the paired fins in the fossil Sarcopterygii: †*Eusthenopteron fordi* (Meunier and Laurin 2012).

At the beginning of the enchondral ossification, the cartilage is invaded by conjunctive buds with blood vessels, chondroclasts and undifferentiated cells. Chondroclasts are at the origin of the formation of an erosion front with medular cavities formation. The undifferentiated cells lay on the surface of these cavities and become osteoblasts that lay thin laminae of enchondral bone. This enchondral bone forms more or less irregular trabeculae (Meunier 1979, Zylberberg and Meunier 2008) that can be eroded and replaced by new bony deposits.

Remark: In the specialized literature, some other notions can be found as the "perichondral ossification" (Fig. 7A) and the "parachondral ossification" (Fig. 7B) (Blanc 1953). In fact, they both are membranous ossifications but they respectively develop in the neighbourhood of the cartilaginous anlagen for the

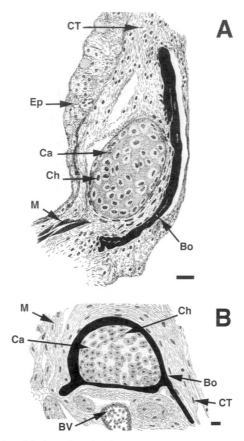

Fig. 7. A. *Salmo salar.* Cross section of the lower jaw showing an example of parachondral ossification. The bone is separated from the cartilage by connective tissue (Scale bar = 25 μm). **B.** *Cottus gobio.* Cross section of a branchial arch showing an example of perichondral ossification (Scale bar = 20 μm) (Bo, bone; BV, blood vessel; Ca, cartilage; Ch, chondrocyte; CT, conjonctive tissue; Ep, epidermis; M, muscle). Drawings from Blanc 1953; Figs. 3, 7.

perichondral ossification or fit exactly the cartilage without its destruction for the parachondral ossification. In the first case, the bone is always separated from the cartilage, at least by the perichondral membrane (the perichondre), and eventually by some other thin mesenchymal components (Blanc 1953).

The histological relationships between bone-cartilage are numerous in the Osteichthyes. The chondro-osseous metaplasia is a progressive transformation of the cartilage to give bony tissue without a previous destruction of the cartilage (Haines and Mohuiddin 1968). Neoplasia is the replacement of an old tissue (cartilage) by a new one (bone) that implicates a destruction of cartilage (Francillon-Vieillot et al. 1990).

Bone Remodelling

Fish bone is a living tissue that can be the object of a partial destruction as a consequence of various solicitations: mechanical constraints, physiological constraints, pathology (Doherty et al. 2014). The main result of this destruction is a release into the internal fluids of minerals (essentially calcium and phosphate) and of organic components (essentially amino-acids coming from the catalysis of collagen). Three modes of bone destruction are known in Osteichthyes: osteoclasy, periosteocytic osteolysis and halastasy (or diffuse demineralization).

Osteoclasy

Osteoclasy is the result of the activity of osteoclasts. They are generally multinucleated cells (Fig. 2E) but mononucleated osteoclasts have also been frequently described in certain teleostean fishes (Sire et al. 1990, Witten 1997, Witten et al. 2001, 2004a). Osteoclastic cells are rich in acid phosphatases and proteolytic enzymes. They are responsible for the formation of more or less wide erosive cavities (or bays of erosion) or canals. At the microscopical level, these clastic cells are obvious when they have several nuclei. They can also de detected owing to the small more or less roundish cavities or alveolae on the resorbed surfaces—the so-called "Howship's lacunae". Chondroclasy is a similar destructive process to osteoclasy but it affects the cartilaginous tissues. The clastic cells are chondroclasts (Fig. 1B).

Periosteocytic Osteolysis

Periosteocytic osteolysis is a cellular activity limited to the osteocytes which, in certain circumstances, function as true osteoclasts but in the vicinity of the osteocytic cell only (Belanger 1969, Lopez 1970, 1973, Kacem and Meunier 2000). Effectively, they are able to demineralize and, subsequently, to destroy bone matrix that just wraps the lacunae where they are embedded. This periosteocytic process is clearly more discrete than osteoclasy, but it can be easily revealed by measuring the mean diameter or the mean surface of the osteocytic lacunae, which significantly enlarges under specific physiological environments (Kacem and Meunier 2000). The authors acknowledge that periosteocytic osteolysis allows very sharp adjustments of calcium homeostasis.

Halastasy

Halastasy is a process of diffuse demineralization which affects the mineral component only, without a clear perceptible cell intervention, and without affecting the organic matrix, contrary to osteoclasy and periosteocytic osteolysis. Halastasy removes small quantities of mineral in the osseous tissue without any destruction of the organic matrix. To reveal a halastatic process, it is necessary to measure the mineralization degree of bone with the quantitative microradiography, a delicate but accurate technique at the histological level (Boivin and Baud 1984, Meunier 1984a, Kacem and Meunier 2003).

These three different modes of partial or total bone destruction are under the control of various physiological factors, e.g., certain hormones can play an important part. Unfortunately, the number of studies in fishes devoted to these biological phenomena is still too scarce, and limited to few taxa belonging to the Anguillidae, Salmonidae and Cichlidae (Lopez 1973, Weiss and Watabe 1979, Yamada and Watabe 1979, Takagi and Yamada 1993, Persson et al. 1998, Kacem et al. 1998, 2000, 2004, 2013, Sbaihi et al. 2007, 2009).

Cyclical Growth of Bone (Fig. 8)

Bone growth is an appositional process that results from the activity of the osteoblasts, which are located in the periosteal membrane (see above). Periost produces primary bone lasts for the life of the fish or can be partially destroyed to be replaced by new bone, which is then considered as secondary bone. The structure of primary growing bony tissue depends on the rate of osteogenesis, i.e., whether it is young or adult, on the one hand, or if it represents a fish with a high or a low metabolic rate, on the other hand. Whatever the factors are that influence the growth rate, the osteoblasts are under the influence of seasonal rhythms. The measurement of periosteal osteogenesis is based on vital labelling techniques (Meunier and Boivin 1974, 1978, Beamish and McFarlane 1983, Babaluk and Craig 1990, Boujard and Meunier 1991, Trébaol et al. 1991). Depending on the length of the fish, its age and physiological stage, periosteal osseous apposition ranges between 0.1 and 20 µm/day (Simmons et al. 1970, Casselman 1974, Boujard and Meunier 1991, Meunier and François 1992, *inter alia*). The rate of new bone deposit in primary bone is responsible for growth marks that can be used for age determination in fishes (Meunier 2002).

Authors who have made use of skeletochronology have generally paid very little attention to the significance of the structural support of the cyclical growth marks. However, knowledge of their histological

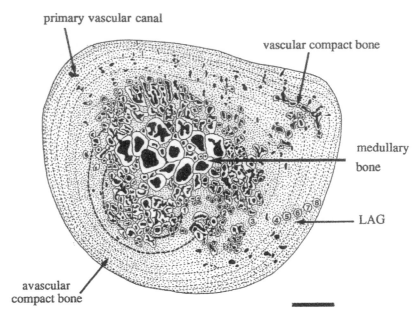

primary vascular canal

vascular compact bone

medullary bone

LAG

avascular compact bone

Fig. 8. *Palinurichthys* sp. (Centrolophidae). Drawing of the cross section of a rib. Note observed lines (1–8) of arrested growth (LAG) in primary bone and numerous secondary osteons, especially in the central area of the rib. The remodelling that has affected the central area has partially destroyed the two first LAG (Scale bar = 2 mm) (From Meunier 1987).

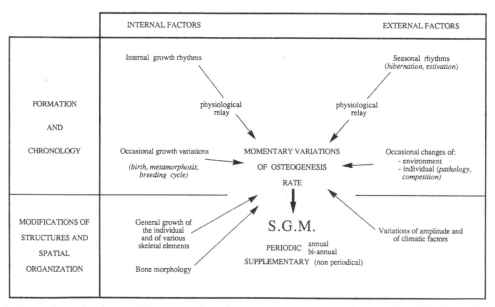

Fig. 9. Various internal and external factors that determine the spatiotemporal organization of the skeletal growth marks (From Castanet et al. 1993).

structure is essential for the understanding of the relationships between the skeleton and genetic and epigenetic factors of growth (Fig. 9) (Castanet et al. 1977, 1993).

Growth marks, that are useful for age determination in fish, are only found in primary bone. This is why the integrity of this type of bone is essential for age estimation studies because remodelling destroys this particular information (Castanet et al. 1992, 1993). Moreover, growth marks have specific properties under the light microscopy, which are used to distinguish them in age estimation studies. There are three categories of growth marks in bone: "zones", "annuli" and "arrested growth lines" (AGL). (i) The "zone" is the thickest growth mark. It appears dark under transmitted light and bright in reflected light. When bone is vascularised, vascular canals and cavities are more numerous in zone marks than in annuli. In fast-growing young animals, opaque layers, i.e., fast-growing layers (zones), are made of woven fibrous bone with isodiametric osteocytes randomly distributed. In adults, the bone growth decreases because of the competition with the reproductive metabolism, the woven fibred bone is replaced by parallel-fibred bone or, eventually by lamellar bone, both of with are more or less flattened cell lacunae (Castanet et al. 1992, 1993). (ii) Annuli or slow-growing layers are clearly less thick than zones. Generally speaking, in bony fishes the annulus consists of one or more bony lamellae, which are weakly hypermineralized in comparison with the mineralization of the fast-growing bone of the zone (Meunier 1988, Castanet et al. 1993, inter alia). Moreover, the density of osteocytes is slightly lower in the annulus (Castanet et al. 1977, 1993). Annuli always appear more translucent than the zones, i.e., white in transmitted light and dark in reflected light. (iii) The "arrested growth lines" are cemented lines that are concordant with the bony layers (see above). In any species, the annulus is lined on its peripheral side by an AGL, sometimes by two (or more) AGL. AGL are rest lines (Castanet 1981, Ricqlès et al. 1991), i.e., they mark a temporary cessation of local osteogenesis (Castanet et al. 1992, 1993). They are generally more translucent and more refringent than other marks. So they appear as the brightest structures when bone is observed with polarized light. They also are very chromophilic (with Haematoxylin and PAS stains) and frequently weakly hypermineralized (Figs. 5A, 6A). The growth of one year corresponds to the combination of a fast-growing zone and an annulus.

Osteichthyans (as Chondrichthyans) are poikilothermic fishes and, as such, they grow all along their life even if growth rate decreases with age. Moreover, the longevity of fishes is highly variable according to the taxa, some of them reaching more than fifty years, some even more than a century (Das 1994, inter alia). The growth rate of bone is time-dependent and it expresses different histological characteristics. The external periodic and aperiodic phenomena, which lead to bone growth variations, interact with general individual growth processes and locally with those of the different skeletal elements. According to bone morphogenesis, the histological structure and the spatial aspect of growth marks changes from place to place in the same bone and between different bones of the same individual. This fact is very important for the choice of suitable bones for ageing studies. Similarly, according to the general evolution of the growth rate throughout life, the structure of growth marks and their sequence will change from birth to death. Until sexual maturity, when body growth rate is high, annuli or AGL are well separated by wide zones of fast-growing tissue. They will become closer and closer during adulthood and very close to each other with ageing (Castanet et al. 1992, 1993, Lord et al. 2007). Moreover, any cyclical anomaly of an external factor can in turn generate more or less fine structural differences in the bony structure and thus creates "false growth marks" or "supernumerary growth marks". If certain supernumerary growth marks are purely accidental (e.g., after food scarcity, dryness, marked disease), others can result from normal acyclical events of life: hatching and/or yolk reduction ("birth mark") (Lecomte et al. 1986, 1989), migration during recruitment (surnumerary acyclical growth marks) (Meunier et al. 1979), breeding ("breeding marks"), and annual adult migrations (Compean-Jiménez and Bard 1980). These supernumerary marks appear as an annulus or AGL, the histological characteristic of which are similar to that of a normal annulus and AGL.

An interesting case is that of the Atipa (*Hoplosternum littorale*), a South American catfish (Callichthyidae), which lives in coastal swamps. This fish builds a nest with grass and bubbles for spawning (Hostache et al. 1992, 1993, Pascal et al. 1994, 1995). At the same time, the male develops secondary sex characteristics on the pectoral spiny rays, whose epithelium thickens (Winemiller 1987) while typical hypervascularized bone is deposited on the external margin of the ray (Boujard and Meunier 1991). When spawning is complete, bony growth falls drastically and the periost deposits avascular pseudo-lamellar bone (Fig. 6C). During the following breeding season, a new area of hypervascularized bone forms, and so on. By counting the number of hypervascularized clusters, the age of the male can thus be estimated so as the number of reproduction instalments. In the female, the only growth marks are alternating zones and hypermineralized annuli (Meunier 2002).

In fact, we may ask if yearly variations of climate directly affect the histological structure of bone so as to deposit the growth marks or if there are relays, and more precisely biological relays, between these external factors (e.g., temperature, rain) and the registering property of bone. Fishes are poikilothermic species. When they live in temperate climates they are exposed to seasonality of an external factor, which induces a yearly cyclical biological rhythm, especially for body growth, i.e., skeleton growth inducing deposition of bone growth marks. Other cyclical biological functions such as reproduction can also induce cyclical growth marks. In many studies of fishes living in temperate countries, a set of growth marks (i.e., zone + annulus or/and AGL) are laid down each year as validated with vital labelling (Casselman 1974, Meunier and Pascal 1981, and others). In tropical climates, there is less contrast in seasonality compared to temperate ones. Nevertheless, some physico-chemical parameters of the environment may show a weak yearly cyclical variation that may be enough to induce and synchronize cyclical growth marks. In tropical Africa as in tropical South America, dry and wet seasons respectively play the same role as winter and spring-summer in temperate areas. The histological expression of these alternated dry and wet seasons is well marked in fish bones (Quick and Bruton 1984, Lecomte et al. 1986, 1989, Meunier et al. 1997, Meunier 2012). In an apparently more or less uniform environment like the marine depth, growth marks are present in bony tissue probably caused by the availability of food coming from the epipelagic production (Meunier and Arnulf 2018, Meunier et al. 2018b).

Contrary to generally accepted views, sclerochronology, and particularly age estimation in fish with bony tissues (= skeletochronology), is not an easy science. It involves a series of process and data-processing sequences that are often complex and time-consuming (Panfili et al. 2002).

Hyperostosis

Various anomalies can affect the skeletal development (Mawdesley-Thomas 1969). Much of them result from ecological contaminations or from difficulties in cultured activities (Kacem et al. 2004, and many others). They are out of our present scope, but one type deserves our interest: the hyperostosis.

Numerous teleostean fishes develop hyperostosis, e.g., typical swelling of a bone as a whole or of part of bone (see Gervais 1875, Chabanaud 1926, Desse et al. 1981, Gauldie and Czochanska 1990, Driesh 1994, Meunier and Desse 1994, Smith-Vaniz et al. 1995, Smith-Vaniz and Carpenter 2006, Chanet 2018). Frequently, these specific bones are swollen whereas the resting skeleton is free of hyperostosis. Firstly, the neurocranium can show several swollen bones. It is generally the occipital crest of the cranium that is affected, the other bones being unmodified, e.g., in Carangidae (Desse et al. 1981, Smith-Vaniz et al. 1995, see also Driesch 1994, Chanet 2018). But other species develop spectacular hyperostosis in cranial bones, for example in various Haemulidae (*Pomadasys hasta* of the Indian Ocean), the large neurocranium of which is so strong that it is resistant to natural burrowing and it is then found in archaeological sites (Meunier and Desse 1994). Secondly, it is noteworthy that hyperostosis can be strictly localized on given caudal vertebra, for example the 21st, with a possible extension to the two nearest ones (20th and 22nd) in the lumptail searobin (*Prionotus stephanophrys*) (Meunier et al. 1999) or the (31st and 32nd) vertebra of the axial skeleton in the black bonite (*Euthynnus lineatus*) (Béarez et al. 2005). Hyperostotic dorsal pterygophores are also frequent in elongate fishes like the scabbard fishes (Olsen 1971, Lima et al. 2002, Meunier et al. 2010, Giaratana et al. 2012), and the oarfish (Paig-Tran et al. 2016).

Among the numerous questions that arose to explain the aetiology of this phenomenon, some of them are: why do some bones constantly develop hyperostosis and some others do not in a given species? Why is this development species-specific? Various studies confirm the taxonomic aspect of the hyperostotic phenomenon because it always affects the same bones in a given species (Driesch 1994) but the phenomenon differs from one species to another (Gauldie and Czochanska 1990, Smith-Vaniz et al. 1995, Smith-Vaniz and Carpenter 2007, Chanet 2018). This suggests that hyperostosis seems to have a genetic origin but probably there are other causes.

The histological characteristics of the swollen bones in hyperostotic extant species are independent of the anatomical localization of the hyperostotic bones. But this histological process affects essentially species with acellular bone (Figs. 4C, 6D). The few description of hyperostosis with cellular bone concerns a fossil cyprinide (Chang et al. 2008) and an extant ariide (Srinivasa-Rao and Lakshini 1986). Hyperostosis was also described in a fossil clupeid of the Miocene (Gaudant and Meunier 1996), the bony tissue of which shows osteocytes. The spectacular swelling of bones results from an activation of osteogenesis that yields primary spongy bone directly, or after an associated remodelling (Fierstine 1968, Desse et al. 1981, Gauldie and Czochanska 1990, Meunier and Desse 1994, Smith-Vaniz et al. 1995, Meunier and Herbin 2014). Hyperostotic bones predominantly show a spongy structure (Fierstine 1968, Desse et al. 1981, Gauldie and Czochanska 1990), except in several fossil taxa which have swollen avascular compact bones, e.g., in the cyprinid *Hsianwenia wui* (Chang et al. 2008) and the cyprinodontid *Aphanius crassicaudus* (Meunier and Gaudant 1987), but these fishes have a small length, not more than 5 to 6 cm (TL).

What are the causality and the biological significance of the hyperostotic phenomenon in various fish? There are no firm and satisfactory explanations yet for any hyperostotic case recorded so far. Various hypotheses have been indeed proposed to explain the development of these curious bones: aid in fin erection, aid in neutral buoyancy, ageing action on bone, reaction to high temperatures, metabolic abnormality, pathogenic phenomena, genetic factors, etc. In fact, the causal factors may be various and the bony response relatively uniform: a similar swelling of the affected bony organ.

Do hyperostotic bones correspond to pathological features? Except for Bhatt and Murti (1960), the authors agree that hyperostosis is not a pathological formation (Olsen 1971, Desse et al. 1981, Gauldie and Czochanska 1990). It also appears that fishes with such swollen bones show a normal behaviour (Johnson 1973). For example, hyperostotic processes seem inescapable in the jack mackerel, *Trachurus trachurus* (Carangidae), since a high number of individuals show swollen bones at the end of their life (Desse et al. 1981). As these fishes do not show abnormal behaviour, we can consider that the phenomenon is not pathologic, at least in this species and, possibly, in the whole carangid family since a lot of carangids have hyperostotic bones (Smith-Vaniz et al. 1995, Smith-Vaniz and Carpenter 2007). It is the same in the triglid *Prionotus stephanophrys* (Meunier et al. 1999) as in the scombrid *Euthynnus alleteratus* (Bearez et al. 2005). In a more general way, the aetiology of the hyperostotic phenomenon in fishes is poorly known. Presently, this phenomenon occurs mainly among marine species and in the majority, among species with acellular bone (Desse et al. 1981, Meunier and Desse 1994). However, hyperostotic bones may have been recorded in some less highly evolved species: for example, the marine Pacific catfish, *Bagre pinnimaculatus* and *B. panamaensis* (Ariidae) (Béarez and Meunier personal observ.). These two species have cellular bone.

Conclusion

The bony tissues of the Osteichthyes are not submitted to progressive structural novelties all along the geological times. There is only a new rearrangement of the various components that were present as soon as the mineralized osseous tissues originated, but there are various adaptative directions. Yet, with the lack of both osteocytes and mineral, there are very derived skeletal tissues, especially in the teleosts (Fig. 10).

The fish bony tissues are true living tissues that provide various functions and maintain closed relations with the other physiological compartments of the fish organism. The knowledge obtained in taxa is considered as relatively closed to the basal teleosts, essentially the Ostariophysi, Salmonidae and Anguillidae, which requires new experimental studies to understand the various aspects of their bone biology as a whole, notably for the factors (as hormonal control) at work in such or such physiological phenomenon. These previously cited taxa are all fishes with cellular bone, two of them being well known as migratory species: the salmon and the eel. It is absolutely necessary to increase efforts to develop physiological studies on species with acellular bone. Among these last ones, cultured species as the Bass, the Sea Bream or the Turbot, for example, the breeding of which is now under control, are good potential models for experimental studies of their bone physiology. At the present time, the first of them, the Bass, is probably the best suitable for experimental works, without totally rejecting the tilapines that are also world wide cultured species and the structure of their acellular bony tissues has been described at several occasions.

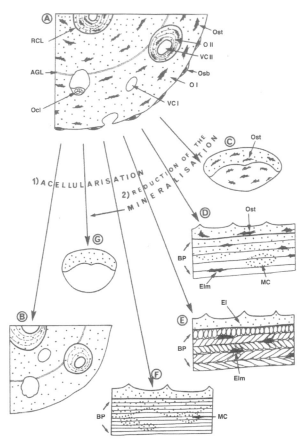

Fig. 10. Recapitulative scheme of the two main evolutionary trends of bony tissues in Osteichthyes (From Meunier 1987): (1) Acellularisation (**B, E** to **G**) and (2) Reduction of mineralization (**C** to **G**). **A.** Compact, cellular, vascularized and pseudo-lamellar bone with localized remodelling. **B.** Compact acellular, vascularized and pseudo-lamellar bone with localized remodelling. **C.** Cellular osseous and cellular unmineralized permanent "preosseous" tissues: camptotrichia of *Neoceratodus forsteri* (Dipnoi, Neoceratodidae). **D.** Cellular bone and partly unmineralized cellular isopedine: scale of *Amia calva* (Amiidae, Holostei). **E.** Acellular bone and partly unmineralized cellular isopedine: scale of *Latimeria chalumnae* (Coelacanthidae, Actinistia, Sarcopterygii) and *Neoceratodus forsteri* (Neoceratodidae, Dipnoi, Sarcopterygii). **F.** Acellular bone and partly unmineralized acellular isopedine: scale of *Hemichromis bimaculatus* (Perciformes, Acanthopterygii, Teleostei). **G.** Acellular bone and unmineralized acellular "preosseous" tissue: camptotrichia of *Protopterus annectens* (Lepidosirenidae, Dipnoi, Sarcopterygii). AGL, arrested growth line; BP, basal plate; El, external layer; Elm, elasmocyte; MC, Mandl's corpuscle; O I, primary bone; O II, secondary bone; Osb, osteoblast; Ocl, osteoclast; Ost, osteocyte; RCL, reversal cementing line; VC I, primary vascular canal; VC II, secondary vascular canal; Dotted areas indicate mineralized areas.

References Cited

Agassiz, L. 1834–43. Recherches sur les Poissons fossiles. 5 volumes and atlas. Imprimerie Petitpierre. Neuchâtel.

Amprino, R. and G. Godina. 1956. Osservazioni sul rinnovamento strutturale dell'osso in Pesci Teleostei. Publ. Staz. Zool. Napoli. 28: 62–71.

Babaluk, J.A. and J.F. Craig. 1990. Tetracycline marking studies with pike, *Esox lucius* L. Aquac. Fish. Manag. 21: 307–315.

Bareille, G., R. Lecomte-Finiger, P. Sasal, N. Mary, F.J. Meunier, M.-H. Deschamps, S. Berail, C. Pecheran and R. Lecomte-Finiger. 2015. Are elemental and strontium isotopic microchemistry of otolith and histomorphometrical characteristics of vertebral bone useful to resolve the eel *Anguilla obscura* status in lake Lalo-Lalo (Wallis Island)? Vie et Milieu—Life and Environment 65(1): 29–39.

Beamish, R.J. and G.A. Mc Farlane. 1983. The forgotten requirement for age validation in fisheries biology. Trans. Am. Fish. Soc. 112: 735–743.

Béarez, P., F.J. Meunier and A. Kacem. 2005. Description morphologique et histologique de l'hyperostose vertébrale chez la thonine noire, *Euthynnus lineatus* (Teleostei, Perciformes, Scombridae). Cah. Biol. Mar. 46: 21–28.

Belanger, L.F. 1969. Osteocytic osteolysis. Calc. Tiss. Res. 4: 1–12.

Benjamin, M. 1988. Mucochondroid (mucus connective) tissues in heads of teleosts. Anat. Embryol. 178: 461–474.

Benjamin, M. 1989. Hyaline-cell cartilage (chondroid) in the heads of teleosts. Anat. Embryol. 179: 285–303.

Benjamin, M. 1990. The cranial cartilages of teleosts and their classification. J. Anat. 169: 153–172.

Benjamin, M., J.R. Ralphs and O.S. Eberewariye. 1992. Cartilage and related tissues in the trunk and fins of teleosts. J. Anat. 181: 113–118.

Beresford, W.A. 1981. Chondroid bone, secondary cartilage and metaplasia. Urban and Schwarzenberg, Baltimore.

Bhatt, Y.M. and N.N. Murti. 1960. Hyperostosis in *Trichiurus haumela* (Forskal). J. Univer. Bombay 28: 84–89.

Blanc, M. 1953. Contribution à l'étude de l'ostéogenèse chez les poisons Téléostéens. Mém. Mus. Nat. Hist. Nat. 7: 1–145.

Boivin, G. and C.A. Baud. 1984. Microradiographic methods for calcified tissues. pp. 391–412. *In*: G.R. Dickson (ed.). Methods of Calcified Tissue Preparation. Elsevier, Amsterdam.

Boujard, T. and F.J. Meunier. 1991. Croissance de l'épine pectorale, histologie osseuse et dimorphisme sexuel chez l'atipa, *Hoplosternum littorale* Hancox, 1828 (Callichthyidae, Siluriforme) Cybium. 15: 55–68.

Brito, P.M., F.J. Meunier, G. Clément and D. Geffard-Kuriyama. 2010. The histological structure of the calcified lung of the fossil coelacanth *Axelrodichthys araripensis* (Actinistia: Mawsoniidae). Palaeontology 53: 1281–1290.

Bystrow, A.P. 1939. Zahnstruktur der Crossopterygier. Acta Zoologica. 20: 283–338.

Cao, L., T. Moriishi, T. Miyazaki, T. Iimura, M. Hamagaki, A. Nakane, Y. Tamamura and T. Komori. 2011. Comparative morphology of the osteocyte lacunocanalicular system in various vertebrates. J. Bone Min. Metab. 29: 662–670.

Casadevall, M., A. Casinos, C. Viladic and M. Ontanon. 1990. Scaling of skeletal mass and mineral content in teleosts. Zool. Anz. 225(3/4), 8: 144–150.

Casselman, J.M. 1974. Analysis of hard tissues of pike *Esox lucius* L. with special reference to age and growth. pp. 13–27. *In*: T.B. Bagenal (ed.). Ageing of Fish. Unwin Brothers.

Castanet, J., F.J. Meunier, C. Bergot and Y. François. 1975. Données préliminaires sur les structures histologiques du squelette de *Latimeria chalumnae*. I—Dents, écailles, rayons de nageoires. pp. 159–168. *In*: Coll. Inter. CNRS, Problèmes actuels de Paléontologie. Evolution des Vertébrés, Paris, 4–9 Juin 1973.

Castanet, J., F.J. Meunier and A. (de) Ricqlès. 1977. L'enregistrement de la croissance cyclique par le tissu osseux chez les Vertébrés poïkilothermes: données comparatives et essai de synthèse. Bull. Biol. Fr. Belg. 111: 183–202.

Castanet, J. 1979. Données comparatives sur la minéralisation des marques de croissance squelettique chez les Vertébrés. Etude par microradiographie quantitative. C. R. Acad. Sci. D. 289: 405–408.

Castanet, J. 1981. Nouvelles données sur les lignes cimentantes de l'os. Arch. Biol. 92: 1–24.

Castanet, J., H. Francillon-Vieillot and F.J. Meunier. 1992. La squelettochronologie à partir des tissus osseux et dentaires des Vertébrés. pp. 257–280. *In*: J.L. Baglinière, J. Castanet, F. Conand and F.J. Meunier (eds.). Tissus durs et Âge individuel des Vertébrés. Colloques et Séminaires, ORSTOM-INRA.

Castanet, J., H. Francillon-Vieillot, F.J. Meunier and A. (de) Ricqlès. 1993. Use of bone growth in aging individuals. pp. 245–283. *In*: B.K. Hall (ed.). Bone, 7, CRC Press.

Chabanaud, P. 1926. Fréquence, symétrie et constance spécifiques d'hyperostoses externes chez divers poissons de la famille des Sciaenidés. C. R. Acad. Sci. 182: 1647–1649.

Chanet, B. 2018. Swollen bones in jacks and relatives (Teleostei: Acantomorphata: Carangidae). Cybium. 42(1): 99–103.

Chang, M., X. Wang, H. Liu, D. Miao, Q. Zhao, G. Wu, J. Liu, Q. Li, Z. Sun and N. Wang. 2008. Extraordinarily thick-boned fish linked to the aridification of the Qaidam Basin (northern Tibetan Plateau). P.N.A.S. 105: 13246–13251.

Compean Jiménez, G. and F.X. Bard. 1980. Utilisation de la squeletto-chronologie chez les Thunnidés. Bull. Soc. Zool. Fr. 105: 329–336.

Cupello, C., F.J. Meunier, M. Herbin, P. Janvier, G. Clément and P.M. Brito. 2017. The homology and function of the lung plates in extant and fossil coelacanths. Scientific Reports. 7: 9244. DOI:10.1038/s41598-017-09327-6 (8 pp).

Das, M. 1994. Age determination and longevity in fishes. Gerontology 40: 70–96.

Davesne, D., F.J. Meunier, M. Friedman, R.B.J. Benson and O. Otero. 2018. Histology of the endothermic opah (*Lampris guttatus*) suggests a structure-function relationship in teleost fish bone. Biol. Letters 14: 20180270.

Davesne, D., F.J. Meunier, A. Schmitt, M. Friedman, O. Otero and R. Benson. 2019. The phylogenetic origin and evolution of acellular bone in teleost fishes: insights into osteocyte function in bone metabolism. Biological Review (in press).

Denison, R.H. 1963. The early history of the vertebrate calcified skeleton. Clin. Orthop. Rel. Res. 31: 141–152.

Desse, G., F.J. Meunier, M. Peron and J. Laroche. 1981. Hyperostose vertébrale chez l'animal. Rhumatologie 33: 105–119.

Doherty, A.H., C.K. Halambor and S.W. Donahue. 2014. Evolutionary physiology of bone: Bone metabolism in changing environments. Physiology 30: 17–29.

Driesch, A. (von den). 1994. Hyperostosis in fish. *In*: W. Van Neer (ed.). Fish Exploitation in the Past, Proc. 7th Meet. ICAZ Fish Remains Working Group, Ann. Mus. Roy. Afrique Centrale, Sci. Zool. 274: 37–45.

Fierstine, H.L. 1968. Swollen dorsal fin elements in living and fossil *Caranx* (Teleostei: Carangidae). Contr. Sci. Los Angeles 137: 1–10.

Fink, W.L. 1981. Ontogeny and phylogeny of tooth attachment modes in Actinopterygian fishes. J. Morph. 167: 167–184.

Francillon, H. 1974. Développement de la partie postérieure de la mandibule de *Salmo trutta fario* L. (Pisces, Teleostei, Salmonidae). Zool. Scripta 3: 41–51.

Francillon, H. 1977. Développement de la partie antérieure de la mandibule de *Salmo trutta fario* L. (Pisces, Teleostei, Salmonidae). Zool. Script. 6: 245–251.

Francillon-Vieillot, H., V. (de) Buffrénil, J. Castanet, J. Géraudie, F.J. Meunier, J.-Y. Sire, L. Zylberberg and A. (de) Ricqlès. 1990. Microstructure and mineralization of Vertebrate skeletal tissues. pp. 471–530. *In*: J.G. Carter (ed.). Skeletal Biomineralization: Patterns, Processes and Evolutionary Trends. Vol. 1, Van Nostrand Reinhold, New-York.

Froese, R. and D. Pauly. 2019. Fishbase: www.fishbase.org. Accessed January 2019.

Gaudant, J. and F.J. Meunier. 1996. Observation d'un cas de pachyostose chez un Clupeidae fossile du miocène terminal de l'ouest Algérien, *Sardina ?crassa* (Sauvage, 1873). Cybium. 20: 169–183.

Gauldie, R.W. and Z. Czochanska. 1990. Hyperostotic bones from the New Zealand snapper *Chrysophys auratus* (Sparidae). Fish. Bull. 88: 201–206.

Germain, D., J. Mondéjar-Fernández and F.J. Meunier. 2016. The detection of plicidentine tooth organisation in teleostean fishes owing to non-destructive tomography. Cybium. 40(1): 75–82.

Germain, D. and F.J. Meunier. 2018. Teeth of Polypteridae and Amiidae have plicidentine organisation. Acta Zoologica. 2017; 00: 1–7. https://doi.org/10.1111/azo.12237.

Géraudie, J. and F.J. Meunier. 1984. Structure and comparative morphology of camptotrichia of lungfish fins. Tissue and Cell. 16: 217–236.

Gervais, P. 1875. De l'hyperostose chez l'homme et chez les animaux. I. J. Zool. 4: 272–284; II. J. Zool. 4: 445–462.

Giarratana, F., A. Ruolo, D. Muscolino, F. Marino, M. Gallo and A. Panebianco. 2012. Occurrence of hyperostotic pterygiophores in the silver scabbardfish, *Lepidopus caudatus* (Actinopterygii : Perciformes : Trichiuridae). Acta Ichthyol. Pisc. 42(3): 233–237.

Giraud-Guilles, M.M. 1988. Twisted plywood architecture of collagen fibrils in Human compact bone osteons. Calc. Tissue Intern. 42: 167–180.

Giraud, M.M., J. Castanet, F.J. Meunier and Y. Bouligand. 1978. The fibrous structure of coelacanth scales: a twisted "plywood". Tissue and Cell. 10: 671–686.

Goodrich, E.S. 1913. On the structure of bone in fishes: a contribution to palaeohistology. Proc. Zool. Soc. Lond. 80–85.

Haines, R.W. 1935. Epiphyseal growth in the branchial skeleton of fishes. Quart. J. Micr. Sci. 77: 79–97.

Haines, R.W. 1937. Posterior end of Meckel's cartilage and related ossifications in bony fishes. Quart. J. Micr. Sci. 80: 1–38.

Haines, R.W. and A. Mohuiddin. 1968. Metaplastic bone. J. Anat. 103: 527–538.

Halstead, L.B. 1963. Aspidin: the precursor of bone. Nature. 199: 46–48.

Halstead, L.B. 1969. Calcified tissues in the earliest vertebrates. Calc. Tis. Res. 3: 107–124.

Halstead, L.B. 1973. The heterostracan fishes. Biol. Rev. 48: 279–332.

Hertwig, O. 1876–1882. Ueber das Hautskelett der Fische. Morph. Jahrb. Z. Anat. Entwik. 2: 328–396; 5: 1–21; 7: 1–42.

Hostache, G., M. Pascal, M. Kernen and C. Tessier. 1992. Temperature et incubation chez l'Atipa, *Hoplosternum littorale* (Teleostei, Siluriforme). Aquat. Liv. Res. 5: 31–39.

Hostache, G., M. Pascal and P. Planquette. 1993. Saisonnalité de la reproduction chez l'Atipa, *Hoplosternum littorale* (Siluriforme, Teleostei), par l'analyse de l'évolution du rapport gonado-somatique. Aquat. Liv. Res. 6: 155–162.

Hughes, D.R., J.R. Bassett and L.A. Moffat. 1994. Histological identification of osteocytes in the allegretto acellular bone of the sea breams *Acanthopagrus australis*, *Pagrus auratus* and *Rhabdosargus sarba* (Sparidae, Perciformes, Teleostei). Anat. Embryol. 190: 163–179.

Huysseune, A. 1986. Late skeletal development at the articulation between upper pharyngeal jaws and neurocranial base in the fish, *Astatotilapia elegans*, with the participation of a chondroid form of bone. Amer. J. Anat. 177: 119–137.

Huysseune, A. and W. Verraes. 1986. Chondroid bone on the upper pharyngeal jaws and neurocranial base in the adult fish *Astatotilapia elegans*. Amer. J. Anat. 177: 527–535.

Huysseune, A. 1989. Morphogenetic aspects of the pharyngeal jaws and neurocranium apophysis in postembryonic *Astatotilapia elegans* (Trewavas 1933) (Teleostei: Cichlidae). Acad. Anal. Brussels. 51: 11–35.

Huysseune, A. and J.Y. Sire. 1990. Ultrastructural observations on chondroid bone in the teleost fish *Hemichromis bimaculatus*. Tissue and Cell. 22: 371–383.

Huysseune, A. 2000. Skeletal system. pp. 307–317. *In*: G.K. Ostrander (ed.). The Laboratory Fish: Part 4. Microscopic Functional Anatomy. Acad. Press, San Diego.

Huysseune, A. 2006. Formation of a successional dental lamina in the zebrafish (*Dano rerio*): support for a local control of replacement tooth initiation. Int. J. Dev. Biol. 50: 637–643.

Ishiyama, M., T. Inage and H. Shimokowa. 1999. An immunocytochemical study of amelogenin proteins in the developing tooth of the gar-pike, Lepisosteus oculatus (Holostei, Actinopterygii). Arch. Histol. Cytol. 62: 191–197.

Jernvall, J. and I. Thesleff. 2012. Tooth shape formation and tooth renewal: evolving with the same signals. Development 139: 3487–3497.

Johnson, C.R. 1973. Hyperostosis in fishes of the genus *Platycephalus* (Platycephalidae). Jap. J. Ichth. 20: 178.

Kacem, A., F.J. Meunier and J.L. Baglinière. 1998. Quantitative study of morphological and histological changes in the skeleton of *Salmo salar* L. (Teleostei: Salmonidae) during its anadromous migration. Preliminary results. J. Fish Biol. 53: 1096–1109.

Kacem, A., S. Gustafsson and F.J. Meunier. 2000. Demineralization of the vertebral skeleton in Atlantic salmon *Salmo salar* L., during spawning migration. Comp. Biochem. Physiol. 125: 479–484.

Kacem, A. and F.J. Meunier. 2000. Mise en évidence de l'ostéolyse périostéocytaire vertébrale chez le saumon atlantique *Salmo salar* L. (Salmonidae, Teleostei), au cours de sa migration anadrome. *In*: 1ères rencontres d'Ichtyologie en France, Soc. Fra. Ichtyol. Cybium. 24(3 suppl.): 105–112.

Kacem, A. and F.J. Meunier. 2003. Halastatic demineralization in the vertebrae of the Atlantic salmon, *Salmo salar* L. (Teleostei, salmonidae), during its anadromous migration. J. Fish Biol. 3: 1122–1130.

Kacem, A., F.J. Meunier, J. Aubin and P. Haffray. 2004. Caractérisation histo-morphologique des malformations du squelette vertébral chez la truite arc-en-ciel (*Oncorhynchus mykiss*) après différents traitements de triploïdisation. *In*: 2èmes rencontres d'Ichtyologie en France, Soc. Fra. Ichtyol. Cybium. 28(1 suppl.): 15–23.

Kacem, A. and F.J. Meunier. 2009. Study of the transformations of the texture and the mineralization of the dentary bone in the Atlantic salmon, *Salmo salar* L. (Teleostei, Salmonidae), during genital maturation. Cybium. 33(1): 61–72.

Kacem, A., J.L. Baglinière and F.J. Meunier. 2013. A quantitative study of scales resorption in Atlantic salmon (*Salmo salar*) during its anadromous migration. Cybium. 37(3): 199–206.

Kemp, N.E. and S.K. Westrin. 1979. Ultrastructure of calcified cartilage in the endoskeletal tesserae of sharks. J. Morph. 160: 75–102.

Kirschbaum, F. and F.J. Meunier. 1981. Experimental regeneration of the caudal skeleton of the glass knifefish, *Eigenmannia virescens* (Rhamphichthyidae, Gymnotoidei). J. Morph. 168: 121–135.

Kölliker, A. 1859. On the different types in the microscopic structure of the skeleton of the osseous fish. Proc. R. Soc. Lond. 9: 656–688.

Lecomte, F., F.J. Meunier and R. Rojas-Beltran. 1986. Données préliminaires sur la croissance de deux Téléostéens de Guyane, *Arius proops* (Ariidae, Siluriformes) et *Leporinus friderici* (Anostomidae, Characoidei). Cybium. 10: 121–134.

Lecomte, F., F.J. Meunier and R. Rojas-Beltran. 1989. Some data on the growth of *Arius proops* (Ariidae, Siluriforme) in the estuaries of French Guyana. Aquat. Liv. Res. 2: 63–68.

Lima, F.C., A.P.M. Souza, E.F.M. Mesquita, G.N. Souza and V.C.J. Chinelli. 2002. Osteomas in cutlass fish, *Trichiurus lepturus* L., from Niteroi, Rio de Janeiro state. Brazil. J. Fish Dis. 25: 57–61.

Lison, L. 1954. Les dents. *In*: P.P. Grassé (éd.). Traité de Zoologie, Masson, Paris 12: 791–853.

Lopez, E. 1970. L'os cellulaire d'un poisson téléostéen "*Anguilla anguilla* L.". I. Étude histologique et histophysique. Z. Zellforsch. 109: 552–565.

Lopez, E. 1973. Étude morphologique et physiologique de l'os cellulaire des poissons téléostéens. Mém. Mus. Nat. Hist. Nat. 80: 1–90.

Lord, C., Y. Fermon, F.J. Meunier, M. Jegu and P. Keith. 2007. Croissance et longévité du Watau yaike, *Tometes lebaili* (Osteichthyes, Teleostei, Serrasalminae) dans le bassin du haut Maroni (Guyane française). Cybium. 31(3): 359–367.

Mawdesley-Thomas, L.E. 1969. Neoplasia in Fish—A bibliography. J. Fish Biol. 1: 187–207.

Meunier, F.J. and G. Boivin. 1978. Action de la fluorescéine, de l'alizarine, du bleu de calcéine et de diverses doses de tétracycline sur la croissance de la truite et de la carpe. Ann. Biol. Anim. Bioch. Biophys. 18: 1293–1308.

Meunier, F.J. 1979. Étude histologique et microradiographique du cartilage hémal de la vertèbre de la carpe, *Cyprinus carpio* L. (Pisces, Teleostei, Cyprinidae). Acta Zool. 60: 19–31.

Meunier, F.J., M. Pascal and G. Loubens. 1979. Comparaison de méthodes squelettochronologiques et considérations fonctionnelles sur les tissus osseux acellulaires d'un Ostéichthyen du Lagon Néo-Calédonien. Aquaculture. 17: 137–157.

Meunier, F.J. and J. Géraudie. 1980. Les structures en contreplaqué du derme et des écailles des Vertébrés inférieurs. Ann. Biol. 19: 1–18.

Meunier, F.J. and Y. François. 1980. L'organisation spatiale des fibres de collagène et la minéralisation des écailles des Dipneustes actuels. Bull. Soc. Zool. Fr. 105: 215–226.

Meunier, F.J. and M. Pascal. 1981. Étude expérimentale de la croissance cyclique des rayons de nageoire de la carpe (*Cyprinus carpio* L.). Résultats préliminaires. Aquaculture. 26: 23–40.

Meunier, F.J. and J. Castanet. 1982. Organisation spatiale des fibres de collagène de la plaque basale des écailles des Téléostéens. Zool. Scripta. 11: 141–153.

Meunier, F.J. 1984a. Étude de la minéralisation de l'os chez les Téléostéens à l'aide de la microradiographie quantitative. Résultats préliminaires. Cybium. 8(3): 43–49.

Meunier, F.J. 1984b. Structure et minéralisation des écailles de quelques Osteoglossidae (Ostéichthyens, Téléostéens). Ann. Sc. Nat. Zool. 13ème Série. 6: 111–124.

Meunier, F.J. 1984c. Spatial organization and mineralization of the basal plate of elasmoid scales in Osteichthyans. Am. Zool. 24: 953–964.

Meunier, F.J. 1987. Os cellulaire, os acellulaire et tissus dérivés chez les Osteichthyens: les phénomènes de l'acellularisation et de la perte de minéralisation. Ann. Biol. 26: 201–233.

Meunier, F.J. and J. Gaudant. 1987. Sur un cas de pachyostose chez un poisson du Miocène terminal du bassin méditerranéen, *Aphanius crassicaudus* (Agassiz), (Teleostei, Cyprinodontidae). C. R. Acad. Sci. Paris 305: 925–928.

Meunier, F.J. 1987–88. Nouvelles données sur l'organisation spatiale des fibres de collagène de la plaque basale des écailles des Téléostéens. Ann. Sci. Nat. Zool. 13ème Sér. 9: 113–121.

Meunier, F.J. 1988. Détermination de l'âge individuel chez les Ostéichthyens à l'aide de la squelettochronologie: historique et méthodologie. Acta Oecolog, Oecol. Gener. 9: 299–329.

Meunier, F.J. and A. Huysseune. 1992. The concept of bone tissue in Osteichthyes. Nether. J. Zool. 42(2-3): 445–458.

Meunier, F.J. and Y. François. 1992. Croissance du squelette chez les Téléostéens. I. Squelette, os, tissus squelettiques. Ann. Biol. 31: 169–184. II. La croissance du squelette. Ann. Biol. 31: 185–219.

Meunier, F.J. and J. Desse. 1994. Histological structure of hyperostotic cranial remains of *Pomadasys hasta* (Osteichthyes, Perciformes, Haemulidae) from archeological sites of the Arabian Golf and the Indian Ocean. pp. 47–53. *In*: W. Van Neer (ed.). Fish Exploitation in the Past, Proc. 7th Meet. ICAZ Fish Remains Working Group, Ann. Mus. Royal Afr. Cent. Sci. Zool. 274.

Meunier, F.J. 1997. Structure et minéralisation des écailles de quelques Characiformes de Guyane française. Rev. Hydrobiol. Trop. (1994) 27(4): 407–422.

Meunier, F.J., R. Rojas-Beltran, T. Boujard and F. Lecomte. 1997. Rythmes saisonniers de la croissance chez quelques Téléostéens de Guyane française. Rev. Hydrobiol. Trop. (1994) 27(4): 423–440.

Meunier, F.J., P. Béarez and H. Francillon-Vieillot. 1999. Some morphological and histological aspects of hyperostosis in the Equatorian marine fish *Prionotus stephanophrys* (Teleostei, Triglidae). pp. 125–133. *In*: B. Séret and J.-Y. Sire (eds.). Proc. 5th Indo-Pacific. Fish Conf., Nouméa. 1997, Paris: Soc. Fr. Ichtyol.

Meunier, F.J. 2002. Skeleton. pp. 65–88. *In*: J. Panfili, H. de Pontual, H. Troadec and P.J. Wright (eds.). Manual of Fish Sclerochronology, Ifremer-IRD coedition, Brest, France.

Meunier, F.J., N. Journiac, S. Lavoué and N. Rabet. 2002. Caractéristiques histologiques des marques de croissance squelettique chez l'Atipa, *Hoplosternum littorale* (Hancock, 1828) (Teleostei, Siluriformes), dans le marais de Kaw (Guyane française). Bull. Fr. Pêche Piscic. 364: 49–69.

Meunier, F.J., M.V. Erdmann, Y. Fermon and R.L. Caldwell. 2008. Can the comparative study of the morphology and histology of the scales of *Latimeria menadoensis* and *L. chalumnae* (Sarcopterygii, Actinistia, Coelacanthidae) bring new insight on the taxonomy and the biogeography of recent coelacanthids? *In*: L. Cavin, A. Longbottom and M. Richter. (eds.). Fishes and the Break-up of Pangaea. Geological Society, London, Special Publications, 295: 351–360.

Meunier, F.J., J. Gaudant and E. Bonelli. 2010. Morphological and histological study of the hyperostosis of *Lepidopus* a fossil Trichiuridae from the Tortonian (Upper Miocene) of Piedmont (Italy). Cybium. 34(3): 293–301.

Meunier, F.J. 2011. The Osteichtyes, from the Paleozoic to the extant time, through histology and paleohistology of bony tissues. C. R., Palevol. 10: 347–355.

Meunier, F.J. 2012. Cyclical growth of freshwater fishes in French Guyana rivers. A skeletochronological approach. Cybium. 36(1): 55–62.

Meunier, F.J. and M. Laurin. 2012. A microanatomical and histological study of the long bones in the Devonian sarcopterygian *Eusthenopteron foordi*. Acta Zool. 93(1): 88–97.

Meunier, F.J., P.M. Brito and M.-E. Leal. 2013. Morphological and histological data on the structure of the lingual tooth plate of *Arapaima gigas* (Osteoglossidae; Teleostei). Cybium. 37(4): 263–271.

Meunier, F.J. and M. Herbin. 2014. La collection histologique du squelette des « poissons » de Paul Gervais. Cybium. 38(1): 23–42.

Meunier, F.J. 2015. New data on the attachment of teeth in the Angler fish *Lophius piscatorius* (Actinopterygii; Teleostei; Lophiidae). Cah. Biol. Mar. 56(2): 97–104.

Meunier, F.J., D. De Mayrinck and P.M. Brito. 2015a. Presence of plicidentine in the labial teeth of *Hoplias aimara* (Erythrinidae; Ostariophysi; Teleostei). Acta Zool. 96(2): 174–180.

Meunier, F.J., J. Mondejar-Fernandes, F. Goussard, G. Clément and M. Herbin. 2015b. Presence of plicidentine in the oral teeth of *Latimeria chalumnae* (Sarcopterygii; Actinistia; Coelacanthidae). J. Struct. Biol. 190(1): 31–37.

Meunier, F.J. and P.M. Brito. 2017. Histological characteristics of lower jaws and oral teeth in the short nose gar, *Lepisosteus platostomus* Rafinesque, 1820 (Lepisosteidae). Cybium. 41(3): 279–286.

Meunier, F.J. and I. Arnulf. 2018. Some histological data of bone and teeth in the Rift Eelpout, *Thermarces cerberus* (Zoarcidae; Perciform; Teleostei). Cybium. 41(1): 83–86.

Meunier, F.J., D. Germain and O. Otero. 2018a. A histological study of the lingual molariform teeth in *Hyperopisus bebe* (Mormyridae; Osteoglossomorpha; Teleostei). Cybium. 41(1): 87–90.

Meunier, F.J., F. Lecomte and G. Duhamel. 2018b. Some histological data on bone and teeth in the grey notothen (*Lepidonotothen squamifrons*) and in the mackerel icefish (*Champsocephalus gunnari*) (Notothenioidei; Perciformes; Teleostei). Cybium. 41(1): 91–97.

Meunier, F.J., O. Otero and M. Laurin. 2018c. A histological study of the jaw teeth in the Devonian actinopterygian = *Cheirolepis canadaensis* (Whiteaves). Cybium. 41(1): 67–74.

Meunier, F. and G. Boivin. 1974. Divers aspects de la fixation du chlorhydrate de tétracycline sur les tissus squelettiques de quelques Téléostéens. Bull. Soc. Zool. Fr. 99: 495–504.

Moss, M.L. 1961. Osteogenesis of acellular teleost fish bone. Amer. J. Anat. 108: 99–110.

Moss, M.L. 1965. The biology of acellular teleost bone. Ann. N.Y. Acad. 109: 337–350.

Olsen, S.J. 1971. Swollen bones in the atlantic cutlass fish *Trichiurus lepturus* Linnaeus. Copeia. (1): 174–175.

Ørvig, T. 1951. Histologic studies of placoderms and fossil elasmobranches. The endoskeleton with remarks on the hard tissues of lower vertebrates in general. Ark. Zool. (ser 2). 2: 321–354.

Ørvig, T. 1967. Phylogeny of tooth tissues: evolution of some calcified tissues in early Vertebrates. pp. 45–110. *In*: A.E.W. Miles (ed.). Structural and Chemical Organization of Teeth, Vol. I, London: Academic Press.

Ørvig, T. 1977. A survey of odontods ("dermal teeth") from developmental, structural, functional, and phyletic points of view. *In*: S.M. Andrews, R.S. Miles and A.D. Walker (eds.). Problems in Vertebrate Evolution. 4: 53–75. Lin. Soc. London.

Paig-Tran, E.W.M., A.S. Barrios and L.A. Ferry. 2016. Presence of repeating hyperostotic bones in dorsal pterygiophores of the oarfish, *Regalecus russellii*. Anat. doi:10.1111/joa.12503 (8pp).

Panfili, J., H. de Pontual, H. Troadec and P.J. Wright. 2002. Manuel de sclérochronologie des poissons. Coédition Ifremer-IRD, Brest, France, 464 p.

Parmentier, E., P. Compère, M. Casadewall, N. Fontenelle, R. Cloots and C. Henrist. 2008. The rocker bone: a new kind of mineralized tissue? Cell Tissue Res. 334: 67–79.

Pascal, M., G. Hostache, C. Tessier and P. Vallat. 1994. Cycle de reproduction et fécondité de l'atipa, *Hoplosternum littorale* (Siluriforme), en Guyane française. Aquat. Liv. Res. 7: 25–37.

Pascal, M., G. Hostache and C. Tessier. 1995. Timing of spawning and location of nests in *Hoplosternum littorale* (Siluriformes, Callichthyidae). Cybium. 19: 143–151.

Patterson, C. 1977. Cartilages bones, dermal bones and membrane bones, or the exoskeleton versus the endoskeleton. pp. 77–121. *In*: S.M. Andrews, R.S. Miles and A.D. Walker (eds.). Problems in Vertebrate Evolution. Lin. Soc., Symp. 4, Acad. Press, London, New-York.

Persson, P. 1997. Calcium regulation during sexual maturation of female salmonid: Estradiol-17b and calcified tissues. Department of Physiology (Ed.). Göteborg University, Sueda.

Persson, P., K. Sundell, B.Th. Björnsson and H. Lundqvist. 1998. Calcium metabolism and osmoregulation during sexual maturation of river running Atlantic Salmon. J. Fish Biol. 52(2): 334–349.

Peyer, B. 1968. Osteichthyes. pp. 80–110. *In*: R. Zangerl (ed.). Comparative Odontology. Univ. Chicago Press.

Qu, Q., S. Sanchez, H. Blom, P. Tafforeau and P.E. Ahlberg. 2013. Scales and tooth whorls of ancient fishes challenge distinction between external and oral "teeth". PLoS One 8: e71890 (doi:10.1371/journal.pone.0071890).

Qu, Q. 2015. Three-dimensional virtual histology of early vertebrates scales revealed by synchrotron X-ray phase-contrast microtomography. Acta Univ. Upsal. Uppsala, 1–49.

Quick, A.J.R. and M.N. Bruton. 1984. Age and growth of *Clarias gariepinus* (Pisces, Clariidae) in the P.K. le Roux Dam, South Africa. South Afr. J. Zool. 19: 37–45.

Ricqlès, A. (de), F.J. Meunier, J. Castanet and H. Francillon-Vieillot. 1991. Comparative microstructure of bone. pp. 1–78. *In*: B.K. Hall (ed.). Bone, 3, CRC Press.

Sanchez, S., P.E. Ahlberg, K.M. Trinajstic, A. Mirone and P. Tafforeau. 2012. Three-dimensional synchrotron virtual paleohistology: a new insight into the world of fossil bone microstructures. Microsc. Microanal. 18: 1095–1105.

Sasagawa, I. and M. Ishiyama. 2005a. Fine structural and cytochemical mapping of enamel organ during stages in gars, *Lepisosteus oculatus*, Actinopterygii. Arch. Oral Biol. 50: 373–391.

Sasagawa, I. and M. Ishiyama. 2005b. Fine structural and cytochemical observations on the dental epithelial cells during cap enameloid formation stages in *Polypterus senegalus*, a bony fish (Actinopterygii). Connect. Tissue Res. 40: 33–52.

Sasagawa, I., M. Ishiyama, H. Yokosuka, M. Mikami and T. Uchida. 2009. Tooth enamel and enameloid in actinopterygian fish. Front. Mater. Sci. China 3: 174–182.

Sasagawa, I., H. Yokosuka, M. Ishiyama, M. Mikami, H. Shimokawa and T. Uchida. 2012. Fine structural and immunochemical detection of collar enamel in the teeth of *Polypterus senegalus*, an actinopterygian fish. Cell Tissue Res. 347: 369–381.

Sasagawa, I., M. Ishiyama, H. Yokosuka and M. Mikami. 2013. Teeth and ganoid scales in *Polypterus* and *Lepisosteus*, the basic actinopterygian fish: an approach to understand the origin of the tooth enamel. J. Oral Biosci. 55: 76–84.

Sasagawa, I., M. Ishiyama, H. Yokosuka, M. Mikami, H. Shimokawa and T. Uchida. 2014. Immunihistochemical and Western blot analysis of collar enamel in the jaw teeth of gars, Lepisosteus oculatus, an actinopterygian fish. Connect. Tissues Res. 55(3): 225–233.

Sbaihi, M., A. Kacem, S. Aroua, S. Baloche, K. Rousseau, E. Lopez et al. 2007. Thyroid hormone-induced demineralization of the vertebral skeleton in the Eel. Gen. Comp. Endocr. 151: 98–107.

Sbaihi, M., A. Kacem, K. Rousseau, F.J. Meunier and S. Dufour. 2009. Cortisol mobilizes mineral stores from vertebral skeleton in the European Eel: an Ancestral origin for glucocorticoid-induced osteoporosis? J. Endocrin. 201: 241–252.

Schmidt, W.J. 1971. The normal tooth tissues. pp. 51–469. *In*: W.J. Schmidt and A. Keil (eds.). Polarizing Microscopy of Normal and Diseased Dental Tissues in Man and other Fertebrates. Pergamon Press, Oxford-New York.

Schönborner, A.A., F.J. Meunier and J. Castanet. 1981. The fine structure of calcified Mandl's corpuscles in teleosts fish scales. Tissue and Cell. 13: 589–597.

Schultze, H.-P. 1966. Morphologische und histologische Untersuchungen an Schuppen mesozoischer Actinoptrygier (Übergang von Ganoid-zu Rundschuppen). N. Jahb. Geol. Paläont. 126: 232–314.

Schultze, H.-P. 1969. Die Faltenzähne der Rhipidisiiden Crossopterygier, der Tetrapoden und der Actinopterygier-Gattung *Lepisosteus* nebst einer Beschreibung der Zahnstruktur von *Onichodus* (Struniiformer Crossopterygier). Palaeontol. Italica, New Series. 35, 65: 63–137.

Schultze, H.-P. 1970. Folded teeth and the monophyletic origin of Tetrapodes. Amer. Mus. Novit. 2408: 1–10.

Schultze, H.-P. 1996. The scales of Mesozoic actinopterygians. pp. 83–93. *In*: G. Arratia and H.-P. Schultze (eds.). Mesozoic Fishes 2—Systematics and Fossil Record. Verlag Dr. F. Pfeil, München.

Simmons, D.J., N.B. Simmons and J.H. Marshall. 1970. The uptake of calcium-45 in the acellular-boned toadfish. Calc. Tis. Res. 5: 206–221.

Sire, J.-Y., A. Huysseune and F.J. Meunier. 1990. Osteoclasts in teleost fish: light- and electron-microscopical observations. Cell Tissue Res. 260: 85–94.

Sire, J.-Y. and F.J. Meunier. 1994. The canaliculi of Williamson in Holostean bone (Osteichthyes, Actinopterygii): a structural and ultrastructural study. Acta Zool. 75: 235–247.

Sire, J.-Y. and F.J. Meunier. 2017. Acellular bone in *Sparus aurata* (Teleostei, Perciformes). A light and TEM study. Cah. Biol. Mar. 58(4): 467–474.

Smith, M.M., G.J. Fraser and T.A. Mitsiadis. 2009. Dental lamina as source of odontogenic stem cells: evolutionary origins and developmental control of tooth generation in Gnathostomes. J. Exp. Zool. 312B: 260–280.

Smith-Vaniz, W.F., L.S. Kaufman and J. Glovacki. 1995. Species-specific patterns of hyperostosis in marine teleost fishes. Mar. Biol. 121: 573–580.

Smith-Vaniz, W.F. and K. Carpenter. 2007. Review of the crevalle jacks, *Caranx hippos* complex (Teleostei : Carangidae), with a description of a new species from West Africa. Fish. Bull. 106: 207–233.

Soukup, V., H.H. Epperhein, I. Horácek and R. Cerny. 2008. Dual epithelial origin of the vertebrate teeth. Nature. 455: 795–U6.

Srinivasa Rao, K. and K. Lakshmi. 1986. Case study of nodular excrescences in *Arius tenuispinis*, hiterto considered as osteoma. Dis. Aquat. Org. 1: 123–130.

Stephan, P. 1900. Recherches histologiques sur la structure du tissu osseux des Poissons. Bull. Scient. Fr. Belg. 33: 281–429.

Suga, S., Y. Taki and K. Wada. 1983. Fluoride concentration in the teeth of Perciform fihes and its phylogenetic significance. Jap. J. Icht. 30(1): 81–93.

Takagi, Y. and J. Yamada. 1993. Changes in metabolism of acellular bone in Tilapia, *Oreochromis niloticus*, during deficiency and subsequent repletion of calcium. Comp. Biochem. Physiol. 105A(3): 459–462.

Texereau, M., D. Germain and F.J. Meunier. 2018. Comparative histology of caniniform teeth in some predatory ichthyophagous teleost. Cybium. 41(1): 75–81.

Trébaol, L., H. Francillon-Vieillot and F.J. Meunier. 1991. Étude de la croissance des mâchoires pharyngiennes chez *Trachinotus teraia* (Carangidae, Perciforme) à l'aide de la technique du marquage vital. Cybium. 15: 263–270.

Weiss, R.E. and N. Watabe. 1979. Studies on the biology of fish bone. III. Ultrastructure of osteogenesis and resorption in osteocytic (cellular) and anosteocytic (acellular) bones. Calc. Tis. Intern. 28: 43–56.

Williamson, W.C. 1849. On the microscopic structure of the scales and dermal teeth of some ganoid and placoid fish. Philos. Trans. R. Soc. Lond. B. 139: 435–475.

Williamson, W.C. 1851. Investigation in the structure and development of the scales and bones of fishes. Philos. Trans. R. Soc. Lond. B. 141: 643–702.

Winemiller, K.O. 1987. Feeding and reproductive biology of the currito, *Hoplosternum littorale*, in the Venezuelan llanos with comments on the possible function of the enlarged male pectoralspines. Env. Biol. Fish. 20: 219–227.

Witten, P.E. 1997. Enzyme histochemical characteristics of osteoblasts and mononucleated osteoclasts in a teleost fish with acellular bone (*Oreochromis niloticus*, Cichlidae). Cell Tis. Res. 287: 591–599.

Witten, P.E., A. Hansen and B.K. Hall. 2001. Features of mono- and multinucleated bone resorfing cells of the zebrafish *Danio rerio* and their contribution to skeletal development, remodelling and growth. J. Morph. 250: 197–207.

Witten, P.E., A. Huysseune, T. Franz-Odendaal, T. Fedak, M. Vickaryous, A. Cole and B.K. Hall. 2004a. Acellular teleost bone/dead or alive, primitive or derived? Paleont. News. 55: 34–41.

Witten, P.E., H. Rosenthal and B.K. Hall. 2004b. Mechanisms and consequences of the formation of a kype (hook) on the lower jaw of male Atlantic salmon (*Salmo salar* L.). Mitt. Hamb. Zool. Mus. Inst. 101: 149–156.

Witten, P.E. and A. Huysseune. 2009. A comparative view on mechanisms and functions of skeletal remodelling in teleost fish, with special emphasis on osteoclasts and their function. Biological Reviews 84: 315–346.

Yamada, J. and N. Watabe. 1979. Studies on fish scale formation and resorption. I. Fine structure and calcification of the scales in *Fundulus heteroclitus* (Atheriniformes: Cyprinodontidae). J. Morph. 159: 49–66.

Zylberberg, L., J. Géraudie, F.J. Meunier and J.Y. Sire. 1992. Biomineralization in the integumental skeleton of the living lower Vertebrates. *In*: B.K. Hall (ed.) Bone, CRC Press. 4: 171–224.

Zylberberg, L. 2004. New data on bone matrix and its proteins. C. R. Palevol. 3: 591–604.

Zylberberg, L. and F.J. Meunier. 2008. New data on the structure and the chondrocyte populations of the haemal cartilage of abdominal vertebrae in the adult carp *Cyprinus carpio* (Teleostei, Ostariophysii, Cyprinidae). Cybium. 32(3): 225–239.

Zylberberg, L., F.J. Meunier and M. Laurin. 2010. A microanatomical and histological study of the postcranial dermal skeleton in the Devonian sarcopterygian *Eusthenopteron foordi*. Acta Palaeont. Polon. 55(3): 459–470.

Chapter **4**

Muscular System

Wincenty Kilarski

INTRODUCTION

Contractility is an intrinsic property of all cells that is necessary for the creation of basic functions involving movements such as cell division and motility. In higher organisms, some cells are specialized to enable movement of tissue or organs. These cells may function as individual contractile units, such as myoepithelial cells or may be aggregated to form special tissue–muscle—for movements of large structures.

The contractile apparatus in all cells consists of fibrillar proteins set in a structured manner in the cytoplasm, and connected by intermolecular bonds. In general, the contractions of all the cells result from the reorganization of the intermolecular bonds with the utilization of chemical energy.

Three different types of muscle tissues may be distinguished on the basis of their morphophysiological characteristics; their embryological origin, microscopic organization and physiological properties.

Skeletal or Striated Muscle

Skeletal or striated muscle has a wide variety of morphological forms, but all have the same basic structure being composed of long multinucleated cells named muscle fibers. Individual muscle fiber comprises sets of fine myofibrils, each composed of discrete series of contractile units termed sarcomeres that run along the myofibril length.

Fibers of striated muscles vary in length from several to several hundred millimeters. The thickness of fibers also varies, on cross sections from several micrometers to several hundred micrometers. These variations are related to the age of animals and the type of muscle. Each individual muscle fiber lies in a thin layer of loose connective tissue–perimysium—that forms the structural framework for nerve fibers and the capillary network. When observing the muscle fiber in cross-section, the more central part of the fiber is occupied by small dots that are uniformly or randomly distributed in the form of polygonal areas named the fields of Cohnheim. These represent the cross section through myofibrils. However, the fields of Cohnheim are a consequence of shrinkage and have no functional significance.

In longitudinal sections, the cross striation of individual myofibrils is well distinguished when histological staining is applied. However, living, isolated muscle fiber examined with the polarizing or phase contrast microscope also shows characteristic birefringence of alternate light and dark bands. The anisotropic A-bands are light in contrast to the dark isotropic I-bands. The A-band's width is constant while

30-147 Kraków, Poland, Wiedeńska 4b.
Email: Wincenty.kilarski@uj.edu.pl

the width of I-bands varies according to the extent of contraction or relaxation of the muscle fiber. I-bands are bisected by a dark fine lined termed the Z-line. Occasionally, when special histological staining is applied or when muscle fibers are observed in the phase contrast microscope, the narrow light band, named the H-band, may be observed, crossing the middle part of the A-band. In the center of the H-band lays the fine dark line, the M-line. The width of the H-band varies due to the state of muscle contraction. It is broad in relaxed muscle while in the contracted muscle it is narrow or may be totally absent. These bands can be verified and interpreted in more details when examining muscle fibers in the electron microscope (Figs. 1A–C).

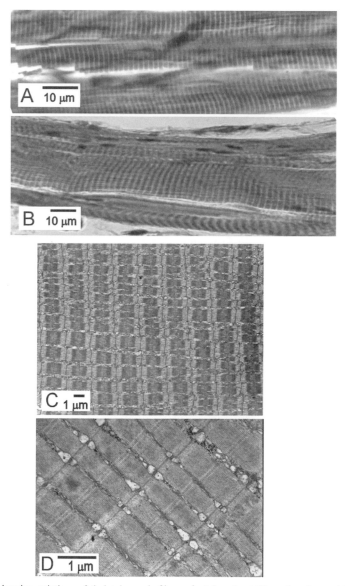

Fig. 1. Micrographs showing striations of skeletal muscle fibers of rainbow trout (*Oncorhynchus mykiss* Walbaum) (**A**) and the guppy (*Poecilia reticulata* Peters) (**B**). The striations are composed of alternating light I-bands (isotropic in polarized light) and dark A-bands (anisotropic in polarized light). Eosin and Hematoxylin stain. **C**. Electron micrograph of small fragment of longitudinally sectioned trunk muscle fibers of tench (*Tinca tinca* L.). The myofibrils are of uniform diameter. Corresponding bands of sarcomeres are in register across the muscle fiber. **D**. Electron micrograph of several myofibrils labeled to indicate the various sarcomere bands (A-, M-, I- and Z-bands, respectively; for details see Fig. 2D) in the normal pattern of cross striation in a relaxed muscle fiber. Individual myofibrils are separated by the cisternae of the sarcoplasmic reticulum.

Visceral or Smooth Muscle

Visceral or smooth muscle forms muscular components of several structures such as the intestinal tract and blood vessels where they construct layers of varying thickness. They are composed of single spindle shape cells, fibers that run parallel to each other. The cytoplasm of smooth muscle fibers is homogenous and lacks the alternating dark and light bands characteristic of striated muscles. Therefore, they appear smooth in the histological preparations. The contractile proteins of the smooth muscle fibres are organized in the form of filaments that run as parallel fine threads of different thickness along the long axis of the fibers. They represent myosin and actin units, respectively, that are similar to those of striated muscle.

Cardiac Muscle

Cardiac muscle is composed of a population of single elongated cells named cardiomyocytes that are linked together by specialized junctions to form physiological units. Cardiac muscle has several structural and functional characteristics of smooth and striated muscle, since its contractile apparatus is regularly organized and forms distinct striations when observed both in light and electron microscopy. Like the visceral muscle, its contractions are strong and utilize a great deal of energy, and their contraction is continuous and initiated by inherent mechanisms though modulated by external autonomic and hormonal stimuli. Cardiac muscle fibers also have a system of T-tubules and sarcoplasmic reticulum analogous to that of skeletal muscle. In the case of cardiac muscle, however, there is a slow leak of calcium ions into the cytoplasm from the sarcoplasmic reticulum after recovery from the preceding contraction; this causes a succession of automatic contractions independent of external stimuli. The rate of this inherent rhythm is then modulated by external autonomic and hormonal stimuli. Cardiac muscle contracts continuously and rhythmically maintaining a permanent blood circulation.

Morphological Basis of Muscle Contraction in Striated Muscle

In order to understand muscle structure organization in detail, we must use a specific nomenclature to describe this highly specialized tissue that leads to the use of a special terminology for some muscle cell components: cell membrane—sarcolemma, cytoplasm—sarcoplasm, endoplasmic reticulum—sarcoplasmic reticulum, mitochondria—sarcosomes. The thick filaments are located in the central part of the sarcomeres, each myosin filament is a collection of myosin molecules (500,000 kDa). In the region of the A-band thick myosin filaments interdigitate with a set of six of thin actin filaments. In specially prepared isolated myofibrils, cross bridges between the myosin and actins filaments occur in a double hexagonal array. In a section across the myofilament bundle, each end of a thick filament is surrounded by a group of six thin filaments.

General Organization of the Contractile System in Striated Muscle

The basic contractile units of the striated muscle are the myofibrils (Fig. 2) composed of a set of linearly arranged contractile units, the sarcomeres. Each sarcomere is demarcated by Z-lines. The Z-line is composed of a characteristic network of proteins, α-actinin. The sarcomeres consist of two kinds of main filaments. Thick myosin filaments (~ 15 nm in diameter), composed of myosin molecules, that form A-bands, and thin actin filaments (~ 5 nm in the diameter) confined to I-bands. The actin filaments extend about 1 μm in either direction from the Z line. Each actin filament approaching the Z line appears to be continuous with four diverging thin strands called Z-filaments. The actin filaments approaching the Z-line from opposite sides are offset, so that in longitudinal sections the cross-connecting Z-filaments give a characteristic zigzag pattern.

Molecular analysis of myosin reveals that each molecule is composed of two long polypeptide chains helically entwined. The individual myosin molecule is shaped like a double headed golf club with a hinged rod. The main shaft of the molecule is about 200 nm long and 2–3 nm thick. Within each bundle, the

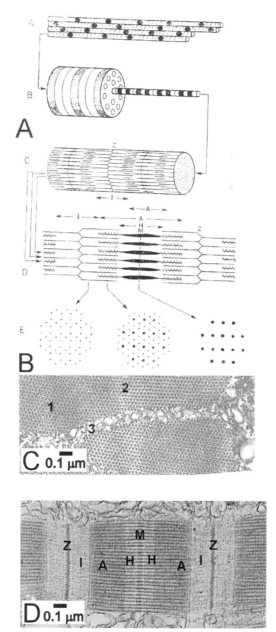

Fig. 2. Diagrams of the organization of skeletal muscle. **A.** Several muscle fibers form the muscle fascicle (small A). A single muscle fiber with one pulled out myofibril (small B). **B.** A fragment of a myofibril (small C) showing a general view of the sarcomere organization on a longitudinal (small D) and cross section (small E) at the levels of I-, A- and M-bands. **C.** Electron micrograph of the cross section of the myofibril at the A- (1) and M-band (2) levels separated by the cisternae of the sarcoplasmic reticulum (3). **D.** Longitudinal section of a single sarcomere (EM-micrograph). All sarcomere bands (I, Z, A, H, M) are indicated.

myosin molecules are arranged in bipolar mode with their protruding club like heads heading for the ends of their shafts and directed towards the middle. For that reason, a myosin bundle has a central, bare zone (H-band) and an array of protruding heads of reverse polarity at each end.

The I-band of the sarcomere is composed of thin filaments that in vertebrate muscle consist of three proteins: actin, tropomyosin and troponin. These last two proteins are regulatory molecules and will be discussed later in more details. Actin filaments are about 1 μm in length and have a beaded appearance. F-actin looks like two long pitch helical strands of globular subunits, G-actins. Globular actin is 5.6 nm in diameter and polymerized to form threads entwined around each other to produce the filamentous polymer, F-actin. In each of the grooves running along the filament helix between the individual threads lie tropomyosin molecules joined end to end, with troponin molecules attached to the tropomyosin filaments at axial intervals of 38.5 nm.

Actins comprise around 20% of the sacomere dry mass, and in the living active state has the form of filaments. The G-actin (70,000 kDa) has a more or less pear shape; however, to simplify the schematic description of actin filaments, G-actins are considered to be regular spheres. In the living state, both G and F actin remain in equilibrium. New actin filaments are formed and regulated by ATP in the presence of Mg^{2+} ions. The G-actin monomers are consistently oriented to give the filament a definite polarity. In the intact sarcomere, the actin filaments on either side of the Z-line have opposite polarity. Polymerization of G-actin to form the F-actin filaments run in one definite direction and is initiated by the protein α-actinin that is associated with the Z-line. The length of the actin filament is controlled by another protein, β-actinin, which becomes attached to the end of the actin filament after it reaches the length of 1 μm and blocks its further elongation.

Regulatory Mechanism in Striated Muscle

The process of striated muscle contraction is regulated at the level of actin filaments, controlled by two proteins already mentioned above, tropomyosin and troponins. The tropomyosin molecule is a dimer about 40 nm long and 2 nm thick, and holds two polypeptide chains, α and β, in α helical conformation. Tropomyosin molecules are put together end to end to form filaments of 70,000 kDa. Each molecule of tropomyosin is associated with seven monomers of G-actin. Tropomyosin is characteristic for animal species, the stage of myogenesis and for the type of muscle. The second regulatory protein is a complex of three molecules named troponin. It consists of three subunits: Tn C (18,000 kDa) with binding sites for Ca^{2+} ions, Tn I (23,500 kDa) inhibits the interaction of actin and myosin and Tn T (38,000 kDa) attaches both troponins to tropomyosin.

Sliding Filament Mechanism of Contraction in Striated Muscle

The sliding filament mechanism is based mainly on electron microscopical observations of muscle fibers in the relaxing and contracting state.

The Myosin-Actin Interaction

The heads of myosin molecules can move towards and bind with the G-actin subunits of the surrounding actin filaments owing to their hinged shaft. The contractile force is produced by the ratchet-like attachment. Because of the preferred polarity in the arrangement of these two groups of interacting molecules, the detachment and reattachment of myosin heads, the thin filaments pull inwards towards the middle of the thick filaments on either side. Since the thin filaments are fixed in the Z-lines, the length of each sarcomere as a consequence is reduced, i.e., contraction occurs. On relaxation, all of the myosin heads become detached and the two packages of filaments slide back to their original position of partial overlap (Figs. 2A–D).

Regulation of the Actin-Myosin Interaction

The regulation of contraction depends upon the availability of Ca^{2+} ions. On stimulation, calcium is released from the sarcoplasmic reticulum cisternae causing its concentration in the milieu of the myofilaments to increase from less than 1.0 to about 10 micromoles.

The sarcoplasmic reticulum (SR) is the main reservoir for calcium ions that is released to stimulate contraction of muscle fibers and is subsequently pumped back into the SR to bring about relaxation. The

SR is composed of an elaborate system of tubules that form a network surrounding each myofibril with periodic swellings forming cisternae in which the calcium ions are stored. The SR is made of smooth membranes containing two major proteins, an ATPase enzyme (around 85%) and calcium channels. The ATPase is responsible for accumulating calcium ions back into the SR after contraction; the calcium channels are responsible for the release of calcium ions into the sarcoplasm to stimulate contraction. The SR is organized in the form of repeated segments of tubules and cisternae enveloping individual sarcomeres. It is compiled from two distinct elements. The longitudinal tubules run along myofibrils and widen into cisternae that abut each sarcomere at both sides of the Z-line. These are named dyads. Longitudinal tubules or canals of the SR that originate from the dyads run along the A-band and they terminate in the wide central cisterna at the level of H-band and are called H-cisterna. In some muscle, however, these cisternae are absent. Between the cisternae of the dyads, a single apparent cross-section transverse tubule (continuous with the sarcolemma that together with two terminal cisternae form named triad) is present. In longitudinal sections of myofibrils, the centrally located 'vesicle' of triads is actually a cross section of a tubule that runs perpendicularly to the long axis of myofibrils and is continuous with the sarcolemma. The longitudinally running tubule is termed transverse tubular system (T-tubule (TS)). In fast contracting muscle, triads are regularly located at the level of the Z-lines while in the extrinsic eye muscle or in striated muscle of the swim bladder they are found at the boundary between the A- and I-bands. The functional significance of this different localization of the transverse tubular system (Figs. 1C and D) is so far not well understood.

Calcium ions are stored in the cisternae of the dyads and are liberated from them when a nerve signal originating in motor—neurons reaches the sarcolemma as action potential. Then the action potential is propagated through the transverse system of the tubules to the dyad cisternae and initiates liberation of Ca^{2+} ions.

In the absence of Ca^{2+} ions in the myofibrils, troponin acts as inhibitor of contraction. This inhibitory action is due to troponin-I. Troponin-I interacts strongly with actin and very weakly with tropomyosin. It is plausible that troponin-I is able to form a stable inhibitory complex with tropomyosin only when it associates along the actin filament in the absence of Ca^{2+} ions. When calcium is released from the SR cisternae into the myofibril environment, calcium ions bind to troponin-C that exhibits high calcium affinity only when reaching a level of $10^{-6} mol^{-1}$. When the concentration of calcium ions in the myofibrils' milieu drops to $10^{-7} mol$, it is released from troponin-C and is pumped back into the longitudinal elements of the SR and is transported from there to the terminal dyads of the cisternae.

The formation of the troponin-C–Ca^{2+} complex results from conformational changes of tropomyosin that moves in the direction of the actin helical groove and unveils active sites of G-actin monomers enabling the interaction with the myosin heads. When the troponin-C–Ca^{2+} complex is split, the tropomyosin filaments return to their previous position and rest on the surface of the actin filament, thus blocking the active sites for the myosin heads.

Contraction and Relaxation of Striated Muscle Fibers

The signal for muscle contraction originates in motor neurons and runs as action potential waves along their axon membrane and reaches the specialized junction region of the motor end plate or synapse. Action potentials arise in the cell body of the neuron as a result of integration of afferent stimuli. Action potentials are then conducted along the axon to stimulate the effectors, in this case the muscle fiber where the axon terminates. The axoplasm of the nerve terminal contains mitochondria and a large number of small vesicles–synaptic vesicles—that are the sites of storage of the neurotransmitter, acetylocholine. It is estimated that each vesicle may contain 10,000 molecules of neurotransmitter. Synaptic vesicles are thought to be derived by budding from the smooth endoplasmic reticulum of the neuron. They then migrate down the axon to the synapse by means of microtubular mechanisms. Synaptic vesicles tend to aggregate towards the presynaptic membrane in places of specialized sites of the presynaptic membrane called active zones.

When the wave of membrane depolarization reaches the end-plate, the calcium ions enter the synapse via calcium channels. These cause the liberation of the neurotransmitter acetylcholine from synaptic vesicles into the presynaptic cleft by means of exocytosis. The neurotransmitter diffuses across the synaptic cleft and is caught by the acetylcholine receptor situated in the postsynaptic membrane. Sarcolemma, facing the motor end-plate, contains several acetylcholine receptors that are oligomeric tetramers and have two binding sites for acetylcholine molecules. When the acetylcholine–receptor complex becomes established, it lasts for 5 milliseconds resulting in a temporal increase in the permeability of the sarcolemma to potassium ions that flow out from the muscle fiber; simultaneously, sodium ions flow into the fiber. In consequence, the sarcolemma becomes depolarized producing an action potential that is spread over the sarcolemma and through the transverse tubular system (TS), activating the SR terminal cisternae, thereby releasing Ca^{2+} ions and triggering muscle contraction.

Acetylcholine released into the postsynaptic cleft is rapidly broken down to choline and acetate by the acetylocholine esterase present in the postsynaptic cleft and these products are then returned back to the synaptic cytoplasm by means of pinocytosis, thereby inactivating the released acetylcholine between successive nerve impulses.

Since there is no direct continuity of the membrane between the transverse tubule and terminal SR cisternae, the action potential is probably transmitted via certain proteins (~ 30,000 kDa) that bridge the membrane of the TS tubules and SR terminal cisternae. These protein bridges were described as "foots" based on their appearance in electron microscopic images.

Smooth Muscle

Smooth or visceral muscle differs from striated muscle characterized above in several respects. Smooth muscle tissue originates from the hypomere, part of the mesoderm situated below the *chorda dorsalis* of the embryo. In places where a layer of smooth muscle will develop later, the presumptive smooth muscle cells begin to stretch out and differentiate into the smooth muscle cell myoblasts. The nuclei of smooth muscle myoblasts are located in the central, thickest part of the cells that become elongated and form fibers (Figs. 3A and C).

Smooth muscles have physiological and pharmacological properties that differ from striated muscle. Their contraction is slower and allows two forms of contractions, rhythmic and tonic contraction. Smooth muscles contract as a whole mass rather than as individual motor units. Contraction of visceral muscles occurs independent of neurological innervations, often in a rhythmic fashion. In the veins or arteries, the smooth muscles respond to vasomotor nerve stimuli with contraction or relaxation, thus changing the diameter of the lumen of the vessels. Smooth muscle cells are able to propagate nerve stimuli from cell to cell via gap junctions. Smooth muscle cells may synthesize collagen and elastin molecules and may function as endocrine cells producing hormones that contribute to the maintenance of blood pressure (renin, angiotensin).

Visceral smooth muscle, e.g., in the intestine, is organized in layers with the cells of one layer arranged at right angle to those of the adjacent layer. This arrangement permits a wave of contraction passing down the tube, thus propelling the content forward. This action is named peristalsis. The visceral muscle activity is boosted by parasympathetic stimulation and controlled by a variety of hormones released in response to changes of the volume of the intestine. Smooth muscle fibers are bound together in irregular branching fascicles that are functional units. Within the fascicle, individual smooth muscle fibers are assembled more or less parallel to one another with the thickest part of one fiber lying against the thinner part of the adjacent cells. The individual smooth muscle fibers may be side connected by means of specialized junctions. Between the individual muscle fibers, there is a supporting network of connective tissue that makes the framework for blood vessels and nerves.

Fine Structure of Smooth Muscle Cells

The contractile machinery of smooth muscle cells contains four types of filaments: thick myosin filaments (14–16 nm), thin actin filaments (6–8 nm), non-contractile intermediate filaments (10 nm) containing desmin (50,000 kDa), also present in striated muscle, and filamin filaments (250,000 kDa). In contrast to striated muscle, the longitudinally running sets of filaments are not assembled into regular sarcomeres. The

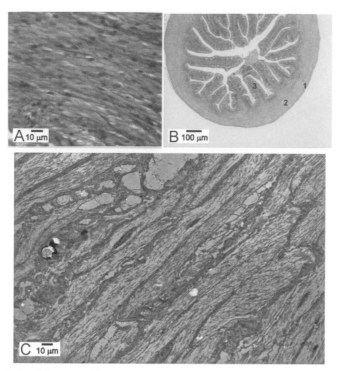

Fig. 3. Visceral muscle. **A.** Small fragment of the smooth muscle layer surrounding the ovary of tench (*Tinca tinca* L.). Visceral muscle consists of elongated, spindle shaped smooth muscle cells with pointed ends which may be occasionally bifurcated. Fibers are generally short and contain only one elongated nucleus centrally located in the cytoplasm. Eosin and Hematoxylin stain. **B.** Cross section through a fragment of the intestine of perch (*Perca fluviatilis*) showing the two layers of the smooth muscle (*muscularis mucosa*) (Circular layer (1) and longitudinal layer (2)) surrounding the intestine. (3) intestinal villi. **C.** Electron micrograph of several smooth muscle fibers from the intestine of the tench (*Tinca tinca* L.). Actin myofilaments running along the long axis of the fibers.

bunch of actin filaments is attached to the cell membrane by means of a dense body that is a plaque-like condensation of glycol proteins. These dense plaques are also present in the cytoplasmic matrix within the cell interior and are used similarly as an attachment for thin filaments. The actin filament, F-actin polymer, is associated with the troponin complex. The intermediate filaments are inserted to the plasma membrane or to the dense plaques that lie in the cell cytoplasm. They play a structural role providing a framework for the attachment of the force-generating components. There is good evidence that filamin filaments can interact with actin to generate a contractile power.

Myosin Linked Regulation of Smooth Muscle Contraction

It is well accepted that the level of calcium ions within the smooth muscle fiber determines the contractile activity of the cell. The mechanism whereby this control is exerted over the contractile apparatus is not well established. There are two theories: (1) Activation results from the phosphorylation of the light chains of myosin (20,000 kDa), (2) Phosphorylation is not required for the activation and a separate regulatory system is involved which has been named leyotonin. Since the leyotonin hypothesis is not well based on experimental evidence, only the first hypothesis will be briefly outlined here.

The phosphorylation has two key components: the myosin light chain kinase activating the system and myosin light chain phosphatase involved in relaxation of the muscle. The myosin light chain kinase is a dimer composed of large, 105,000 kDa and small, 17,000 kDa subunits identified as being calmodulin. The sequence of events thought to occur during the activation of the myosin light chain kinase is that calcium is bound to calmodulin, and that the Ca^{2+}—calmodulin complex forms the active tertiary complex with the larger kinase subunit. It is therefore accepted that calcium activates the myosin light chain kinase by its interaction with calmodulin. This activation leads to the activation of the myosin molecule as an ATPase. Consequently, only when activated by calcium can myosin hydrolyze ATP and combine to actin by means of myosin heads that oscillate and move the actin filaments forward. This mechanism is independent of any regulation by the actin filaments.

Smooth Muscle as an Auxiliary Component of Particular Organs

The Bulbus Arteriosus

The structural architecture of the bulbus arteriosus of fish hearts makes it the most important modulator of several cardiovascular activities. It is a highly distensible chamber that dilates during ventricular systole and may transiently store 25–100% of the stroke volume (Bushnell and Jones 1994). Therefore, it is generally described as an elastic reservoir that attenuates the extremes of systolic and diastolic pressure and produces a virtually continuous flow of blood through the ventral aorta (Stevens and Randal 1967). Therefore, it seems reasonable to suggest that the presence of smooth muscle and elastic fibers in the tunica media of the wall of the bulbus arteriosus are the most important elements for the "windkessel" function that may modulate bulbar elasticity. Thus, the presence of the smooth muscle appears to be actively involved in the bulbus arteriosus wall dynamics (Figs. 4A and B).

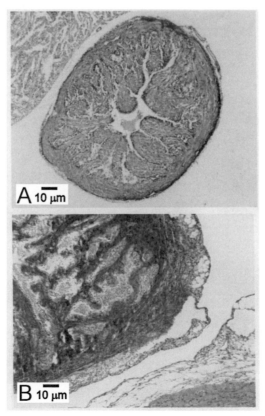

Fig. 4. Comparison of the two types of bulbi arteriosi; they differ in the form and thickness of their trabeculae. **A.** Bulbus arteriosus of the sunblake (*Leucaspius delineatus* Heckel). **B.** The bulbus arteriosus of the whitefish (*Coregonus lavaretus* L., Coregonidae) contains a profusion of trabeculae across the entire width of its lumen. The three layered structure of the bulbular wall contains a smooth muscle layer in its center. The bulbus arteriosus is covered by a layer of tunica adventitia (the epicardium). Eosin and Hematoxylin stain.

Color version at the end of the book

The Gastrointestinal Tract

The gastrointestinal tract is another organ whose task strongly depends on the proper function of smooth muscle tissue. Food is propelled along the gastrointestinal tract by two main mechanisms: (1) The voluntary muscular action in the mouth, pharynx and partly the oesophagus (striated muscle fibers) is (2) succeeded by involuntary waves of smooth muscles contraction called peristalsis modulated by the autonomic nervous system and a variety of hormones.

The tissue that is responsible for the dynamic activity of the gastrointestinal tract comprises three layers of smooth muscle cells. These are: (1) the muscularis mucosae, a thin layer of smooth muscle, which produces local movements and folding of the mucosa, (2) the muscularis propria that consists of smooth muscle cells organized into two separate layers: an inner circular layer and an outer longitudinal layer. The contraction of the circular musculature constricts the lumen of the digestive tract, but in the absence of a compensatory mechanism, it would also extend the portion of the tract in which it is located. Such a lengthening could be compensated by a contraction of the longitudinal smooth muscle layer (Fischer and Seesemann 1961). The action of the smooth muscle layers, opposed at right angles to one another, is the basis of peristaltic action of the gastrointestinal tract. A peculiar muscle system is found in the digestive tract of some teleosts. An example among freshwater fish is the tench (*Tinca tinca* L.). The tunica muscularis of the tench is composed of three layers: an external layer consisting of striated muscle fiber organized in the form of separate bundles running longitudinally; an inner layer, also made up from striated muscle fiber but running longitudinally; and the most inner layer built of smooth muscle (Kilarski and Bigaj 1971). The replacement of the smooth musculature by striated muscle causes rhythmic reflectory peristaltic waves during the action of the alimentary tract. Based on morphological criteria which can be applied to the muscle of all vertebrates (the structure of the Z line, presence or absence of the M band), the muscle fibers of the tunica muscularis of the tench could be classified as slow muscle.

An essential function of the muscular tube built up from striated muscle is the induction of pressure increase inside the tube, thus compressing the gas contained in it, as in the case of the swim-bladder of some fish, e.g., in the burbot (*Lota, lota* L.) which has a muscular layer composed of striated muscle fibers (Kilarski and Bigaj 1964).

The Ovary

In the majority of fish, the ovary is enclosed in an ovarian sac. The eggs pass into this sac as soon as they are released from the follicle, and are led from there to the genital pore to the body cavity (*Salmonidae*). The ovarian sac is provided with smooth muscle cells that are arranged into two layers, a circular and a longitudinal layer. The waves of the muscular contraction run cranio-caudally, pushing the eggs out of the abdominal cavity.

Cardiac Muscles

Cardiac muscle shows many structural and functional qualities of skeletal and visceral muscles. Contractions of heart muscle are strong and require a great amount of energy. The continuous pulsation of the whole organ is initiated and maintained by an inherent mechanism regulated by the autonomous nervous system and hormonal stimuli.

The cardiac tissue is composed of elongated muscle cells–cardiomyocytes—containing mainly one, or rarely two, nuclei found in the central part of the cell. Each cardiomyocyt contains no more then 2–3 myofibrils. The muscle fibers are branched and at the borders between two fibers they join by means of specialized intercellular junctions located at intercalated discs. Between the adjacent cardiomyocytes, special low resistance junctions (gap junctions) are located responsible for the propagation of action potentials from one cell to another and for propagation of some metabolites. Thus, adjacent cells are forced to contract simultaneously, thereby acting as a functional syncytium (Fig. 5).

The cardiomyocytes comprising the main constituents of the walls of cardiac compartments form bundles called trabeculae or sheets of the structural units of the heart. The spongy like structure is filled

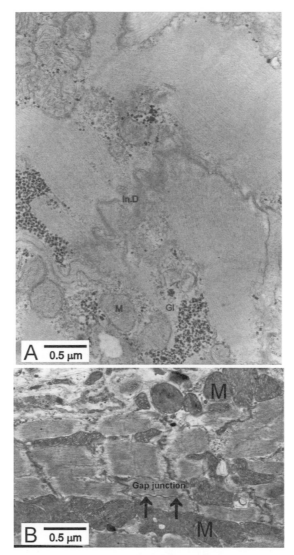

Fig. 5. A. Electron micrograph of ventricular myocardial cells of the lamprey (*Petromyzon marinus* L.) illustrating the junction between two cardiomyocytes termed intercalated disc (In. D). M, mitochondria; Gl, glycogen granules. **B.** Small fragment of the ventricular part of the lamprey illustrating the presence of a gap junction (arrow) (EM micrograph). M, mitochondrium.

with blood. Coordinated and rhythmical contraction of the cardiac cells squeezes out blood from the heart ventricle and pushes it to the gills.

The following description of heart elements will concentrate mainly on the teleost heart but in later parts of this chapter, the description of hearts of remaining taxa of fish will also be considered. The structure of the fish heart and the blood supply appears to present considerable diversity in different fishes. Characteristically, the fish heart pumps only venous blood to the gills and is therefore defined as gills heart. However, there are some exceptions present in dipnoan fish and in some air breathing fish that possess lungs or air sacs which receive blood directly from the heart. In general, the fish heart is comprised of four anatomical units: (1) the sinus venosus, (2) the atrium, (3) the ventricle and (4) conus or bulbus arteriosus. The bulbus arteriosus is not part of the tissue of the heart muscle but it is a local enlargement of the aorta. In elasmobranch fishes and in Holostei, the bulbus arteriosus is well developed, but is hardly discernible in teleosts.

General Organization

The blood circulates in a closed system comprising veins and arteries and it is propagated by means of constant rhythmic pulsation of the heart muscle. The blood drained from the whole fish body enters the first heart chamber, the sinus venosus to which the blood is delivered by the paired ductus Cuvieri and one hepatic vein. Then the blood flows to the atrium and further to the ventricle and subsequently the blood is ejected through the bulbus arteriosus (Figs. 4A and B) to the gills.

The heart of teleosts is relatively small in comparison with the heart of mammals. Its mass ranges from 0.08 to 0.11% and depends on the swimming activity of the fish. The mass of fast and continuously swimming mackerel (*Scomber scombrus* L.) or flying fish (*Cheilopogonidae*) is 0.196% while the mass of the heart of a fish with a sedentary life (e.g., the plaice, *Pleuronectes platessa* L.) is barely 0.06%.

The fish heart can adapt its size temporarily to a changing environment. A good example is the heart of the salmon (*Salmo salar* L.). While it inhabits the sea, the heart is large but it becomes reduced in its mass when the salmon migrate up river for spawning.

The localization of the heart is ventral and lies behind the gills in the pericardial cavity that consists of two layers: an outer layer, the parietal pericardium; and the second layer, the visceral pericardium or epicardium. The pericardial cavity is roughly conical in outline with the apex facing frontally. The pericardium is perforated for blood vessels leaving and entering the heart. The heart moves synchronously with the inner layer of the pericardium that lies firmly on it during contractions. However, the narrow cleft that separates the pericardium from myocardium is filled with serous fluid that lubricates the surface of the myocardium allowing the heart to move smoothly within the pericardial cavity during contraction. The main part of the epicardium is composed of a thin layer of loose connective tissue composed of fibroblasts and collagenous reticular fibers. In that layer, dispersed lipocytes may be found the number of which varies according to the age and physical condition of the fish (Figs. 6A–D).

The sinus venosus is a thin walled sac about 60–90 μm thick. It has the shape of an inverted cone with tapering apex. Its tip joins the next heart chamber, the atrium, with the sino-atrial channel equipped with a pair of sino-atrial valves that prevent backflow of blood into the sinus venosus. The wall of the sinus venosus consists of three layers: (1) The external, single layer consisting of visceral epicardium. (2) The internal layer built of endocardial cells; and (3) The middle layer containing collagen and elastic fibrils among which a few myocardial cells are present. These make the sinus venosus walls slightly contractile, particularly in elongated fish, e.g., eels (*Anguilla anguilla* L.), that effectively helps the ejection of the blood from the sinus to the atrium (Figs. 6C and D).

The atrium has the shape of a pyramid whose top is directed towards the front of the body. The bottom of the atrium lies on the ventricle and is connected with it by a funnel—a shaped passage. The atrio-ventricular opening in this passage is surrounded by a muscular ring and equipped with two atrio-ventricular valves in the crucian carp (*Carassius carassius* L.), four in the sunfish (*Mola mola* L.) and six in the garfish (*Lepisosteus* sp.). The real shape of the atrium is difficult to describe in fixed histological preparations. In the living heart, the shape and size of the atrium varies in consequence of the cyclic changes of the whole heart volume during its rhythmic pulsation. In general, the shape and the volume of the atrium depend on the shape of the ventricle, and on the distance of the atrium from the bulbus arteriosus. When this distance is reduced (*Clupeidae*, *Scombroidei*, *Salmonidae*), then the atrium lies close to the bulbus arteriosus. In elongated fish like the eel (*Anguilla anguilla* L.), the atrium lies directly on the ventricle and its auricles wrap up around the ventricle.

The atrium is a thin walled sack composed of trabecular myocardium that contains cardiomyocytes located centrally and interlaced with collagen fibrils. The spatial organization of the trabecular myocardium is very complicated. It also differs in thickness. Fast swimming fish like the mackerel (*Scomber scombrus* L.) have a narrow trabecule ranging from 15 to 35 μm while small fish like the fresh water sunblakes (*Leucaspius delineatus* L.) have a large trabecule of up to 105 μm. The entire inner surface of the atrium chamber is lined with a single layer of epithelial tissue—the mesocardium. The cells of the mesocardium are generally flat but in some fish, e.g., in the atrium of the stickleback (*Gasterosteus aculeatus* L.), they are roundish and contain monoamine granules and in large fish even lipid droplets (*Opsanus tau* L.). These mesothelial cells may express mitotic activity when the heart is injured. The atrium is connected

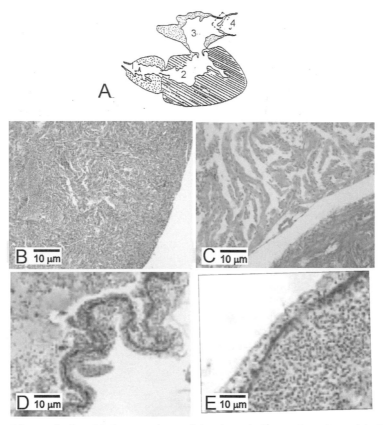

Fig. 6. A. Scheme of the organization of the heart seen in a medial section. 1, bulbus arteriosus; 2, ventricle; 3, atrium; 4, sinus venosus. Sino-atrial, atrio-ventricular and bulbo-ventricular valves are visible. Pictures **B–E** illustrate cross sections through small fragments of the sinus venosus of the crucian carp (*Carassius aureatus gibelio* Bloch) (**B**), the sunblake (*Leucaspius delineatus*) (**C**); the roach (*Rutilus rutilus*) (**D**), and the rudd (*Scardinius erythrophtalmus*) (**E**). The surface of the sinus is covered by a single layer of visceral pericardium and a deeply stained layer of collagen and fibrocytes.

with the next heart chamber–the ventricle—by means of the atrio-ventricular sulcus guarded by a pair of atrio-ventricular valves that stop backflow of blood into the atrium.

The ventricle is commonly described as a pyramid but, in fact, it differs significantly from species to species. According to the schematic description of the ventricle, its apex is directed anteriorly; the roof of the ventricle is flat and is covered by the atrium. The flat sides of the ventricle are directed to its base forming the narrow keel.

Having in mind the diversity of the ventricle shape, Santer (1989) distinguished three categories of ventricles: (1) A tubular one (cod, *Merluccius merluccius* L.), (2) A sack-like one (plaice, *Pleuronectes platessa* L.), and (3) A pyramid like one (weeverfish, *Trachinus draco* L.).

The myocardium is organized in the form of trabeculae containing striated myocytes or cardiac fibers. The cardiac tissue forms two layers, the cortical dense layer and spongy like inner layer (spongiosa). In large fast swimming fish, the heart has a distinct dense cortical layer but in small pelagic forms, the cortical dense layer does not exist and the ventricular wall consists only of a spongy like myocardium. Both layers are separated by thin connective tissue. The cortical layer consists of longitudinally, circularly and spirally assembled bundles of cardiac muscle fibers. If the cortical layer is present, it is furnished with blood capillaries that nourish the surface layer of the ventricle. The blood supplying the dense layer comes from the coronary circulatory system. Blood which sustains and nourishes the cortical layer of the ventricle is brought to it by the artery originating from the last epibranchial arteries. The spongy part of the myocardium has no extra capillary system and the trabeculae are bathed in the venous blood.

Innervation of the Fish Heart

The fish heart is myogenic which means that it has its own independent specific excitation system. This system is developed out of specifically modified muscle fibers that in mammalian hearts are known as Purkinie fibers. However, these fibers are not easily discerned in fish hearts; we may only assume that this type of excitatory system may exist in fish hearts. Based on histological and histochemical studies, the innervation is evidently present in the heart of different species. Nerve impulses are transmitted by specific fibers which are located in the muscle rings. The greatest density of cardiac innervation observed in the sino-atrial region suggests that it is the dominant pacemaker region of the teleost heart. This morphological observation has strong electrophysiological evidence. Von Skramlik (1935) suggested that the prime pacemaker region in teleosts is situated in the basal wall of the atrium adjacent to the sino-atrial junction and that a secondary automatism exists in the region of the atrio-ventricular junction. The elegant physiological studies on the developing heart in the last larval stage of the eel (glass-eel, *Anguilla anguilla* L.) and in the trout embryo (*Salmo trutta* L.) by Grodzinski (1954a, b) showed the presence of the three pacemaker centers. On the base of the distribution of the pacemaker center, von Skramlik (1935) has distinguished three types of fish hearts: (1) Three pacemaker centers exist. The first is located in the sinus venosus and the dusctus Cuvieri, the second at the base of the atrium in the close vicinity of the atrio-ventricular sulcus and the third one in the atrio-ventricular funnel. This type is rare and found so far in eels (*Anguilliformes*). (2) The second type was described mainly in Elasmobranchii where two pacemakers were described, one in the wall of the sinus venosus and the second in the atrio-ventricular sulcus. (3) The third type was described in several groups of teleosts where one pacemaker is situated in the conus auricularis (a part of the atrium) and the second one in the atrio-ventricular junction.

The stimulation of the heart's pulsation is not directly depending on the nervous system; nevertheless, the heart is abundantly innervated and its automatic system is regulated by the parasympathetic and sympathetic systems which form far-reaching plexuses at the base of the heart. There is a marked difference in the density of innervations of different species and in the different parts of the heart when silver-stained sections of different heart regions are observed. It was noticed that the density of innervations is greatest in the sino-atrial region, but noticeably lower in the atrial wall and only scarcely seen in the ventricle.

Histological and electron microscopical observation demonstrate numerous cross sections through nerves, both myelinated and unmyelinated. They were found generally in the sino-atrial region. Two categories of myelinated fibers occur in the vagus nerve. The large diameter nerve fibers are most probably sensory afferents originating from the wall of the atrium and from the epicardial and endocardial receptors of the ventricle. Over the sinus venosus and sino-atrial margin, the cardiac plexuses are formed by small diameter myelinated nerve fibers. In the midst of this plexus are scattered unipolar ganglion cells upon which synaptic formations build up, derived from small diameter myelinated nerves. The unmyelinated nerves innervate post-ganglionic axons of the atrium, atrio-ventricular region and the proximal part of the ventricle. The nerve endings that come from the nervus vagus are cholinergic and their neurotransmitters cause deceleration of the heart contraction. In contrast to the adrenergic synapses that are very numerous in the wall of the atrium, the ventricle originates from the nervus sympathicus and liberates the neurotransmitter adrenalin which stimulates acceleration of the heart contraction frequency.

Chévrel (1887) detected the presence of adrenergic innervation in the fish heart. His concept relies on the basis of an anatomical study and suggested that the fish heart has sympathetic innervations. Subsequently, Young (1931) confirmed this observation and stated that nerve fibers from the sympathetic ganglia join the extracranial branch of the nervus vagus.

The development of formaldehyde-induced fluorescence and other more sophisticated histochemical techniques for the intracellular visualization of biogenic monoamines has made possible the demonstration of monoamine-containing nerves in the hearts of fish. Since then, adrenergic nerves were reported in the hearts of several teleost species. Gannon (1971) subsequently verified, electrophysiologically and pharmacologically, that the trout (*Salmo trutta* L.) atrium and ventricle contain functional adrenergic innervations. There are, however, species that do not possess adrenergic innervation. Santer (1972, 1977) and Cobb and Santer (1973) failed to detect any evidence of adrenergic nerves in bottom-dwelling species like the plaice (*Pleuronectes platessa* L.). Therefore, it may be assumed that the absence of adrenergic innervations may be widespread among different fish species.

The Heart of Agnatha (Myxinomorphi and Petromyzontimorphi)

The superclass, Myxinomorphi, contains the order Myxiniformes that includes two families, the Myxinidae and Bdellostomatidae. The fish that belongs to these groups are entirely sea-dwellers. One of the first elegant descriptions of the heart of myxine (*Myxine glutinosa* L.) can be found in the monograph "Comparative anatomy of the Myxinoids" (Müller 1839). The heart of the myxine is enclosed in a cartilaginous pericardium that communicates with the peritoneal cavity by means of the pericardio-peritoneal canal.

Myxine

The heart of *Myxine glutinosa* consists of three chambers but the ventricle and atrium are distinctly separated from each other and are only bridged by a narrow canal. The heart is situated posterior to the last pair of gill pockets and lies asymmetrically such that the ventricle is located medially, positioned slightly to the right. But the atrium and the sinus venosus are situated on the left side of the body. The sinus venosus consists of two partitions, an anterior one and a posterior one. They are separated from each other by a small groove that grows out from the left sidewall of the sinus venosus. The atrium and the sinus venosus communicate by a canal that is secured by the two large valves of the semi-lunar type. These prevent the backflow of blood from the atrium to the sinus venosus. An analogous valve secures the doorway between the atrium and the ventricle and prevents regurgitation of blood into the atrium. The blood from the ventricle is pumped through the truncus arteriosus to the bulbus arteriosus and consequently to the ventral aorta which sequentially sends off branches, the afferent branchial arteries to the six pairs of gills.

The atrial myocardium has a loose trabeculate organization composed of thin (6.1–0.71 μm) myocardial cells. The ventricular myocardium has a dense, trabeculate, sponge-like structure similar to that already described in teleost hearts. The trabeculae of the myxine heart are on average 10 μm in diameter. In addition to the cardiomyocytes, endothelial and interstitial cells are also present. The distinctive feature of the agnathan heart is the presence of cells that contain a quite large population of catecholamine granules and, as a consequence, the catecholamine level of the heart is very high.

The presence of a portal heart is a unique structure among the whole vertebrate phylum. It was first described by Retzius (1822) who pointed out the similarity between the structures of the atrium of the branchial heart. The portal heart of myxine can be considered the true heart in the sense that its myocardium is built of striated cells and has an intrinsic spontaneous rhythm. The portal heart is equipped with three blood vessels: (1) The right anterior cardinal vein, (2) The supraintestinal portal vein and (3) The common portal vein. Three apertures from these vessels are all secured by valves that guarantee that blood enters the heart through the first two and leaves the heart through the common portal vein. Another peculiarity is the presence of the caudal heart in the circulatory system of Myxine. This was first described by Retzius (1890) and he suggested that the blood from the caudal sinuses returns into the veins.

An innervation of the systemic heart of the Myxiniformes was often questioned and Greene in 1902 reported in his elegant study that the hagfish (*Eptatretus stouti* L.) is free from all nervous regulation (aneural heart). He reached his conclusion from the fact that he was unable to change the rate of heart stroke when the vagus nerve was electrically stimulated. Greene's conclusion was reinforced later by similar findings of Augustinsson et al. (1956).

More sophisticated methods used for histological and chemical analysis as well as electron microscopical techniques subsequently questioned the old point of view that hagfish hearts are aneural. One of the main points that contradicted the old idea was the presence catecholamines in the cardiac tissue. This appears to be dependent upon the integrity of cardiac innervation.

Applying special silver staining histological techniques revealed the presence of large nerves with myelinated fibers and ganglion cells in the proximity of the heart. Recognized was also a system of ganglion cells with fibers distributed along the aorta, the epicardium, and the endocardium. The myocardium has a plexus of coarse and fine argyrophilic fibers and fibril twigs similar to that which occurs in the heart of teleost fish.

These findings contradict the original statement of Greene (1902) that the heart of the hagfish is aneural (Hirsch et al. 1964).

Petromyzontiformes

The heart of petromyzons consists of four anatomically distinct parts: (1) The sinus venosus, (2) The atrium or auricle, (3) The ventricle and (4) The bulbus arteriosus. The heart is outstandingly large and in the sea lamprey (*Petromyzon marinus* L.), the heart ratio ×100 (the relation of heart weight to body weight) is higher than that found in most poikilothermic vertebrates and approaches the average value for mammals (0.59 versus 0.64). This high heart index may reflect the high demand for oxygen (Pupa and Ostadál 1969) of an efficient cardiac pump which is required for the muscular performance during active searching for pray, anadromous migration, and subsequent spawning activity.

The heart of species of the family petromyzontidae is completely enclosed by a cartilaginous pericardium. To secure the passage of the blood, the backflow between the compartments is controlled by valves. Electron microscopical studies (Bloom 1962, Kilarski 1964) revealed the fine structure of the myocardial cardiomyocytes. Their structural organization is very similar to that already described for the hearts of teleosts as well as of the hearts of higher vertebrates. The cardiac cells of *petromyzon* contain one or two myofibrils with well distinguishable striation with all the bands and lines-Z, A, I, M and N-present. The sarcoplasmic reticulum is also relatively well developed. Myocardial cells are connected with each other by means of modifications of the plasma membrane forming junctions like desmosomes organized at intercalated discs. However, the latter are not as complex as those observed in the hearts of higher vertebrates.

Owsiannikof (1883) and Ransom and Thompson (1886) observed nervous elements (ganglion cells) in the heart of lampreys. They suggested that the heart receives nerve fibers from the nervus vagua which branches to the heart and runs from the epibranchial trunk of the vagus nerve. Tretiakoff (1927), after using methylene blue stain, demonstrated a dense network of varicose fibers present in the wall of the sinus venosus of *Lampetra* sp. All parts of the lampetra heart contain numerous cells which resemble ganglion cell bodies; these cells are located mainly along the interior surface of the heart cavities. Augustinsson et al. (1956), studying innervations of the lamprey heart, described two types of ganglion cells. A large cell type has a diameter around 14 µm and a small cell type averages 5–7 µm. These ganglionic cells were present in the walls of all cardiac compartments.

In 1912, Gaskell noticed the presence of chromaffin cells in the wall of the sinus venosus but Augustinsson et al. (1956) pointed out that these cells are present in all parts of the heart resulting in a distinct chromaffin reaction. They form chains along the muscle trabeculae and according to Fänge et al. (1963), the chromaffin cells may be considered as a primitive type of adrenergic nerve cell that, together with the larger ganglion cell, creates the innervation system of the lamprey heart. There is strong evidence of the existence of chromaffin cells since they can be visualized after treatment of the specimens with formaldehyde and then visualized in ultraviolet illumination. This technique is used for the detection of monoamines and their granules exhibit yellow green fluorescence.

The Heart of Elasmobranchs

The heart of elasmobranchs varies in its weight as a percentage of body weight ranging from 0.061% of body weight in the small deep-water shark (*Etmopterus spinax* L.) to 0.146% in the shallow-water shark (*Galeorhinus galeus* L.) (Hesse 1921). A similar relation exists in teleosts and it may be speculated that the reduction of heart weight is an effect of the reduced skeleton and somatic musculature of deep-water dwellers. The heart of elasmobranchs lies within its cartilaginous pericardium and is composed of a thin walled atrium or auricle and a thick walled ventricle. The atrium receives blood from the sinus venosus and pumps it to the ventricle which propels the blood forward to the gills via the truncus arteriosus. The sinus venosus is a delta-shape collecting sac that communicates with the atrium. Dorsally, the sinus venosus is fused with the posterior part of the roof of the pericardial cavity. Laterally, each angle of the sinus extends to the right or left as a large vessel, the ductus Cuvieri.

The blood enters the atrium from the sinus through the sinu-auricular aperture which is secured by the sinu-auricular valves, double folds of the endothelial lining of the atrium. The valves prevent blood regurgitation.

In elasmobranchs, the atrium is in general a thin-walled sack and lies over the ventricle. The architecture of the walls of the atrium is changeable; they are folded but smooth out during the contraction rhythm. In the extended form, the inner surface of the walls of the atrium are smooth but in some species they are extensively folded to such a degree that they may even give the appearance of possessing two compartments. The atrium communicates with the ventricle by the auro-ventricular opening that has two pocket-like flaps—valves whose concavities are directed toward the ventricle cave. The atrium is composed of thin randomly arranged myocardial trabeculae. The trabeculae are composed of striated fibers lined by thin endothelium. A single layer of cubical cells covers the epicardium. The epicardium layer is separated from the myocardium by a narrow sub-epicardial space in which lies thin bundles of collagen fibers and a sparse capillary network.

The ventricle of the elasmobranch heart is small and has a pyramidal shape whose base is directed posteriorly. The wall of the ventricle is thick and its surface is exceedingly jagged and irregular. In the ventricle occur special tendinous cords (chordae tendineae), extended from one end of the muscular trabecule to the opposite wall; this prevents the ventricle chamber from increasing beyond its capacity. The myocardium of the ventricle is divided into two parts; the compact layer-*pars compacta*—which occupies from 17% of the total area of the ventricle in Rajidae (*Raja hypoborea* L.) to 5% in Holocephali (*Chimera monstrosa* L.) (Santer and Greer Walker 1980). The second part, the pars spongiosa, is the inner trabecular part of the ventricular myocardium and in fixed preparations the ventricle appears to have a completely spongy organization lacking a ventricular lumen. However, when fixation is done during diastole a small ventricular lumen may be observed which communicates on one side with the atrium and, on the other, with the conus arterious.

There are, however, examples in which the compact layer is composed of three layers of myocardial cells independently oriented and separated by well-defined inter fascicular connective tissue, as in the fresh-water shark *Alopias vulpinus* L. (Cumini et al. 1977).

The base of the ventricle is continued anteriorly by the conus arteriosus and is confluent with the ventral aorta. Its shape is tubular in cross section. Internally, the conus arteriosus is universally provided with three longitudinal rows of semi lunar valves that separate the conus arteriosus from the ventral aorta. From the outside, the conus arteriosus is covered by epicardium beneath which lie distinct bundles of collagen fibers surrounding the region of the cardiac muscle. In small rays and sharks, the cardiac muscle is circumferentially oriented. In large sharks, an outer circular layer surrounds the alternating thinner bands of longitudinal and circularly arranged cardiac muscle. Beneath the cardiac muscle is situated the region containing collagen and elastin fiber layers which run circularly. These provide an anchorage for the valves (Santer 1989).

The innervation of the heart of elasmobranchs originates from the nervus vagus (Young 1933, Pick 1970). From the nervus vagus two branches arise, one from the visceral branch of the nervus vagus and the other from the post-branchial branch of the fourth branchial division of the nervus vagus (Short et al. 1977, Taylor et al. 1977).

There is no anatomical or histological evidence for the existence of sympathetic innervation of the heart, and no connections between the paravertebral sympathetic ganglia and the vagal cardiac branches have been detected. There are, however, a few reports that claim the existence of few fluorescent varicose axons following formaldehyde-induced fluorescence in the wall of the sinus venosus of *Heterodontus portus-jacksoni* L. (Gannon et al. 1972). However, the lack of any catecholamine fluorescence in the nervus vagus and the absence of any identifiable sympathetic nerves in the heart make the existence of such innervation most unlikely. The vagal branches contain entirely myelinated nerve fibers. Unmyelinated axons have not been found in elasmobranch hearts. Therefore, it may be assumed that they are devoid of cardiac sympathetic innervations.

Myomeric Structure of the Somatic Musculature

The somatic musculature of fish comprises axial muscle and extremity muscle. Axial muscle may, for analytical purposes, be divided into (i) trunk and tail musculature and (ii) extrinsic eye muscle. These have different origins, special innervation and very characteristic anatomical organization. For histological analysis, we shall here concentrate mainly on the axial muscle and briefly on the extrinsic eye muscle.

The fish body is composed predominantly of muscle. Axial musculature comprises from 40 to 60 percent of the total body weight and contains only a few percent of connective tissue (2–5% of the muscle) and takes the form of delicate membranes, separating the long muscle into segments (Dunajaski 1979). For that reason, fish meat is considers as real meat. The axial muscles of fish comprises layers rather than bundles as in other vertebrates.

The body musculature of fish is arranged in segmental myotomes (fish embryos) or myomeres that have a complex three-dimensional organization. The myomeres form piled cones in which the muscle fibers follow a complex helical trail from one myomere to the next along the fish. Muscle fibers are oriented at angles of up to 40 degrees to the longitudinal axis of the fish. This complex arrangement enables all fibers across a body section to undergo similar tensions during body bending (Alexander 1969). Each myomere contains a superficial region lying directly beneath the skin, where the muscle fibers run parallel to the body axis (Figs. 7A–C).

Each myomere is separated from its neighbor by a thin layer of connective tissue. Other layers of connective tissue, called septa, occur along the vertical midline of the body separating the muscles of the left and right sides of the body, and horizontally separating the muscles of the upper (epaxial) and lower

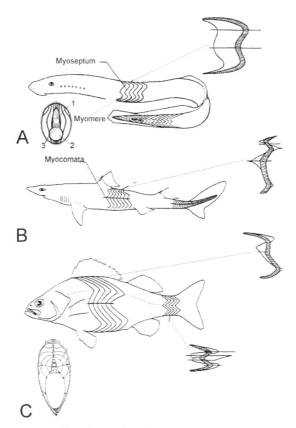

Fig. 7. Schematized myomere patterns of lateral musculature in the major living fish groups and the details of a lateral view of a single myomere (right part of pictures). **A.** Lampreys (*Petromyzon marinus*). Its lateral view and on the cross section (below): 1, chorda dorsalis; 2, body cavity; 3, dorsal vessels (dorsal aorta and dorsal veins). **B.** Shark (*Squalus acantias* L., squalidae). **C.** Perch (*Perca fluviatilis* L.) and its cross section at the level of its dorsal fin (after Nursdall 1956, modified).

halves of the body (hypaxial). The horizontal septum is called myocomata. The epaxial musculature was described as musculus lateralis superior and the hypaxial muscle as musculus lateralis inferior. The ventral musculature of the abdominal region can be classified differently due to the direction of muscle fiber orientation and is described as musculus obliquus externus and musculus obliquus internus. Ventral to these runs the musculus rectus lateralis whose muscle fibres run parallel to the long axis (Figs. 8A–C).

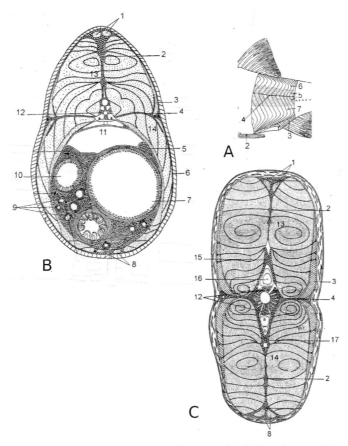

Fig. 8. Myomere pattern of the main trunk muscle as seen in cross and sagittal sections. **A.** Myomere pattern in the tail muscle after removal of the skin of the roach (*Rutilus rutilus*). 1, anal portion of intestine; 2, fragment of the ventral fin; 3, muscles of the anal fin; 4, myocomata; 5, musculus rectus lateralis; 6, musculus lateralis magnus pars dorsolateralis; 7, musculus rectus lateralis. **B, C.** Myomere pattern of the main trunk muscle as seen in cross sections of the rainbow trout (*Oncorhynchus mykiss*) at the level of the body cavity (**B**) and in the tail region (**C**). 1, musculi recti dorsalis; 2, septum sagitale; 3; musculus rectus lateralis; 4, myocomata, septum horizontale; 5, fat tissue surrounding stomach; 6, skin; 7, cavity of the stomach; 8, musculi infrarcarinales anteriores; 9, cross section through the appendices pyloricae; 10, intestine cavity; 11, air bladder; 12, vertebral column with fat deposit; 13, musculus lateralis magnus, parts dorsolateralis; 14, musculus lateralis magnus pars ventrolateralis; 15, dorsal spinal process containing fat tissue; 16, spinal cord; 17, ventral spinal process with the vascular arch (hemapophysis).

Muscle Fiber Types

Fish muscles occur in three different types depending on the concentration of myoglobin, amount of mitochondria and density of the capillary network and they are classified as red, pink or intermediate and white. Most fish have a mixture of two, or all three types of muscle fibers. Generally, these are kept in discrete groupings; however, in some fish, e.g., in the Salmonidae, the red and white muscle types are mixed to form a mosaic type of muscle.

Analysis of muscle composition and histological diversity has mainly been conducted on the main axial musculature. That is understandable from a commercial reason as the main axial musculature is the major edible part of the fish. It is also the easiest part for biochemical and for histological studies and also for physiological reasons because the axial musculature is principally responsible for the locomotion of the fish.

Myomers are composed of individual fibers whose length varies from a tenth of a centimeter to several centimetres. This depends on the size, age and location of the myomere. Myomeres located near the head have longer muscle fibers than those positioned closer to the tail. Our histological analysis will start from the description of the muscle fiber variety situated just below the skin (musculus rectus lateralis, superficialis) (Figs. 9A–C).

Slow-Red Fibers

Slow-red fibers (Fig. 9A) are the chief component of the musculus rectus lateralis that is confined to a narrow strip along the middle line of the body which divides this into a dorsal and a ventral part. Red muscle fibers represent less than 10% of the lateral musculature and are small in diameter (25–45 μm). This is, however, species specific and depends also on the condition of the individual. The red muscle fibers are frequently identified as slow fibers and are used mainly for sustained energy-efficient swimming. The characteristics of these fibers are a good capillary supply, large numbers of mitochondria, lipid droplets, glycogen stores, and a high concentration of myoglobin and cytochromes. The energy metabolism of these fibres is almost entirely aerobic, based mainly on lipid as a fuel supplemented with carbohydrates (Sänger and Stoiber 2001).

Intermediate or Pink Fibers

Intermediate or pink fibers (Fig. 9B) occupy the narrow margin that separates the red muscle fibre zone from the white muscle fiber zone. Their name reflects not only their intermediate position between red and white muscle fiber zones but also many other characteristics. Their mean diameter lies between those of red and white fibers. Intermediate muscle fibers are relatively fast contracting fibers but are also characterized by intermediate resistance to fatigue and intermediate speed of contraction between red and white muscle fibers (Bone 1978).

White or Fast Fibers

White or fast fibers (Fig. 9C) are of the glycolytic type and compose the major part of the myomere occupying less than 70% of it (Sanger and Stoiber 2001). The white fibers are the largest fibers in diameter, spanning between 50 and 100 μm or even more according to age, size and physical condition of the fish. The proportion of the cross-sectioned area of the axial musculature that is comprised of white muscle fibers varies along the length of the fish, being greatest in the anterior part and declining in the direction of the tail. Generally, white musculature is used at high swimming speed initiating fast start burst swimming for prey capture and escape response but for a relatively short time. White muscle fibers are tightly packed with myofibrils occupying between 75 and 95% of the fiber volume. They contain a low number of mitochondria and lipid droplets. Myoglobin and cytochromes also are in low concentration in most species so far examined.

Vascularization of the white portion of the myomere is poor. Glycogen content is also low. It was noticed, however, that the amount of glycogen varies among the white muscle fibers. A marked heterogeneity in glycogen content between different sized white fibers, with a significantly higher content in the smaller fibers, has been observed (Kiessling and Ostrowski 1997).

White muscle mass decreases dramatically in the caudal region because of the necessary streamlining of the fish in general. During fast swimming, the bulk of the fast muscle may be active (Johnston 1977) and the anterior muscle must be the major source of power (Altringham et al. 1993). Decreased muscle mass towards the tail, and a power-transmitting role for the posterior fast muscle seems likely (Altringham

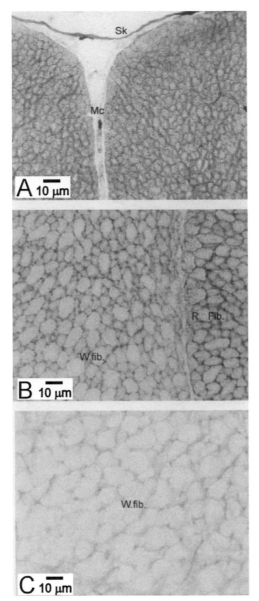

Fig. 9. Cryostat sections of the trunk muscle of the crucian carp (*Carassius aureatus gibelio* Bloch) displaying a high activity of succinate dehydrogenase (SDH). **A.** A superficial sheet of red muscle fibers covering the main mass of the lateral musculature located just under the skin (Sk). They exhibit highly positive SDH staining. Mc, myocomata. **B.** In contrast, intermediate muscle fibers located below the red muscle fibers (R. fib.) exhibit moderate SDH activity. The intermediate fibers form a mosaic with the white muscle fibers (W. fib.). **C.** White anaerobic muscle fibers (W. fib.) are the largest in the cross section. These muscle fibers have low amounts of mitochondria and therefore exhibit weak SDH activity and staining.

Color version at the end of the book

and Ellerby 1999). The energy for white muscle fibers, operating in a nearly closed system, is supplied by anaerobic breakdown of intramuscular glycogen with a small contribution from cytosolic phosphocreatine and ATP (Kiesseling et al. 2004).

The different fiber types can, for example, be identified due to the concentration their enzymatic components, e.g., the SDH enzyme as shown in Fig. 9, or based on their ATPase activity or the myofibril characteristics (Fig. 10). See for comparison the last paragraph below, muscles of Elasmobranchii.

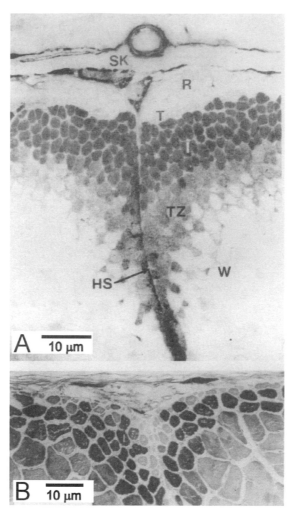

Fig. 10. The amount of ATPase activity can be used to determine the relative proportion of different fiber types. **A.** Cryostat section of the stone loach (*Noemacheilus barbatulus* L.) at the level of myocomata (HS, horizontal septum) stained for ATPase at pH 10.4. Red fibers (R) located just under the skin (SK) are ATPase negative. The thin layer of "tonic" fibers (T) demonstrates very low staining for ATPase. The intermediate fibers (I) display highest ATPase activity. Below lies the transitional zone (TZ) whose muscle fibers show intermediate staining and the main part of the trunk muscle is composed of white anaerobic muscle fibers (W) that are negative for ATPase staining. **B.** Paraffin section of freeze substituted trunk muscle of the stickleback (*Gasterosteus aculeatus* L.) incubated with anti-intermediate fibers serum and immunoperoxidase stained showing a positive reaction in the intermediate fibers. The stickleback trunk muscular system apparently does not have a red muscle fiber zone.

The Extrinsic Eye Muscles

The eye muscles of fish are comprised of two pairs of straight and one pair of oblique muscles. The muscle system that is responsible for the eye ball movements is very precise and is composed of red and white muscle fibers which occupy distinct portions of the muscle. The smallest red fibers, with a diameter not exceeding 10 µm, occupy the marginal zone of the muscle. Large white fibers ranging from 18.4 to 80 µm reside opposite to the red fiber zone (Kilarski 1965). It should be noted that all the values cited for muscle fiber size depend on the mass of the animal and the species. Eye muscles are innervated by three cranial nerves: III (nervus oculomotorius), IV (nervus trochlearis) and VI (nervus abducens). The high sensitivity of muscle fibers is reflected in the ratio of the number of axons that innervate and control the group of muscle fibers and this structure is named motor-neural units. In general, one axon innervates between 4 to 10 muscle fibers, while in the axial musculature one axon innervates up to 100 fibers. The histology of these muscles has rarely been investigated but their ultrastructural organization is well characterized (Kilarski 1973).

Muscles of Agnatha

The phylogenetically lowest representatives of the Craniata are the Myxinomorphi (Hagfish) and the Petromyzontomorphi (Lampreys) (Nelson et al. 2016). Their axial musculature is comprised of three main muscle masses: (1) The parietal muscle of the trunk, which forms the dorsal and parietal parts left and right of the musculature; and (2) The ventral part of the musculature which contains two oblique muscles and (3) One straight muscle. All the muscles are divided into myotomes by means of connective tissue forming myosepts. A myocomata is missing.

In their elegant study, Korneliessen and Nicolaysen (1973) described three main types of muscle fibers: white, red and intermediate. Additionally, they have described the ultrathin red muscle fibers in the velum muscle (musculus craniovelaris) of the hagfish (*Myxine glutinosa* L.). Histological as well as ultrastructure characteristics of these fibres are almost identical as those described previously for teleost muscles.

The axial musculature of lampreys is organized in compartments that are composed of 3–5 plate like "central" muscle fibers that correspond to white muscle fibers of other fish. The "central" fibers are encircled by a single layer of "parietal" fibers that have the characteristics of red fibers of teleosts and other vertebrates (Lie 1974). Stanius (1851) was the first to recognize the presence of these fibers in lampreys. Terävainen (1971) described some ultrastructural and electrophysiological features of the lamprey's body muscle. He concluded that the "central" muscle fibers have a twitch function while the "parietal" fibers have a tonic function. However, Lie (1974) in his elegant study based on histochemical analysis demonstrated the presence of three muscle fiber types: a tubular parietal corresponding to red fibers, plate-like white fibres and plate-like central fibers. In this study, he demonstrated the existence of the tri partition of the muscle mass of Agnatha as seen in other vertebrates as far as the muscle fiber types are concerned.

Muscles of Elasmobranchii

Several species of elasmobranch fish were investigated concerning their myotomal organization. These studies included the *Chimaera monstrosa* L. (Kryvi and Totland 1978), the dogfish (*Scyliorhinus canicula* L.) (Bone 1966, 1978) and sharks (*Etmopterus spinax* L. and *Galeus melastomus* L.) (Kryvi 1977). See also the exhaustive paper by Bone and Chubb (1978). They were able to distinguish five types of muscle fibers based on the staining of succinic dehydrogenase (SDHase), Ca^{2+} activated myofibrillar ATPase and glucose phosphate-isomerase.

The outer surface of the myotome facing the skin is composed of relatively thick fibers forming the interrupted layer. These fibers are SDHase and ATPase negative but show an intense staining for glucose phosphate-isomerase. So far the function of these fibers is unclear. These fibers were described by Bone et al. (1986) as superficial large diameter fibers. Very similar fibers were also described in the myotome of the golomianka (*Comephorus baicalensis* Pall.), the endemic, deep-water dwellers of Lake Bajkal (Kilarski et al. 1992). Two further types of red fibers can be discriminated based upon the above criteria. Outer red fibers, which are relatively large, display high SDHase activity and low ATPase activity, and inner red fibers display high SDHase activity. The white muscle fibers represent two classes: outer white fibers located close to the inner red fiber zone and inner white fibers positioned below the outer white fiber layer. Both white muscle fibers are characterized by high Ca^{2+} stimulated myofibrillar ATPase activity and a low mitochondrial volume.

Sharks have a special muscle mechanism which modulates their tail operation. In contrast to teleosts, which have a complex array of intrinsic muscle that adjusts fin ray orientation in a three-dimensional range of movement (Flammang and Lauder 2008, 2009), sharks have only a single band of red muscle, radial muscle, located deep inside the dermis and separated from the axial myotomes. The caudal fin of sharks is the most important handling surface for locomotion, controlling the angle of attack of the fin by leading the tail beat with the dorsal tip of the caudal fin. This vital function of the shark's body has special anatomical adaptation not found in other fish. The axial myotomes extend over the length of the vertebral column, ending in the dorsal lobe of the caudal fin. In the caudal fin, five layers of orthogonally arrayed subdermal collagen fibers—the stratum compactum—are found. The radial muscle is a thin strip of red muscle fiber which lies ventral to the axial musculature of the caudal fin. The individual muscle fibers originate at the haemal arches of the caudal vertebrae and insert posteriorly into the fifth deepest layer of collagen fibers of the stratum compactum.

The anatomical position of the radialis muscle with regard to the collagen fibers of the stratum compactum provides a stiffening mechanism for the shark tail. This is accomplished solely by the radial muscle at slower swimming speed, as inferred from electromyographic and kinomatic analyses (Flammang 2010).

Acknowledgements

I would like to thank Dr. Nicolas J. Severs, Professor of the Imperial College of London, for invaluable advice in the preparation of this manuscript. I also acknowledge Dr. Danuta Semik and Dr. Grzegorz Tylko for helping with the preparation of the illustrations.

References Cited

Alexander, M.McN. 1969. Orientation of muscle fiber in the myomeres of fishes. J. Mar. Biol. Ass. U.K. 49: 263–290.

Altringham, J.D., C.S. Wardle and C.I. Smith. 1993. Myotomal muscle function at different location in the body of a swimming fish. J. Exp. Biol. 182: 191–206.

Altringham, J.D. and D.J. Ellerby. 1999. Fish swimming: Patterns in muscle function. J. Exp. Biol. 202: 3397–3403.

Augustinsson, K.B., R. Fänge, A. Johnels and A. Ostlund. 1956. Histological, physiological and biochemical studies of the heart of two cyclostomes, hagfish (*Myxine*) and lamprey (*Lampetra*). J. Physiol. 131: 257–276.

Bloom, G.D. 1962. The fine structure of cyclostomes cardiac muscle cells. Z. Zellforsch. Mikrosk. Anat. 57: 213–239.

Bone, Q. 1966. On the function of the two types of myotomal muscle fibers in elasmobranch fish. J. mar. biol. Ass. U.K. 46: 321–349.

Bone, Q. and A.D. Chubb. 1978. The histochemical demonstration of myofibrillar ATPase in elasmobranch muscle. Histochem. J. 10: 489–494.

Bone, Q. 1978. Locomotor muscle. pp. 361–424. *In*: W.S. Hoar and D.J. Randal (eds.). Fish Physiology, Vol. 7. New York, London, Acad. Press.

Bone, Q., I.A. Johnston, A. Pulsford and K.P. Rayan. 1986. Contractile properties and ultrastructure of three types of muscle fiber in the dogfish myotome. J. Muscle. Res. Cell Mot. 7: 47–56.

Brodal, A. and R. Fänge. 1963. The biology of myxine. Universitetsforlaget. Oslo, 124–136.

Bushnell, P.G. and D.R. Jones. 1994. Cardiovascular and respiratory physiology of yuna and billfishes: adaptations and for support of exceptionally high metabolic rates. Environ. Biol. Fishes. 40: 303–318.

Chévrel, R. 1887. Sur l'anatomie du systeme nerveux grand sympathique des elasmobranchs et des poissons osseux. Arch. Zool. Exp. Gen. 5: 1–96.

Cobb, J.S.L. and M.R. Santer. 1973. Electrophysiology of cardiac function in teleosts: cholinergically mediated inhibition and rebound excitation. J. Physiol. (London) 230: 561–573.

Cumini, V., A. Maresca, G. Tajana and B. Tola. 1977. On the heterogeneity of the fish heart ventricle: 1. Preliminary morphological observations. Boll. Soc. Ital. Biol. Sper. 53: 543–548.

Dunajaski, E. 1979. Texture of fish muscle. J. Texture Stud. 10: 301–318.

Ebashi, S. 1980. *In*: S. Ebashi, K. Maruyama and M. Endo (eds.). Muscle Contraction its Regulatory Mechanism. Japan Scientific Societies Press, Tokyo.

Fänge, R., A.G. Johnels and P.S. Enger. 1963. The autonomic nervous system. pp. 124–136. *In*: A. Brodal and R. Fänge (eds.). The Biology of Myxine. Universitetsforlaget. Oslo.

Flammang, B.E. and G.V. Lauder. 2008. Speed-dependent intrinsic caudal fin muscle recruitment during steady swimming in bluegill sunfish, *Lepomis macrochirus*. J. Exp. Biol. 211: 587–598.

Flammang, B.E. and G.V. Lauder. 2009. Caudal fin shape modulation and control during acceleration, breaking and backing maneuvers in bluegill sunfish *Lepomis macrochirus*. J. Exp. Biol. 212: 277–286.

Flammang, B.E. 2010. Functional morphology of the radialis muscle in shark tails. J. Morph. 271: 340–352.

Gannon, B.J. 1971. A study of dual innervations of teleost heart by a field stimulation technique. Comp. Gen. Pharmacol. 2: 175–183.

Gannon, B.J., G.D. Campbell and G.H. Satchell. 1972. Monoamine storage in relation to cardiac regulation in the Port Jackson shark *Heterodontus portus-jacksoni*. Z. Zellforsch. Mikrosk. Anat. 131: 437–450.

Gaskell, J.F. 1912. The distribution and physiological action of the suprarenal medullary tissue in *Petromyzon fluviatilis*. J. Physiol. 44: 59–67.

Greene, C.W. 1902. Contribution to the physiology of the California hagfish *Polistotrema stouti*. II. The absence of regulative nerves for the systemic heart. Am. J. Physiol. 6: 318–324.

Greer Walker, M., R.M. Santer, M. Benjamin and D. Norman. 1985. Heart structure of some deep-sea fish (Teleostei–Macruridae). J. Zool. (London) 195: 275–324.

Grodzinski, Z. 1954a. Contraction of the isolated heart of the European glass eel (*Anguilla anguilla* L.). Bull. Acad. Polon. Sci. Let. Cl. Sci. Math. Nat. Ser. B. II. 2: 19–22.

Grodzinski, Z. 1954b. Initiation of contraction in the heart of the sea trout (*Salmo trutta* L.) embryo. Bull. Acad. Polon. Sci. Let. Cl. Sci. II. 2: 127–130.

Hesse, R. 1921. Das Herzgewicht der Wirlbeltiere. Zool. Jahrb. Abt. Allg. Zool. Physiol. Tiere. 38: 243–364.

Hirsch, E.F., M. Jelinek and T. Cooper. 1964. Innervation of the systemic heart of the California hagfish. Circ. Res. 14: 212–217.

Johnston, I.A. 1977. A comparative study of glycolysis in red and white muscles of the trout (*Salmo gairdneri*) and mirror carp (*Cyprinus carpio*). J. Fish Biol. 11: 575–588.

Kiessling, A. and A. Ostrowski. 1997. A comparison of the relative importance of hyperplasia and hypertrophy for muscle growth in the extremely fast swimming fish Mahi mahi. Martinique Proceedings 2.

Kiesseling, A., K. Lindah-Kiesseling and K.-H. Kiesseling. 2004. Energy utilisation and metabolism in spawning migrating Early Stuart sockeye salmon (*Oncorhynchus nerka*): The migratory paradox. Canadian. J. Fish. Aquatic Sci. 61(3): 452–456.

Kilarski, W. 1964. The organization of the cardiac muscle cell of the lamprey (*Petromyzon marinus* L.). Acta Biol. Cracov. Ser. Zool. 7: 75–87.

Kilarski, W. and J. Bigaj. 1964. Organization of the sarcoplasmic reticulum in skeletal muscle of fishes. I. The sarcoplasmic reticulum of striated muscles of the swim-bladder of the burbot (*Lota lota* L.). Acta Biol. Cracov. 7: 111–125.

Kilarski, W. 1965. The organization of the sarcoplasmic reticulum in skeletal muscle of fishes. Part II. The perch (*Perca fluviatilis* L.). Acta Biol. Crac. 8: 51–57.

Kilarski, W. and J. Bigaj. 1970. Organization and fine structure of extraocular muscle in *Carassius* and *Rana*. Z. Zellforsch. 94: 194–204.

Kilarski, W. and J. Bigaj. 1971. The fine structure of striated muscle fibres of tunica muscularis of the intestine in some teleosts. Z. Zellforsch. 113: 472–489.

Kilarski, W. 1973. Cytomorphometry of sarcoplasmic reticulum in extrinsic eye muscles of the teleost (*Tinca tinca* L.). Z. Zellforsch. 136: 535–544.

Kilarski, W., M. Kozlowska and M.G. Martynova. 1992. The ultrastructure of myotomal muscles of the golomianka, *Comephorus baikalensis* Pallas. J. Fish Biol. 40: 489–495.

Korneliessen, H. and K. Nicolaysen. 1973. Ultrastructure of four types of striated muscle fibers in the Atlantic Hagfish (*Myxine glutinosa*, L.). Z. Zellforsch. 143: 273–290.

Kryvi, H. 1977. Ultrastructure of the different fiber types in axial muscles of the sharks *Etmopterus spinax* and *Galeus melastomus*. Cell Tissue Res. 184: 287–300.

Kryvi, H. and G.K. Totland. 1978. Fiber types in locomotory muscles of the cartilaginous fish *Chimaera monstrosa*. J. Fish. Biol. 12: 257–256.

Lie, H.R. 1974. A quantitative identification of three muscle fiber types in the body muscles of *Lampetra fluviatilis*, and their relation to blood capillaries. Cell Tiss. Res. 1954: 109–119.

Murray, J.M., A. Weber and A. Wegner. 1980. Tropomyosin and the various states of the regulated actin filament. pp. 221–236. *In*: S. Ebashi, K. Maruyama and M. Endo (eds.). Muscle Contraction its Regulatory Mechanism. Japan Scientific Societies Press, Tokyo.

Müller, 1839. Comparative anatomy of the Myxinoids. Berlin.

Nelson, J.S., T.C. Grande and M.V.H. Wilson. 2016. Fishes of the World. 5th ed. John Wiley and & Sons, New Jersey.

Nursdall, J.R. 1956. The lateral musculature and the swimming fish. Proc. Zool. Soc. London. 126: 137–148.

Ohtsuki, I. 1980. *In*: S. Ebashi, K. Maruyama and M. Endo (eds.). Muscle Contraction and its Regulatory Mechanism. Japan Scientific Societ. Press, Tokyo.

Owsiannikof, T. 1883. Über das sympatische Nervensystem der Flussneunauge, nebst einigen histologischen Notizen über anderes Gewebe desselben Thieres. Bull. Acad. Sci. St.-Petersb. 28: 440–448.

Pick, J. 1970. The autonomic nervous system. Morphological, comparative and surgical aspects. Lippincott Philadelphia.

Poupa, O. and B. Ostadál. 1969. Experimental cardiomegalies and "cardiomegalies" in free-living animals. Ann. NY Acad. Sci. 165: 445–468.

Ransom, W.B. and D'A.W. Thompson. 1886. On the spinal and visceral nerves of Cyclostomata. Zool. Anz. 9: 421–426.

Retzius, A.A. 1822. Vascular system of myxine. K. Vet. Akad. Handlgr. Stockholm.

Retius, G. 1890. Eine soganntes Caudalherz bei Myxine glutinosa. Biol. Untersuchungen (Neue Folge), Stockholm. 1: 94–96.

Santer, R.M. 1972. Ultrastructural and histochemical studies on the innervations of the heart of a teteost *Pleuronectes platessa* L. Z. Zellforsch. Mikrosk. Anat. 131: 519–528.

Santer, M.R. 1977. Monoaminergic nerves in the central and peripheral nerves of fishes. Gen. Pharmacol. 8: 157–172.

Santer, R.M. and M. Greer Walker. 1980. Morphological studies on the ventricle of teleost and elasmobranch hearts. J. Zool. (London) 190: 259–272.

Santer, R.M. 1989. Morphology and innervations of the fish heart. Adv. Anat. Embyol. Cell Biol. 89: 1–102.

Sänger, A.M. and W. Stoiber. 2001. Muscle fiber diversity and plasticity. pp. 187–250. *In*: Ian A. Johnston (ed.). Acad. Press, London.

Stanius, H. 1851. Über den Bau der Muskeln bei *Petromyzon fluviatilis*. Göttingsche Gelehrt. Az. Nachricht. 17.

Short, S., P.J. Butler and E.W. Taylor. 1977. The relative importance of nervous. humoral and intrinsic mechanism in regulation of heart rate and stroke volume in the dogfish *Scyliorhinus canicula*. J. Exp. Biol. 70: 77–92.

Skramlik, E. von. 1935. Über den Kreislauf bei den Fische. Ergebn. Biol. 11: 1–130.

Stevens, E.D. and D.J. Randal. 1967. Changes in blood pressure, heart rate and breathing rate during moderate swimming activity in rainbow trout. J. Exp. Biol. 46: 307–310.

Taylor, E.W., S. Short and P.J. Butler. 1977. The role of cardiac vagus in the response of the dogfish *Scyliorhinus canicula* to hypoxia. J. Exp. Biol. 70: 57–75.

Teräväinen, H. 1971. Anatomical and physiological studies on muscles of lamprey. J. Neuro-physiol. 84: 954–973.

Tretiakoff, D. 1927. Das periphere Nervensystem des Flussneunauges. Z. Wiss. Zool. 129: 359–452.

Young, J.Z. 1931. On the autonomic nervous system of the teleostean fish *Uranoscopus scaber*. Q. J. Micros. Sci. 74: 491–535.

Young, J.Z. 1933. On the anatomic nervous system of selachians. Q. J. Micros. Sci. 75: 571–624.

Chapter 5

Structure and Function of Electric Organs

Frank Kirschbaum

INTRODUCTION

Vertebrates are characterized by a segmented body organisation. This can, for example, be seen in the axial skeleton, in particular the vertebral column (vertebrae) and the spinal cord segmentation comprised of the ventral and dorsal roots. This segmentation is more clearly seen in lower vertebrates like fishes in particular at the level of the axial muscle (myomeres) (Harder 1975, see also Chapter Muscular System). If muscle fibres are transformed into electric cells (electrocytes), a segmental arrangement of these elements is therefore already present. If these electrocytes are activated via nerves synchronously, small potentials of about 100 mV of the individual electrocytes (Bennett 1971a) can add resulting in larger potentials. Thus, electric fish producing electric fields—weak or strong electric—can emerge (Moller 1995).

Evolution of Electric Organs

As seen in the introduction, fishes process a precondition for the development of electric organs based on the metameric organisation of their axial muscle. It is, thus, not surprising that electric organs have evolved several times independently from each other (convergent evolution). Electric fishes have evolved twice during the evolution of cartilaginous fishes (skates and electric rays) and six times in bony fishes (Teleostei) (Fig. 1): in the African mormyroids belonging to the osteoglossomorpha; in the South American electric eels (Gymnotiformes); in the African catfish families Malapteruridae, Mochokidae, and Clariidae; and in the marine family Uranoscopidae (*Astroscopus*) of the order Perciformes.

Ontogenetic data based on the reproduction of mormyrid and gymnotiform fishes in captivity (Kirschbaum and Schugardt 2002) have shown that larval and adult electric organs exist in the mormyroids (Kirschbaum 1977, 1995) and in some gymnotiforms (Kirschbaum 1977, Kirschbaum and Schwassmann 2008) representing two steps in the evolution of electric organs (Kirschbaum 1995).

Humboldt University, Faculty of Life Sciences, Unit of Biology and Ecology of Fishes, Philippstr. 13, Haus 16, D-10115 Berlin, Germany.
Email: frank.kirschbaum@staff.hu-berlin.de

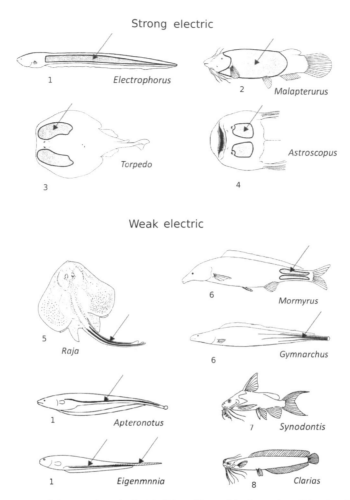

Fig. 1. Representative genera of strong and weak electric fishes. The position/extension of the electric organs is indicated (black or hatched, see also arrows). Four genera of the strong electric fish are shown. The monospecific South American *Electrophorus* (ca. 600 V) belongs to the Gymnotiformes. The African *Malapterurus* (ca. 300 V) represents a catfish (Malapteruridae, Siluriformes). Both are freshwater fish. The marine *Torpedo* (ca. 60 V) is a cartilaginous fish (Torpedinidae, Chondrichthyes). The marine *Astroscopus* (5–50 V according to author) belongs to the Perciformes (Uranoscopidae). Weak electric fish produces discharges of about 1 Volt. Some species of the marine scates of the genus *Raja* (Rajidae) are electric. *Mormyrus* represents one of the many genera of the African mormyrid fish (Mormyridae, Osteoglossomorpha). The African monospecific genus *Gymnarchus* is related to the mormyrids (combined in the superfamily Mormyroidea). The South American weak electric gymnotiforms (knifefishes) are represented here by the genus *Apteronotus* (Apteronotidae). All species of this family possess neurogenic electric organs. Species of the other knifefish families all possess myogenic electric organs. An example is the fishes of the genus *Eigenmannia* (Sternopygidae). Some species of the genus *Synodontis* of the African family Mochokidae as well as species (*Clarias*) of the African family Clariidae possess weak electric organs (their position is not indicated in this Figure). In total, electric organs have evolved eight times independently (see numbers).

Function of Electric Organs

Strong electric fish use their discharge for prey capture or defence. In the night active *Malapterurus electricus*, the discharge volleys of the predatory behaviour can be elicited by gustatory stimuli (Bauer 1968): large numbers of taste buds are not only found in the epithelium of the oral cavity but also distributed over the whole body (Herrick 1901, Bauer 1968). But also water current stimuli elicit predatory behaviour (Bauer 1968, Kastoun 1972). Weber (1982) investigated the response to mechanical stimuli. The presence of ampullary electroreceptors (for electroreception see Chapter Sensory Organs) (Bennett 1971b) in

Malapterurus electricus suggests that predatory behaviour is also guided by passive electroreception. Concerning the behaviour of *Malapterurus electricus* in the field, see review by Moller (1995).

Electrophorus electricus emits both strong and weak electric discharges (Moller 1995). Laboratory studies (e.g., Bauer 1970, 1979, Westby 1988, Catania 2015) have investigated the significance of the strong electric discharges in prey capture behaviour. Westby (1988) described the hunting behaviour in a coastal stream in French Guiana. The weak electric discharges are apparently used during social behaviour (Coates et al. 1937, Cox 1938, Lüling 1975, Lamarque 1979) and probably during reproductive behaviour (Assuncao and Schwassmann 1995).

Torpedo marmorata is a well studied species concerning biology, ecology and reproduction (Mellinger 1969, 1971, 1973, 1974, Mellinger et al. 1978, review: Whitaker 1992, Belbenoit 1970, 1974, 1979, review: 1986, Belbenoit and Bauer 1972). *T. marmorata* is a viviparous species with the females growing much bigger than the males (63 vs. 37 cm). The benthic *T. marmorata* is in general buried in the sand waiting for prey. Strong electric discharge volleys are elicited by nearby fast moving fish or by mechanical stimulation (Belbenoit 1970, 1986, Belbenoit and Bauer 1972). The discharge volleys (3.2 to 24.5 s duration) paralyze the prey which is subsequently covered and then swallowed by the *Torpedo*. Newborn *Torpedo* of about 10 cm body length is capable of discharging during prey capture behaviour (Belbenoit 1974, Michaelson et al. 1978).

There are several species of electric stargazers (*Astroscopus* spp.) (Moller 1995). These marine fish in general hide buried in the sand. The prey capture behaviour in *Astroscopus y-graecum* was investigated by Pickens and McFarland (1964): as soon as small prey fish enter the visual field of the fish, it leaps out of the sand, opens its large mouth and catches the prey fish. Opening of the mouth is accompanied by electric organ discharge of 100 ms to 19 s duration. However, prey capture behaviour is also successful after silencing of the electric organ via sectioning the electric nerve (reviewed by Moller 1995).

The weak electric catfishes of the families Mochokidae (*Synodontis* spp.) and Clariidae (*Clarias gariepinus*) use their discharges during social, aggressive interactions (Hagedorn et al. 1990, Baron et al. 1994a, b). The different species of the weak electric marine skates produce weak discharges of different durations (70–217 ms). The discharges are emitted during social interactions (Bratton and Ayers 1987, Bratton et al. 1993, New 1994, see also review by Moller 1995).

There has grown a vast literature on the function of the electric organs of the weak electric freshwater fishes of the two African families Mormyridae and Gymnarchidae and of the neotropical order Gymnotiformes (see reviews by Bennett 1971a, Heiligenberg 1977, Bullock and Heiligenberg 1986, Kramer 1990, Moller 1995, Bullock et al. 2005) since the discovery of the function in 1958 (Lissmann 1958, Lissmann and Machin 1958): the three main functions are prey detection, object location (orientation) and electrocommunication. Some recent papers on electrocommunication shall be cited here: Nagel et al. (2018) investigated the possible role of the electric discharge in species recognition in two species of the mormyrid genus *Campylomormyrus*; Worm et al. (2017) described electrocommunication and locomotor behaviour of *Mormyrus rume proboscirostris* (Mormyridae) towards a mobile dummy fish; Henninger et al. (2018) investigated electrocommunication during social and reproductive behaviour of the knifefish *Apteronotus rostratus* in the wild in a rain forest creek in Panama.

Anatomy and Fine Structure of Electric Organs

In this section, the anatomy and the fine structure of the electric organs of the various taxa in which electric organs have evolved independently will be treated in a phylogenetic order, according to Nelson et al. (2016).

Cartilagenous Fishes

Skates (Rajidae)

The family Rajidae comprises more than 200 species and many genera (Nelson et al. 2016). In the genus *Raja*, several species possess electric organs in the tail (Fig. 2A) (Ewart 1892). These organs are found in the centre of the most lateral bundle of longitudinally running muscle fibres (Figs. 2B, C).

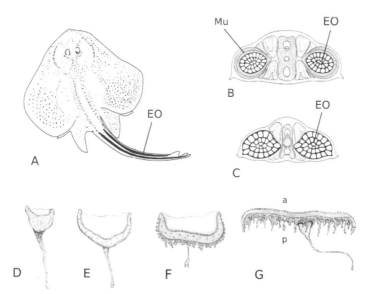

Fig. 2. Anatomy and histology of the electric organ of scates (*Raja*). **A.** Position of the electric organ (EO) in the tail of *Raja erinacea*. **B, C.** Cross sections of the tail of *R. batis* showing two aspects of the anatomy of the electric organs (EO), either still surrounded by muscle (Mu) (**B**) or devoid of muscle (**C**). **D–E.** Different anatomies of electrocytes of full grown skates representing different steps in the differentiation of electrocytes from muscle tissue. **D.** *R. radiata*. **E.** *R. circularis*; the muscular striation has entirely disappeared. **F.** *R. fullonica*; villous-like projections extend from the outer surface of the electrocytes. **G.** *R. batis*; the electrocyte is rather flat and possesses on the posterior face (p) long villous-like projections. a, anterior face. **A, B, C.** After: P. GRASSE, "Traité de zoologie, Tome 13 Volume 2—Agnathes poissons Anatomie ethologie systematique", © Armand Colin, Paris, 1958. With permission, modified. **D–G.** After Ewart 1892. With permission, modified.

There are two types of electrocytes: cup-type electrocytes of different modification found in the species *R. radiata* (Fig. 2D), *R. circularis* (Fig. 2E), and *R. fullonica* (Fig. 2F). These different cup-type electrocytes seem to present different steps in the evolution of muscle fibres towards electrocytes. Disc-shaped electrocytes were found in *R. batis* (Fig. 2G). Similar disc-shaped electrocytes occur in the species *R. macrorhynchus*, *R. alba*, *R. oxyrhynchus*, *R. clavata*, *R. maculate*, and *R. microcellata* (Ewart 1892). Both types of electrocytes are oriented antero-posteriorly; the innervation is found at the anterior face. Often short processes (stalks) are found as remnants of the muscle fibres from which the electrocytes develop. An abundance of striated myofibrils is seen indicative of the myogenic origin of the electrocytes (syncytium). The non-innervated, posterior faces possess large number of protuberances tens of microns in diameter and length (Figs. 2F, G). Each electrocyte lies in a small connective tissue compartment.

Electric Rays (Families Torpedinidae and Narcinidae)

In the family Torpedinidae 23 species and in the family Narcinidae 42 species are known (Nelson et al. 2016). The fishes of the family Torpedinidae are all strong electric. In some species of the family Narcinidae, two electric organs are present: one producing strong and the other weak electric discharges. If all the species of this family are electric is not well established. The best studied electric ray is *Torpedo marmorata* (concerning references see paragraph Function of Electric Organs). The electric organ is derived from gill muscles. The two organs on each side of the body are located lateral to the gills (Figs. 3A, B) and they extend in the ventro-dorsal direction as seen in a cross section (Fig. 3B). They are composed of about 450 hexagonal columns (Fig. 3C); in adult rays, they are 4–8 mm in diameter. In each column, about 1000 thin plates (electrocytes), ca. 10–30 μm thick, are packed on top of each other, filling entirely the space of the column. Nerves innervate the electrocytes on their ventral surface (Fig. 3C). The individual electrocytes are spaced by about 5–10 μm. No striated myofibrils are found in the electrocytes. The differentiation of the electrocyte plate from muscle fibre like structures is shown in Figs. 3D to G.

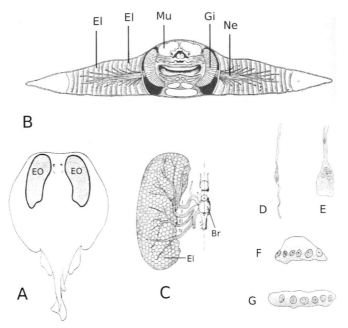

Fig. 3. Anatomy and histology of the electric organ of *Torpedo marmorata*. **A.** Position of the two electric organs (EO) near the eyes. **B.** Cross section at the level of the electric organs showing the extension of the electric organ and the innervating nerves (Ne). Each organ consists of about 450 columns (**C**) extending in the ventro-dorsal direction. **D–E.** Differentiation of an electrocyte plate (**G**) from muscle fibre precursors (**D**). Br, brain; El, electrocyte; Gi, gills; Mu, muscle. After: P. GRASSE, "Traité de zoologie, Tome 13 Volume 2—Agnathes poissons Anatomie ethologie systematique", © Armand Colin, Paris, 1958. With permission, modified.

Bony Fishes (Teleostei)

Mormyroidei

The Mormyroidei belong to the subdivision Osteoglossomorpha and comprise two families, the Mormyridae with more than 200 species and the monospecific family Gymnarchidae (Nelson et al. 2016). In this chapter, we describe the electric organ of the mormyrid *Pollimyrus isidori*, which is the best studied species of the mormyrid family concerning structure and development of electric organs due to the reproduction of this species in captivity (Kirschbaum 1987).

Pollimyrus isidori possesses two distinct, functional electric organs which succeed each other during ontogeny (Fig. 4A) (this situation applies to other species as well) (Kirschbaum 1977, 1981, 1982, 1995, Kirschbaum and Westby 1975, Westby and Kirschbaum 1977, 78, Denizot et al. 1978, 1982). There is a short phase during early ontogeny when both organs are functional. During the larval phase, up to 15 mm total length in the medial part of the axial muscle modified muscle fibres–electrocytes—are seen which are arranged metamerically (Fig. 4B) and represent cells which are about 60 µm in length and 30–40 µm in width. They possess a caudally oriented appendage—a stalk—which receives the innervation (via electromotor neurones situated in the spinal cord). The stalks extend into a space filled with connective tissue limiting the different rows of electrocytes (Fig. 4C). The electrocytes contain many nuclei which also extend into the stalk. The plasma of the electrocytes contains densely packed striated myofibrils; however, these are not arranged in order and thus do not seem to be functional (Denizot et al. 1978). The electrocytes of the larval electric organ will disappear (Kirschbaum 1981) at the beginning of the juvenile phase when the adult electric organ starts to become functional (Denizot et al. 1982).

The adult electric organ is found in the caudal peduncle filling nearly the entire space (Figs. 4A, D). It comprises two dorsal and two ventral columns of about 100 electrocytes arranged in the rostro-caudal direction of the caudal peduncle (Fig. 4E). Each electrocyte has a kidney-like shape (Fig. 4F) and possesses

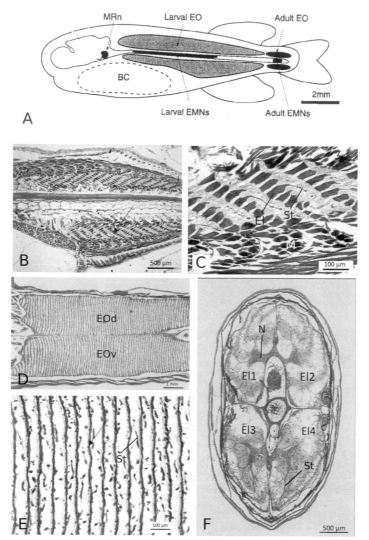

Fig. 4. Topographical, anatomical and histological data of the larval and the adult electric organs of *Pollimyrus isidori*. **A–C.** Larva of 15 mm length. **D–F.** Juvenile of 25 mm length. **A.** Extension of the larval and adult electric organ (EO) and electromotor neurons (EMNs). BC body cavity, MRn medullary relay nucleus. **B.** Medial sagittal section showing the maximal extension of the larval electric organ. Metamerically arranged electrocytes (arrow) are seen both in the epaxial and hypaxial muscle. **C.** At higher magnification, individual electrocytes (El) and the caudally extending stalk (St) emerge. **B and C.** Rostral left. M, muscle. **D.** Sagittal section through the caudal peduncle showing the dorsal (EOd) and ventral column (EOv) of ca. 100 individual electrocytes aligned in the rostro-caudal direction. Rostral left. **E.** Detail of **D** showing 14 individual electrocytes (arrow) and the caudally emerging stalk (St). **F.** Cross section of the caudal peduncle indicating the anatomy and position of four columns of electrocytes, two dorsal (El1 and El2) and two ventral columns (El3 and El4) of the adult electric organ. The kidney shaped individual electrocytes possess protuberances on the caudal face that join together to become the large stalk (St) that receives innervation (N). **A** after Kirschbaum 1995, with permission.

many protuberances (stalks) (Fig. 4F) on the caudal face which join together and finally become a big stalk which receives the innervation (via electromotor neurones situated in the spinal cord in the middle of the caudal peduncle) (Fig. 4F). Some of the stalks penetrate the electrocyte, run for some distance on the rostral face, and again penetrate the electrocyte to appear back on the caudal face (Fig. 4E). Each electrocyte lies in a compartment surrounded by a thin layer of connective tissue.

Gymnotiformes (Knifefishes and Electric Eel)

The gymnotiforms comprise the strong electric eel (*Electrophorus electricus*) and between 200 and 250 weakly electric species (Crampton and Albert 2006). Due to the great diversity in the anatomy and histology of the electric organs of this monophyletic group, three species will be presented here: the strong electric *E. electricus* possessing three electric organs; the weakly electric *Eigenmannia virescens* with a weak electric organ originating from muscle tissue and the weakly electric *Apteronotus leptorhynchus* with a neurogenic organ, i.e., that is derived from spinal nerves. The neurogenic organ presents the only exception concerning the origin of electric organs; all other electric organs are derived from muscle tissue. Interestingly, in the early ontogeny of this fish a myogenic organ is found that degenerates later on after the differentiation of the neurogenic organ (Kirschbaum 1983). Some knifefishes possess accessary electric organs in the head region (Bennett 1971a) the significance of which is not well understood.

Electrophorus electricus (Electric eel). The strong electric eel possesses three electric organs (Fig. 5A). The main organ extends over most of the length of the fish. At its caudal end, about one third in length compared to the main organ Sach's organ is present, separated by the former by a connective tissue border. The electrocytes of these two organs fill most of the space as seen in a cross section (Fig. 5B). Below the main organ in between the anal fin ray muscles, the electrocytes of Hunter's organ are situated. During ontogeny, the electrocytes originate from embryonic tissue situated below the hypaxial muscle (Figs. 5D, E) (Schwassmann et al. 2014). The electrocytes of the main organ split off dorsad (Figs. 5D, E), the older the fish the more electrocytes have differentiated (compare Figs. 5D and E). The electrocytes of Hunter's organ split off ventrad later during ontogeny (Schwassmann et al. 2014). The innervation of the electrocytes is found on the posterior face; each electrocyte is enclosed in a sheet of connective tissue (Fig. 5C).

Eigenmannia virescens. The electric organ of *Eigenmannia virescens* extends over nearly the whole length of the fish (Fig. 6A). In the most rostral part, the electrocytes are situated below the hypaxial muscle in

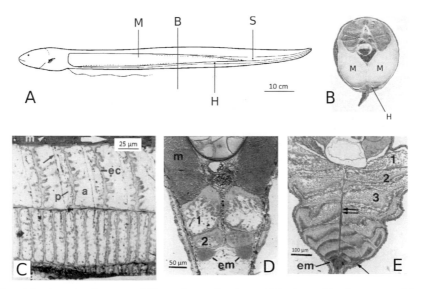

Fig. 5. Position of the three electric organs of *Electrophorus electricus* and histology of the electrocytes of the main organ of larvae of different size (**C, D**, 15 mm long; **E**, 25 mm long). **A.** Position and extension of the three electric organs: Main Organ (M), Sach's Organ (S), and Hunter's Organ (H). **B.** Anatomy of the Main (M) and Hunter's Organ (H) seen in a cross section at a level indicated in **A**. **C.** The first two rows (upper row is first row) of electrocytes (ec) seen in a sagittal section. Each electrocyte is limited by a connective tissue sheet (arrow). a, anterior; m, muscle; p, posterior. **D.** Cross section showing the first two rows (1, 2) of electrocytes. em, electromatrix; m, muscle. **E.** Cross section depicting eight columns (1, 2, 3 ..) of electrocytes. Double arrows point to the medium septum. Single arrow indicates the lateralis imus muscle. em, electromatrix. **A, B**: After: P. GRASSE, "Traité de zoologie, Tome 13 Volume 2—Agnathes poissons Anatomie ethologie systematique", © Armand Colin, Paris, 1958. With permission, modified. **D, E:** After Schwassmann et al. 2014, with permission, modified.

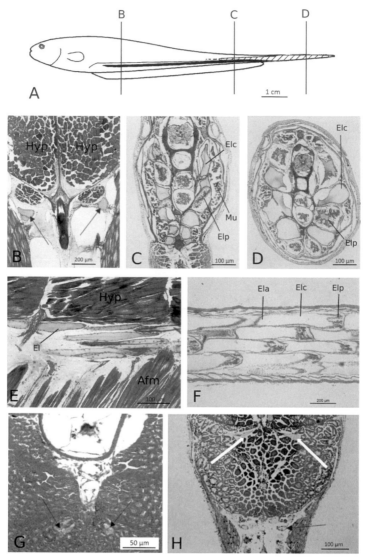

Fig. 6. Anatomy and histology of the larval (**G, H**) and adult electric organs (**A–F**) of *Eigenmannia virescens*. **A.** Anatomy of the electric organs in lateral view. Black, anal fin organ; hatched, hypaxial organ. Level of cross sections of Figures **B–D** are indicated. **B.** Position of the electrocytes (arrows) of the anal fin organ below the hypaxial muscle (Hyp) seen in a cross section. **C–D.** Cross sections at levels C and D (**A**) showing the anatomy of the electric organs. **C.** Ca. seven to eight rows of electrocytes are seen on each side of the body, five rows are present in the hypaxial organ (**D**). Elc, cross section of the centre of the electrocyte; Elp, caudal end of electrocyte (see **F**). **E.** Sagittal section at level B (**A**). Several rows of electrocytes (El) are present below the hypaxial muscle (Hyp) and above the anal fin muscles (Afm). **F.** Sagittal section at level D (**A**). Five rows of electrocytes are seen. Ela, rostral face of electrocyte; Elc, central face; Elp, posterior face. Rostral left in **E** and **F**. **G.** Cross section of the hypaxial muscle of a 17 day old larva showing the electrocytes of the larval electric organ (arrows) inserted inside the mass of muscle fibres. **H.** Cross section at level C (**A**) of a 41 day old juvenile presenting the electrocytes of the larval organ (white arrows) and those of the anal fin organ (adult organ, black arrow).

Color version at the end of the book

between and below the anal fin muscles (Fig. 6B). There are just one to two electrocytes on either side seen in this cross section. In a cross section before the end of the anal fin (Fig. 6C), it can be seen that the electric organ is composed of 7–8 rows of electrocytes on each side of the body. A cross section of the caudal appendage (Fig. 6D) reveals that the number of electrocytes has decreased to 5. A sagittal section in the rostral part of the electric organ (Fig. 6E) shows several thin, several hundred micrometer long electrocytes in between the hypaxial muscle and the muscle of the anal fin. The electrocytes of the caudal appendage have a clear polar anatomy, the rostral part has some prologations, whereas the caudal face, which receives the innervation, shows many ramifications (Fig. 6F). This polar organisation is also visible in the two cross sections (Figs. 6C, D). Each electrocyte is embedded in a connective tissue compartment. Figure 6G shows the electrocytes of the larval electric organ in the hypaxial muscle of a 17 day old larva. In a 41 day old juvenile, electrocytes of both larval and adult electric organ are present (Fig. 6H). The electrocytes of the larval electric organ will completely degenerate later during ontogeny.

Apteronotus leptorhynchus. All species of the knifefish family Apteronotidae possess neurogenic electric organs, i.e. they are derived from blind ending nerve fibres originating from modified motoneurons-electromotor neurones—extending over nearly the whole length of the spinal cord (de Oliveira Castro 1955, Bennett 1971a). This is the only example of a group of electric fish which possesses an electric organ that is not of myogenic origin. However, the apteronotids possess as larvae a larval electric organ that is derived from muscle tissue (Kirschbaum 1983, Kirschbaum and Schwassmann 2008) that will degenerate later during ontogeny.

Here we describe the neurogenic organ of *A. leptorhynchus*; the development of its electric organs has been studied in detail (Kirschbaum 1983, Kirschbaum and Schwassmann 2008). The neurogenic organ extends over nearly the whole length of the fish from behind the operculum up to the small caudal fin (Fig. 7A) typical for all Apteronotid fish. The myelinated nerve fibres leave the spinal cord via spinal nerves

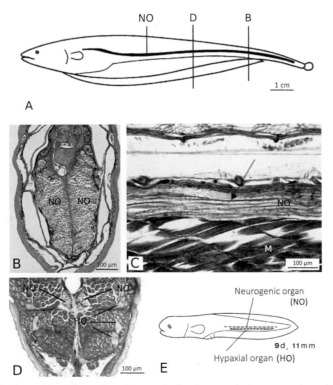

Fig. 7. Anatomy and histology of the neurogenic and myogenic electric organs in *Apteronotus leptorhynchus*. **A.** The position of the neurogenic organ (NO) is indicated as well as the level of the cross sections (B, D). **B.** Cross section showing the large extension of the neurogenic organ (NO). **C.** Medial sagittal section at level D (see Fig. **A**) depicting the modified nerve fibres (arrow) of the neurogenic organ (NO) running in the rostro-caudal direction. M, muscle. **D.** Cross section at level D (see Fig. **A**) of a 22 day old larva. Note the medial position of the neurogenic organ (NO) and the electrocytes of the myogenic (hypaxial, HO) organ. **E.** Position of the future neurogenic organ and that of the hypaxial organ (myogenic) in a 9 day old larva are indicated.

and enter the organ situated in a medial location below the vertebral column (Figs. 7B, C). In a 22 day old larva, electrocytes of the myogenic organ in the hypaxial muscle and neurocytes of the neurogenic organ are both present (Fig. 7D). Both organs are functional (Kirschbaum 1983). The position of the myogenic organ (hypaxial organ) and the future position of the neurogenic organ are indicated in a 9 day old larva (Fig. 7E).

Siluriformes

Malapterurus electricus. The genus *Malapterurus* was originally a monospecific genus (*M. electricus*). Meanwhile, 10 species are known (Norris 2003). Here the electric organ of *M. electricus* is presented. It comprises a ca. 2 mm thick layer of electrocytes located directly below the skin. The organ begins rostrally at the level of the pectoral fin and ends at about the middle of the anal fin (Fig. 8A). The individual electrocytes are arranged in parallel at 90 degree to the longitudinal axis of the fish and are innervated at their rostral faces (Fig. 8B). They represent about 1 mm large cells with rounded edges and a central innervation point (Fig. 8C).

Mochokidae and Clariidae. The structure and anatomy of the electric organs of species of the families Mochokidae and Clariidae are not well known. In the *Synodontis* species (Mochokidae), the electric organ is apparently derived from a dorsally located sonic muscle (Hagedorn et al. 1990) which is comprised of modified striated muscle fibres. The anatomical origin of the weak electric discharges of *Clarias gariepinus* has not yet been revealed.

Perciformes

Astroscopus y-graecum (Stargazer). The electric organs of the several species of the genus *Astroscopus* (Nelson et al. 2016) are situated behind the eye (Fig. 9) and occupy the eye chamber and seem to have developed from eye muscles. In *Astroscopus y-graecum,* there are about 150 to 200 layers of electrocytes

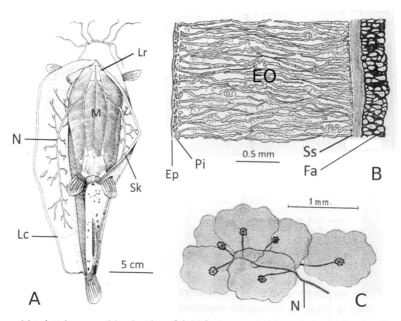

Fig. 8. Anatomy of the electric organ of the electric catfish *Malapterurus electricus.* **A.** *In situ* situation showing the cut open skin (Sk). Rostral (Lv) and caudal (Lc) limits of the electric organ are indicated. N, nerve innervating the electric organ. M, muscle. **B.** Cross section of the epidermis (Ep) and the electric organ (EO) showing the extension of the EO in between the epidermis and the fat tissue (Fa). Pi, pigment cell layer; Ss, stratum subelectricum. **C.** Anatomy of individual electrocytes. N, nerve innervating the electrocytes. After: P. GRASSE, "Traité de zoologie, Tome 13 Volume 2—Agnathes poissons Anatomie ethologie systematique", © Armand Colin, Paris, 1958. With permission, modified.

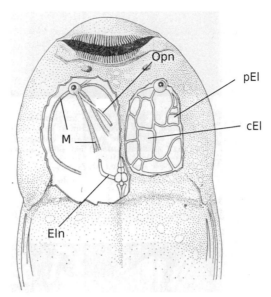

Fig. 9. Head of *Astroscopus y-graecum* seen from above showing the position of the right electric organ and individual electrocytes and on the left, after removal of the electric organ, the eye muscles and the nerves innervating the eye and the electric organ. cEl, central electrocyte; pEl, peripheral electrocyte. After: P. GRASSE, "Traité de zoologie, Tome 13 Volume 2—Agnathes poissons Anatomie ethologie systematique", © Armand Colin, Paris, 1958. With permission.

arranged in a horizontal plane. Each layer is composed of about three to five large central and eight to twelve small electrocytes located peripherally (Fig. 9).

References Cited

Assuncao, M.I.S. and H.O. Schwassmann. 1995. Reproduction and larval development of *Electrophorus electricus* on Marajó island (Pará, Brazil). Ichthyol. Explor. Freshwaters. 6(2): 175–184.

Baron, V.D., K.S. Morshnev, V.M. Olshansky and A.A. Orlov. 1994a. Electric organ discharges of two species of African catfish (*Synodontis*) during social behaviour. Anim. Behav. 48: 1472–1475.

Baron, V.D., A.A. Orlov and A.S. Golubtov. 1994b. African *Clarias* catfish elicits long-lasting weak electric pulses. Experientia. 50: 644–647.

Bauer, R. 1968. Untersuchungen zur Entladungstätigkeit und zum Beutefangverhalten des Zitterwelses *Malapterurus electricus* Gmelin 1789 (Siluroidea, Malapteruridae, Lacép. 1803). Z. Vergl. Physiol. 59: 371–402.

Bauer, R. 1970. La décharge électrique pendant le comportement alimentaire de *l'Electrophorus electricus*. J. Physiol. Paris 62: 341–341.

Bauer, R. 1979. Electric organ discharge (EOD) and prey capture behaviour in the electric eel, *l'Electrophorus electricus*. Behav. Ecol. Sociobiol. 4: 311–319.

Belbenoit, P. 1970. Comportement alimentaire et décharge électrique associée chez *Torpedo marmorata* (Selachii, Pisces). Z. Vergl. Physiol. 67: 205–216.

Belbenoit, P. and R. Bauer. 1972. Video recordings of prey capture behaviour and associated electric organ discharge of *Torpedo marmorata* (Chondrichthyes). Mar. Biol. 17: 93–99.

Belbenoit, P. 1974. Hérédité des coordinations sensorimotrices et électrique (CSME) dans le comportement prédateur de *Torpedo marmorata*. J. Physiol. Paris. 69: 187A.

Belbenoit, P. 1979. Electric organ discharge of *Torpedo* (Pisces); basic pattern and ontogenetic changes. J. Physiol. Paris 75: 435–441.

Belbenoit, P. 1986. Fine analysis of predatory and defensive motor events in *Torpedo marmorata* (Pisces). J. Exp. Biol. 121: 197–226.

Bennett, M.V.L. 1968. Neural control of electric organs. pp. 147–169. *In*: D. Ingle (ed.). The Central Nervous System and Fish Behaviour. University of Chicago Press, Chicago, Illinois.

Bennett, M.V.L. 1971a. Electric organs. pp. 347–484. *In*: W.S. Hoar and D.J. Randall (eds.). Fish Physiology, Vol. 5, Academic Press, New York.

Bennett, M.V.L. 1971b. Electroreception. pp. 493–574. *In*: W.S. Hoar and D.J. Randall (eds.). Fish Physiology, Vol. 5, Academic Press, New York.

Bratton, B.O. and L.J. Ayers. 1987. Observations on the electric organ discharge of two skate species (Chondrichthyes: Rajidae) and its relationship to behaviour. Env. Biol. Fishes. 20: 241–254.

Bratton, B.O., A. Christiano, N. McClennan, M. Murray, E. O'Neill and K. Ritzen. 1993. The electric organ discharge of the skate (Rajidae): waveform, occurrence and behaviour relationships. Soc. Neurosci. Abstr. 19: 374.

Bullock, T.H. and W. Heiligenberg. 1986. Electroreception. John Wiley & Sons, New York, USA.

Bullock, T.H., C.D. Hopkins, A.N. Popper and R.R. Fay. 2005. Electroreception. Springer. New York.

Catania, K.L. 2015. Electric eels use high-voltage to track fast-moving prey. Nat. Commun. 6: 8638 doi: 10.1038/ncomms9638.

Coates. C.W., R.T. Cox and L.P. Granath. 1937. 1. The electric discharge of the electric eel, *Electrophorus electricus* (Linneaus). Zoologica (N.Y.) 22: 1–34.

Cox, R.T. 1938. The electric eel at home. Bull. N.Y. Zool. Soc. 41: 59–65.

Crampton, W.G.R. and J. Albert. 2006. Evolution of electric signal diversity in gymnotiform fishes. pp. 647–731. *In*: F. Ladich, S.P. Collin, P. Moller and B.G. Kapoor (eds.). Communication in Fishes. Science Publishers. Enfield (NH), Jersey, Plymouth.

Denizot, J.P., F. Kirschbaum, G.W.M. Westby and S. Tsuji. 1978. The larval electric organ of the weakly electric fish *Pollimyrus (Marcusenius) isidori*. (Mormyridae, Teleostei). J. Neurocytol. 7: 165–181.

Denizot, J.P., F. Kirschbaum, G.W.M. Westby and S. Tsuji. 1982. On the development of the adult electric organ in the mormyrid fish *Pollimyrus isidori* (with special focus on the innervation). J. Neurocytol. 11: 913–934.

De Oliveira Castro, G. 1955. Differentiated nervous fibres that constitute the electric organ of *Sternarchus albifrons* Linn. Anais Acad. Brasil. Cien. 27: 557–570.

Ewart, J.C. 1892. The electric organ of the skate: Observations on the structure, relations, progressive development, and growth of the electric organ of the skate. Phil. Trans. Roy. Soc. London B 183: 389–420.

Fessard, A. 1958. Les organs electriques. pp. 1143–1238. *In*: P.P. Grasse (ed.). Traite de Zoologie—Anatomie, Systematique, Biologie. Vol. XIII. Agnathes et Poissons. Anatomie, Ethologie, Systematique. Fascicule II. Masson et Companie. Paris.

Hagedorn, M., M. Womble and T.E. Finger. 1990. Synodontid catfish: a new group of weakly electric fish. Brain Behav. Evol. 35: 268–277.

Harder, W. 1975. Anatomy of Fishes. E. Schweizerbartsche Verlagsbuchhandlung. Nägele and Obermiller, Stuttgart.

Heiligenberg, W. 1977. Principles of electrolocation and jamming avoidance in electric fish. A Neuroethological Approach. Springer, Berlin.

Henninger, J., R. Krahe, F. Kirschbaum, J. Greve and J. Benda. 2018. Statistics of natural communication signals observed in the wild identify important yet neglected stimulus regimes in weakly electric fish. J. Neurosc. 2018 Jun. 13; 38(24): 5456–5465. doi: 10.1523/JNEUROSCI.0350-18.2018. Epub 2018 May 7.

Herrick, C.J. 1901. The cranial nerves and cutaneous sense organs of the North American siluroid fishes. J. comp. Neurol. 11: 179–249.

Kastoun, E. 1972. Das Verhalten des Zitterwelses, *Malapterurus electricus* Gmelin, im elektrischen Feld. PhD thesis, Universität Köln, 56 pp.

Kirschbaum, F. and G.W.M. Westby. 1975. Development of the electric discharge in Mormyrid and Gymnotid fish (*Marcusenius* sp. and *Eigenmannia virescens*). Experientia. 31: 1290–1293.

Kirschbaum, F. 1977. Electric-organ ontogeny—distinct larval organ precedes adult organ in weakly electric fish. Naturwissenschaften. 64: 387–388. doi: 10.1007/BF00368748.

Kirschbaum, F. 1981. Ontogeny of both larval electric organ and electromotoneurones in *Pollimyrus isidori* (Mormyridae, Teleostei). pp. 129–157. *In*: T. Szabo and G. Czeh (eds.). Advances in Physiological Sciences. Vol. 31, Sensory Physiology of Aquatic Lower Vertebrates. Akadémiai Kiadió, Budapest.

Kirschbaum, F. 1982. Die Entwicklung des "adulten" elektrischen Organs bei *Pollimyrus isidori* (Mormyridae, Teleostei). Verh. Dtsch. Zool. Ges. 75: 242.

Kirschbaum, F. 1983. Myogenic electric organ precedes the neurogenic organ in apteronotid fish. Naturwissenschaften. 70: 305–307.

Kirschbaum, F. 1995. Reproduction and development in mormyriform and gymnotiform fishes. pp. 267–301. *In*: P. Moller (ed.). Electric Fishes—History and Behavior. Fish and Fisheries Series. Chapman and Hall, London, New York.

Kirschbaum, F. and C. Schugardt. 2002. Reproductive strategies and developmental aspects in mormyrid and Gymnotiform Fishes. J. Physiol. (Paris) 96(5-6): 557–566.

Kirschbaum, F. and H.O. Schwassmann. 2008. Ontogeny and Evolution of electric organs in gymnotiform fish. J. Physiol. (Paris) 102: 347–356.

Kramer, B. 1990. Electrocommunication in Teleost Fishes: Behavior and Experiments. Springer, New York.

Lamarque, P. 1979. Le Gymnote *Electrophorus electricus* et la peche à l'électricité. La pisciculture francaise. 55: 22–26.

Lissmann, H.W. 1958. On the function and evolution of electric organs in fish. J. Exp. Biol. 35: 156–191.

Lissmann, H.W. and K.E. Machin. 1958. The mechanism of object location in *Gymnarchus niloticus* and similar fish. J. Exp. Biol. 35: 457–486.

Lüling, K. 1975. Ichthyologische Untersuchungen an der Yarina Cocha. Zool. Beiträge. 21: 29–96.

Michaelson, D.M., D. Sternberg and L. Fishelson. 1978. Observations on feeding, growth and electric discharge of newborn *Torpedo ocellata* (Chondrichthyes, Batoidei). J. Fish Biol. 15: 159–163.

Moller, P. 1995. Electric Fishes—History and Behavior. Chapman and Hall. Fish and Fisheries Series, 17.

Nagel, R., F. Kirschbaum, J. Engelmann, V. Hofmann, F. Pawelzik and R. Tiedemann. 2018. Male-mediated species recognition among African weakly electric fishes. R. Soc. Open Sci. 5: 170443. http://dx.doi.org/10.1098/rsos.170443.

Nelson, J.S., T.C. Grande and M.V.H. Wilson. 2016. Fishes of the World. 5th ed. John Wiley and & Sons, New Jersey.

Norris, S.M. 2003. Malapteruridae. pp. 175–194. *In*: D. Paugy, C. Leveque and G.G. Teugels (eds.). The Fresh and Brackish Water Fishes of West Africa. Volume II. Collection Faune et Flore Tropicales. Paris.

Pickens, P.E. and W.N. McFarland. 1964. Electric discharge and associated behaviour in the stargazer. Anim. Bchav. 12: 362–367.

Schwassmann, H.O.S., M.I.S. Assuncao and F. Kirschbaum. 2014. Ontogeny of the electric organs in the electric eel, *Electrophorus electricus*: Physiological, histological, and fine structural investigations. Brain Behav. Evol. 84: 288–302. DOI: 10.1159/000367884.

Westby, G.W.M. and F. Kirschbaum. 1977. Emergence and development of the electric organ discharge in the mormyrid fish, *Pollimyrus isidori*. I. The larval discharge. J. Comp. Physiol. 122: 251–251.

Westby, G.W.M. and F. Kirschbaum. 1979. Emergence and development of the electric organ discharge in the mormyrid fish, *Pollimyrus isidori*. II. Replacement of the larval by the adult discharge. J. Comp. Physiol. 127: 45–59.

Westby, G.W.M. 1988. The ecology, discharge diversity and predatory behaviour of gymnotiform electric fish in the coastal streams of French Guiana. Behav. Ecol. Sociobiol. 22: 341–354.

Worm, M., F. Kirschbaum and G. von der Emde. 2017. Social interactions between live and artificial weakly electric fish: Electrocommunication and locomotor behavior of *Mormyrus rume proboscirostris* towards a mobile dummy fish. PLOS ONE 12(9): e0184622.

Chapter **6**

Digestive System

*Ostaszewska Teresa** and *Kamaszewski Maciej*

INTRODUCTION

The gastrointestinal tract of fish larvae is incompletely developed, morphologically and functionally compared to that of juvenile stages (Ostaszewska 2005, Sysa et al. 2006, Sánchez-Amaya et al. 2007, Yúfera and Darias 2007). At the moment of hatching, there are no histological differences along its whole length (O'Connell 1981). At the beginning of exogenous feeding, the gastrointestinal tract is still less differentiated structurally and functionally than in the adult fish (Ostaszewska 2002). In sharp contrast to other vertebrates, fish consume a great variety of food and there are many modes of feeding. Based on the nature of the food taken, there are often overlapping feeding types, namely (I) herbivores and detritophages, (II) omnivores, consuming small invertebrates, and (III) carnivores, consuming fishes and bigger invertebrates.

General Anatomy of the Digestive System

According to Harder (1975) and Wilson and Castro (2011), the gastrointestinal tract can be divided into four parts considering its topography: headgut, foregut, midgut, and hindgut (Fig. 1A). The headgut covers the anterior section of the digestive tract which includes the mouth and the pharynx, whose functions involve catching of the food and its mechanical processing (Clements and Raubenheimer 2005). The foregut consists of the esophagus and the stomach in which chemical digestion of feed begins. However, in agastric fish, like the Cyprinidae, but also in fish from the families Triglidae, Lophiidae, and Syngnathidae, the foregut is a functional part of the digestive tract. The bile and pancreatic ducts join to it, just behind the esophagus.

The foregut epithelium is of ectodermal and the midgut of endodermal origin. The midgut is the largest part of the digestive tract, where digestion is continued and nutrients are absorbed from the gut lumen. It is located between the pyloric valve and the anus, and can be divided into two sections: (1) The middle section being an equivalent of the duodenum and small intestine of mammals, and (2) The posterior section being an equivalent of the colon.

The last part of the digestive tract is constituted of the hindgut which includes the rectum. In some species, however, there is no clear morphological distinction between the midgut and the hindgut.

Department of Ichthyobiology, Fisheries and Biotechnology in Aquaculture, Faculty of Animal Science, Warsaw University of Life Sciences (WULS-SGGW), Poland.
* Corresponding author: teresa_ostaszewska@sggw.pl

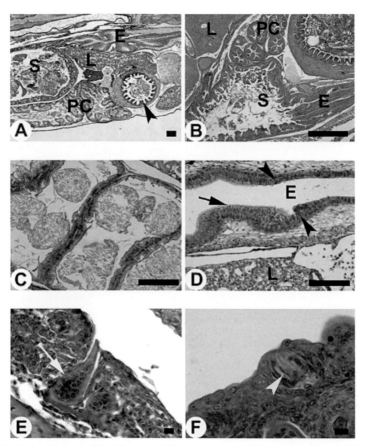

Fig. 1. A–D. Longitudinal sections. **A.** Digestive tract of Russian sturgeon (*Acipenser gueldenstaedtii*). **B.** Digestive tract of the pacu (*Piaractus mesopotamicus*). Abbreviations for **A** and **B**: E, esophagus; L, liver; S, stomach; PC, pyrolic caeca. Arrowhead, non-glandular stomach. **C.** Spiral valve intestine of the Russian sturgeon. **D.** Oral cavity and esophagus (E) of Russian sturgeon. Mucous membrane of the esophagus with multiple mucous cells (arrowhead) and ciliated cells (black arrow). L, liver. **E.** Cross-section of a tooth (white arrow) of Russian sturgeon. **F.** Taste bud (yellow arrowhead) in the pharyngeal mucosa of Russian sturgeon. AB/PAS staining. Scale bars: **A, C** and **D**—100 μm, **B**—500 μm, **E** and **F**—10 μm.

Color version at the end of the book

There are also some other divisions of the intestine, including, among others, the sections "duodenum", foregut, hindgut, and rectal section. In all cases, the divisions between particular sections vary. The beginning of the midsection or the duodenal section of the gut might be determined based on the orifices of bile and pancreatic ducts, as well as the orifice of the pyloric caeca, if they occur in a given species (Kilarski 2012).

Pyloric caeca are located near the junction of the stomach and the midsection of the gut (Fig. 1B). Their presence and number are usually species-specific, e.g., agastric fish have no pyloric caeca. The physiological role of these caeca is still uncertain, they probably serve supporting functions during feed digestion and absorption.

Another modification in the anatomy of the gastrointestinal tract is the spiral intestine, formed by the spiral valve (Fig. 1C), which unfolds from the hindgut and has been described in Acipenseridae (Wegner et al. 2009, Dabrowski and Ostaszewska 2009, Ostaszewska et al. 2011a, Kamaszewski et al. 2014), lungfish (Hassanpour and Joss 2009), sharks (Jhaveri et al. 2015, Leigh et al. 2017), and stingrays (Chatchavalvanich et al. 2006). The spiral valve intestine does not occur in teleosts. Its walls are lined with a ciliated cylindrical epithelium. In addition, its mucosa contains mucus-secreting and endocrine cells (Jhaveri et al. 2015). Considering its histology, the spiral valve is supposed to retard digesta passage through the gastrointestinal tract and to aid in the process of digestion (Leigh et al. 2017).

The length of the gastrointestinal tract, and of the intestine in particular, depends on the feed ingested. Herbivore fish, like the common carp (*Cyprinus carpio*), usually have a long intestine with numerous loops, whereas the intestine of predatory fish is usually short and can be arranged in a single loop (like in pike) or in multiple loops (like in perch). An exception is the garfish (*Belonidae*) with a straight pipe-shaped intestine (Bočina et al. 2017).

Esophagus

The oral cavity and pharynx are lined with a stratified squamous epithelium containing abundant mucus-secreting cells and often sensory cells forming taste buds, which turns into a monolayer cylindrical epithelium near the orifice of the stomach. The mucosa of the esophagus forms long and spiral folds. In some species (e.g., of the families Percidae and Acipenseridae), it can contain ciliated cells apart from the mucous cells (Figs. 1D and E). In a few species, the pharynx epithelium is lined with a horny layer. Mucosa development and appearance of the mucous cells in the esophagus are correlated with the opening of the oral cavity and beginning of mixed endo-exogenous feeding. Among epithelial cells, numerous mucous cells secrete acidic carboxyl and sulfate mucins (Fig. 1D) (Ostaszewska 2005). In most fishes, the tongue is poorly developed and lacks specific muscles. It often comprises only connective tissue covered with an epithelium that contains many unicellular glands. Cartilage is sometimes found, particularly in the Chondrichthyes.

The shape and position of the oral cavity, dentition of jaws, the bucco-pharynx and gills are determined by the feeding habits and type of ingested feed (Kapoor et al. 1975). The oral cavity can be superior or inferior, placed either on the bottom or in front of the head in the horizontal or upper position. In the elasmobranch fishes, lungfishes and some osteichthyes (Labridae and Scaridae), a skinfold appears on the ridge of the mouth which seals the closed oral cavity. It forms "lips" which have no muscles and differ from those found in mammals. Their form varies considerably, even within the same family like the Cyprinidae, where the lips are protruding and brightly-colored as in carp (*Cyprinus carpio*), while inconspicuous and hardly recognizable as lips in asp (*Aspius aspius*).

The fish tongue is poorly developed except in lampreys in which the tongue is a special lamprey apparatus of different origin than that of gnathostome fish. It has no muscles and is formed by a fold of mucosa having a significantly thickened connective tissue. It strongly adheres to the skeletal elements which merge with both halves of the hyoid arch and the adjacent branchial arches. The tongue moves passively following the respiratory movements of the mandible and gills. In the Teleostomi, the hyoid arch is strongly bound to the mandibular arch, and the tongue may aid in food transport to the pharynx. In many species, the hyoid arch terminus possesses some folds which are built of connective tissue and mucosa. The folds protrude to a various extent and are often completely separated from the bottom of the oral cavity. Often, bones located inside these folds are armed with teeth, whereas the epithelium of the esophagus mucosa is multi-layered and contains goblet cells as well as papillae and sensory buds.

Smooth muscles have been found in the tongue of the Nile bichir (*Polypterus* sp.), and striated muscles in lungfish (*Protopterus* sp.). The Polypteridae have a well-developed tongue which originates on the ossa hyoidae of both sides of the body; its free end extends into the oral cavity (Marcus 1934). The tongue of *Plecoglossus altivelis* (Salmoniformes) is a complicated structure and is located between the tip of the glossohyale and the angle of the jaw (Iwai 1962).

Cyprinids have a special organ in their oral cavity, called pharyngeal pad. It is a muscular bulge of the roof of the oral cavity and is located anterior to the pharyngeal teeth. This organ—resembling a pillow—is constituted of a multi-layered mucosal epithelium, placed on a thick layer of connective tissue. The epithelium is folded and contains multiple goblet and mucous cells as well as taste buds, whereas the connective tissue contains both collagen and elastic fibers. A layer of striated muscles is located underneath it.

The edge of the oral cavity is usually armed with teeth (Fig. 1E), which can appear on the palate, on the tongue fold, and on gill arches, like in the case of pike (*Esox lucius*). There are several kinds of teeth in the Osteichtyes. They are found on the jaw, hyal, palatine, vomer, and pharyngeal. All of them are joined by connective tissue to their respective bones (tooth-bearing bones) and possess a similar histological structure.

The pulp is composed mainly of connective tissue and occupies the center of the tooth. The odontoblasts are arranged at the outermost region of the pulp and secrete dentin. Dentin is composed mainly of collagen fibers, and is very similar to bone. This part of the tooth contains calcium salts. Protoplasmic projections from the odontoblasts direct the dentin through the dentinal tubules. The surface of the dentin is usually covered with enamel (or an enameloid substance). Teeth of rainbow trout (*Oncorhynchus mykiss*) and ayu (*Plecoglossus altivelis*), for example, have an enamel covering. The structure of the enamel is very complex. Ground sections of the teeth exhibit fiber-like striations extending from the dentinal tubules, which are arranged radially from the superficial to the intermediate layers. Enamel is secreted by the enamel organ, which is composed of two layers of epithelial cells, while the tooth is still in the tooth germ.

The teeth (see also chapter Fish Skeletal Tissues) occur in various shapes, including filiform, pin-like, awl-like, conical, nodular or cube-like teeth. Their number fluctuates in a wide range, e.g., catfish have around 10,000 teeth which are small, filiform, and tightly packed one next to another. In many predatory fish, teeth are located only on the dental, pharyngeal, tongue, maxillary, submaxillary, and palatinal bones. Many species of fish, among others from the Cyprinidae and the Pleuronectiformes, have so-called pharyngeal teeth located on the faucial bones. The pharyngeal teeth are arranged in one row (*Tinca tinca*, *Rutilus rutilus*), in two rows (*Leuciscus leuciscus*, *Leuciscus cephalus*) or in three rows (*Cyprinus carpio*, *Barbus barbus*).

Taste buds (TB) (see also chapter Sensory Organs) are distributed around and in the mouth, in the oropharyngeal cavity, the gills and often also on the fish's external skin (Fig. 1F) (Whitear 1971, Gomahr et al. 1992). The cells associated with TB are not of neural origin but are modified epithelial cells. The signals are transferred from the TB to the brain stem by the facial (VII) nerve from the anterior part of the mouth, and by the glossopharyngeal (IX) and vagus (X) nerves from the middle and posterior part of the oropharyngeal cavity and gills (Kinnamon 1987). Studies have shown that in fishes the gustatory sense is extremely well-developed, forming an important part in the individual's interpretation of its 'Umwelt'. The functional physiology of the gustatory sense has been summarized by several authors (Hara 1992, 1994, Tagliafierro and Zaccone 2001). Taste bud cytology has been studied in catfishes, poeciliids and lungfishes (Storch and Welsch 1970, Reutter 1978, 1991); in a freshwater mrigal carp (*Cirrhinus mrigala*) (Sinha 1975); in holostean fishes (Reutter et al. 2000), and recently in cardinal fishes (Fishelson et al. 2004). Most of the older literature on this subject was compiled by Kapoor et al. (1975). As demonstrated by Fishelson (1997) and Iwai (1980), similar to other organs vital for orientation, the TB evolves very early in the embryos and newly hatched larvae of fishes, providing an early tool for chemoreception.

Stomach

In terms of morphology, stomachs of fish can be classified by shape as straight (I), siphonal (U or J) or cecal (Y) (Harder 1975). The straight stomach is rare; it occurs, e.g., in pike (*Esox lucius*) (Smith 1989); the J-shaped stomach occurs in South American catfish (Hernández et al. 2009), channel catfish (Sis et al. 1979), and rainbow trout (Ezeasor and Stokoe 1981); the U-shaped stomach was found in *Salmo* sp., *Clupea* sp. (Smith 1989), and *Coregonus lavaretus* (Ostaszewska et al. 2018), whereas the cecal stomach occurs in (Y-shaped) in pike-perch (*Sander lucioperca*) (Ostaszewska 2005, Hamza et al. 2015) and Nile tilapia (*Oreochromis niloticus*) (Caceci et al. 1997). In various fish species, the stomach has one to three regions, usually a cardiac, fundic, and pyloric one (Fig. 2A).

The mucous epithelium of each stomach region is mono-layered and folded. In the cardiac stomach, the folds are shallow (short), whereas in the fundic and pyloric regions they are deeper. Epithelial cells are cuboidal in the cardiac region and cylindrical in the other regions. The nucleus is located at the cell base. In the cardiac and fundic regions, gastric cells are located in the lamina propria and surrounded by loose connective tissue and located in recesses formed by the fold of the mucosa. Fundus of glands is formed by secretory cells of polygonal shape. Nuclei of these cells are round and the cell cytoplasm contains acidophil granules observed after staining with hematoxylin and eosin (Fig. 2B).

Gastric glands of fish contain only one type of secretory cells without division into the pepsinogen-secreting and hydrochloric acid-secreting cells. These oxyntopeptidic cells are adapted to secrete both pepsinogen and hydrochloric acid. Acid secretion is associated with a well-developed intracytoplasmic

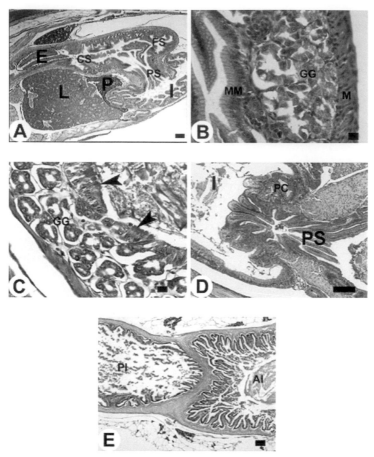

Fig. 2. Longitudinal sections. **A.** Digestive tract of pike-perch. CS, cardiac stomach; E, esophagus; FS, fundic stomach; I, intestine; L, liver; P, exocrine pancreas; PS, pyloric stomach. **B.** Stomach of the lavaret (*Coregonus lavaretus*). GG, gastric glands; M, muscle layer; MM, mucous membrane; **C.** Russian sturgeon stomach, glandular region. GG, gastric glands; arrowhead, mucous cells secreting neutral glycoprotein. **D.** Russian sturgeon stomach, pyloric stomach. I, intestine; PC, pyloric caeca; PS, pyloric stomach; **E.** Anterior (AI) and posterior intestine (PI) of the sunshine bass (*Morone chrysops* x *Morone saxatilis*). AB/PAS staining (**A, C–E**) and H/E staining (**B**); scale bars: **A, D** and **E**—100 μm, **B** and **C**—10 μm.

Color version at the end of the book

system consisting of a tubulo-vesicular network of smooth membranes (Noaillac- Depeyre and Gas 1978, Ostaszewska 2005, Hamza et al. 2015). Activation of these cells by distending the stomach prior to fixation results in the fusion of these tubulovesicles with the apical plasma membrane creating an apical labyrinth (Ezeasor 1981).

The glands open into crypts of folds of the mucosa. In many fish species, gastric glands are located mainly in the cardiac and fundic regions, like in the sisorid catfish *Glyptosternum maculatum* (Xiong et al. 2011), the yellowhead catfish *Tachysurus fulvidraco* (Yang et al. 2010), the white sturgeon *Acipenser transmontanus* (Domeneghini et al. 1999), the sterlet *Acipenser ruthens* (Wegner et al. 2009), and the pike-perch *Sander lucioperca* (Hamza et al. 2015). A similar arrangement of gastric glands has been observed in the Atlantic bluefin tuna *Thunnus thynnus* and in the striped dwarf catfish *Mystus vittatus* (Kozaric et al. 2007, Chakrabarti and Gosh 2014). In the spotted sand bass *Paralabrax maculatofasciatus* larvae, they are found only in the anterior part of the stomach (Pena et al. 2003). In the pyloric region, mucous glands are found in only a few fish species from the genera *Salmo*, *Esox*, *Anguilla*, *Perca*, and *Tilapia* (Harder 1975, Osman and Caceci 1991).

The mucosa of the pyloric stomach is built of a monolayered columnar epithelium, the lamina propria, and the muscular layer of the mucosa. Long folds of the mucosa form fan-shaped branches. The apical region of the cytoplasm of cells of the columnar epithelium of the cardiac, fundic and pyloric stomach stain PAS-positive, which is indicative of the presence of neutral glycol-derived compounds (Fig. 2C) (Ikpegbu et al. 2013, Hamza et al. 2015). The presence of only neutral glycoproteins, which might be related to the conduction of food, may provide efficient protection against proteolysis and mechanical injury (Petrinec et al. 2005), and has buffering effects on acidic stomach digesta (Kozarić et al. 2008). In gobioid fishes, however, mucus is not detected on the surface columnar epithelium (Jaroszewska et al. 2008) suggesting that acid-peptic digestion is not significant since a protective mucous coat is not required. The columnar cells in the chondrosteans are ciliated (Radaelli et al. 2000), while in other fishes the surface may be smooth or lined with microvilli. The presence of microvilli in the mucosa of the stomach indicates potential absorption of nutrients therein; however, this process has not been confirmed so far (Buddington and Christofferson 1985).

The stroma of the mucosa contains numerous lymphocytes and eosinophilic granular cells. A muscularis mucosae is found and consists almost entirely of smooth muscle cells disposed longitudinally. The submucosa contains nerves, arteries, veins, and lymphatics and coarse eosinophilic granulocytes. The muscular coat is composed of inner circular and outer longitudinal coats of smooth muscle cells with sometimes an additional inner oblique layer.

Generally, the same three layers that form the interior of the digestive tract from the esophagus to the anus are also found in the stomach. They include the mucous sheath with its epithelium, the muscle layer (consisting of orbicular muscles arranged internally to the longitudinal muscles), and serosa which covers this system from the outside.

The stomach ends at the pylorus, which may be present as a muscular sphincter created from the thickening of circular smooth muscle and/or a mucous membrane fold that serves as a valve-like structure (Fig. 2D).

Histology of the Intestine

The digestive tract is a long tube built of four layers: mucosa (mucous membrane), sub-mucosa (sub-mucous membrane), muscular coat, and adventitia. The histological structure of the middle part of the digestive tract (intestine) corresponds to the histology of the entire gastrointestinal tract.

The mucosa is formed from the epithelium overlaying two layers of connective tissue (basal lamina and lamina propria) and a thin layer of smooth muscles of the mucosa (muscularis mucosa). It is the outermost layer. The mucous membrane forms longitudinal folds which never form villi typical of the mucosa of mammals. Folds of fish intestine are covered with a monolayered cylindrical epithelium. Cells of the epithelium are located on the connective tissue layer of the sub-mucosa membrane, separated from the epithelium by a thin basement membrane. On the surface of intestinal epithelium cells (enterocytes), there is a brush border—a palmate outgrowths of the apical part of the enterocytes, which serves absorptive functions. In the mucous membrane, goblet cells are located between erythrocytes which secrete acidic and neutral mucopolysaccharides into the lumen of the intestine.

The structure of the mucous membrane of the posterior intestine differs from that of the mucosa of the midgut (Fig. 2E). It is characterized by the presence of lower intestinal folds compared to the midgut, which are formed by the monolayer cylindrical epithelium with sparse goblet cells between the enterocytes. In addition, the mucous membrane has no or has only a very thin muscularis mucosa.

The sub-mucosa consists of two layers: dense and granular, which are composed of fibroblasts and different types of collagen. It contains blood and lymphatic vessels, and nerves.

The muscular layer is composed of two layers of smooth muscles: an inner circular layer and an outer longitudinal layer of myocytes. In some fish (e.g., in tench), the muscular layer is constituted (throughout the digestive tract) by striated voluntary muscles overlying the smooth muscle layer.

The adventitia is composed of cells of connective tissue and is directly connected to the mesentery. It is the outermost layer of the gastrointestinal tract. Its wall contains large blood vessels and nerves. The outer part of the adventitia's wall surface is covered with a squamous epithelium called endothelium.

In the white-edge freshwater whipray (*Himantura signifer*), the wall of the digestive tract from the mouth to the esophagus and the wall of the posterior intestine are composed of three layers (mucosa, muscularis and adventitia), whereas the fourth layer–submucosa—was reported to occur in the wall of the stomach and anterior intestine (Chatchavalvanich et al. 2006).

Cells of the Mucous Membrane of the Intestine

Enterocytes

Enterocytes are cells of the monolayered cylindrical epithelium which line the mucous membrane of the intestine and are responsible for the absorption and transport of nutrients from the gastrointestinal tract lumen to blood vessels. Palmate outgrowths called microvilli are located on their apical part (Figs. 3A and B). These microvilli form a brush border on the surface of the intestinal mucosa which enlarges the surface of digestive enzyme secretion and nutrients absorption in the gut lumen. A typical trait of enterocytes is also the location of the cell nucleus close to the basal region of the cell (Figs. 3A and B).

Goblet Cells

Goblet cells which secrete both neutral and acidic glycoproteins occur in the mucous membrane of the intestine (Figs. 3C and D) (Domeneghini et al. 2005, Díaz et al. 2008). Glycoproteins serve multiple functions in the gastrointestinal tract, e.g., protecting the intestine surface, enhancing effectiveness of digestion and absorption of nutrients, buffering intestinal fluid, or facilitating defecation (Domeneghini et al. 1998, 2005, Díaz et al. 2008, Ostaszewska et al. 2018). As suggested by de Silva and Anderson (1995), the number of goblet cells in fish intestines may fluctuate as affected by feeding habits and starvation status.

Endocrine Cells

The digestive tract of vertebrates is the largest organ of endocrine secretion (Holst et al. 1996, Buddington and Krogdahl 2004). Multiple hormones secreted by endocrine cells change the metabolism of the alimentary tract and other organs, thereby allowing fish to adapt to changes in feeding regime (both quantitative and qualitative) and in environmental conditions (Buddington and Krogdahl 2004). The endocrine cells located in the mucosa of the gastrointestinal tract (Fig. 3E) synthesize various hormones which significantly affect functions of this tract (Bell 1979). They are claimed to be anatomic units responsible for the synthesis of digestive tract hormones, whereas changes in their density reflect production capabilities of these hormones (El-Salhy and Sitohy 2001). Hormones secreted by endocrine cells affect gut peristalsis, and secretion of enzymes by the intestine, liver and pancreas (Deveney and Way 1983). Distribution of endocrine cells depends on fish species and on the type of ingested feed (Solcia et al. 1975, Andreozzi et al. 1997). Examples of hormones secreted by neuroendocrine cells of the intestine are described below.

Cholecystokinin (CCK) is synthesized in mammals and fish by endocrine cells in the first segment of the small intestine (Kamaszewski and Ostaszewska 2014) and by neurons of the enteric nervous system, spinal cord, and brain. CCK stimulates secretion of pancreatic juice proteins and contractions of gallbladder, as well as feed ingestion and release of insulin and somatostatin (Koven et al. 2002, Volkoff 2006). In the intestine of percid fish, cholecystokinin was observed in the pyloric stomach, in pyloric caeca and in the initial segment of the anterior intestine. No CCK-positive endocrine cells were reported in the posterior intestine (Kamaszewski and Ostaszewska 2014). Similar localization of cholecystokinin-secreting cells was observed in turbot (*Scophthalmus maximus*) (Bermúdez et al. 2007), aju (*Plecoglossus altivelis*) (Kamisaka et al. 2003), white halibut (*Hippoglossus hippoglossus*) (Kamisaka et al. 2001), *Coreoperca herzi* (Lee et al. 2004), sea trout (*Salmo trutta*) (Bosi et al. 2004), and rainbow trout (*Oncorhynchus mykiss*) (Ostaszewska et al. 2010a). The presence of these cells in the epithelium of the entire intestine

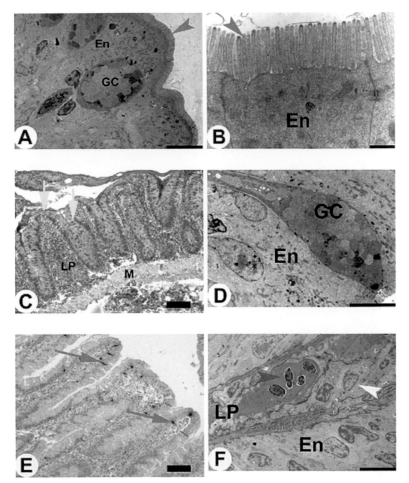

Fig. 3. Cross (**E**) and longitudinal sections (**A–D, F**) of the anterior intestine. **A.** Mucous membrane of the intestine of zebrafish (*Danio rerio*). Cells of monolayered cylindrical epithelium of the intestinal mucosa with microvilli (red arrow) and a goblet cell (GC) and an enterocyte (En). **B.** Microvilli (red arrow) on the surface of enterocytes (En). **C.** Anterior intestine of common carp with AB-positive goblet cells (yellow arrow). Lp, lamina propria; M, muscular layer. **D.** Anterior intestine of zebrafish with goblet cell (GC) and enterocyte (En). **E.** Anterior intestine of common carp. Immunohistochemical detection of neuroendocrine cells secreting gastrin (red arrows). **F.** Anterior intestine of zebrafish. Lamina propria (LP) with blood vessels containing erythrocytes (blue arrowhead) and macrophages (white arrowhead). En—enterocyte. TEM photos (**A, B, D, F**), AB/PAS staining (**C**), Immunohistochemical staining using anti-gastrin antibody (**E**). Scale bars: **A, D** and **F**—5 μm, **B**—0,5 μm, **C** and **E**—100 μm.

Color version at the end of the book

was described by Ku et al. (2004) in the pale chub (*Zacco platypus*), by Rombout et al. (1986) in the rosy barb (*Puntius conchonius*), and by Kiliaan et al. (1992) in the goldfish (*Carassius auratus auratus*) and in the Mozambique tilapia (*Oreochromis mossambicus*).

Gastrin (GAS) is secreted by endocrine cells of the stomach and duodenum of mammals and fish (Kurokawa et al. 2003, Krzymowski and Przała 2005, Ostaszewska et al. 2010a, Kamaszewski and Ostaszewska 2014, Leigh et al. 2017). It is responsible for the stimulation of gastric juice secretion and modulates somatostatin secretion. In mammals, it regulates gastric juice secretion and growth of the digestive tract epithelium (Kinoshita and Ishihara 2000). As in mammals, gastrin stimulates secretion of hydrochloric acid and pepsinogen by adenocytes in the fish stomach (Bjenning and Holmgren 1988, Kurokawa et al. 2003). In the intestine of pike-perch, the GAS-secreting endocrine cells were detected in the pyloric stomach, in pyloric caeca, and in the initial sections of the anterior intestine (Kamaszewski and Ostaszewska 2014). Analogous to this species, the presence of GAS-positive cells was observed in the intestine of rock bass (*Ambloplites rupestris*), sunshine bass (*Lepomis gibbosus*), and yellow perch (*Perca flavescens*) (Reifel et al. 1983, Ostaszewska et al. 2013). Bosi et al. (2004) also reported on the presence of gastrin-secreting cells in the epithelium of the anterior intestine and pyloric caeca of trout. In turn, in juvenile turbots, secretion of gastrin was observed in the stomach and initial section of the anterior intestine (Reinecke et al. 1997). In the agastric fishes, like the Cyprynidae, GAS-positive cells were detected only in the initial section of the anterior intestine (Rombout et al. 1986, Ostaszewska et al. 2010b).

Chromogranin A (CgA) is a glycoprotein secreted by the diffuse neuroendocrine system (Winkler and Fisher-Colbrie 1992). It is located in secretory granules of endocrine cells and is used as a marker of histopathological lesions (Lloyd and Wilson 1983). In addition, owing to its common presence in endocrine cells, chromogranin A is used as a marker of all endocrine cells (Ostaszewska et al. 2008). Chromogranin A is a precursor of such regulatory peptides as vasostatin, pancreastatin and parastatin (Helle and Aunis 2000). No chromogranin was detected in the intestine of some fish species, like in *Coreoperca herzi* and the pale chub (*Zacco platypus*) (Lee et al. 2004, Ku et al. 2004). In turn, in the case of vimba bream (*Vimba vimba*), Ostaszewska et al. (2008) observed CgA-positive cells in the oral cavity, taste buds, the brush border and in the supranuclear regions of gut enterocytes. A study conducted by Reinecke et al. (1997) demonstrated co-localization of chromogranin A with neurostetin, serotonin, gastrin and cholecystokinin in endocrine cells of the digestive tract of turbot.

Leptin is one of the key factors involved in the regulation of feed intake and homeostasis in fish (de Pedro et al. 2006, Ostaszewska et al. 2010a). The presence of this hormone in the gastrointestinal tract of rainbow trout was confirmed in the intestine, pyloric caeca, stomach epithelium, and gastric glands (Ostaszewska et al. 2010a). Its presence in different sections of the gastrointestinal tract, e.g., in the cytoplasm of epithelial gastric gland cells of the Salmonidae, was also reported by other authors (Bosi et al. 2004). Leptin was detected in the mucous membrane of the stomach and the intestine of the European sea bass (*Dicentrarchus labrax*), and also in the intestine of agastric goldfish (*Carassius auratus auratus*) (Russo et al. 2011). In chondrichthyans, like in the small-spotted catshark (*Scyliorhinus canicula*), expression of this hormone was also confirmed in goblet cells and in neuroendocrine cells of the intestine (Gambardella et al. 2010), whereas in the European sea bass–Leptin was mainly found in mucous cells of the esophagus, neuroendocrine cells of the stomach and intestine, and in eosinophilic granular cells of the gastrointestinal lamina propria (Gambardella et al. 2010a, b). Latest studies have also demonstrated the expression of various forms of leptin in the gastrointestinal tract of zebrafish (*Danio rerio*) (Mania et al. 2017).

As demonstrated by Hindlet et al. (2007) based on analyses carried out in mammals, leptin may induce translocation of peptide transporters and regulate both the activity and expression of intestinal PepT1. An investigation conducted in rainbow trout confirmed the role of leptin in fish digestion physiology. Ostaszewska et al. (2010a) pointed out that leptin secreted by the fish stomach can regulate protein absorption from feed through its modulating effect of PepT1 activity, as in mammals.

Additional hormones secreted by enteric endocrine cells include: somatostatin, grelin, gastric inhibitory peptide (GIP), or vasoactive intestinal peptide (VIP). These hormones affect modulation of the hormonal response in the gastrointestinal tract and may influence the secretion of gastric juice, intestinal juice and bile, as well as motor activity of the digestive tract (Buddington and Krogdahl 2004, Volkoff et al. 2009, Rønnestad et al. 2013).

Rodlet Cells

Rodlet cells are found in the epithelium of the gastrointestinal tract of various species of teleosts (Laura et al. 2012, Bosi et al. 2018). They are usually ovoid with a basally-located nucleus. In addition, they are characterized by a wide fibrous layer beneath the plasma membrane and by the presence of characteristic large rod-shaped cytoplasmic granules. Relatively less is known about functions they play in the gastrointestinal tract. Usually, they are supposed to serve secretory and immunological functions (Reite 2005). Laura et al. (2012) described changes in the localization and cytology of rodlet cells during maturation. Immature rodlet cells were described by other authors near the basal epithelium membrane. They were characterized by a well-developed rough endoplasmic reticulum with extended cisternae. It seems that maturating rodlet cells migrate into the middle section of the epithelium and their cell organelles undergo reorganization. In turn, mature rodlet cells placed near the free surface show a thick subplasmalemmar fibrillar coat (Laura et al. 2012).

Cells of the Immune System

The gastrointestinal tract is part of the lymphoid system in fish; hence, immune cells form gut-associated lymphoid tissue (GALT) in it (Abelli et al. 1997). Multiple lymphoid tissues were observed in blood vessels present in the sub-mucous membrane and in blood vessels of the lamina propria of intestinal folds (Figs. 3F, 4A and B) (Uran et al. 2008, Ostaszewska et al. 2013). In addition, infiltration of the enteric mucosa by lymphocytes, labeled by the receptor CD-3 (cluster of differentiation 3), has been reported in many fish species during enteritis (Figs. 3E, 4A and B) (Bakke-McKellep et al. 2007, Ostaszewska et al. 2013).

Physiology—Epithelial Turnover

The proper functioning of cells of the gastrointestinal tract epithelium is regulated by epithelial turnover (Potten and Booth 1997). In fish, the proliferating cells are localized in the basal region of intestinal folds, as the mother cells in crypts in the intestine of mammals (Leung et al. 2005). In turn, the apoptotic cells are usually found in the apical region of intestinal folds (Berntssen et al. 2001). Continuity of the enteric epithelium is ensured by maintained balance between processes of cell proliferation in the basal part and processes of cell apoptosis on top of the intestinal folds (Potten et al. 1997, Ostaszewska et al. 2010b, Kamaszewski and Ostaszewska 2014). Under exposure to pathological factors, changes in cell proliferation in the intestine are often the first symptom of homeostasis disruption (Berntssen et al. 2004, Sanden et al. 2005, Ostaszewska et al. 2010b). The intensity of apoptosis processes is also perceived as an early and

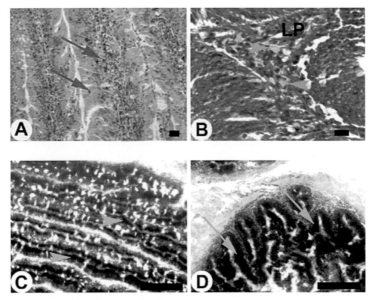

Fig. 4. Cross sections of the anterior intestine. **A.** Yellow perch. CD-3-positive lymphocytes (blue arrows) infiltrating the enteric mucosa. Immunohistochemical detection of CD-3-positive cells. **B.** Common carp. Granulocytes with basophilic granules (green arrowheads) in a blood vessel. LP, lamina propria. **C.** Pike-perch. Histochemical detection of non-specific esterase (blue arrowheads). **D.** Pike-perch. Histochemical detection of aminopeptidase M (blue arrows). **A,** immunohistochemical staining using anti-CD-3 antibody, **B,** Crossman's staining, **C** and **D,** Histochemical detection of enzymes. Scale bars: **A, B**—10 μm, **C** and **D**—100 μm.

sensitive indicator of chemical stress, activity of pathogens and disorders in body homeostasis (Berntssen et al. 1999, Sweet et al. 1999).

Physiology of Digestion

Gland cells (principal cells), localized in the gastric mucosa, secrete pepsin—an inactive form of pepsinogen. It is the main enzyme engaged in the proteolytic activity in the acidic environment of the stomach. Pepsin has been reported in many species of gastric fish after the period of metamorphosis and has been claimed to be the marker of stomach functionality (Zambonino Infante and Cahu 1994, Yúfera et al. 2012, Hamza et al. 2015). As demonstrated by many authors, in predatory and omnivorous gastric fish, the activity of acid proteases, determined in the stomach, is usually higher than the activity of alkaline proteases, which results from the activity of different proteases, like trypsin, chymotrypsin, aminopeptidase and others, in the intestine (Wu et al. 2007, Xiong et al. 2009). This confirms a phenomenon observed in other vertebrates, i.e., the stomach being the main site of protein digestion.

Pepsin is probably responsible for the earliest stage of protein digestion, which consists in degradation of large polypeptide chains in the acidic environment of the stomach (Tengjaroenkul et al. 2000, Xiong et al. 2009). Successive stages of digestion are continued in the intestine, at a higher AS alkaline pH of intestinal digesta.

During fish ontogenesis, particularly at the moment of differentiation and physiological maturation of the intestine, enhancement is observed in the activity of enzymes of the brush border of enterocytes but suppression in the activity of intracellular enzymes (Zambonino Infante et al. 1997). Development and differentiation of enterocytes may be evaluated based on the ratio of aminopeptidase N activity in the brush border to the activity of such intracellular enzymes as leucine-alanine peptidase. The value of this ratio reflects the digestion process in the brush border compared to the intracellular digestion and indicates the stage of fish larvae development (Cahu and Zambonino Infante 1994, Buchet et al. 2000, Hamza et al. 2008).

As in other vertebrates, protein digestion in the alimentary tract of fish proceeds extracellularly and intracellularly at the entire length of the intestine (Hirji and Courtney 1982). Extracellular digestive enzymes are secreted by cells of the pancreas, stomach epithelium and intestine to the lumen of the gastrointestinal tract. Proteins are digested in the stomach and intestine, whereas final stages of the digestion process proceed with enzymes of the brush border of enterocytes (Rhodes et al. 1967). The extracellular enzymes include peptidases: pepsin, trypsin, chymotrypsin, aminopeptidase, carboxypeptidase, and elastase. In turn, the intracellular enzymes—like non-specific esterases (Fig. 4C) or acid phosphatase—are associated with lysozymes and are responsible for the digestion inside cells and for maintaining the balance between the content of protein and products of its hydrolysis inside cells. Lysozymes contain a set of hydrolases which take part in decomposition of macromolecules delivered to a cell. Phosphatases are responsible for detaching a phosphate residue from carbohydrates, lipids, and proteins (Lojda et al. 1979). Finally, the non-specific esterase dissociates glycerides and fatty acid esters having chains not longer than eight atoms of carbon (Kamaszewski and Ostaszewska 2013).

Enzymes of the brush border of the intestine play a significant role in protein digestion (Zambonino Infante and Cahu 2007, Hamza et al. 2008, Kamaszewki et al. 2014). Aminopeptidase—being an exopeptidase—may serve as an example in this case (Fig. 4D) (Lojda et al. 1979). It is an enzyme associated with the surface of enterocyte cell membrane which decomposes peptide bonds of N-terminal amino acids. The free amino acids and small molecules of peptides are then actively transported through the cell membrane of enterocytes (Gawlicka et al. 1995). Likewise, digestion of carbohydrates initiated by exocrine amylase is continued by enzymes of the brush border of enterocytes, including, e.g., disaccharidase and maltase (Krzymowski and Przała 2005, Krogdahl et al. 2005).

The teleosts include many agastric species. However, this morphological lesion does not impair trophic capabilities of these fish; therefore, various trophic adjustments are observed in the group of fish with a similar morphology of the digestive tract (Day et al. 2011). In the case of agastric fish, the main proteases active in the alkaline environment include trypsin, chymotrypsin, and aminopeptidase. This may point to an adaptation in these fish to digest feed characterized by rather low protein content and high carbohydrate content. As demonstrated in a study conducted by Hidalgo et al. (1999), the total amylolytic: total proteolytic activity ratio was higher in the omnivorous fish whose diet is characterized by a significant contribution of plant-derived feed compared to the carnivorous fish. In addition, as reported by Day et al. (2011), the high proteolytic activity determined in the digestive tract of some agastric herbivore fish may point to ineffective protein digestion resulting presumably from the lack in the activity of acid proteases.

The intake of wood is a rare adaptation among animals. In the case of fish, some species of the family Loricariidae may be wood-eating taxa or detritivores. Physiological analyses conducted in these fish demonstrated that digestion, which includes relatively effective hydrolysis of polysaccharides, proceeds in their proximal and mid intestine. Nevertheless, in-depth enzymatic and microbiological analyses did not demonstrate the formation of a typical fermenter in the digestive tract and did not confirm the presence of symbiotic microflora that could effectively decompose wood (German 2009, German and Bittong 2009). These authors reported that the *Loricariidae* were not true xylivore animals, like, e.g., termites, but in fact were detritivore fish. However, as indicated by Maiuta et al. (2013), the major role of putative cellulolytic bacteria found in feces of fish from the genus *Panaqua* sp. is the metabolic activity, rather than altering the microbial diversity once an increased amount of polysaccharide, such as cellulose, is available depending on the accessible food source, which does not exclude the cellulolytic activity in feces and gastrointestinal microbiome. McDonald et al. (2015) demonstrated nitrogen uptake in the gastrointestinal tract of fish from the genus *Panaqua* sp. based on the reported activity of nitrogenase iron protein (NifH). Sequence analysis pointed to a high likelihood of the occurrence of bacteria of the genus *Rhizobiales* (McDonald et al. 2015); however, these observations require further analyses, as concluded by these authors.

Histological lesions in the anatomy of the gastrointestinal tract, including the number and height of folds of the intestinal epithelium or decreased surface of enterocytes, may reflect improper fish feeding (Ostaszewska et al. 2006, Przybył et al. 2006, Chen et al. 2007, Bowzen et al. 2011, Ostaszewska et al. 2011b, 2013, Kamaszewski and Ostaszewska 2014, Kamaszewski et al. 2014). In turn, Ostaszewska et al. (2006) and Chen et al. (2007) claim that a low number of intestinal folds and reduced surface of enterocytes may be indicative of fish starvation. Likewise, Gisbert and Doroshov (2003) report lower height of intestine folds and decreased surface of the intestinal epithelium cells to be symptoms of fish starvation.

The anterior intestine is the site of absorption of lipid substances from the lumen of the gastrointestinal tract (Noallac-Depeyre and Gas 1973, Rombout et al. 1985, Kjørsvik et al. 1991). The presence of lipid vacuoles in supranuclear regions of the cytoplasm of anterior intestine enterocytes is interpreted as a temporary form of storage of esterified fatty acids (Fontagné et al. 1998). As suggested by Sire and Vernier (1981), Kjørsvik et al. (1991), and by Cabarello et al. (2003), accumulation of lipid vacuoles in the supranuclear regions of anterior intestine enterocytes may be due to the inhibited transport of lipids from intestinal epithelium cells to the cardiovascular system. A high number of lipid vacuoles in enterocytes may also result from fish feeding with high-fat feed mixtures (Luizi et al. 1999). According to Caballero et al. (2002), accumulation of lipid vacuoles in enterocytes of fish may be due to inappropriate ratio of fatty acids in their feed.

In turn, the posterior intestine is the site of absorption of amino acids, peptides and also proteins (Noaillac-Depeyre and Gas 1973, Rombout et al. 1985, Sire and Vernier 1992). Vacuoles containing PAS-positive inserts in the supranuclear regions of enterocytes of the posterior intestine were described in *Chaenogobius annularis* (Watanabe 1985), the summer flounder (*Paralichthys dentatus*) (Bisbal and Bengtson 1995), *Prochiludus scrofa* (Nachi et al. 1998), ide (*Leuciscus idus*) (Ostaszewska et al. 2003), pike-perch (Ostaszewska et al. 2005a), common carp (*Cyprinus carpio*) (Przybył et al. 2006), and yellowtail kingfish (*Seriola lalandi*) (Chen et al. 2007). Vacuoles containing PAS-positive granules result from pinocytotic uptake of protein from the intestine lumen (Govoni et al. 1986). The presence of absorptive vacuoles is indicative of proper digestion and absorption of nutrients (Ostaszewska et al. 2005b). A lack of these vacuoles in the posterior intestine and decreased height of enterocytes may indicate either starvation or improper feeding of fish (Watanabe 1985, Oozeki et al. 1989, Crespo et al. 2001, Chen et al. 2007). In turn, as suggested by Deplano et al. (1991), large vacuoles present in the supranuclear absorption centers of enterocytes may contain multiple acidic hydrolases that can induce changes in cellular structure and lead to cell death.

The Role of the Gastrointestinal Tract in Gas Exchange in Fish

In the course of evolution, fish have independently developed the capability to breathe with atmospheric air through organs of the gastrointestinal system, including, among others, the intestine (Nelson 2014). Usually, these fish facultatively breathe atmospheric air under deficit of water-dissolved oxygen in the aquatic environment. Fish, which respire with atmospheric air using the intestine, have been observed to develop many morphological adaptations. Gut section involved in the respiratory process is usually separated from the section engaged in the digestion process with multiple sphincters (Nelson 2014). Typical traits of the mucous membrane of this section of the intestine include decreased thickness of the intestinal wall, including the mucosa, and an expanded network of capillary vessels.

Liu and Wang (2017) described the histology of the digestive tract of *Paramisgurnus dobryanus*, being an air-breathing species, and found a thin epithelium composed of squamous cells and a dense blood capillary network underneath in their posterior intestine devoid of folds. As suggested by Liu and Wang (2017), epithelial cells present in the posterior intestine of the investigated species are characterized by ultrastructural lesions typical of the pneumocyte types I and II localized in the alveoli of mammals. Likewise, structures enabling gas exchanges were also reported in the posterior intestine of the fish from the genus *Corydoras* (Podkowa and Goniakowska-Witalińska 2002). Histological lesions were observed in these fish from the family Callichthyidae in the posterior section of their intestine, including, among others, penetration of capillaries into the intestine's thick epithelial layer, which is a typical structure of the respiratory intestine.

The loricariids also breathe atmospheric air; however, oxygen uptake proceeds in their stomach (Nelson 2014). As demonstrated by Satora (1998), the stomach of *Ancistrus multipinnis* may be divided into three typical parts: cardia, corpus and pylorus; however, the anatomy of cardia and pylorus is typical of other fish species. In turn, the corpus of the stomach is characterized by decreased thickness of the blood-air interface, which enables oxygen diffusion across walls of the stomach. In addition, Satora (1998) observed that part of epithelial cells in the corpus of the stomach contain multiple lamellar bodies, which may probably produce substances with functions analogous to those of lung surfactants in mammals or some amphibians.

References Cited

Abelli, L., S. Picchietti, N. Romano, L. Mastrolia and G. Scapigliati. 1997. Immunohistochemistry of gut-associated lymphoid tissue of the sea bass *Dicentrarchus labrax* (L.). Fish & Shell. Immunol. 7: 235–245.

Andreozzi, G., P. de Girolamo, C. Affatato, P. Russo and G. Gargiulo. 1997. VIP-like immunoreactivity in the intestinal tract of fish with different feeding habits. European J. Histochem. 41: 57–64.

Bakke-McKellep, A.M., M.K. Frøystad, E. Lilleeng, F. Dapra, S. Refstie, A. Krogdahl and T. Landsverk. 2007. Response to soy: T-cell-like reactivity in the intestine of Atlantic salmon, *Salmo salar* L. J. Fish Dis. 30: 13–25.

Bell, F.R. 1979. The relevance of the new knowledge of gastrointestinal hormones to veterinary science. Vet. Sci. Commun. 2: 305–314.

Bermúdez, R., F. Vigliano, M.I. Quiroga, J.M. Nieto, G. Bosi and C. Domeneghini. 2007. Immunohistochemical study on the neuroendocrine system of the digestive tract of turbot, *Scophthalmus maximus* (L.), infected by *Enteromyxum scophthalmi* (*Myxozoa*). Fish Shell. Immun. 22: 252–263.

Berntssen, M.H.G., O.Ø. Aspholm, K. Hylland, S.E. Wendelaar Bonga and A.K. Lundebye. 2001. Tissue metallothionein, apoptosis and cell proliferation responses in Atlantic salmon (*Salmo salar* L.) parr fed elevated dietary cadmium. Comp. Bioch. Physiol. C 128: 299–310.

Berntssen, M.H.G., K. Hylland, K. Julshamn, A.K. Lundebye and R. Waagbø. 2004. Maximum limits of organic and inorganic mercury in fish feed. Aquac. Nutrit. 10: 83–97.

Bisbal, G.A. and D.A. Bengtson. 1995. Development of digestive tract in larval summer flounder. J. Fish Biol. 47: 277–291.

Bjenning, C. and S. Holmgren. 1988. Neuropeptides in the fish gut. An immunohistochemical study of evolutionary patterns. Histochemistry 88: 155–163.

Bočina, I., Ž. Šantić, I. Restović and S. Topić. 2017. Histology of the digestive system of the garfish *Belone belone* (Teleostei: *Belonidae*). Europ. Zool. J. 84: 89–95.

Bosi, G., A. Di Giancamillo, S. Arrighi and C. Domeneghini. 2004. An immunohistochemical study on the neuroendocrine system in the alimentary canal of the brown trout, *Salmo trutta*, L., 1758. Gen. Comp. Endocrinol. 138: 166–181.

Bosi, G., J.A. DePasquale, M. Manera, G. Castaldelli, L. Giari and B. Sayyaf Dezfuli. 2018. Histochemical and immunohistochemical characterization of rodlet cells in the intestine of two teleosts, *Anguilla anguilla* and *Cyprinus carpio*. J. Fish Dis. 41: 475–485.

Bowzer, J., K. Dabrowski, K. Ware, T. Ostaszewska, M. Kamaszewski and M. Botero. 2011. Growth, survival, and body composition of sunshine bass after a feeding and fasting experiment. North Amer. J. Aquacult. 73: 373–382.

Buchet, V., J.L. Zambonino Infante and C.L. Cahu. 2000. Effect of lipid level in a compound diet on the development of red drum (*Sciaenops ocellatus*) larvae. Aquaculture 184: 339–347.

Buddington, R. and J. Christofferson. 1985. Digestive and feeding characteristics of the chondrosteans. Env. Biol. Fish 14: 31–41.

Buddington, R.K. and A. Krogdahl. 2004. Hormonal regulation of the fish gastrointestinal tract. Comp. Biochem. Physiol. Part A 139: 261–271.

Caballero, M.J., A. Obach, G. Rosenlund, D. Montero, M. Gisvold and M.S. Izquierdo. 2002. Impact of different dietary lipid sources on growth, lipid digestibility, tissue fatty acid composition and histology of rainbow trout, *Oncorhynchus mykiss*. Aquaculture 214: 253–271.

Caballero, M.J., M.S. Izquierdo, E. Kjørsvik, D. Montero, J. Socorro, A.J. Fernández and G. Rosenlund. 2003. Morphological aspects of intestinal cells from gilthead seabream (*Sparus aurata*) fed diets containing different lipid sources. Aquaculture 225: 325–340.

Caceci, T., H.A. El-Habback, S.A. Smith and B.J. Smith. 1997. The stomach of *Oreochromis niloticus* has three regions. J. Fish Biol. 50: 939–952.

Cahu, C.L. and J.L. Zambonino Infante. 1994. Early weaning of sea bass (*Dicentrarchus labrax*) larvae with a compound diet: effect on digestive enzymes. Comp. Biochem. Physiol. 109A: 213–222.

Chakrabarti, P. and S.K. Ghosh. 2014. A comparative study of the histology and microanatomy of the stomach in *Mystus vittatus* (Bloch), *Liza parsia* (Hamilton), and *Oreochromis mossambicus* (Peters). J. Microsc. Ultrastruct. 2: 245–250.

Chatchavalvanich, K., R. Marcos, J. Poonpirom, A. Thongpan and E. Rocha. 2006. Histology of the digestive tract of the freshwater stingray *Himantura signifer* Compagno and Roberts, 1982 (*Elasmobranchii, Dasyatidae*). Anat. Embryol. (Berl) 211: 507–518.

Chen, B.N., J.G. Qin, J.F. Carragher, S.M. Clarke, M.S. Kumar and W.G. Hutchinson. 2007. Deleterious effect of food restrictions in yellowtail kingfish *Seriola lalandi* during early development. Aquaculture 271: 326–335.

Clements, K.D. and D. Raubenheimer. 2005. Feeding and nutrition. pp. 47–82. *In*: D.H. Evans and J.B. Claiborne (eds.). The Physiology of Fishes. CRC Press, Boca Raton.

Crespo, S., M. de Mareo, C.A. Santamaría, R. Sala, A. Grau and E. Pastor. 2001. Histopathological observations during larval rearing of common dentex *Dentex dentex* L. (*Sparidae*). Aquaculture 192: 121–132.

Day, R.D., D.P. German and I.R. Tibbetts. 2011. Why can't young fish eat plants? Neither digestive enzymes nor gut development preclude herbivory in the young of a stomachless marine herbivorous fish. Comp. Biochem. Physiol. Part B 158: 23–29.

de Pedro, N., R. Martinez-Alvarez and M.J. Delgado. 2006. Acute and chronic leptin reduces food intake and body weight in goldfish (*Carassius auratus*). J. Endocrinol. 188: 513–520.

Deplano, M., R. Connes and J.P. Diaz. 1991. Postvalvular enterocytes in feral and farm-reared sea bass Dicentrarchus labrax; hypervacuolization related to artificial feed. Dis. Aquat. Org. 11: 9–18.

Deveney, C.W. and L.W. Way. 1983. Regulatory peptides of the gut. pp. 479–499. *In*: F.S. Greenspan and P.H. Forsham (eds.). Basic and Clinical Endocrinology. Asian End. Singapore: Maruzen.

Díaz, A.O., A.M. García and A.L. Goldemberg. 2008. Glycoconjugates in the mucosa of the digestive tract of *Cynoscion guatucupa*: A histochemical study. Acta Histoch. 110: 76–85.

Domeneghini, C., R. Pannelli Straini and A. Veggetti. 1998. Gut glycoconjugates in *Sparus aurata* L. (Pisces, Teleostei). A comparative histochemical study in larval and adult ages. Histol. Histopathol. 13: 359–372.

Domeneghini, C., S. Arrighi, G. Radaelli, G. Bosi and S. Mascarello. 1999. Morphological and histochemical peculiarities of the gut in white sturgeon, *Acipenser transmontanus*. Eur. J. Histochem. 43: 135–145.

Domeneghini, C., S. Arrighi, G. Radaelli, G. Bosi and A. Veggetti. 2005. Histochemical analysis of glycoconjugate secretion in the alimentary canal of *Anguilla anguilla* L. Acta Histoch. 106: 477–487.

El-Salhy, M. and B. Sitohy. 2001. Abnormal gastrointestinal endocrine cells in patients with diabetes type 1: relationship to gastric emptying and myoelectrical activity. Scand. J. Gastroenterol. 36: 1162–1169.

Ezeasor, D.N. 1981. The fine structure of the gastric epithelium of the rainbow trout, *Salmo gairdneri* Richardson. J. Fish Biol. 19: 611–627.

Ezeasor, D.N. and W.M. Stokoe. 1981. Light and electron microscopic studies on the absorptive cells of the intestine, caeca and rectum of the adult rainbow trout, *Salmo gairdneri*, Rich. J. Fish Biol. 18: 527–544.

Fishelson, L. 1997. Comparative ontogenesis and cytomorphology of the nasal organs in some species of cichlid fish (*Cichlidae, Teleostei*). J. Zool. London 243: 281–294.

Fishelson, L., Y. Delarea and A. Zverdling. 2004. Taste bud form and distribution on lips and in the oropharyngeal cavity of cardinal fish species (*Apogonidae, Teleostei*), with remarks on their dentition. J. Morphol. 259: 316–327.

Fontagné, S., I. Geuden, A.M. Escaffre and P. Bergot. 1998. Histological changes induced by dietary phospholipids in intestine and liver of common carp (*Cyprinus carpio* L.) larvae. Aquaculture 161: 213–223.

Gambardella, C., L. Gallus, S. Ravera, S. Fasulo, M. Vacchi and S. Ferrando. 2010a. First evidence of a leptin-like peptide in a cartilaginous fish. Anat. Rec. 293: 1692–1697.

Gambardella, C., S. Ferrando, T. Ferrando, S. Ravera, L. Gallus, S. Fasulo and G. Tagliafierro. 2010. Immunolocalisation of leptin in the digestive system of juvenile European sea bass (*Dicentrarchus labrax*). Ital. J. Zool. 77: 391–398.

Gawlicka, A., S.J. Teh, S.S.O. Hung, D.E. Hinton and J. de la Noüe. 1995. Histological and histochemical changes in digestive tract of white sturgeon larvae during ontogeny. Fish Physiol. Bioch. 14: 357–371.

German, D.P. 2009. Inside the guts of wood-eating catfishes: can they digest wood? J. Comp. Physiol. Part B 179: 1011–1023.

German, D.P. and R.A. Bittong. 2009. Digestive enzyme activities and gastrointestinal fermentation in wood-eating catfishes. J. Comp. Physiol. Part B 179: 1025–1042.

Gisbert, E. and S.I. Doroshov. 2003. Histology of developing digestive system and the effect of food deprivation in larval green sturgeon (*Acipenser medirostris*). Aquat. Liv. Res. 16: 77–89.

Gomahr, A., M. Palzenberg and K. Kotrschal. 1992. Density and distribution of external taste buds in cyprinids. Environm. Biol. Fish. 33: 125–134.

Govoni, J.J., G.W. Boehlert and Y. Watanabe. 1986. The physiology of digestion in fish larvae. Envir. Biol. Fishes 16: 59–77.

Hamza, N., M. Mhetli, I.B. Khemis, C. Cahu and P. Kestemont. 2008. Effect of dietary phospholipid levels on performance, enzyme activities and fatty acid composition of pikeperch (*Sander lucioperca*). Aquaculture 275: 274–282.

Hamza, N., T. Ostaszewska and P. Kestemont. 2015. Development and functionality of the digestive system in percid fishes early life stages. pp. 239–264. *In*: P. Kestemont, K. Dabrowski and R.C. Summerdelt (eds.). Biology and Culture of Percid Fishes: Principles and Practices. Springer, London.

Hara, T.J. 1992. Fish Chemoreception. London: Chapman & Hall.

Hara, T.J. 1994. Olfaction and gustation in fish: an overview. Acta Physiol. Scand. 152: 207–217.

Harder, W. 1975. Anatomy of Fishes, Schweizerbart, Verlagsbuchhandlung, Stuttgart.

Hassanpour, M. and J. Joss. 2009. Anatomy and histology of the spiral valve intestine in juvenile Australian lungfish, *Neoceratodus forsteri*. Open Zool. J. 2: 62–85.

Helle, K.B. and D. Aunis. 2000. A physiological role for the granins as prohormones for homeostatically important regulatory peptodes? Adv. Exp. Med. Biol. 482: 389–397.

Hernández, D.R., M. Pérez Gianeselli and H.A. Domitrovic. 2009. Morphology, histology and histochemistry of the digestive system of South American catfish (*Rhamdia quelen*). Int. J. Morphol. 27: 105–111.

Hidalgo, M.C., E. Urea and A. Sanz. 1999. Comparative study of digestive enzymes in fish with different nutritional habits. Proteolytic and amylase activities. Aquaculture 170: 267–283.

Hirji, K.N. and W.A.M. Courtney. 1982. Leucine aminopeptydaze activity in the digestive tract of perch, *Perca fluviatilis* L. J. Fish Biol. 21: 615–622.

Holst, J.J., J. Fahrenkrug, F. Stadil and J.F. Rehfeld. 1996. Gastrointestinal endocrinology. Scand. J. Gastroenterol. 216: 27–38.

Ikpegbu, E., D.N. Ezeasor, U.C. Nlebedum and O. Nnadozie. 2013. Morphological and histochemical observations on the oesogaster of the domesticated African catfish (*Clarias gariepinus*, Burchell, 1822). B.J.V.M. 16: 88–95.

Iwai, T. 1962. Studies on the *Plecoglossus altivelis* problems: Embryology and histophysiology of digestive and osmoregulatory organs. Bull. Misaki Mar. Biol. Inst. Kyoto Univ. 2: 1–101.

Iwai, T. 1980. Sensory anatomy and feeding of fish larvae. In: Fish behavior and its use in capture and culture of fishes. ICLARM Conference Proceedings 5: 124–145.

Jaroszewska, M., K. Dabrowski, B. Wilczynska and T. Kakareko. 2008. Structure of the gut of the racer go by *Neogobius gymnotrachelus* (Kessler, 1857). J. Fish Biol. 72: 1773–1786.

Jhaveri, P., Y.P. Papastamatiou and D.P. German. 215. Digestive enzyme activities in the guts of bonnethead sharks (*Sphyrna tiburo*) provide insight into their digestive strategy and evidence for microbial digestion in their hindguts. Comp. Biochem. Physiol. A Mol. Integr. Physiol. 189: 76–83.

Kamaszewski, M. and T. Ostaszewska. 2013. The effect of feeding on aminopeptidase and non-specific esterase activity in the digestive system of pike-perch (*Sander lucioperca* L.). Ann. Warsaw Univ. of Life Sci. SGGW, Anim. Sci. 52: 49–57.

Kamaszewski, M. and T. Ostaszewska. 2014. The effect of feeding on morphological changes in intestine of pike-perch (*Sander lucioperca* L.). Aquacult. Int. 22: 245–258.

Kamaszewski, M., T. Ostaszewska, M. Prusińska, R. Kolman, M. Chojnacki, J. Zabytyvskij, B. Jankowska and R. Kasprzak. 2014. Effects of *Artemia* sp. enrichment with essential fatty acids on functional and morphological aspects of the digestive system in *Acipenser gueldenstaedtii* larvae. Tur. J. Fish Aquat. Sc. 14: 929–938.

Kamisaka, Y., G.K. Totland, M. Tagawa, T. Kurokawa, T. Suzuki, M. Tanaka and I. Rønnestad. 2001. Ontogeny of cholecystokinin—immunoreactive cells in the digestive tract of Atlantic halibut, *Hippoglossus hippoglossus*, larvae. Gen. Comp. Endocrinol. 123: 31–37.

Kamisaka, Y., Y. Fujii, S. Yamamoto, T. Kurokawa, I. Rønnestad, G.K. Totland, M. Tagawa and M. Tanaka. 2003. Distribution of cholecystokinin-immunoreactive cells in the digestive tract of the larval teleost, ayu, *Plecoglossus altivelis*. Gen. Comp. Endocrinol. 134: 116–121.

Kapoor, B.G., H. Smit and I.A. Verighina. 1975. The alimentary canal and digestion in teleosts. Adv. Mar. Biol. 13: 109–239.

Kilarski, W. 2012. Anatomia ryb. Powszechne Wydawnictwo Rolnicze i Leśne, Poznań (in Polish).

Kiliaan, A., S. Holmgren, A.C. Jonsson, K. Dekker and J. Groot. 1992. Neurotensin, substance P, gastrin/cholecystokinin, and bombesin in the intestine of the tilapia (*Oreochromis mossambicus*) and the goldfish (*Carassius auratus*): immunochemical detection and effects on electrophysiological characteristic. Gen. Comp. Endocrinol. 88: 351–363.

Kinoshita, A.J. and S. Ishihara. 2000. Mechanism of gastric mucosal proliferation induced by gastrin. J. Gastroenterol. Hepatol. 15: D7–D11.

Kjørsvik, E., T. van der Meeren, H. Kryvi, J. Arnfinnson and P.G. Kvenseth. 1991. Early development of the digestive tract of cod larvae, *Gadus morhua* L., during start-feeding and starvation. J. Fish Biol. 38: 1–15.

Koven, W., C.R. Rojas–Garcia, R.N. Finn, A. Tandler and I. Rønnestad. 2002. Stimulatory effect of ingested protein and/or free amino acids on the secretion of the gastro-endocrine hormone cholecystokinin and tryptic activity, in early-feeding herring larvae, *Clupea harengus*. Marine Biology 140: 1241–1247.

Kozarić, Z., S. Kuzir, Z. Petrinec, E. Gjurevic and N. Baturina. 2007. Histochemistry of complex glycoproteins in the digestive tract mucosa of Atlantic bluefin tuna (*Thunnus thynnus* L.). Vet. Arh. 77: 441–452.

Kozarić, Z., S. Kužir, Z. Petrinec, E. Gjurčević and M. Božić. 2008. The development of the digestive tract in larval European catfish (*Silures glanis* L.). Anat. Histol. Embryol. 37: 141–146.

Krogdahl, A., G.I. Hemre and T.P. Mommsen. 2005. Carbohydrates in fish nutrition: digestion and absorption in postlarval stages. Aquacult. Nutr. 11: 103–122.

Krzymowski, T. and J. Przała. 2005. Fizjologia zwierząt. Powszechne Wydawnictwo Rolnicze i Leśne, Warszawa (in Polish).

Ku, S.K., J.H. Lee and H.S. Lee. 2004. Immunohistochemical study on the endocrine cells in the gut of the stomachless Teleost, *Zacco platypus* (*Cyprinidae*). Anat. Histol. Embryol. 33: 212–219.

Kurokawa, T., T. Suzuki and H. Hashimoto. 2003. Identification of gastrin and multiple cholecystokinin genes in teleost. Peptides 24: 227–235.

Laura, R., G.P. Germana, M.B. Levanti, M.C. Guerrera, G. Radaelli, F. de Carlos, A.A. Suarez, E. Ciriaco and A. Germana. 2012. Rodlet cells development in the intestine of sea bass (*Dicentrarchus labrax*). Microsc. Res. Tech. 75: 1321–1328.

Lee, J.H., S.K. Ku, K.D. Park and H.S. Lee. 2004. Immunohistochemical study of the gastrointestinal endocrine cells in the Korean aucha perch. J. Fish Biol. 65: 170–181.

Leigh, S.C., Y. Papastamatiou and D.P. German 2017. The nutritional physiology of sharks. Rev. Fish Biol. Fisheries 27: 561–585.

Leung, A.Y.H., J.C.K. Leung, L.Y.Y. Chan, E.S.K. Ma, T.T.F. Kwan, K.N. Lai, A. Meng and R. Liang. 2005. Proliferating cell nuclear antigen (PCNA) as a proliferative marker during embryonic and adult zebrafish hematopoiesis. Histochem. Cell Biol. 124: 105–111.

Liu, Y.Q. and Z.J. Wang. 2017. Study on structural characteristics of intestinal tract of the air-breathing loach, *Paramisgurnus dabryanus* (Sauvage, 1878). Pakistan J. Zool. 49: 1223–1230.

Lloyd, R.V. and B.S. Wilson. 1983. Specific endocrine tissue marker defined by a monoclonal antibody. Science. 222: 628–630.

Lojda, Z., R. Gossran and T.H. Schebler. 1979. Enzyme histochemistry. A laboratory manual. Springer-Verlag. Berlin, Heidelberg, New York.

Luizi, F.S., B. Gara, R.J. Shields and N.R. Bromage. 1999. Further description of the development of the digestive organs in Atlantic halibut (*Hippoglossus hippoglossus*) larvae, wiyh notes on differential absorption of copepod and *Artemia* prey. Aquaculture 176: 101–116.

Maiuta, N.D., P. Schwarzentruber, M. Schenker and J. Schoelkopf. 2013. Microbial population dynamics in the faeces of wood-eating loricariid catfishes. Lett. Appl. Microbiol. 56: 401–407.

Mania, M., L. Maruccio, F. Russo, F. Abbate, L. Castaldo, L. D'Angelo, P. de Girolamo, M.C. Guerrera, C. Lucini, M. Madrigrano, M. Levanti and A. Germanà. 2017. Expression and distribution of leptin and its receptors in the digestive tract of DIO (diet-induced obese) zebrafish. Ann. Anat. 212: 37–47.

Marcus, H. 1934. Zur stammesgeschichte der Zunge. II. Uber Muskulatur in der Polypteruszunge. Anat. Anz., lxxvii.

McDonald, R., F. Zhang, J.E.M. Watts and H.J. Schreier. 2015. Nitrogenase diversity and activity in the gastrointestinal tract of the wood-eating catfish *Panaque nigrolineatus*. The ISME Journal 9: 2712–2724.

Nachi, A.M., F.J. Hernandez-Blazquez, R.L. Barbieri, R.L. Leite, S. Ferri and M.T. Phan. 1998. Intestinal histology of a detrivorous (iliophagous) fish *Prochilodus scrofa* (*Characiformes, Prochilodontidae*). Annal. Scienc. Naturelles. 2: 81–88.

Nelson, J.A. 2014. Breaking wind to survive: fishes that breathe air with their gut. J. Fish Biol. 84: 554–576.

Noaillac-Depeyre, J. and N. Gas. 1973. Absorption of protein macromolecules by the enterocytes of the carp (*Cyprinus carpio* L.). Z. Zellforsch. 146: 525–541.

Noaillac-Depeyre, J. and N. Gas. 1978. Ultrastructural and cytochemical study of the gastric epithelium in a freshwater teleostean fish (*Perca fluviatilis*). Tiss. Cell 10: 23–37.

O'Connell, C.P. 1981. Development of organ systems in the northern anchovy, *Engraulis mordax*, and other teleosts. Amer. Zool. 21: 429–446.

Oozeki, Y., T. Ishii and T. Hirano. 1989. Histological study of the effects of starvation on reared and wild-caught larval stone flounder, *Kareius bicoloratus*. Mar. Biol. 100: 269–275.

Osman, A.H.K. and T. Caceci. 1991. Histology of the stomach of *Tilapia nilotica* (Linnaeus, 1758) from the River Nile. J. Fish Biol. 38: 211–223.

Ostaszewska, T. 2002. Zmiany morfologiczne i histologiczne układu pokarmowego i pęcherza pławnego w okresie wczesnej organogenezy larw sandacza (*Stizostedion lucioperca* L.) w różnych warunkach odchowu. Roprawy Naukowe i Monografie. SGGW, Warszawa (in Polish).

Ostaszewska, T., A. Wegner and M. Węgiel. 2003. Development of the digestive tract of ide *Leuciscus idus* (L.) during the larval stage. Arch. Pol. Fish 11: 181–195.

Ostaszewska, T. 2005. Developmental changes of digestive system structure in pike-perch (*Sander lucioperca* L.). Electr. J. Ichthyol. 2: 65–78.

Ostaszewska, T., K. Dabrowski, K. Czumińska, W. Olech and M. Olejniczak. 2005a. Rearing of pike-perch larvae using formulated diets—first success with starter feeds. Aquacult. Res. 36: 1167–1176.

Ostaszewska, T., K. Dabrowski, M.E. Palacios, M. Olejniczak and M. Wieczorek. 2005b. Growth and morphological changes in the digestive tract of rainbow trout (*Oncorhynchus mykiss*) and pacu (*Piaractus mesopotamicus*) due to case in replacement with soybean proteins. Aquaculture 245: 273–286.

Ostaszewska, T., M. Korwin-Kossakowski and J. Wolnicki. 2006. Morphological changes of digestive structures in starved tench *Tinca tinca* (L.) juveniles. Aquacult. Inter. 14: 113–126.

Ostaszewska, T., K. Dabrowski, P. Hliwa, P. Gomółka and K. Kwasek. 2008. Nutritional regulation of intestine morphology in larval cyprinid fish, silver bream (*Vimba vimba*). Aquacult. Res. 39: 1268–1278.

Ostaszewska, T. and K. Dabrowski. 2009. Early development of *Acipenseriformes* (*Chondrostei, Actinopterygii*). pp. 171–230. *In*: Y.W. Kunz, C.A. Luer and B.G. Kapoor (eds.). Development of Non-Teleost Fish. Science Publishers Inc.

Ostaszewska, T., M. Kamaszewski, P. Grochowski, K. Dabrowski, T. Verri, E. Aksakal, I. Szatkowska, Z. Nowak and S. Dobosz. 2010a. The effect of peptide absorption on PepT1 gene expression and digestive system hormones in rainbow trout (*Oncorhynchus mykiss*). Comp. Biochem. Physiol. Part A 155: 107–114.

Ostaszewska, T., K. Dabrowski, M. Kamaszewski, P. Grochowski, T. Verri, M. Rzepkowska and J. Wolnicki. 2010b. The effect of plant protein based diet, supplemented with dipeptide or free amino acids on PepT1, PepT2 expression and digestive tract morphology in common carp (*Cyprinus carpio* L.). Comp. Biochem. Physiol. Part A 157: 158–169.

Ostaszewska, T., R. Kolman, M. Kamaszewski, G. Wiszniewski, D. Adamek and A. Duda. 2011a. Morphological changes in digestive tract of Atlantic sturgeon *Acipenser oxyrinchus* during organogenesis. Int. Aquat. Res. 3: 101–105.

Ostaszewska, T., K. Dabrowski, K. Kwasek, T. Verri, M. Kamaszewski, J. Sliwinski and L. Napora-Rutkowski. 2011b. Effects of various diet formulations (experimental and commercial) on the morphology of the liver and intestine of rainbow trout (*Oncorhynchus mykiss*) juveniles. Aquacult. Res. 42: 1796–1806.

Ostaszewska, T., K. Dąbrowski, M. Kamaszewski, K. Kwasek, M. Grodzik and J. Bierla. 2013. The effect of dipeptide, Lys-Gly, supplemented diets on digestive tract histology in juvenile yellow perch (*Perca flavescens*). Aquacult. Nutr. 19: 100–109.

Ostaszewska, T., A.T. Karczewska, D. Adamek-Urbańska, K. Krajnik, M. Rzepkowska, M. Luczynski, R. Kasprzak and K. Dabrowski. 2018. Effect of feeding strategy on digestive tract morphology and physiology of lake whitefish (*Coregonus lavaretus*). Aquaculture (under revision).

Pena, R., S. Dumas, M. Villalejo-Fuerte and J.L. Ortız-Galindo. 2003. Ontogenetic development of the digestive tract in reared spotted sand bass *Paralabrax maculatofasciatus* larvae. Aquaculture 219: 633–644.

Petrinec, Z., S. Nejedli, S. Kuzir and A. Opacak. 2005. Mucosubstances of the digestive tract mucosa in northern pike (*Esox lucius* L.) and European catfish (*Silurus glanis* L.). Vet. Arh. 75: 317–327.

Podkowa, D. and L. Goniakowska-Witalińska. 2002. Adaptations to the air breathing in the posterior intensine of the catfish (*Corydoras aeneus, Callichthyidae*). A histological and ultrastructural study. Fol. Biol. 50: 69–82.

Potten, C.S. and C. Booth. 1997. The role of radiation-induced- and spontaneous apoptosis in the homeostasis of the gastrointestinal epithelium: a brief review. Comp. Bioch. Physiol. B 118: 473–478.

Potten, C.S., C. Booth and D.M. Pritchard. 1997. The intestinal steam cell: the mucosal governor. Int. J. Exp. Pathol. 78: 219–243.

Przybył, A., T. Ostaszewska, J. Mazurkiewicz and A. Wegner. 2006. The effect of experimental starters on growth and morphological changes in the intestine and liver of common carp larvae reared under controlled conditions. Arch. Pol. Fish 14: 67–83.

Radaelli, G., C. Domeneghini, S. Arrighi, M. Francolini and F. Mascarello. 2000. Ultrastructural features of the gut in the white sturgeon, *Acipenser transmontanus*. Histol. Histopathol. 15: 429–439.

Reifel, C.W., M. Marin-Sorensen and I.M. Samloff. 1983. Gastrin immunoreactive cells in the gastrointestinal tracts from four species of fish. Can. J. Zool. 61: 1464–1468.

Reinecke, M., C. Müller and H. Segner. 1997. An immunohistochemical analysis of the ontogeny, distribution and coexistence of 12 regulatory peptides and serotonin in endocrine cells and nerve fibers of the digestive tract of the turbot, *Scophthalmus maximus* (Teleostei). Anat. Embryol. 195: 87–102.

Reite, O.B. 2005. The rodlet cells of teleostean fish: their potential role in host defence in relation to the role of mast cells/eosinophilic granule cells. Fish Shellfish Immunol. 19: 253–267.

Reutter, K. 1978. Taste organs in the Bullhead (Teleostei). Adv. Anat. Embryol. Cell Biol. 55: 7–97.

Reutter, K. 1991. Ultrastructure of taste buds in the Australian lungfish, Neoceratodus forsteri. Chem. Senses 16: 404.

Reutter, K., F. Boudriot and M. Witt. 2000. Heterogeneity of fish taste bud ultrastructure as demonstrated in the holosteans *Amia calva* and *Lepisosteus oculatus*. Philosoph. Trans. Royal Soc. London B 355: 1225–1228.

Rhodes, J.B., A. Eichholz and R.K. Crane. 1967. Studies on the organization of the brush border in intestinal epithelial cells. IV. Aminopeptidase activity in microvillus membranes of hamster intestinal brush borders. Biochim. Biophys. Acta 135: 959–965.

Rombout, J.H.W.M., C.H.J. Lamewrs, M.H. Helfrich, A. Dekker and J.J. Taverne-Thiele. 1985. Uptake and transport of intact macromolecules in the intestinal epithelium of carp (*Cyprinus carpio* L.) and the possible immunological implications. Cell Tiss. Res. 239: 519–530.

Rombout, J.H.W.M., C.P.M. van der Grinten, F.M. Peeze Binkhorst, J.J. Taverne-Thiele and H. Schooneveld. 1986. Immunocytochemical identification and localization of peptide hormones in the gastro-entero-pancreatic (GEP) endocrine system of the mouse and a stomachless fish, *Barbus conchonius*. Histochemistry 84: 471–483.

Rønnestad, I., M. Yúfera, B. Ueberschär, L. Ribeiro, Ø. Sæle and C. Boglione. 2013. Feeding behaviour and digestive physiology in larval fish: current knowledge, and gaps and bottlenecks in research. Rev. Aquacult. 5: 59–98.

Russo, F., P. De Girolamo, S. Neglia, A. Gargiulo, N. Arcamone, G. Gargiulo and E. Varricchio. 2011. Immunohistochemical and immunochemical characterization of the distribution of leptin-like proteins in the gastroenteric tract of two teleosts (*Dicentrarchus labrax* and *Carassius auratus* L.) with different feeding habits. Microsc. Res. Tech. 74: 714–719.

Sánchez-Amaya, M.I., J.B. Ortiz-Delgado, A. García-López, S. Cárdenas and C. Sarasquete. 2007. Larval ontogeny of redbanded seabream *Pagrus auriga* Valenciennes, 1843 with special reference to the digestive system. A histological and histochemical approach. Aquaculture 263: 259–279.

Sanden, M., M.H.G. Berntssen, Å. Krogdahl, G.-I. Hemre and A.M. Bakke-McKellep. 2005. An examination of the intestinal tract of Atlantic salmon, *Salmo salar* L., parr fed different varieties of soy and maize. J. Fish Diseases 28: 317–330.

Satora, L. 1998. Histological and ultrastructural study of the stomach of the air-breathing *Ancistrus multispinnis* (Siluriformes, Teleostei). Can. J. Zool. 76: 83–86.

Silva, S.S. and T.A. Anderson. 1995. Fish Nutrition in Aquaculture. Chapman and Hall, London, UK.

Sinha, G.M. 1975. On the origin, development and probable functions of taste buds on the lip and bucco-pharyngeal epithelia of an Indian freshwater major carp, *Cirrhinus mrigala* (Hamilton) in relation to food and feeding habits. Zeitsch. Mikroskop. Anatom. Forsch. 89: 294–304.

Sire, M.F. and J.M. Vernier. 1981. Étude ultrastructurale de la synthèse de chylomicrons au cours de l'absorption des lipids chez la Truite. Influence de la nature des acides gras ingérés. Biol. Cell 40: 47–62.

Sire, M.F. and J.M. Vernier. 1992. Intestinal absorption of protein in teleost fish. Comp. Bioch. Physiol. A 103: 771–781.

Sis, R.F., P.J Ives, D.M. Jones, D.H. Lewis and W.E. Haensly. 1979. The microscopic anatomy of the esophagus, stomach, and intestine of the channel catfish, *Ichtalurus punctatus*. J. Fish Biol. 14: 179–186.

Smith, L.S. 1989. Digestive functions in teleost fishes. In: Fish Nutrition. 2nd ed. London: Academic Press, 331–421.

Solcia, L.R., R. Capella, R. Vassallo and E. Buffa. 1975. Endocrine cells of the gastric mucosa. Int. Rev. Cytol. 42: 223–286.

Storch, V.N. and U.N. Welsch. 1970. Electron microscopic observations on the taste buds of some bony fishes. Arch. Histolog. Japan 32: 145–153.

Sweet, L.I., D.R. Passinoreader, P.G. Meier and G.M. Omnan. 1999. Xenobiotic-induced apoptosis: significance and potential application as a general biomarker of response. Biomarkers 4: 237–253.

Sysa, P., T. Ostaszewska and M. Olejniczak. 2006. Development of digestive system and swim bladder of larval nase (*Chondrostoma nasus* L.). Aquacult. Nutrit. 12: 331–339.

Tagliafierro, G. and G. Zaccone. 2001. Morphology and immunochemistry of taste buds in bony fish. pp. 335–345. *In*: B.G. Kapoor and T.J. Hara (eds.). Sensory Biology of Jawed Fishes. IBH Publishing Company, New Delhi, Calcutta.

Tengjaroenkul, B., B.J. Smith, T. Caceci and S.A. Smith. 2000. Distribution of intestinal enzyme activities along the intestinal tract of cultured Nile tilapia, *Oreochromis niloticus* L. Aquaculture 182: 317–327.

Uran, P.A., A.A. Gonçalves, J.J. Taverne-Thiele, J.W. Schrama, J.A. Verreth and J.H. Rombout. 2008. Soybean meal induces intestinal inflammation in common carp (*Cyprinus carpio* L.). Fish Shellfish Immunol. 25: 751–760.

Volkoff, H. 2006. The role of neuropeptide Y, orexins, cocaine and amphetamine-related transcript, cholecystokinin, amylin and leptin in the regulation of feeding in fish. Comp. Bioch. Physiol. A 144: 325–331.

Volkoff, H., M. Xu, E. MacDonald and L. Hoskins. 2009. Aspects of the hormonal regulation of appetite in fish with emphasis on goldfish, Atlantic cod and winter flounder: Notes on actions and responses to nutritional, environmental and reproductive changes. Comp. Biochem. Physiol. Part A 153: 8–12.

Watanabe, Y. 1985. Histological changes in the liver and intestine of freshwater goby larvae during short-term starvation. Biull. Japan. Soc. Scient. Fish 51: 707–709.

Wegner, A., T. Ostaszewska, and W. Rożek. 2009. The ontogenetic development of the digestive tract and accessory glands of sterlet (*Acipenser ruthenus* L.) larvae during endogenous feeding. Rev. Fish Biol. Fish 19: 431–444.

Whitear, M. 1971. Cell specialization and sensory function in fish epidermis. J. Zool. London 63: 237–264.

Wilson, J.M. and L.F.C. Castro. 2011. Morphological diversity of the gastrointestinal tract in fishes. pp. 1–44. *In*: M. Grosell, A.P. Farrell and C.J. Brauner (eds.). The Multifinctional Gut of Fish. Elsevier, London.

Winkler, H. and R. Fisher-Colbrie. 1992. The chromogranins A and B: the first 25 years and future perspectives. Neuroscience 49: 497–528.

Wu, W., Z. Xi-xun, M. Xu-zhou and L. Wei-chun. 2007. Effect of feeding frequency on the growth and protease activities of *Pelteobagrus vachelli*. J. Shanghai Fish Univ. 16: 224–229.

Xiong, D.M., C.X. Xie, H.J. Zhang and H.P. Liu. 2009. Digestive enzymes along digestive tract of a carnivorous fish *Glyptosternum maculatum* (*Sisoridae, Siluriformes*). J. Anim. Physiol. Anim. Nutr. 95: 56–64.

Yang, R., C. Xie, Q. Fan, C. Gao and L. Fang. 2010. Ontogeny of the digestive tract in yellow catfish *Pelteobagrus fulvidraco* larvae. Aquaculture 302: 112–123.

Yúfera, M. and M.J. Darias. 2007. The onset of exogenous feeding in marine fish larvae. Aquaculture 268: 53–63.

Yúfera, M., F.J. Moyano, A. Astola, P. Pousao-Ferreira and G. Martínez-Rodríguez. 2012. Acidic digestion in a teleost: Postprandial and circadian pattern of gastric pH, pepsin activity, and pepsinogen and proton pump mRNAs expression. PLoS ONE 7: e33687.

Zambonino Infante, J.L. and C.L. Cahu. 1994. Development and response to a diet change of some digestive enzymes in sea bass (*Dicentrarchus labrax*) larvae. Fish Physiol. Biochem. 12: 399–408.

Zambonino Infante, J.L., C.L. Cahu and A. Péres. 1997. Partial substitution of di- and tripeptides for native proteins in sea bass diet improves *Dicentrarchus labrax* larval development. J. Nutr. 127: 608–614.

Zambonino Infante, J.L. and C.L. Cahu. 2007. Dietary modulation of some digestive enzymes and metabolic processes in developing marine fish: Applications to diet formulation. Aquaculture 268: 98–105.

Chapter 7

Glands of the Digestive Tract

Bogdana Wilczyńska[1,*] and *Katarzyna Wołczuk*[2]

LIVER

INTRODUCTION

The liver performs both exocrine and endocrine functions (Ross et al. 1989, Chakrabati and Gosh 2015). The exocrine function is related to the production (secretion) of the bile which is passed through a duct system into the duodenum. Many liver products are directly disposed into the blood stream which is directly linked with the endocrine function (Ross et al. 1989, Sternberg 1997, Van Dyk 2003). An important function of the liver is the synthesis and secretion of lipoproteins and proteins such as albumins, fibrinogen and yolk vitellogenin (Hamlett 1988, Bruslé and Anandon 1996, Bertolucci et al. 2008, Genten et al. 2009). The liver stores lipids, carbohydrates, vitamin A and iron (Bruslé and Anandon 1996). Many digestion products are neutralised in the liver and hormones and biogenic amines are deactivated here. These processes constitute the detoxification function of the organ (Channa and Mir 2009, Tao and Peng 2009, Mir et al. 2011). The liver is also involved in immune responses of the organism due to its ability to identify and neutralise micro-organisms. All biochemical functions of the liver occur in the parenchymal cells of the liver, i.e., hepatocytes, and depend on the complex relations between the vascularisation system (branches of the liver artery and the common portal vein, sinusoids and central veins), hepatocytes and the bile duct system (bile ductules and intrahepatic bile ducts). The structure of the liver reflects its function. The liver is composed of parenchyma, which is made of epithelial cells performing the main functions of the organ and the stroma of the liver, consisting of blood vessels and connective tissue (Eurell and Haensly 1982, Ross et al. 1989, Biagianti-Risbourg 1991, Sternberg 1997, Diaz et al. 1999, Van Dyk 2003, Atamanalp et al. 2008, Bertolucci et al. 2008, Genten et al. 2009, Mir et al. 2011).

Histology and Morphology of the Fish Liver

Fish liver is a dense organ vantrally located in the cranial region of the visceral cavity. Its shape, size and volume are adapted to the space available among other internal organs (Bruslé and Anandon 1996, Datta

[1] Faculty of Biology and Environment Protection, Nicolaus Copernicus University, Toruń, Poland.
[2] Department of Vertebrate Zoology, Faculty of Biology and Environment Protection, Nicolaus Copernicus University, Toruń, Poland, Lwowska 1, 87-100 Toruń.
Email: k.wolczuk@gmail.com

Munshi and Dutta 1996, Sternberg 1997, Vicentini et al. 2005). The liver is divided into two (Chondrichthyes and Dipnoi) or three lobes (many teleost species); however, no lobulation was recognized in some teleost fish species, e.g., in *Oncorhyncus mykiss*, *Serranus cabrilla* (Gingerich 1982, Ferguson 1989, Gonzalez et al. 1993, Vicentini et al. 2005, Bertolucci et al. 2008). The surface of the liver is covered by a serosa and a thin connective tissue capsule (Hamlett 1988, Seyrafi et al. 2009). Streaks of connective tissue penetrate the parenchyma of the liver; however, they do not divide it into clear lobuli (Hamlett 1988, Rocha et al. 1994, Alboghobeish and Khaksar Mahabady 2005, Vicentini et al. 2005, Seyrafi et al. 2009, Sayrafi et al. 2011, Ikpekbu et al. 2012, Miura et al. 2012/2013). The lack of hexagonal division of the liver parenchyma makes it difficult to identify the liver acini. The porta hepatis is a place of connection between the hepar artery and the portal vein; it is also a place from which the bile duct extends (Eurell and Haensly 1982, Ross et al. 1989, Hinton and Laurén 1990, Munshi and Dutta 1996, Petkoff et al. 2006). Inside the liver, the blood vessels ramify into smaller ones. The branches of the portal vein and artery transport the blood to sinusoids where the substances are exchanged between blood and hepatocytes. From the sinusoids, the blood drains into the central vein finally leading to the liver vein (Rocha et al. 1994, Ross et al. 2003, Petkoff et al. 2006). The central veins in the fish liver parenchyma are scattered without forming any specific system and are surrounded by the liver parenchyma or pancreatic cells; they sometimes appear in the vicinity of the artery or bile duct (Eurell and Haensl 1982, Rocha et al. 1994, Hinton and Laurén 1990, Datta Munshi and Dutta 1996, Petkoff et al. 2006, El-Bakary and El-Gammal 2010). The fish liver is composed of two tissue compartments: parenchyma (the epithelial cells that perform the organs' major functions) and stroma (blood vessels and connective tissue). The parenchyma includes various cells situated within the liver as well as the respective extra-cellular spaces. Figure 1 provides the structural stratification of the fish liver (Hinton and Laurén 1990, van Dyk 2003).

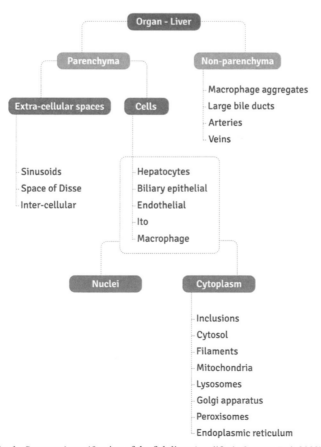

Fig. 1. Structural stratification of the fish liver (modified after van Dyk 2003).

Hepatocytes

Hepatocytes constitute ca. 80–85% of the liver cells (van Dyk 2003, Kapoor and Khanna 2004, Nejedli and Gajger 2013). Hepatocytes are arranged as tubules or cords (Hampton et al. 1989, Rocha et al. 1994, Rosety et al. 2001, Petkoff et al. 2006, El-Bakary 2010). In this arrangement, hepatocytes have their bases directed towards sinusoids, mainly for absorption, and their tapered apices form the wall of the initial portion of the biliary system, the bile canaliculi, mainly for excretion (Hinton et al. 1984, Hampton et al. 1985, Schär et al. 1985, Hampton et al. 1988, Hinton and Laurén 1990, Robertson and Bradley 1992). Between the neighbouring sinusoids, the hepatocytes are arranged as plates, usually two cells thick, but branching and anastomoses of cords can result in four or more cell layers per plate (Geyer 1989, van Dyk 2003, El-Bakary and El-Gammal 2010). This cord-like structure is not always clearly visible (Kendall and Hawkins 1975, Hinton and Pool 1976, Gonzalez et al. 1993, Bruslé and Anadon 1996, van Dyk 2003, Vincentini et al. 2005). Hepatocytes are polygonal-shaped cells, appearing hexagonal with a clearly marked cell membrane easily visible with the light microscope (Geyer 1989, Petkoff et al. 2006, Genten et al. 2009, Monsefi et al. 2010, Sayrafi et al. 2011). Polynucleated cells do not occur in the fish liver. Each hepatocyte contains a large, centrally located nucleus with a clearly marked dark nucleolus. The nucleolus is homogeneous and has a high electron density. The chromatin is granular, with more condensed heterochromatin located at the periphery of the nucleus (Rocha et al. 1994, Rosety et al. 2001, van Dyk 2003, Vicentini et al. 2005, Petkoff et al. 2006, Bertolucci et al. 2008, El-Bakary and El-Gammal 2010, Mir et al. 2011, Sayrafi et al. 2011). Hepatocyte cytoplasm is subject to continuous changes associated with the activity of the stored substances—mainly glycogen and lipids (Hibiya 1982, Petcoff et al. 2006, Genten et al. 2009, Monsefi et al. 2010, Mir et al. 2011, Sayrafi et al. 2011). Fat deposition in fish is affected by many factors, including dietary lipid content, which has been positively related to the fat content in tissue (Bruslé and Anadon 1996, Kapoor and Khanna 2004, Grigorakis 2007). In females of fish, the amount of lipid present is related to their reproductive condition (Kapoor and Khanna 2004). Particularly rich in lipids is the liver of sharks, a little less lipid is present in the liver of skates. According to Hamlett (1988), the large accumulation of low-density lipids in the liver of elasmobranchs may have evolved as an analogue of the teleost swim bladder. The cytoplasm of hepatocytes contains a considerable amount of rough endoplasmic reticulum (rER) situated parallel to the nuclear membrane and free ribosomes (Ross et al. 1989, Rocha et al. 1994, van Dyk 2003, Vicentini et al. 2005, Mir et al. 2011). In the hepatocytes of the majority of bony fish, the smooth endoplasmic reticulum (sER) is not present. In some Salmonidae and Gadidae, there are areas of smooth endoplasmic reticulum and in some places they are continuous with the rough endoplasmic reticulum rER (Ross et al. 1989, Rocha et al. 1994, van Dyk 2003). The canaliculi and the vesicles of the rER are connected to the Golgi apparatus. The Golgi apparatus is usually situated in the neighbourhood of the bile canaliculi. It takes an active part in the secretion of bile and its transport to the bile canaliculi. In the cytoplasm, in the neighbourhood of the bile canaliculi, there are abundant lysosomes responsible for the breakdown of cellular organella (van Dyk 2003). Mitochondria have various shapes from rounded to elongated and are randomly scattered in the cytoplasm (Rocha et al. 1994, Vicentini et al. 2005, Bertolucci et al. 2008, El-Bakary and El-Gammal 2010, Mir et al. 2011). Other cytoplasmic structures include microfilaments, which make up the cytoskeleton ensuring rigidity and flexibility of the cells, and cytostol in which the organella are suspended (Leeson and Leeson 1976, Tanuma et al. 1982, Solomon et al. 1993). Hepatocytes have three surfaces (sinusoidal surface, canalicular surface and basolateral surface), which are of great importance as they participate in the exchange of substances between hepatocytes, blood vessels and bile ductules (Sayrafi et al. 2011). A considerable area of hepatocytes is oriented towards sinusoids and takes part in the exchange of substances with the blood. Another area on the surface forms bile canaliculi—a place where the bile secreted by the hepatocyte is emptied into the bile ducts. The basal and lateral surfaces occur between the adjacent hepatocytes in the places where they are not in touch with the sinusoids or bile canaliculi (Sayrafi et al. 2011).

Melano-Macrophage Cells

The fish parenchyma contains specific macrophages (melano-macrophage (MM)), the size, number and content of which varies depending on the species, age and condition (Rocha et al. 1994, van Dyk 2003,

Vicentini et al. 2005, Bertolucci et al. 2008, El-Bakary and El-Gammal 2010, Mir et al. 2011). They are usually situated in the vicinity of the hepar artery, portal vein and bile ducts and make up the aggregates of melano-macrophages (MMC). They collect heterogeneous materials such as lipofuscin, melanin, or hemosyderyna. The numbers and morphological characteristics of MMC can be influenced by environmental toxins and by a range of pathological conditions (Borucinska et al. 2009).

Ito Cells

Other cell types include Ito cells (lipocytes, perisinusoidal cells, fat-storing cells). The Ito cells are of various shapes and are rich in free ribosomes (Nopanitaya et al. 1979, Fujita et al. 1980, 1986, van Dyk 2003). Their cellular organella are poorly developed. They are situated around sinusoids (perisinusoidal cells) in the perisinusoidal space (space of Disse). They are with long cytoplasmic processes oriented towards the interparenchymatous and perisinusoidal space. The Ito cells are closely linked through long (0.2–1.5 um in length) desmosomes. The Ito cell's cytoplasm contains a number of filaments and lipid droplets (0.5–1.0 µm in diameter), where vitamin A is stored (Nopanitaya et al. 1979, Sakano and Fujita 1982, Fujita et al. 1986, Miura et al. 2012/2013). The Ito cells are not present in elasmobranchs (Hamlett 1988, Kapoor and Khanna 2004).

Space of Disse and Sinusoids

Free surfaces of hepatocytes are oriented towards sinusoids and have multiple microvilli, the majority of which are immersed in the perivascular liquid, with only few touching the wall of the sinusoid. These surfaces contain multiple membrane receptors where the hepatocyte-blood and blood-hepatocyte exchange takes place. Free perisinusoidal spaces (space of Disse), which are very well developed and are the place of origin of the lymphatic vessels, separate the hepatocyte capillaries and sinusoidal endothelium (Rocha et al. 1994, Datta Munshi and Dutta 1996, Sternberg 1997). The Disse space, which is an area where rapid intercellular exchange occurs, contains little connective tissue and perisinusoidal cells such as Ito cells (*Salmo trutta fario*; Nopanitaya et al. 1979, Fujita et al. 1980, Sakano and Fujita 1982, Fujita et al. 1986, Rocha et al. 1994, Sternberg 1997).

Sinusoids are irregularly dilated capillary vessels, with the diameter exceeding the size of the regular capillary diameter (El-Bakary and El-Gammal 2010). The wall of the sinusoid is comprised of a single layer of flat cells called endothelium (Ross et al. 1989, Petkoff et al. 2006, Monsefi et al. 2010, Sayrafi et al. 2011). The endothelial cells are poor in organella, but rich in microfilaments. They have multiple pores (fenestrae) and do not contain a basal lamina (Bruslé et al. 1996, Genten et al. 2009). The endothelial cells rest only on small microvilli of hepatocytes and on a delicate network of reticular fibres. The presence of the pores and the lack of a basal lamina make it easier to penetrate the blood-hepatocyte barrier (Ferri and Sesso 1981, Rocha et al. 1994).

In sinusoids of some teleost species, such as *Salmo gairdneri*, *Ictalurus punctatus*, *Careius bicoloratus*, *Carrasitus auratus*, *Oreochromis niloticus*, *Pimelodus maculatus*, *Pangasius hypothalamus*, Kupffer cells are present (Feri and Sesso 1981, Tanuma et al. 1982, Hampton et al. 1987, 1988, Wolf and Wolfe 2005, Seyrafi et al. 2009, Abdel-Warith et al. 2011). Kupffer cells are phagocytic cells (macrophages) originating from the monocytes circulating in the blood. They contain more cytoplasm and are larger than the perikaryon of the endothelial cells (Ferri and Sesso 1981). Kupffer cells have not been identified in cartilaginous fish (Hamlett 1988, Kapoor and Khanna 2004).

Bile Ducts

The biliary tree is formed by canaliculi, preductules, ductules and ducts (Hampton et al. 1988, Rocha et al. 1994, van Dyk 2003, Monsefi et al. 2010). The canaliculi are small tubular channels on the apical surfaces of the adjacent hepatocytes (Bertolucci et al. 2008). They extend along the cell membrane of hepatocytes inside hepatocyte cords. Research conducted using the electron microscope showed that bile canaliculi stem from the extended sections of pericellular spaces, and their surfaces contain microvilli penetrating into the lumen of the canaliculi (Rocha et al. 1994; Fig. 2). The bile canaliculi are not visible in

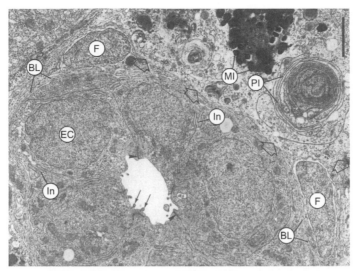

Fig. 2. TEM micrograph of a small bile duct. The epithelial biliary cells (EC) assume a somewhat cuboidal shape and are linked via apical junctions (arrowheads) and interdigitations (In). Short microvilli project from the surface of the EC (arrows). A basal lamina (BL) lies at the base of the EC. Flattened fibrocyte-like cells (F) entrap the channel contacting directly the BL. A melanomacrophage (outlined by block arrows) lies adjacent to the channel; its cytoplasm contains melanin-like granules (MI) and a putative phagolysosome (Pl). Scale bar: 2 μm (adapted from Rocha et al. 1994).

the preparations stained using the standard methods. They can be observed after impregnation with silver nitrate (Bertolucci et al. 2008). In some cyprinids, intracellular canaliculi have been observed (Gingerich 1982, Ferguson 1986).

In addition to hepatocytes, preductules contain epithelial cells. These cells do not contain the basal lamina and are connected to the hepatocytes through intercellular junctions and desmosomes. The apical parts of the epithelial cells contain a delicate fibrillar network.

In the ductules, the epithelium is made up of 2 to 4 layers of cells interconnected through thin tight junctions, desmosomes and interdigitations and, additionally, using desmosomes with the hepatocyte basis (Rocha et al. 1994).

The common bile ducts are situated in the vicinity of the hepatic artery and portal vein and are lined with single-layered cylindrical epithelium. The epithelial cells have a spherical nucleus and a PAS-positive brush border. In *Ictalurus punctatus*, some of the larger ducts contain a few goblet cells scattered among the columnar cells (Morgans 1972). In some fish (*Salmo trutta fario, Seranus cabrilla*), rodlet cells are occasionally found between epithelial cells (Gonzalez et al. 1993, Rocha et al. 1994). Underneath the epithelium, there is a fibrous connective tissue containing multiple elastic fibres, fibroblasts and lymphocytes and a layer of smooth myocytes in a circular arrangement (Petkoff et al. 2006, Genten et al. 2009, Monsefi et al. 2010).

PANCREAS

INTRODUCTION

With regard to structure and function, fish pancreas may be divided into two parts: exocrine and endocrine parts. The exocrine part produces digestive enzymes (including elastase, amylase, trypsin, chymotrypsin, phospholipase A, carboxypeptidase A, bile salt-dependent lipase) that pass through the pancreatic duct into the anterior intestine. These enzymes break down proteins, carbohydrates, fats and nucleotides (Field et al. 2003, Genten et al. 2009, Seyrafi et al. 2009, Miura et al. 2012/2013). In the stomachless holocephalan fish *Chimera monstrosa*, the chitinase is also produced in the pancreas (Hamlett 1988). The endocrine part of the pancreas secretes peptide hormones such as insulin, glucagon, somatostatin and pancreatic polypeptide, which influence the level of lipolysis, glicogenolysis and gluconeogenesis (Al-Mahrouki and Youson 1999, Krogdahl and Sundby 1999, Hoehne-Reitan and Kjorsvik 2004).

Histology and Morphology of the Fish Pancreas

In cartilaginous fish, the pancreas is a large discrete organ with both exocrine and endocrine function (Hamlett 1988, Kapoor and Khanna 2004). It lies along the intestine near the spleen (Kapoor and Khanna 2004). The pancreas of the teleosts may have a compact (e.g., *Esox lucius*, *Silurus glanis*, *Anguilla anguilla*, *Pleuronectes americanus*), or dispersed (e.g., Belone, Perca, Cyprinus, Gadus) structure (Hoar et al. 1979, Murray et al. 2004, Kozaric et al. 2008). In the latter case, the pancreatic tissue is usually scattered along the blood vessels and bile ducts of the liver forming an organ called hepatopancreas (*Astyanax altiparanae*, *Cetonopharingodon idella*, *Cynoscion gatucupa*, *Cyprinus carpio*, *Hydrocinus forskahlii*, *Ictalurus punctatus*, *Micropogon undulates*, *Oreochromis niloticus*, *Oligosarcus jenynsii*, *Pangasius sanitwongsei*, *Schizotorax curvifrons*, *Sparus aurata*; Hinton and Pool 1976, Eurell and Haensly 1982, Geyer et al. 1996, Diaz et al. 1999, Alboghobeish and Khaksar Mahabady 2005, Vicentini et al. 2005, Petcoff et al. 2006, Bertolucci et al. 2008, El-Bakary and El-Gammal 2010, Mir et al. 2011, Sayrafi et al. 2011; Fig. 3). It may also occur in the adipose tissue and mesentery connecting the small intestine, stomach, liver and gallbladder (*Gadus morhua*, *Mystus gulio*, *Paralychthys olivaceus*, *Serranus cabrilla*, *Tinca tinca*; Hoar et al. 1979, Stipp et al. 1980, Beccaria et al. 1992, Gonzalez et al. 1993, Kurokawa and Suzuki 1995, Ostaszewska et al. 2006, Genten et al. 2009, Ghosh and Chakrabarti 2016). The pancreatic tissue is infrequently found in the spleen (*Barbus pectoralis*, *Labeo rohita*; Khaksary Mahabady et al. 2012, Chakrabarti and Ghosh 2015).

The Exocrine Part of the Pancreas

The exocrine part of the pancreas constitutes the main part of the organ. The parenchyma of the exocrine part of the pancreas displays an acinar and tubular structure lacking division into lobules typical of the mammalian pancreas (Petcoff et al. 2006). The acini perform secretory functions while the ducts remove the secretion from the pancreas (pancreatic ducts).

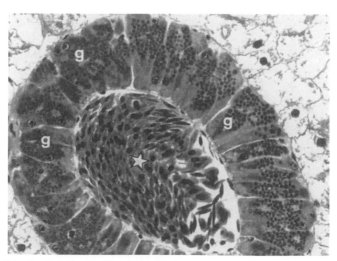

Fig. 3. Organization of the intrahepatic exocrine pancreatic tissue around a blood vessel (star). Note the distribution of zymogen granules (g) in the exocrine cells, H/E x 400 (adapted from Vicentini et al. 2005).

The Secretory Acini

The wall of the secretory acini is made of a single layer of pyramidal cells (acinar cells) whose broader side rests on the basal lamina. The acinar cells are filled with basophilic cytoplasm (except for the apical part) and display a round, euchromatic basal nucleus with a distinct nucleolus (Stipp et al. 1980, Beccaria et al. 1992, Roberts and Ellis 2001, Vicentini et al. 2005, Genten et al. 2009, Seyrafi et al. 2009, El-Bakary and El-Gammal 2010, Mir et al. 2011, Khaksary Mahabady et al. 2012, Miura et al. 2012/2013, Chakrabarti and Ghosh 2015). The cytoplasm contains abundant mitochondria, Golgi apparatus and an extensive rough endoplasmic reticulum (rER) taking the form of broad concentric cisternae (Stipp et al. 1980, Vicentini et al. 2005, Bertolucci et al. 2008, Gilloteaux et al. 2008). The apical part of the cell reveals multiple acidophilic zymogen granules containing proenzymes responsible for the digestion of proteins, carbohydrates, fats, and nucleotides (Petcoff et al. 2006, Bertolucci et al. 2008, Gilloteaux et al. 2008, Seyrafi et al. 2009, El-Bakary and El-Gammal 2010, Mir et al. 2011, Sayrafi et al. 2011, Khaksary Mahabady et al. 2012, Miura et al. 2012/2013, Chakrabarti and Ghosh 2015). The concentration of proenzymes in the zymogen granules varies with the dietary intake (Valaroutsou et al. 2013). The proenzymes are synthesised in the rough ER, and then disposed into the Golgi apparatus, where they are modified and enveloped in a membrane. The so formed secretory granules are collected in the apical part of the cell where the proenzymes are released into the lumen of the acinus. From here, the secretion is directed through the ducts into the anterior intestine, where it is activated in the alkaline environment.

The Excretory Ducts

The pancreatic secretion collected in the lumen of the acini is released into the intestine through the excretory duct system. The system starts in the acini, where the centroacinar cells making up the centroacinar intercalated ducts are present. The intercalated ducts are lined by a single-layered cuboidal epithelium, which in the further sections of the excretory ducts takes the form of single layered cylindrical epithelium (Genten et al. 2009, Sayrafi et al. 2011). This type of epithelial cells also lines the main pancreatic duct, which makes up the final section of the excretory ducts of the pancreas. The main pancreatic duct opens in the direct vicinity of or after connecting to the main bile duct and empties into the front part of the intestine.

The Endocrine Part of the Pancreas

The endocrine part of the pancreas is made up of the accumulations of endocrine cells forming the so-called pancreatic islets (Islets of Langerhans or Brockmann bodies, Lima et al. 1991). In teleost fish, the pancreatic islets are dispersed among various elements of the exocrine part or gathered into separate structures called Brockmann bodies that lie adjacent in the mesentery (Slack 1995, Yang and Wright 1995, Wright et al. 2015). In some cases (e.g., *Gillichthys mirabilis*; Kelly 1993), there is a single large Brockmann body (see also chapter Endocrine Organs), in others there may be several. The cartilaginous fish have islets scattered in the exocrine pancreas, but in holocephalans (e.g., ratfish) they contain only insulin, glucagon and somatostatin cells, while pancreatic polypeptide cells are found in the gut. In Selachii (sharks and rays), all four types are found in the islets (Slack 1995). In addition to the islets, the pancreatic apparatus of fish contains abundant lymphoid follicles usually situated in the vicinity of large blood vessels (Danilova and Steiner 2002).

References Cited

Abdel-Warith, A.A., E.M. Younis, N.A. Al-Asgah and O.M. Wahbi. 2011. Effect of zinc toxicity on liver histology of Nile tilapia, *Oreochromis niloticus*. Sci. Res. Essays 6: 3760–3769.

Alboghobeish, N. and M. Khaksar Mahabdy. 2005. Histological study of the liver and pancreas *Cetenopharyngodon idella*. Iran. J. Vet. Med. 11: 25–34.

Al-Mahrouki, A.A. and J.H. Youson.1999. Ultrastructure and immunocytochemistry of the islet organ of Osteoglossomorpha (Teleostei). Gen. Comp. Endocr. 116: 409–421.

Atamanalp, M., T. Sisman, F. Geyikoglu and A. Topal. 2008. The histopathological effect of copper sulphate on rainbow trout liver (*Oncorhyncus mykiss*). J. Fisheries and Aquatic Sci. 3: 291–297.

Beccaria, C., J.P. Diaz and R. Connes. 1992. Effects of dietary conditions on the exocrine pancreas of the sea bass, *Dicentrarchus labrax L.* (Teleostei). Aquacult. 101: 163–76.

Bertolucci, B., C.A. Vicentini, I.B. Franceschini Vicentini and M.T. Siqueira Bombonato. 2008. Light microscopy and ultrastructure of the liver of *Astyanax altiparanae* Garutti and Britski, 2000 (Teleostei, Characidae). Acta Sci. Biol. Sci. 1: 73–76.

Biagianti-Risbourg, S. 1991. Fine structure of hepatocytes in juvenile grey mullets: *Liza saliens* Risso, *L. ramada* Risso and *L. aurata* Risso (Teleostei, Mugilidae). J. Fish Biol. 39: 221–234.

Borucinska, J.D., K. Kotran, M. Shackett and T. Barker. 2009. Melanomacrophages in three species of free-ranging sharks from the northwestern Atlantic, the blue shark *Prionacae glauca* (L.), the shortfin mako, *Isurus oxyrhinchus Rafinesque*, and the thresher, *Alopias vulpinus* (Bonnaterre). J. Fish Dis. 32(10): 883–891.

Bruslé, J. and G.G. Anadon. 1996. The structure and function of fish liver. pp. 77–93. *In*: J.S.D. Munshi and H.M. Dutta (eds.). Fish Morphology. Science Publishers Inc., Enfield, USA.

Channa, A. and I.H. Mir. 2009. Distribution of neutral lipids in the intestinal tract and liver of *Schizothorax curvifrons* Heckel: A histological study. Indian J. Applied Pure Biol. 24: 285–288.

Chakrabarti, P. and S.K. Ghosh. 2015. Comparative histological and histochemical studies on the pancreas of *Labeo rohita* (Hamilton, 1922), *Mystus vittatus* (Bloch, 1790) and *Notopterus notopterus* (Pallas, 1796). Int. J. Aquat. Biol. 3(1): 28–34.

Danilova, N. and L.A. Steiner. 2002. B cells develop in the zebrafish pancreas. PNAS 99: 13711–13716.

Datta Munshi, J.S.D and H.M. Dutta. 1996. Fish morphology: horizon of the new research. Science Publishes, Inc. USA.

Diaz, A.O., C.M. Gonzalez, A.M. Garcia, C.V. Devincenti and A.L. Goldemberg. 1999. Morphological and histochemical characterization of liver from stripped weakfish, *Cynoscion guatucupa* (Cuvier, 1830). Biociencias 7: 67–78.

El-Bakary, E.R.N. and H.L. El-Gammal. 2010. Comparative histoloigical, histachemical and ultrastructural studies on the liver of flathead grey mullet (*Mugil cephalus*) and sea bram (*Sparus Aurata*). Global Vet. 4: 548–553.

Eurell, J.A. and W.E. Haensly. 1982. The histology and ultrastructure of the liver of Atlantic croacker *Micropogon undulatus* L. J. Fish Biol. 21: 113–125.

Ferguson, H.W. 1989. Systemic pathology of fish: a text and atlas of comparative tissue responses in diseases of teleosts. Iowa state University Press, Ames, IA.

Ferri, S. and A. Sesso. 1981. Ultrastructural study of Kupffer cells in teleost liver under normal and experimental conditions. Cell Tissue Res. 220: 387–391.

Field, H.A., P.D. Si Dong, D. Beis and D.Y.R. Stainier. 2003. Formation of the digestive system in zebra fish. ii pancreas morphogenesis. Dev. Biol. 261: 197–208.

Fujita, H., T. Tamaru and J. Miyagawa. 1980. Fine structural characteristics of the hepatic sinusoidal walls of the goldfish (*Carassius auratus*). Arch. Histol. Jpn. 43: 265–273.

Fujita, H., H. Tatsumi, T. Ban and S. Tamura. 1986. Fine structural characteristics of the liver of the cod (*Gadus morhua macrocephalus*), with special regard to the concept of a hepatoskeletal system formed by Ito cells. Cell Tissue Res. 244: 63–67.

Genten, F., E. Terwinghe and A. Danguy. 2009. Atlas of fish histology. Science Publishers, Enficld, NH, USA.

Geyer, H.J. 1989. Die morfologie, histologie en ultrastruktur van die pankreas, lewer, en galblaas van die algvoeder *Oreochromis mossambicus* (Reters). M. Sc. thesis, Rand Afrikaans Universitety, South Afica.

Geyer, H.J., M.N. Nel and J.H. Swanepoel. 1996. Histology and ultrastructure of the hepatopancreas of the tigerfish, *Hydrocynus forskahlii*. J. Morphol. 227: 93–100.

Gilloteaux, J., R. Kashouty and N. Yono. 2008. The perinuclear space of pancreatic acinar cells and synthetic pathway of zymogen in *Scorpaena scorofa* L.: ultrastructural aspects. Tiss. Cell 40: 7–20.

Gingerich, W.H. 1982. Hepatic toxicology of fishes. pp. 55–105. *In*: L.J. Weber (ed.). Aquatic Toxicology. Raven Press, New York, USA.

Gonzalez, G., S. Crespo and J. Bruslé. 1993. Histo-cytological study of the liver of the cabrilla seabass, *Serranus cabrilla* (Teleostei, Serranidae), an available model for marine fish experimental studies. J. Fish Biol. 43: 363–373.

Grigorakis, K. 2007. Compositional and organolptic quality of farmed and wild gilthead sea bream (*Sparus aur*ata) and sea bass (*Dicentrarchus labrax*) and factors affecting it. A review. Aquacult. 272(1-4): 55–75.

Hamlett, W.C. 1988. The elasmobranch liver. A model for chemical carcinogenesis studies in a 'naturally resistant' vertebrate. pp. 177–190. *In*: P.M. Motta (ed.). Biopathology of the liver. Springer Science + Business Media BV, Dordrecht, Netherlands.

Hampton, J.A., P.A. Mccuskey, R.S. Mccuskey and D.E. Hinton. 1985. Functional units in rainbow trout (*Salmo gairdneri*, Richardson) liver. I. Arrangement and histochenlical properties of hepatocytes. Anat. Rec. 213: 166–175.

Hampton, J.A., J.E. Klaunig and P.J. Goldblatt. 1987. Resident sinusoidal macrophages in the liver of the brown bullhead (*Ictalurus nebulosus*): an ultrastructural, functional and cytochemical study. Anat. Rec. 219: 338–346.

Hampton, J.A., R.C. Lantz, P.J. Goldblatt, D.J. Laurén and D.E. Hinton. 1988. Functional units in rainbow trout (*Salmo gairdneri*, Richardson) liver. II. The biliary system. Anat. Rec. 221: 619–634.

Hampton, J.A., R.C. Lantz and D.E. Hinton. 1989. Functional units in rainbow trout (*Salmo gairdneri*, Richardson) liver: III. Morphometric analysis of parenchyma, stroma, and component cell types. Am. J. Anat. 185: 58–73.

Hibiya, T. 1982. An atlas of fish histology: normal and pathological features. Gustav Fisher Verlag, Stuttgart, New York.

Hinton, D.E. and C.H. Pool. 1976. Ultrastructure of the liver in channel catfish *Ictalurus punctatus* (Rafinesque). J. Fish Biol. 8: 209–219.

Hinton, D.E., R.C. Lantz and J.A. Hampton. 1984. Effect of age and exposure to a carcinogen on the structure of the medaka liver: a morphometric study. Natl. Cancer Inst. Monogr. 65: 239–249.

Hinton, D.E. and D.J. Laurén. 1990. Integrative histophatological approaches to detecting effects of enviromental stressors on fishes. Am. Fish. Symp. 8: 51–66.

Hoar, W.S., J. David and J.R. Brett (eds.). 1979. Fish Physiology. Vol. 8 Bioenergetics and Growth. Academic Press, New York, San Francisco, London.

Hoehne-Reitan, K.A.T.J.A. and E. Kjorsvik. 2004. Functional development of the liver and exocrine pancreas in teleost fish. *In*: J.J. Govoni (ed.). The Proceedings of the 40 American Fisheries Society Symposium, The Morphological Development and Physiological Function in Fish, Bergen, Norway.

Khaksary Mahabady, M., H. Morovvati, A. Arefi and M. Karamifar. 2012. Anatomical and histomorphological study of spleen and pancreas in Berzem (*Barbus pectoralis*). World J. Fish Mar. Sci. 4: 263–267.

Kapoor, B.G. and B. Khanna. 2004. Ichtyology Handbook, Springer, Berlin, Germany.

Kendall, M.W. and W.E. Hawkins. 1975. Hepatic morphology and acid phosphatase localization in the channel catfish (*Ictalurus punctatus*). J. Fish. Res. Bd. Can. 32: 1459–64.

Krogdahl, A. and A. Sundby. 1999. Characteristics of pancreatic function in fish. pp. 437–458. *In*: S. Pierzynowski and R. Zabielski (eds.). Biology of the Pancreas in Growing Animals. Elsevier Science, Amsterdam, Holand.

Kurokawa, T. and T. Suzuki. 1995. Structure of exocrine pancreas of flounder (*Paralichthys olivaceus*): immunological localization of zymogen granules in the digestive tract using antitrypsinogen antibody. J. Fish Physiol. 46: 292–301.

Lima, F.J.A., F.C.L.B. Lima, M.H. Kireger-Azzolini and A.C. Boschero. 1991. Topography of the pancreatic region of the Pacu, *Piaratus mesopotamicus*, Holmberg, 1887. Bol. Tec. CEPTA 4: 47–56.

Lesson, C.R. and T.S. Lesson. 1976. Histology. Third edition. W.B. Saunders Company, Philadelphia, USA.

Mir, H.I., A. Channa and S. Nabi. 2011. Ultrastructural analysis of the liver of the snow trout, *Schizothorax curvifrons* Heckel. Int. J. Zool. Res. 7: 100–106.

Miura, M., Y. Mezaki, M. Morii, T. Hebiguchi, H. Yoshino, K. Kawatsu, M. Fujiwara, K. Imai and H. Senoo. 2012/2013. Histology of the hepatopancreas of puffer fish (*Takifugu rubripes*) in relation to the localization of tetrodotoxin. Arch. Histol. Cytol. 74: 59–70.

Monsefi, M., Z. Gholami and H.R. Esmaeili. 2010. Histological and morphological studies of digestive tube and liver of the Persian tooth-carp, *Aphanius persicus* (Actinopterygii: Cyprinodontidae). IUFS J. Biol. 69: 57–64.

Morgans, L.F. 1972. Histological study of channel catfish *Ictalurus punctatus*. Proc. Arkansas Acad. Sci. 26: 67–69.

Nejedli, S. and I.T. Gajger. 2013. Hepatopancreas in some sea fish different species and the structure of the liver in teleost fish, common pandora, *Pagellus erythinus* (Linnaeus, 1978) and whiting, *Merlangius merlangus euxinus* (Nordmann, 1840). Vet. Archiv. 83(4): 441–452.

Nopanitaya, W., J. Aghajanian, W.J. Grisham and L.J. Carson. 1979. An ultrastructural study on a new type of hepatic perisinusoidal cell in fish. Cell Tissue Res. 198: 35–42.

Ostaszewska, T., M. Korwin-Kossakowski and J. Wolnicki. 2006. Morphological changes of digestive structures in starvedtench *Tinca tinca* (L.) juveniles. Aquaculture Int. 14: 113–126.

Petcoff, G.M., A.O. Diaz, A.H. Escalante and A.L. Goldemberg. 2006. Histology of the liver *oligosarcus jenynsii* (Ostariophsi, Characidae) from Los Pades Lake, Argentina. Ser. Zool. 96: 205–208.

Roberts, R.J. and A.E. Ellis. 2001. The anatomy and physiology of teleosts. *In*: R.J. Roberts (ed.). Fish Pathology. W B Saunders, Philadelphia, USA.

Robertson, J.C. and T.M. Bradley. 1992. Liver ultrastructure of juvenile Atlantic salmon (*Salmo salar*). J. Morphol. 211: 41–54.

Rocha, E., R.A.F. Monterio and C.A. Pereira. 1994. The liver of the brown trout, *Salmo trutta fario*: a light and electron microscope study. J. Anat. 185: 241–249.

Rosety, M., F.J. Ordoñez, A. Ribelles, M. Rosety-Rodriguez, A. Dominguez, C. Carrasco and J.M. Rosety. 2001. Morpho-histochemical changes in the liver and intestine of young giltheads (fish-nursery), *Sparus aurata*, L., induced by acute action of the anionic tensioactive alkylbenzene sulphonate. Eur. J. Hist. 45: 259–265.

Ross, M.H., E.J. Reith and L.J. Rombell. 1989. Histology: a text and atlas. Second edition. Williams and Wilkins, Baltimore, USA.

Ross, M., G. Kaye and W. Pawlina. 2003. Histology: a text and atlas with cell and molecular biology. Fourth edition. Lippincott Williams and Wilkins, Philadelphia, USA.

Sakano, E. and H. Fujita. 1982. Comparative aspects on the fine structure of the teleost liver. Okajimas Folia Anat. 58: 501–520.

Sayrafi, R., G. Najafi, H. Rahmati-Holasoo, A. Hooshyari, R. Akbari, S. Shokrpoor and M. Ghadam. 2011. Histological study of hepatopancreas in Hi Fin Pangasius (*Pangasius sanitwongsei*). Afr. J. Biotechnol. 17: 3463–3466.

Seyrafi, R., G. Najafi, H. Rahmati-Holasoo, B. Hajimohammadi and A.S. Shamsadin. 2009. Histological study of hepatopancreas in iridescent shark catfish (*Pangasius hypothalmus*). J. Anim. Vet. Adv. 8(7): 1305–1307.

Schär, M. and S.D. Malyip. 1985. Histochemical studies on metabolic zonation of the liver in the trout (*Salmo gairdneri*). Histochemistry 83: 147–151.

Slack, J.M.W. 1995. Developmental biology of the pancreas. Development 121: 1569–1580.

Stipp, A.C., S. Ferri and A. Sesso. 1980. Fine structural analysis of a teleost exocrine pancreas cellular components. A freeze fracture and transmission electron microscopic study. Anat. Anz. 147: 60–70.

Solomon, E.P., L.R. Berg, D.W. Martin and C. Ville. 1993. Biology. Third edition. Saunders College Publlishing, Philadelphia, USA.

Sternberg, S.S. 1997. Histology for pathologist. Second edition. Raven Press, New York, USA.

Tao, T. and J. Peng. 2009. Liver development in zebrafish (*Danio rerio*). J. Genet. Genomics 36: 325–34.

Tanuma, Y., M. Ohata and T. Ito. 1982. Electron microscopic study on the sinusoidal wall of the liver in the flatfish, Kareius bicoloratus: demonstration of numerous desmosomes along the sinusoidal wall. Arch. Histol. Jpn. 45: 453–457.

Valaroutsou, E., E. Voudanta, E. Mente and P. Berillis. 2013. A microscope and image analysis study of the liver and exocrine pancreas of sea bream *Sparus aurata* fed different diets. Int. J. Zool. Res. 3(1): 54–58.

Van Dyk, J. 2003. Histological changes in the liver of *Oreochromis mossambicus* (Cichlidae) afer exposure to cadmium and zinc. M. Sc. Thesis, Aquatic Health, Faculty of Science, Rank. Africans University, Johannesburg.

Vicentini, C.A., I.B. Franceschini-Vicentini, M.T.S. Bombonato, B. Bertolucci, S.G. Lima and A.S. Santos. 2005. Morphological study of the liver in the teleost *Oreochromis niloticus*. Int. J. Morphol. 23: 211–216.

Wolf, J.C. and M.J. Wolfe. 2005. A brief overview of nonneoplastic hepatic toxicity in fish. Toxicol. Pathol. 33: 75–85.

Wright, J.R., A. Bonen, J.M. Conlon and B. Pohajdak. 2015. Glucose homeostasis in the teleost fish Tilapia: insights from Brockmann body xenotransplantation studies 1. Integr. Comp. Biol. 40(2): 234–245.

Yang, H. and J.R. Wright. 1995. A method for mass harvesting islets (Brockmann bodies) from teleost fish. Cell transplant.

Chapter 8

Swim Bladder

Ostaszewska Teresa * and *Kamaszewski Maciej*

INTRODUCTION

The swim bladder originates from an outgrowth of the anterior part of the alimentary canal. A diverticulum is formed in the esophagus wall during embryonic development which grows backwards under the spine and kidneys, and forms a thin-walled bladder. The wall of the bladder consists of two layers: the tunica interna and tunica externa. The tunica interna, which lines the bladder lumen, is subdivided into two layers: a simple flat or columnar epithelium and an underlying layer of connective tissue which contains some smooth muscle fibers. In turn, the tunica externa is composed of connective tissue, contains collagenous and elastic fibers, and is supplied with numerous capillaries.

Unlike any other organ of the vertebrates, the swim bladder serves multiple functions (Allen et al. 1976). It is a hydrostatic organ, it enables respiration (*Umbra* sp., *Arapaima gigas*, *Gymnarchus niloticus*) and transport of sound waves to the labyrinth (*Cyprinidae*, *Siluridae*), and also is an element of sound-producing organs in fish from, e.g., *Pogonias* and *Triglidae* genera (Kilarski 2012).

Swim Bladder Development

Swim bladder filling up with gas at the stage of larval development is a very important moment, as it allows the fish to control their buoyancy. The mechanism of swim bladder filling up has not been explicitly recognized so far in the case of the *Osteichthyes*. It may proceed via gulping atmospheric air from above the water surface (Doroshev and Cornacchia 1979), swallowing air bubbles suspended underneath the water surface (McElman and Balon 1979), producing gas through gas glands (Schwarz 1971), and by gas transport across the respiratory and cardiovascular system (Doroshev and Cornacchia 1979). Most freshwater fish fill up their swim bladder by gulping air from above the water surface. The swallowed air bubbles are transported across the gut owing to peristaltic movements and then with the pneumatic duct to the swim bladder lumen (Rieger and Summerfelt 1998). Single large air bubbles are disintegrated into small bubbles (ca. 15 μm in diameter) presumably through the surface action of the gallbladder and peristalsis of the foregut. In the soft-finned fish, the pneumatic duct functions throughout their life (Fig. 1A) (Hunter and Sanchez 1976). These fish species are called physostomes (Physostomi) and include Acipenseridae, Cyprinidae, Salmonidae, and many other bony fishes.

Department of Ichthyobiology, Fisheries and Biotechnology in Aquaculture, Faculty of Animal Science, Warsaw University of Life Sciences (WULS-SGGW), Poland.

* Corresponding author: teresa_ostaszewska@sggw.pl

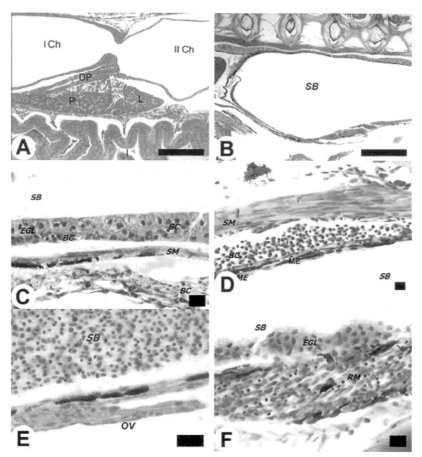

Fig. 1. Longitudinal sections. **A.** Swim bladder of the ide (*Leuciscus idus*) showing the two swim bladder chambers (ICH and IIC) and the pneumatic duct (DP). I, intestine; L, liver; P, pancreas. **B.** Swim bladder (SB) of pike perch (*Sander lucioperca*). **C.** Lower wall of the swim bladder (SB) of pike perch. Note the glandular epithelium (EGL), collagenous and elastic fibers (BC), and smooth muscle fibers (SM). **D.** Upper wall of the swim bladder (SB) of pike perch showing smooth muscle fibers (SM), the flat epithelium (ME), and blood vessels (BV). **E.** Diaphragm flat epithelium of the Oval (OV). **F.** Cells of the gas gland of pike perch. EGL, glandular epithelium; RM, rete mirabile; SB, swim bladder. AB/PAS staining. Scale bars: **A, B**—100 μm, **C–F**—10 μm.

Color version at the end of the book

In most of the hard-finned fish, the pneumatic duct functions only at the larval stage—these fish are physoclist species (Physoclisti) which include, among others, Perciformes, Gadiformes, and Gasterosteiformes (Ostaszewska 2003, Kilarski 2012). Larvae of some species of the Cichlidae (Physoclisti), like *Tilapia mossambica*, lack a pneumatic duct and fill up their swim bladders without accessing the water surface (Doroshev and Cornacchia 1979). The physostomes have no difficulties with swim bladder inflation (Tait 1960). The filling up of the swim bladder at early developmental stages of these fish is initiated by gulping atmospheric from above the water surface. A research conducted by Tait (1960) has demonstrated that after hatching the brown trout (*Salmo trutta lacustris* L.), rainbow trout (*Salmo gairdneri*), and sea trout (*Salmo trutta trutta* L.) remain at the bottom of an aquifer till the content of their yolk sac is substantially resorbed. Only then they become more motile and swim up to the water surface to fill up their swim bladder. A different behavior was observed in larvae of *Coregonus clupeaformis*, whose swim bladders were filled up with gas two months post hatching. Until then, the young fish were able to function with their non-filled swim bladder by using the air contained in their intestine (Tait 1960). The same author demonstrated that when deprived of the possibility of filling up their swim bladders for a longer period of time, the larvae of rainbow trout (*Salmo gairdneri*) and those of *Coregonus clupeaformis* fill them up as soon as the factors disturbing the filling process subside.

In physostomes, the orifice of the pneumatic duct is found at the terminal section of the esophagus near the site where the esophagus enters the intestine, and is closed with a special sphincter muscle. In salmonids, the swim bladder takes the form of a long sac connected with the esophagus with a relatively wide pneumatic duct. In cyprinids, a narrowing divides the swim bladder into two chambers, while the pneumatic duct runs along the intestine and ends with an orifice at the posterior chamber of the swim bladder (Fig. 1A). The posterior chamber differentiates first at the early stage of the postembryonic development when the larvae have still large reserves of the yolk sac, but begin to ingest feed. The anterior chamber of the swim bladder differentiates and fills up with air after yolk sac resorption when the larvae ingest exclusively exogenous feed (Ostaszewska et al. 1997, 2003). Both chambers serve different functions. The posterior chamber contracts under the influence of adrenaline, whereas the anterior one remains unchanged. The anterior chamber is connected with the labyrinth via Weber's organs. Changes in the tension of the swim bladder wall and wall vibrations induced by sound waves are transferred to the inner ear. The size of the posterior chamber changes depending on circumstances because it serves a hydrostatic function.

In the case of the physoclist species, it is hypothesized that the pneumatic duct functions only at the larval stage as a necessary element to initiate swim bladder filling and that the filling up is possible only in the first days of larval life. Successive regulation of gas volume in the bladder proceeds via metabolism of the gas gland and venous gases (Pelster 1995). The presence of the pneumatic duct in larvae of the physoclist species was confirmed in *Sander vitreum* (Marty et al. 1995) and *Morone saxatilis* (Walbaum) (Doroshev et al. 1981).

Swim Bladder Morphology

Observation of swim bladder anatomy in pike perch (*Sander lucioperca*) revealed the presence of a gas gland, a rete mirabile, and an oval. The swim bladder wall consists of two layers: tunica interna and tunica externa (Ostaszewska 2002). The tunica interna lines the swim bladder lumen and is composed of a simple flat epithelium in the upper wall and a glandular epithelium in the bottom wall. A layer of connective tissue containing smooth muscle fibers is located underneath the epithelium. The tunica externa is composed of connective tissue which contains collagenous and elastic fibers. The collagenous fibers are arranged in layers between fibroblasts and pigment cells. The tunica externa of the bottom abdominal wall of the swim bladder is lined with a glandular epithelium at two-thirds of its length. The glandular epithelium is multilayered and contains crypts and niches. Multiple blood vessels run between epithelial cells which are filled with large amounts of cytoplasm (Figs. 1B–E).

The *rete mirabile* ("wonderful net") is found in the cephalic-abdominal wall of the swim bladder and consists of a few layers: a serous membrane which constitutes the tunica externa, a fibrous tunica, a muscular membrane, and an epithelial tunica interna (Fig. 1F). The serous membrane is very thin and includes a layer of epithelium, connective tissue, blood vessels, and nervous fibers. In turn, the fibrous tunica is made up of dense collagenous fibers of the connective tissue and its internal wall is lined with a well-vascularized muscular coat. The muscular layer with smooth muscle bundles is separated by a layer of pigment cells from the submucosa, which in turn is composed of the connective tissue and is the major constituent of the wall of the rete mirabile. The complicated system of the rete mirabile is connected to the cuboid glandular epithelium of the tunica interna (Ostaszewska 2002).

In anatomical terms, the upper wall of the swim bladder differs from its bottom wall. The tunica interna of the upper wall is lined with a simple flat epithelium. The oval is located in the dorsal-posterior section of the upper wall of the swim bladder and is composed of a thin flat epithelium. The occluding layer of the oval is just opposite the heavily vascularized wall. The rete mirabile and the glandular epithelium are located in the cephalic-abdominal wall of the swim bladder, whereas the oval is located in the dorso-posterior section of the upper wall of the swim bladder.

In pike perch, as in other Perciformes, two organs regulate the volume of gases in the swim bladder, i.e., the gas gland which secretes gases into the bladder, and the oval which absorbs them. Both the rete mirabile and the gas gland play a significant role in gas secretion (Scheid et al. 1990, Pelster and Scheid

1992). No oval has been reported in some of the Physoclisti species, e.g., in *Morone saxatilis* (Groman 1982), in which gas pressure in the bladder is regulated via the gas gland and the rete mirabile.

References Cited

Allen, J.M., J.H.S. Blaxter and E.J. Denton. 1976. The functional anatomy and development of the swimbladder-inner ear-lateral line system in herring and sprat. J. Mar. Biol. Assoc. U.K. 56: 471–486.

Doroshev, S.I. and J.W. Cornacchia. 1979. Initial swim bladder inflation in larvae of *Tilapia mossambica* (Peters) and *Morone saxatilis* (Walbaum). Aquaculture 16: 57–66.

Doroschev, S.I., J.W. Cornacchia and K. Hogan. 1981. Initial swim bladder inflation in the larvae of physoclistous fishes and its importance for larval culture. Rapp. P.-v. Réun. Cons. Int. Explor. Mer. 178: 495–500.

Groman, D.B. 1982. Histology of the striped bass. American Fisheries Society.

Hunter, J.R. and S. Sanchez. 1976. Diel changes in swim bladder inflation of the larvae of the northern anchovy, *Engraulis mordax*. Fish Bull. 74: 847–855.

Kilarski, W. 2012. Anatomia ryb. Powszechne Wydawnictwo Rolnicze i Leśne, Poznań (in Polish).

Marty, G.D., D.E. Hinton and R.C. Summerfelt. 1995. Histopathology of swimbladder moninflation in walleye (*Sizostedion vitreum*) larvae: role of development and inflammation. Aquaculture 138: 35–48.

McElman, J.F. and E.K. Balon. 1979. Early ontogeny of walleye *Stizostedion vitreum*, with steps of salutatory development. Environ. Biol. Fish 4: 309–348.

Ostaszewska, T., R. Wojda and M. Mizieliński. 1997. Swim bladder development in vimba (*Vimba vimba* L.) larvae. Arch. Pol. Fish 5: 247–257.

Ostaszewska, T. 2002. The morphological and histological development of digestive tract and swim bladder in early organogenesis of pike-perch larval (*Stizostedion lucioperca* L.) in different rearing environments. SGGW, Warszawa [in Polish with English summary], pp. 1–96.

Ostaszewska, T. 2003. Histopathological changes during pikeperch *Sander lucioperca* swim bladder development. Proceedings of PERCIS III, The Third International Percid Fish Symposium, University of Wisconsin, Madison, Wisconsin, U.S.A., July 20–24, 2000; Barry, T.P. and J.A. Malison (eds.). pp. 111–112.

Ostaszewska, T., A. Wegner and M. Węgiel. 2003. Development of the digestive tract of ide, *Leuciscus idus* (L.) during the larval stage. Arch. Pol. Fish 11: 2.

Pelster, B. and P. Scheid. 1992. Countercurrent concentration and gas secretion in the fish swim bladder. Physiol. Zool. 65: 1–16.

Pelster, B. 1995. Metabolism of the swimbladder tissue. pp. 101–118. *In*: P.W. Hochachka and T.P. Mommsen (eds.). Biochemistry and Molecular of Biology of Fishes. 4. Elsevier, Amsterdam.

Reiger, P.W. and R.C. Summerfet. 1998. Microvideography of gas bladder inflation in larval walleye. J. Fish Biol. 53: 93–99.

Scheid, P., B. Pelster and H. Kobayashi. 1990. Gas exchange in the fish swim bladder. pp. 735–742. *In*: J. Piiper (ed.). Oxygen Transport to Tissue. Plenum Press, New York.

Schwarz, A. 1971. Swim bladder development and function in haddock, *Melanogrammus aeglefinus* L. Biol. Bull (Woods Hole) 141: 176–188.

Tait, J.S. 1960. The first filling of the swim bladder in salmonids. Can. J. Zool. 38: 179–187.

Chapter 9

Kidney

Frank Kirschbaum

INTRODUCTION

Teleost kidneys regulate water balance and ion concentration and are therefore part of the osmoregulatory (see also Chapter Gills: Respiration and Ionic-Osmoregulation) and excretory system; however, most of the protein metabolism is excreted by the gills. In addition, kidneys comprise hematopoietic (differentiation of red blood cells), phagocytotic, endocrine elements (see Chapter Endocrine Organs) and lymphoid tissue. The structures which have osmoregulatory and excretory function are the nephrons. Ontogenetically, they arise metamerically, i.e., one nephron per body segment. Later during ontogeny, additional nephrons develop; however, they do no longer originate metamerically. The nephrons lie outside the coelomic cavity and below the vertebral column (Harder 1975) (Figs. 1A–E). In most fishes, the rostral part of the kidney is the head kidney (Fig. 1A) (derived from the pronephros), comprising adrenocortical and chromaffin endocrine elements, hematopoetic and lymphoid tissues. The body kidney, derived from the mesonephros, is comprised of the nephrons (Figs. 1B, C) and, in addition, contains hematopoietic and lymphoid tissues. The form of the kidney varies in different species (Hibiya 1986); Ogawa (1962) distinguishes five configurational types of kidneys of marine teleosts. The kidneys of freshwater teleosts can be grouped into some of these categories (Ogawa 1961). The head kidney can be missing in some species. The kidney is supplied with blood by the caudal vein (Fig. 1E). However, most freshwater teleosts lack a true renal portal system (Moore 1933).

The nephrons, the functional elements of the kidney, consist of the renal corpuscle (Malphigian body) and a renal tubule (urinary tubule) (see below). The Malphigian body is composed of the glomerulus and a surrounding capsule (Bowman's capsule). The urinary tubules are connected to two collecting ducts (also called Wolffian ducts, see, e.g., Harder 1975) which join together (Fig. 1F) in the caudal part of the kidney and become the ureter (Fig. 1G). The ureter either opens directly to the outside by a urinary pore or ends into the bladder (Fig. 1H). The bladder is lined by a thin epithelium (Fig. 1I). The ureter near the urinary pore (Fig. 1J) is lined by a columnar epithelium (Figs. 1K–L).

Humboldt University, Faculty of Life Sciences, Unit of Biology and Ecology of Fishes, Philippstr. 13, Haus 16, D-10115 Berlin, Germany.
Email: frank.kirschbaum@staff.hu-berlin.de

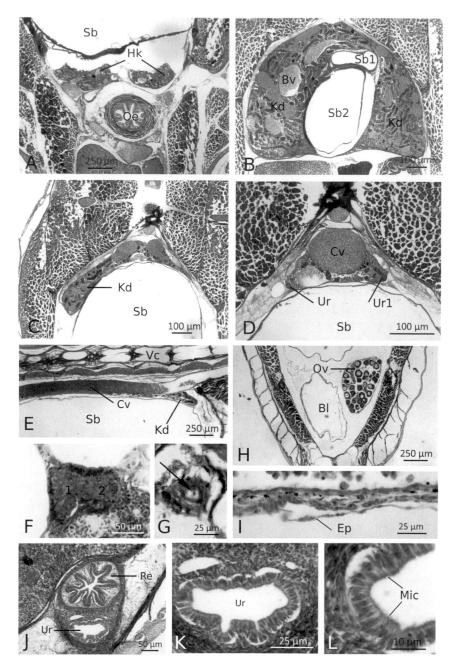

Fig. 1. Anatomy and histology of the kidney and the appropriate ducts in *Eigenmannia virescens* (Sternopygidae, Gymnotiformes) shown in cross sections and a sagittal section (E). Azan stain. **A.** Head kidney (Hk) below the swim bladder (Sb) and above the oesophagus (Oe). **B.** Body kidney (Kd) filling the large space between the anterior (Sb1) and posterior (Sb2) swim bladder. **C.** Posterior part of the body kidney (Kd) with just a few nephrons and tubules and two large veins. **D.** Proximal part of the body kidney showing the large caudal vein (Cv) and two ureters (Ur and Ur1) at the periphery of the kidney. **E.** Caudal vein (Cv) entering the posterior part of the body kidney (Kd). **F.** The two ureters (Wolffian ducts) (1, 2) joining together and joined together shown in **G, H.** The urinary bladder (Bl) in the rear part of the body cavity. **I.** The flattened epithelium (Ep) lining the bladder. **J.** Position of the ureter (Ur) below the rectum (Re) just before the opening to the outside. **K.** Anatomy of the ureter (Ur). **L.** The epithelium of the ureter is composed of columnar cells lined with microvilli (Mic). Bv, blood vessel; Ov, ovary; Sb, swim bladder.

Color version at the end of the book

The anatomy of the nephrons is different in stenohaline marine teleosts which are characterized by hypo-osmoregulation and freshwater teleosts which perform hyper-osmoregulation (Marshall and Grosell 2006). In marine teleosts, the nephron is comprised of the renal corpuscule and the renal tubule. The renal tubule contains an early and a late proximal tubule. Secretion of Na^+, Cl^-, Mg^{+2}, SO_4^{2-}, and water occurs in the early proximal tubule. Reabsorption of Na^+, Cl^-, and water occurs in the late proximal tubule, and also in the urinary bladder (Marshall and Grosell 2006) and leads to a low-average urine flow rate. The contribution of the glomerulus in the secretion is small (Beyenbach and Baustian 1989, Beyenbach and Liu 1996). Therefore, it is not surprising to find nephrons in marine teleosts lacking the glomerulus, so called aglomerulur tubules. Aglomerular nephrons occur in fish of 6 families, 13 genera and 23 species of marine teleosts (Hickman and Trump 1969). Examples of fishes with aglomerular tubules are *Hippocampus coronatus* (seahorse), *Syngnathus schlegeli* (pipefish), and *Antennarius tridens* (Frogfish) (Hibiya 1986). Although fishes with aglomerular kidneys are rare, the loss has apparently occurred three times during evolution (Beyenbach 2004). Interestingly, aglomerular nephrons are also found in fish occurring in brackish water and even freshwater (Collette 1995, 1996, Beyenbach 2004).

In freshwater teleosts, the nephrons consist of the Malphigian body and tubules comprised of a proximal tubule, a distal tubule and a collecting tubule. Urine flow and composition in freshwater teleosts differs considerably from that of marine fishes (Marshall and Grosell 2006). Urine flow rates in freshwater teleosts in general range between 2 and 10 ml $kg^1 h^1$ (Braun and Dantzler 1997, Curtis and Wood 1991, Grosell et al. 2004, 1998, Hickman and Trump 1969, McDonald and Wood 1998). These high flow rates indicate the need to eliminate excess water due to hyper-osmoregulation. In addition, high urine flow allows the fish to eliminate metabolic waste products such as creatin and creatinine (Hickman and Trump 1969). The dominant electrolytes in freshwater teleost urethral urine are Na^+ and Cl^- with a substantial reabsorption of these monovalent ions across the tubular epithelia (proximal, distal, and collecting tubule) (Marshall and Grosell 2006).

The Nephron of Freshwater Teleosts: Cytological Features

Here we restrict our histological description to the nephrons of various freshwater teleosts. The glomeruli of freshwater teleosts are supplied by branches of segmented arteries (Figs. 2A and B) arising from the dorsal aorta (Hickman and Trump 1969) and they form capillary loops (Fig. 2E). The glomerulus is covered by specific epithelial cells, the podocytes (Figs. 2C to E) which produce the filtrate which is directed towards the neck segment of the proximal tubule (Fig. 2D). In the wall of the afferent arteriols, juxtaglomerular cells (Fig. 2E) are found which are supposed to produce the hormone renin (Hibiya 1982). The glomerulus is surrounded by Bowman's capsule (e.g., Figs. 2C to F) composed of two layers of single flattened epithelia. Lymphocytes (Figs. 2E and G) and hematopoietic tissue (Fig. 2H) are also seen in the vicinity of the nephrons. In the knife fishes, *Gymnotus carapo* (Fig. 2F) and *Rhamphichthys* sp. (Fig. 2H), the Malphigian body and proximal (Pt) and distal tubules (Dt) are found grouped together. The different parts of the renal tubules are cytologically different. They all consist of a lumen surrounded by a singular epithelium of, in general, cuboidal/columnar cells. The proximal tubule is characterized by a wide lumen and the presence of microvilli constituting the brush border protruding into the lumen. The distal tubule in general possesses a small lumen lacking the brush border of the proximal tubule (Genthen et al. 2009). In some species, the distal tubule is quite large and the singular epithelium has a lightly stained plasma, as seen in the Zebrafish *Danio rerio* (Fig. 2G).

Acknowledgements

I thank Prof. Rüdiger Krahe for his offer to use his microscope-digital-camera device for taking the photomicrographs.

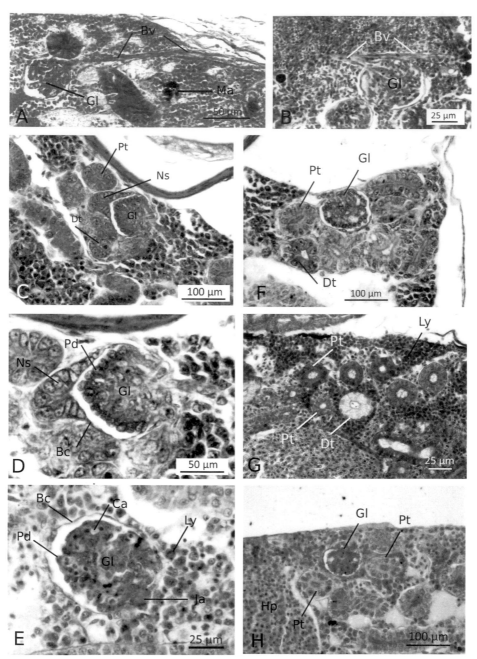

Fig. 2. Details of the nephrons of various freshwater teleosts. Azan stain. **A, B.** Malphigian bodies and the connecting blood vessel in the knife fish *Eigenmannia virescens*. An aggregatioin of macrophages (Ma) is also seen in **A. C.** Glomerulus (Gl) and the nearby neck segment (Ns) and proximal (Pt) and distal tubules (Dt) in the knife fish *Gymnotus carapo*. **D.** Detail of a glomerulus (Gl), the neck segment (Ns), podocytes (Pd), and Bowman's capsule (Bc) in the knife fish *Gymnotus carapo*. **E.** Detail of a glomerulus (Gl), podocytes (Pd), and Bowman's capsule (Bc) in the knife fish *Rhamphichthys* sp. A juxtaglomerular cell (Ja) and lymphocytes are also present. Capillaries (Ca), part of the capillary loops, are also seen in the glomerulus. **F.** An arrangement of a Malphigian body, proximal (Pt) and distal tubules (Dt) in *Gymnotus carapo*. **G.** Proximal (Pt) and distal tubules (Dt) in the Zebrafish *Danio rerio*. Note the distal tubule with small lumen and large cuboidal cells with light cytoplasm. Lymphocytes (Ly) are also seen. **H.** Arrangement of a glomerulus (Gl) with proximal (Pt) and distal tubules (Dt) in *Rhamphichthys* sp. Bl, blood vessel; Gl, glomerulus; Hp, hematopoietic tissue.

References Cited

Beyenbach, K.W. and M.D. Baustian. 1989. Comaparative physiology of the proximale tubule. pp. 104–142. *In*: R.K.H. Kinne, E. Kinne-Saffran and K.W. Beyenbach (eds.). Comparative Physiology, Structure and Function of the Kidney. Krager, Basel.

Beyenbach, K.W. and P.L. Liu. 1996. Mechanism of fluid secretion common to aglomerular and glomerular kidneys. Kidney Int. 49: 1543–1548.

Beyenbach, K.W. 2004. Kidneys sans glomeruli. Amer. J. Physiol. 286: F811–F827.

Braun, E.J. and W.H. Dantzler. 1997. Vertebrate venal system. pp. 481–576. *In*: W.H. Dantzler (ed.). Handbook of Physiology—Comparative Physiology I.

Collette, B.B. 1995. *Potamobatrachius trispinosus*, a new freshwater toadfish (Batrachoididae) from the Rio Tocantins, Brazil. Ichthyol. Explor. Freshwaters 6: 333–336.

Collette, B.B. 1996. A review on the venomous toadfishes, subfamily *Thalassophryninae*. Copeia 4: 846–864.

Curtis, B.J. and C.M. Wood. 1991. The function of the urinary-bladder *in vivo* in the freshwater rainbow-trout. J. Exp. Biol. 155: 567–583.

Genten, F., E. Terwinghe and A. Danguy. 2009. Atlas of Fish Histology. Science Publishers, Enfield, Jersey, Plymouth.

Grosell, M., C. Hogstrand and C.M. Wood. 1998. Renal Cu and Na excretion and hepatic Cu metabolism in both Cu acclimated and non acclimated rainbow trout (*Oncorhynchus mykiss*). Aqu. Tox. 40: 275–291.

Grosell, M., M.D. McDonald, C.M. Wood and P.J Walsh. 2004. Effects of prolonged copper exposure in the marine gulf toadfish (*Opsanus beta*). I. Hydromineral balance and plasma nitrogenous waste products. Aqu. Tox. 68: 249–262.

Harder, W. 1975. Anatomy of Fishes. E. Schweizerbartsche Verlagsbuchhandlung. Nägele und Obermiller, Stuttgart.

Hibiya, T. (ed.). 1982. An Atlas of Fish Histology. Normal and Pathological Features. Kodansha Ltd., Tokyo. Gustav Fischer Verlag, Stuttgart, New York.

Hickman, C.P. Jr and B.J. Trump. 1969. The kidney. pp. 91–239. *In*: W.S. Hoar and D.J. Randall (eds.). Fish Physiology Vol. I. Academic Press, New York, London.

Marshall, W.S. and M. Grosell. 2006. Ion transport, osmoregulation, and acid-base balance. pp. 177–230. *In*: D.H. Evans and J.B. Claiborne (eds.). The Physiology of Fishes. CRC Taylor & Francis, Boca Raton, London, New York.

McDonald, M.D. and C.M. Wood. 1998. Reabsorption of urea by the kidney of the freshwater rainbow trout. 18: 375–386.

Moore, R.A. 1933. Morphology of the kidneys of Ohio fishes. Contrib. Stone Lab. Ohio Univ. 5: 1–34.

Ogawa, M. 1961. Comparative study of the external shape of the teleostean kidney with relation to phylogeny. Sci. Rep. Tokyo Kyoiku Daigaku B 10: 61–68.

Ogawa, M. 1962. Comparative study on the internal structure of the teleostean kidney. Sci. Rep. Saitama Univ. B 4: 107–129.

Chapter **10**

Ovaries and Eggs

Mari Carmen Uribe,[1],* Harry J. Grier[2,3] and Arlette Amalia Hernández Franyutti[4]

INTRODUCTION

This chapter considers the structure of the female reproductive system of teleosts which differs in several features compared to all other vertebrates. Teleosts lack Müllerian ducts and consequently oviducts. The caudal region of the ovary is the gonoduct, which lacks germinal cells and opens at the genital pore. The cystovarian type of ovary due to the saccular structure and the germinal epithelium bordering its lumen leads to the release of ovulated eggs to the ovarian lumen rather than into the coelom (Turner 1947, Wourms 1981, Dodd and Sumpter 1984, Schindler and Hamlett 1993); both features characterize unique morphological and physiological aspects of the reproduction in this diverse group of vertebrates. This morphology excludes some "lower" fishes such as bowfin, trout and paddlefish that have the ovarian lumen open to the coelom, developing a gymnovarian type of ovary. Consequently, in these species, the ovarian lamellae project into the ovarian lumen connected to the coelom, and ovulation occurs into the coelom. Both types of teleost ovaries, cystovarian and gymnovarian, contrast the internal position of the germinal epithelium with the superficial position of the germinal epithelium of the ovary that borders the coelom of all other vertebrates and defines a unique teleost characteristic as ovulation into the ovarian lumen (Brummett et al. 1982, Grier 2012, Sales et al. 2013, Grier et al. 2009, 2016).

Oogenesis, the essential function of the ovary, is a morphological and functional process by which oogonia transform into fertilizable eggs. In addition to describing the current, basic knowledge on the morphology of the ovary and oogenesis of teleosts, this chapter applies a terminology for oocyte staging and follicular development proposed by Grier et al. (2009) as a guide for analysis of teleost oogenesis. The foundation of this staging is the cell division processes, mitosis and meiosis, from which oocyte stages and their subdivisions, called steps, were erected.

[1] Laboratorio de Biología de la Reproducción Animal, Departamento de Biología Comparada, Facultad de Ciencias, Universidad Nacional Autónoma de México, Ciudad Universitaria, Ciudad de México 04510, México.
[2] Division of Fishes, Department of Vertebrate Zoology, National Museum of Natural History, Smithsonian Institution, Washington, DC 20013-7012, USA.
[3] Florida Fish and Wildlife Research Institute, 100 8th Avenue, SE, St. Petersburg, FL 33701-5020, USA.
[4] Laboratorio de Acuicultura Tropical, División Académica de Ciencias Biológicas, Universidad Juárez Autónoma de Tabasco, Villahermosa, Tabasco, 86150, México.
  Emails: harry.grier1@verizon.net; arhefr@hotmail.com
* Corresponding author: mari3uribe3@gmail.com

The evolution of viviparity in teleosts has resulted in unique types of insemination and gestation that, because the lack of oviducts, occur inside the ovary (intraovarian gestation) differing from all other vertebrates. In viviparous teleosts, ovarian morphology is clearly modified because teleost viviparity, invariably, involves the ovary in a gestational role. Consequently, gestation implies specific morphological and physiological characteristics of the ovary that vary among viviparous fishes and change according to the stages of gestation. As a pre-requisite for viviparity, the ovary receives the spermatozoa and the fertilization is intraovarian (Turner 1947, Wourms et al. 1988, Wourms and Lombardi 1992, Burns and Weitzman 2005, Uribe et al. 2005). Some species store spermatozoa in fluids within the ovarian lumen, generating adequate conditions to maintain viable the spermatozoa for a period of time (Turner 1933, 1947, Wourms et al. 1988, Potter and Kramer 2000, Uribe and Grier 2011). Therefore, the ovary of viviparous teleosts differs from those of all other vertebrates because it is the site, not only for production of eggs, but also for internal sperm storage and fertilization that is followed by gestation (Amoroso 1960, Wourms 1981, Uribe et al. 2005, 2009). The intraovarian fertilization occurs mainly into the ovarian follicle, later the gestation may follow two different pathways: (a) intrafollicular gestation, in which the embryo remains into the follicle and move into the lumen just before birth, as in poeciliids; or, (b) intraluminal gestation, in which the embryo moves, during early development (cleavage or neurula), from the follicle into the ovarian lumen where it continues to develop, as in goodeids (Wourms et al. 1988, Uribe et al. 2009).

Therefore, the structural characteristics of the fish ovaries, mentioned here, are analyzed, described and illustrated in the different sections of this chapter.

Ovarian Structure

In most fishes, the female reproductive system consists of two ovaries. They are located in the abdominal cavity and suspended dorsally within the coelomic cavity by the mesovarium. The ovaries lie symmetrically on either side of the midline of the body, between the swim bladder and the intestine, in parallel position to the kidneys. Most teleost ovaries are saccular structures or cystovaries with a central lumen or ovocoel (Fig. 1A), elongated and irregular in shape. In some species, there is a partial fusion of the ovaries along the medial-caudal portion, as in the Dusky Grouper *Epinephelus marginatus* (Mandich et al. 2002); in other species, there is only one ovary as the result of the total fusion of paired ovaries during embryological development. The single ovary was described in species such as *Fundulus heteroclitus* (Brummett et al. 1982), *Oryzias latipes* (Strüssmann and Nakamura 2002), *Synbranchus marmoratus* (Ravaglia and Maggese 2002, Lo Nostro et al. 2003), and most viviparous teleosts as poeciliids and goodeids (Bailey 1933, Turner 1947, Mendoza 1965, Wourms 1981, Lombardi 1998, Uribe et al. 2009) and in Osteoglossomorpha (Yanwirsal et al. 2017).

The ovarian wall includes the mucosa, smooth muscle subjacent to the mucosa, and peritoneum located over the ovarian external surface (Figs. 1A, B, 2A, B). The mucosa is formed by (a) the germinal epithelium and (b) the stroma that is formed by loose and vascularized connective tissue. The germinal epithelium, lining the ovarian lumen, consists of somatic epithelial cells that are commonly cuboidal or columnar. The character that distinguishes the germinal epithelium from all other epithelia is the presence of oogonia that give rise to oocytes. A basement membrane supports the germinal epithelium and separates the epithelium from the stroma (Hoar 1969, Dodd 1977, Nagahama 1983, Begovac and Wallace 1987, Grier 2000, 2012, Grier et al. 2009, 2016). The stroma contains follicles in different stages of development, blood vessels, nerves, reticular fibers, and diverse types of somatic cells such as undifferentiated mesenchymal cells, fibroblasts, cells of the immune system and sometimes melanocytes. The smooth muscle is formed by layers in circular and longitudinal orientation around the periphery of the ovary (Fig. 1A). The peritoneum consists of squamous epithelium which is bordered by the coelomic cavity (Grier et al. 2009).

The mucosa of the ovarian wall forms numerous and irregular folds, the ovarian lamellae, which are extended into the lumen (Figs. 1A, B, 2A, B, 3A–C). The lamellae extend as large and abundant rows that are highly branched and sometimes anastomosed, as in the Common Snook, *Centropomus undecimalis* (Grier 2000), the Red Drum, *Sciaenops ocellatus* (Grier 2012), the Rainbow Trout, *Oncorhynchus mykiss* (Grier et al. 2007). The lamellae are composed also of two histological components (a) the germinal

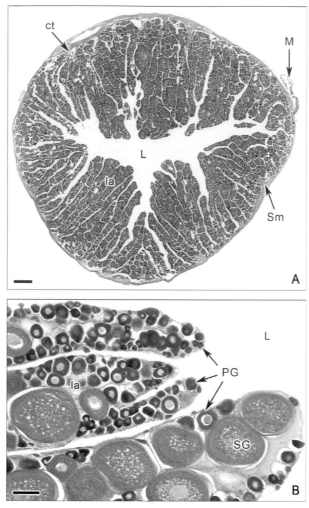

Fig. 1. Structure of the ovary. Ovary of Common Snook *Centropomus undecimalis*. **A.** The ovary is saccular, of cystovarian condition with a central lumen (L). A portion of the mesovarium (M) is seen at the dorsal side of the ovary. The ovarian wall forms numerous and irregular folds, the ovarian lamellae (la). Stroma of connective tissue (ct) and smooth muscle (Sm) are seen at the periphery. Bar = 0.3 mm. **B.** Detail of an ovarian lamellae (la) containing follicles in Primary Growth (PG) and Secondary Growth (SG) Stages. The central lumen (L) is seen. Scale bar = 100 μm.

Color version at the end of the book

epithelium, and (b) the stroma which contains follicles in various stages of development (Figs. 1B, 2B, 3C) (Brummett et al. 1982, Muñoz et al. 2001, Grier et al. 2009, 2016).

The structure of the ovary is deeply transformed in viviparous teleosts during intraovarian gestation (Figs. 4A, B, 5A), during intraluminal gestation (Fig. 4B), as in goodeids, or during intrafollicular gestation (Figs. 5B, C), as in poeciliids (Wourms 1981, Wourms et al. 1988, Uribe et al. 2009), and the hemirhamphid *Dermogenys pusillus* (Mendoza 1943, Amoroso 1960). These changes are progressively observed not only because of the gestation condition but also due to the stage of development of the embryos. Additionally, in some species the development of embryos of different ages occurs at the same time in the ovary because some embryos initiate the gestation and before the birth of these embryos, the other group of embryos initiates their development, being two groups in different embryonic stages. This complex process is called superfoetation. The poeciliid *Heterandria formosa* may contain upto 6 groups of embryos in different stage of development at the same time (Fig. 5C).

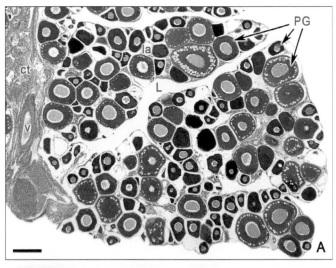

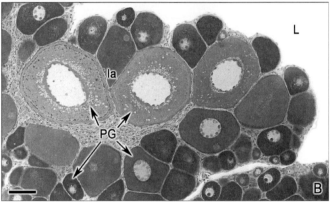

Fig. 2. Structure of the ovarian lamellae. Ovaries of Red Drum *Sciaenops ocellatus*. **A, B.** Ovaries containing lamellae (la) bordering the ovarian lumen (L). The lamellae contain oocytes in Primary Growth Stage (PG) with basophilic ooplasm and some with oil droplets. Stroma of connective tissue (ct) surrounds the follicles containing blood vessels (v). **A.** Scale bar = 100 μm. **B.** Scale bar = 50 μm.

Color version at the end of the book

Gonoduct

Female teleosts do not have oviducts because Müllerian ducts, from which oviducts develop during embryogenesis in other vertebrates, do not develop; this is an exclusive feature of teleosts. Consequently, communication of the ovary to the exterior occurs through the caudal portion of the ovary, the gonoduct, to the genital pore (Fig. 6). The gonoduct is a single structure, even in species that have double or partially fused ovaries, as in the Dusky Grouper *Epinephelus marginatus* (Mandich et al. 2002). The gonoduct is characterized by lack of germinal cells. This aspect differentiates the morphology and physiology of the gonoduct from the rest of the ovary where germinal cells are present and oogenesis occurs (Fig. 6) (Hoar 1969, Wourms et al. 1988, Lombardi 1998, Uribe et al. 2005, Campuzano-Caballero and Uribe 2014). The gonoduct is lined internally by a mucosa formed by epithelium and stroma, and surrounding the mucosa there are smooth muscle and peritoneum. The epithelium is cuboidal or columnar and has both ciliated and non-ciliated cells; the caudal region of the gonoduct has exocrine glands (Campuzano-Caballero and Uribe 2014). Brummett et al. (1982) suggested that secretory cells of this region in *Fundulus heteroclitus* may produce the jelly coat that covers the chorion of ovulated eggs. Gelatinous material surrounding

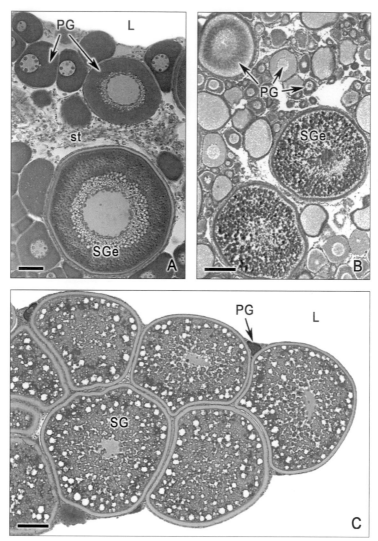

Fig. 3. Structure of the ovarian lamellae. The lumen (L) surrounds the lamellae. **A.** In the Red Drum *Sciaenops ocellatus*, the lamellae contain Primary Growth oocytes (PG) and one oocyte at early Secondary Growth (SGe). Stroma (st) of connective tissue surrounds the follicles. Scale bar = 50 μm. **B.** In the Goldfish *Carassius auratus*, the lamellae contain oocytes at Primary Growth Stage (PG) and two oocytes at early Secondary Growth Stage (SGl). Scale bar = 100 μm. **C.** In the Striped Mullet *Mugil cephalus*, a synchronous spawning fish, the lamellae contain abundant oocytes in the Full-Grown of Secondary Growth Stage (SG). At the periphery of the lamellae, some oocytes in Primary Growth Stage (PG) are observed. The lumen (L) surrounds the lamellae. Scale bar = 100 μm.

Color version at the end of the book

eggs at spawning were observed in several species as *Sebastolopus alascanus* (Erickson and Pikitch 1993), *Aspitrigla obscura* (Muñoz et al. 2001), *Genypterus blacodes* (Freijo et al. 2009), *Helicolenus d. dactylopterus* (Sequeira et al. 2011).

During embryogenesis of gymnovaries, such as in salmonids, as the Rainbow Trout, the gonoduct has been secondarily lost in whole or in part; it remains as a discontinuous caudal funnel. This funnel is opened, at the cephalic edge, to the coelomic cavity by the *ostium* and at the caudal edge, it communicates to the exterior via the genital pore (Kendall 1921, Lombardi 1998).

In species with internal fertilization, the gonoduct receives spermatozoa during insemination, where they may be stored for some time or move to the interior of the ovary where the fertilization occurs

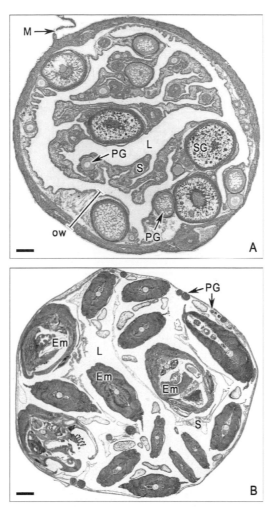

Fig. 4. Structure of the ovary in viviparous teleosts. A. Transverse sections of the ovary of the goodeid *Goodea atripinnis* during non-gestation. The ovarian wall (ow) forms several irregular folds which contain oocytes in Primary Growth Stage (PG) and Secondary Growth Stage (SG). The lumen (L) is divided by a highly folded septum (S). Scale bar = 50 μm. **B.** Transverse sections of the ovary of the goodeid *Xenotoca eiseni* during gestation. Several embryos (Em) are developing in the ovarian lumen (L). The septum (S) is observed. The folds of the ovarian wall contain oocytes in Primary Growth (PG) and Secondary Growth (SG) Stages. Scale bar = 200 μm.

(Campuzano-Caballero and Uribe 2014). Burns and Weitzman (2005) affirm that the introduction of sperm into the gonoduct and the fusion of male and female gametes within the female system are essential requirements for the evolution of vertebrate viviparity.

Development of the Ovary

During embryogenesis of most teleosts, the ovary develops in a unique process among vertebrates; it occurs by an invagination of the genital ridges from the coelomic epithelium. The genital ridges are thickenings of coelomic epithelium in both right and left mesonephric regions of the developing embryo. The genital ridges contain the oogonia, located among the coelomic epithelial cells. During their invagination, the genital ridges become enclosed by elongation of lateral outgrowths that extend ventrally, fusing with each other to form the ovarian lumen. This embryological process establishes the saccular ovarian structure that defines the cystovarian condition, common to the majority of teleosts (Turner 1947, Wourms 1981, Dodd

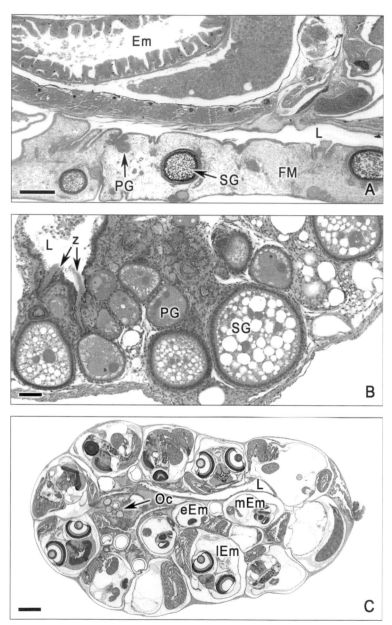

Fig. 5. Structure of the ovary in viviparous teleosts. A. Detail of the ovary of the goodeid *Ilyodon whitei* during gestation. The embryo (Em) is developing in the ovarian lumen (L), near the ovarian fold of maternal tissue (FM) where oocytes in Primary Growth Stage (PG) and Secondary Growth Stage (SG) Scale bar = 20 µm. **B.** Ovary of the poeciliid *Heterandria formosa* where spermatozoa (z) are located in deep folds of the mucosa. The ovarian wall contains oocytes in Primary Growth (PG) and Secondary Growth (SG) Stages. Lumen (L) Scale bar = 50 µm. **C.** Panoramic view of the single ovary of *Heterandria formosa* during gestation. The ovary contains oocytes (Oc) in different stages of development and embryos in different ages of gestation as early embryos (eEm), middle embryos (mEm) and late embryos (lEm) in evident condition of superfoetation. Lumen (L). Scale bar = 0.3 mm.

Color version at the end of the book

and Sumpter 1984, Schindler and Hamlett 1993). Consequently, as ovarian lamellae develop from the genital ridges, their surface epithelium becomes the germinal epithelium that borders the ovarian lumen. However, in fishes, such as the Holostei, lungfishes, paddlefishes, sturgeons, and bowfins (Kendall 1921, Robertson

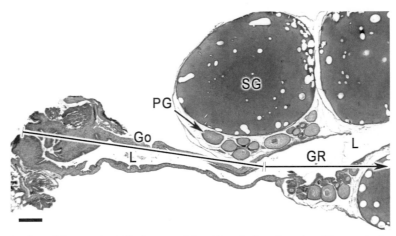

Fig. 6. Panoramic view of the gonoduct. Single ovary of *Poecilia reticulata.* A portion of the germinal region (GR) of the ovary and the gonoduct (Go) are seen. The germinal region of the ovary contains oocytes in Primary Growth Stage (PG) and Secondary Growth Stage (SG); in contrast, the gonoduct lacks germinal cells. The lumen (L) of both parts of the ovary is continuous. Scale bar = 0.3 mm.

1953, Kunz 2004, Grier et al. 2007), during the invagination of the genital ridges, lateral outgrowths do not develop, and then the ovary maintains an open side to the lumen; consequently, the lamellae project laterally into the coelom. This is known as the gymnovarian condition and it is considered the primitive condition (Kendall 1921, Hoar 1969, Dodd 1977, Grier et al. 2009). Therefore, teleost ovaries may be of two types: the cystovaries possess a lumen that has continuity with the gonoduct, and the gymnovaries project to the coelom. Thus, in the cystovarian condition, during ovulation the oocytes are released into the ovarian lumen, becoming eggs, and they always exit through the gonoduct and the genital pore. In the gymnovarian condition, because there is a discontinuous gonoduct, during ovulation, the oocytes are released directly into the coelom; from there, they enter into the *ostium* of the caudal funnel and to the exterior at the genital pore (Kendall 1921).

Cyclic Ovarian Changes

The ovarian structure and their cyclic changes are controlled by essential functions, as in other vertebrates, including: (1) the process of oogenesis, when oogonia develop into mature oocytes; (2) ovulation; and (3) the hormonal activity associated with the secretion of female sex hormones, indispensable for reproduction. The ovaries of fishes, similar to those seen in other vertebrates, show seasonal cyclic changes based on variations of climatic patterns such as photoperiod, temperature, rainfall and humidity through the year (Bailey 1933, Wallace and Selman 1981, Selman and Wallace 1989, Kirschaum and Schugardt 2002, Parenti et al. 2015, Uribe et al. 2016). These variations provide signals that influence the morphology and physiology of the ovaries via the central nervous system and the hypophyseal hormones that stimulate oogenesis and control the reproduction (Coward and Bromage 2000, Patiño and Sullivan 2002, Lubzens et al. 2010). An adequate supply of nutrients is an essential factor for the reproductive process, particularly in females, where vitellogenesis and maturation of the oocytes require abundant energy reserves. The nutrients could also be obtained by the utilization of those stored in the adipose tissues of the fish.

Ovarian size varies seasonally related with the sequential stages of oogenesis and progressively increases in size as a consequence of oocyte development during previtellogenesis and vitellogenesis, when abundant vitellogenic oocytes occupy most of the abdominal cavity. Vitellogenesis is characterized by protein yolk formation, when the oocytes accumulate the nutrients needed for the development of embryo and larva. Wallace and Selman (1981), and Selman and Wallace (1986) described three types of ovarian development in fishes during the breeding season: (1) asynchronous, when oocytes at different stages of development and having different sizes are present in the ovary at the same time; consequently, clutches of eggs ovulate multiple times during a breeding season, as occurs in the ovary of *Fundulus heteroclitus* (Selman and Wallace 1986); (2) group-synchronous, when there are two or more clutches of oocytes at different stages of development in the ovary. Then, clutches of eggs are successively ovulated several times during the breeding season, as occurs in the European sea bass, *Dicentrarchus labrax* (Asturiano et al. 2002); (3) synchronous, when oocytes of similar size and stage are present in the ovary. During the breeding season, the eggs are ovulated at the same time or over a short period of time, as occurs in the ovary of the Pacific salmon, *Oncorhynchus gorbuscha* (Patiño and Takashima 1995), and the Rainbow Trout *Oncorhynchus mykiss* (Tyler et al. 1996, Grier et al. 2007). Previtellogenic follicles are always present in the ovary regardless of the type of ovarian development (Mandich et al. 2002, Grier et al. 2007, 2009).

In viviparous species, the ovarian size increases progressively during the development of oocytes, but also during gestation with the development of the intraovarian development of embryos (Hoar 1969, Amoroso 1960, Mendoza 1965, Wourms 1981, Nagahama 1983, Schindler and Hamlett 1993).

Oogenesis

Oogenesis goes through the same basic pattern of development in all fishes, occurring in a sequence of oocyte differentiation that involves morphological, physiological and molecular changes implicated in the transformation of oogonia into eggs (Wallace and Selman 1981, 1990, Selman and Wallace 1989, Tyler and Sumpter 1996, Muñoz et al. 2001, Grier et al. 2009, Viedma et al. 2011).

Oogonia are the earliest stage of female germ cells. They are scattered in the germinal epithelium as individual cells or in small groups forming cell nests (Figs. 7A–C). The cell nests, besides the germinal cells, also contain somatic epithelial cells; they are the prefollicle cells (Fig. 7B) derived from the somatic cells of the celomic epithelium. During folliculogenesis, at the beginning of meiosis, the primordial follicles are formed when meiotic oocytes, in cell nests, begin to be enclosed by a single layer of prefollicle cells. Folliculogenesis is completed when an oocyte and its surrounding prefollicle cells are completely encompassed by a basement membrane. Then, prefollicle cells become follicle cells. The follicle cells form a single layer of squamous or cuboidal cells, surrounding the oocyte membrane (Begovac and Wallace 1987, Grier 2000, 2012, Quagio-Grassiotto et al. 2011, Wildner et al. 2013, Grier et al. 2007, 2009, 2016). In some viviparous fishes, such as *Heterandria formosa*, the follicle cells become columnar or pseudostratified during oocyte Secondary Growth (Uribe and Grier 2011). The follicle cells remain as a monolayer throughout the entire period of oocyte development, even if they become columnar during vitellogenesis. Subsequently, a thin theca, derived from the stroma, develops around the basement membrane of the follicle. The theca has been divided into interna and externa. The theca interna is composed of vascularized and fibrous connective tissue and steroids secretory cells; the theca externa is a single, squamous layer of cells that rests upon reticular fibers, separating it from the theca interna. The basement membrane separates the follicle cells from the theca (Patiño and Takashima 1995, Patiño and Sullivan 2002) as in *Centropomus undecimalis* (Grier 2000), *Oncorhynchus mykiis* (Grier 2007), *Heterandria formosa* (Uribe and Grier 2011), *Sciaenops ocellatus* (Grier 2012) and *Chlorophthalmus agassizi* (Parenti et al. 2015). At this point, the structure comprised of the follicle (oocyte and follicle cells) and surrounding elements (basement membrane and theca) is referred as "follicle complex", a term introduced by Grier (2000), emphasizing the separation of structure derived from an epithelium (the follicle) and that which is derived from stroma (the theca), separated from each other by a basement membrane. The follicle and the theca form a morphological and physiological unit that ends upon ovulation when the oocyte becomes an egg.

Grier et al. (2009) described and defined six stages of oocyte development that characterize oogenesis in teleosts: (1) Oogonial Proliferation (OP), (2) Chromatin Nucleolus (CN), (3) Primary Growth (Previtellogenesis) (PG), (4) Secondary Growth (Vitellogenesis) (SG), (5) Oocyte Maturation (OM) and (6) Ovulation (OV). The stages are divided into steps. Letters designating stages are capitalized, whereas letters designating steps are lower case. CN is divided into four steps: Leptotene (CNl), Zygotene (CNz), Pachytene (CNp), and early Diplotene (CNed). PG is divided into five steps: One Nucleolus (PGon), Multiple Nucleoli (PGmn), Perinucleolar (PGpn), Oil Droplets (PGod) and Cortical Alveoli (PGca). The Secondary Growth Stage (SG) is divided into three steps: Early Secondary Growth (SGe) with small yolk globules, Late Secondary Growth (SGl) with yolk globules at their maximum diameter, and Full-Grown

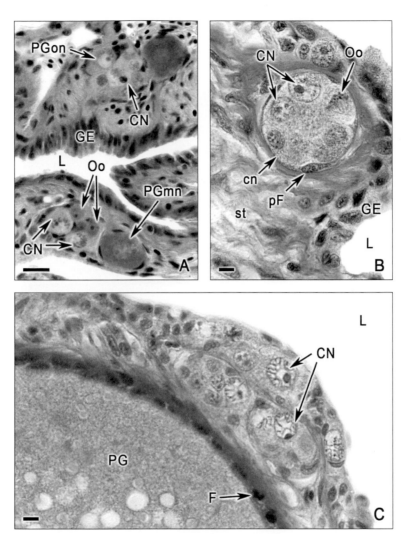

Fig. 7. Oogenesis. Oogonia and oocytes in Primary Growth stage. A. Ovary of the poeciliid *Heterandria formosa*. Germinal epithelium (GE) borders the ovarian lumen (L). Oogonia with one nucleolus (Oo), oocytes during Chromatin Nucleolus Stage (CN) with chromosomes appearing as fine filaments, oocytes that also have one nucleolus during One Nucleolus of Primary Growth Stage (PGon) and during Multiple Nucleoli Step of Primary Growth (PGmn), are seen. Scale bar = 20 μm. **B.** Ovary of Rainbow Trout *Oncorhynchus mykiss*. A cell nest (cn) containing one oogonium (Oo) during metaphase of mitosis, and oocytes during Chromatin Nucleolus Stage (CN) are surrounded by prefollicle cells (pF). Stroma (st), germinal epithelium (GE) and ovarian lumen (L). Scale bar = 10 μm. **C.** Ovary of Rainbow Trout, *Oncorhynchus mykiss*. Cell nest appears in the germinal epithelium at the border of the lumen (L), containing oocytes during the Chromatin Nucleolus Stage (CN), when the chromosomes are evident. A portion of an oocyte during the Primary Growth Stage (PG) and the surrounding follicle cells (F) are seen. Scale bar = 10 μm.

Color version at the end of the book

(SGfg) when the oocyte attains its maximum size. OM is divided into four steps: Eccentric Germinal Vesicle (OMegv), Germinal Vesicle Migration (OMgvm), Germinal Vesicle Breakdown (OMgvb) and Meiosis Resumes (OMmr). The staging is very useful for understanding cellular development during oogenesis and for defining similarities and differences of this essential process in diverse species. The staging is easily adapted to different species of fishes, for example, ostariophysan fishes lack oil droplets and globules in their developing oocytes. To adapt the staging, the Oil Droplets Step is omitted as a step in Primary Growth.

Oogonial Proliferation Stage (OP)

Oogonia are the source of renewing germ cell population in the ovary when they divide mitotically (Figs. 7A, B), increasing in number in the germinal epithelium (Grier et al. 2007, Uribe et al. 2014). Oogonia transform into oocytes when they enter in meiosis (Dodd and Sumpter 1984, Guraya 1986, Begovac and Wallace 1988, Grier et al. 2016). In Rainbow Trout, *Oncorhynchus mykiss* (Grier et al. 2007) and Tilapia, *Oreochromis mossambicus* (Smith and Haley 1987), oogonial proliferation is very active during and after ovulation when mitotic oogonia are frequently seen. In viviparous teleosts such as the goodeid *Ilyodon whitei*, proliferation of oogonia occurs during late gestation (Uribe et al. 2014) and in the poeciliid, *Heterandria formosa*, occurs throughout gestation, when several clutches of oocytes form and several broods of embryos, at different stage of embryogenesis, develop at the same time (Uribe and Grier 2011).

Chromatin Nucleolus Stage (CN)

The Chromatin Nucleolus Stage is the beginning of meiosis and occurs within the germinal epithelium, as in *Centropomus undecimalis* (Grier 2000), *Sciaenops ocellatus* (Grier 2012) and *Ilyodon whitei* (Uribe et al. 2014). Oocytes contain a spherical nucleus and the ooplasm is hyaline (Figs. 7A–C). Oocytes in this Stage progress through leptotene, zygotene, pachytene and early diplotene of prophase I of meiosis. Similar to that described in other vertebrates, oocytes become arrested in late diplotene while the lampbrush chromosomes develop (Fig. 7C). During this Stage, folliculogenesis advances as the prefollicle cells progressively surround the oocyte (Mazzoni et al. 2010, Quagio-Grassiotto et al. 2011, Sales et al. 2013, Uribe et al. 2014). The process of folliculogenesis is documented in a wide analysis of various species (Grier et al. 2016). Folliculogenesis is completed when the oocyte and follicle cells, comprising the ovarian follicle, are encompassed by a basement membrane that demarcates them from the theca (Guraya 1986, Selman and Wallace 1989). At least in *Centropomus undecimalis* (Grier 2000) and *Sciaenops ocellatus* (Grier 2012), folliculogenesis does not end until after the initiation of primary growth. Prior to that, the final step during CN is the early diplotene oocyte that has one nucleolus, a lucent ooplasm and developing lampbrush chromosomes.

Primary Growth Stage (PG) or Previtellogenesis

During this Stage, the nucleus of the oocyte, now called germinal vesicle, enlarges in volume and initially includes one nucleolus and the distinctive lampbrush chromosomes. Subsequently, the germinal vesicle contains multiple nucleoli, an indication of increased activity in the synthesis of rRNAs (Fig. 8A). Progressively, the nucleoli become located around the periphery of the germinal vesicle (Figs. 8B, C) (Selman and Wallace 1989, Thiry and Poncin 2005, Wildner et al. 2013).

Ooplasm growth involves an enormous increase in the number of cell organelles and molecules associated with synthetic activities, as ribosomes, mitochondria, endoplasmic reticulum, Golgi apparatus, glycogen, enzymes and diverse precursors of macromolecular synthesis. During this growth, the ooplasm becomes basophilic (Figs. 8B, C), mainly due to the increase in number of ribosomes. Irregular structures of the ooplasm formed adjacent to the germinal vesicle, known as Balbiani bodies, are evident during Primary Growth (Figs. 8B, C). Balbiani bodies are evolutionarily conserved aggregates of organelles in early oocytes of all vertebrates examined. They contain mitochondria, endoplasmic reticulum, Golgi apparatus and associated RNAs (Anderson 1968, Guraya 1986, Marlow and Mullins 2008, Grier et al. 2009). With the advance of the Primary Growth Stage, the Balbiani bodies disperse throughout the ooplasm, and subsequently, fragment into small portions and disperse in the vegetal pole ooplasm (Grier 2012, Grier et al. 2016). Additionally, the morphology of the Balbiani body is highly variable, even between closely related teleost species (Begovac and Wallace 1988). Kloc and Etkin (2005) analyzed the role of the irregular distribution of RNAs during the early stages of embryogenesis, affirming that the embryological patterns are dependent on the proper spatial and temporal distribution of the vast supply of maternal molecules deposited during oogenesis. Among these molecules, the irregular localization of RNAs in the ooplasm is involved in the generation of oocyte polarity, which defines the development of the embryo.

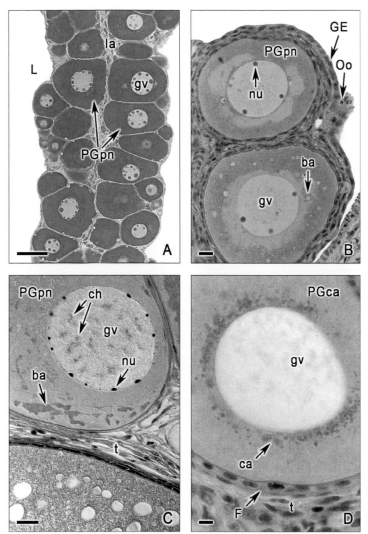

Fig. 8. Oogenesis. Oocytes in Primary Growth stage. A. Longitudinal section of a lamella of the ovary of Red Drum *Sciaenops ocellatus.* Oocytes during the Perinucleolar Step of Primary Growth (PGpn) with central germinal vesicle (gv) and deep basophilic ooplasm. Lumen (L). Scale bar = 50 μm. **B.** Ovary of the goodeid *Xenotoca eiseni* with oocytes during the Perinucleolar Step of Primary Growth (PGpn), the germinal vesicle (gv) encloses nucleoli (nu). Balbiani bodies (ba) around the germinal vesicle. Germinal epithelium (GE) with an oogonium (Oo) borders the ovarian lumen. Scale bar = 20 μm. **C.** Ovary of Rainbow Trout *Oncorhynchus mykiss* with an oocyte in Perinucleolar Step of Primary Growth (PGpn). The ooplasm contains Balbiani bodies (ba) and the germinal vesicle (gv) contains lampbrush chromosomes (ch) and peripheral nucleoli (nu). Theca (t) surrounding the adjacent follicle. Scale bar = 10 μm. **D.** Ovary of the goodeid *Xenotoca eiseni* with an oocyte during Cortical Alveolar Step of the Primary Growth Stage (PGca). Cortical alveoli (ca) surround the germinal vesicle (gv). The follicle cells (F) form a squamous epithelium. The theca (t) surrounds the follicle. Scale bar = 10 μm.

Color version at the end of the book

Cortical alveoli begin to form near the germinal vesicle (Fig. 8D) but later they move to the peripheral ooplasm (Figs. 9A–D). Their appearance correlates with the formation of endoplasmic reticulum and Golgi apparatus. The temporal appearance of cortical alveoli varies among species. In *Syngnathus fuscus* (Anderson 1968), *Fundulus heteroclitus* (Anderson 1968, Selman and Wallace 1986, 1989, Selman et al. 1988), *Dicentrarchus labrax* (Mayer et al. 1988), *Cichlasoma urophthalmus* (Viedma et al. 2011) and *Heterandria formosa* (Uribe and Grier 2011), cortical alveoli appear early during the Primary Growth

Stage and prior to the appearance of oil droplets. In *Dicentrarchus labrax* (Mayer et al. 1988), they appear after the beginning of vitellogenesis. The cortical alveoli have different sizes according to the species, for example, 50 µm in diameter in *Fundulus heteroclitus* (Anderson 1968, Selman et al. 1988, Wallace and Selman 1990) and *Atractosteus tropicus* (Méndez-Marín et al. 2012); 25 µm in the Chum Salmon *Oncorhynchus keta* (Kunz 2004); 12 µm in the pipefish *Syngnathus fuscus* (Anderson 1968).

Gradually, during the Primary Growth Stage, oil droplets are also deposited in the ooplasm. Oil droplets begin to form also near the germinal vesicle (Figs. 9D, 10A–C). Progressively, they increase in number and move to the periphery as in *Poecilia latipinna* (Fig. 10D).

During the Primary Growth Stage (PG), as the zona pellucida begins to develop between the oocyte and the follicle cells, two layers are seen as morphologically different: a thin layer of amorphous material that begins to accumulate, forming the *homogeneous layer*; and the *zona radiata* which is developed by numerous interdigitated microvilli of oocyte and follicle cells. The microvilli provide an active and ample surface for the interchange of diverse materials between follicle cells and the oocyte. Glycoproteins are essential components of the zona pellucida combined with other carbohydrates and proteins (Anderson 1968, Kunz 2004). The follicle cells, initially being a single layer of squamous cells around the oocyte, become thicker during progression of the Primary Growth Stage when these cells become cuboidal in some species; even in other species as *Centropomus undecimalis* (Grier 2000) and *Sciaenops ocellatus* (Grier 2012), they are always squamous. The zona pellucida continues to develop during the Secondary Growth Stage. A basement membrane and a vascularized theca completely surround the follicle (Grier 2000, 2012, Grier et al. 2009).

Secondary Growth Stage (SG) or Vitellogenesis

This Stage is characterized by the occurrence of the process of yolk accumulation (Figs. 10A–D, 11A–D). Similar to that described in other vertebrates, yolk is the fundamental and more abundant material stored during oogenesis in the ooplasm for the nutrition and the high metabolic activity required during embryonic and larval development (Guraya 1986, Wallace and Selman 1990, Patiño and Sullivan 2002, Ponce de León et al. 2011). The yolk is composed of about 45% phosphoproteins, 25% lipids, and 8% glycogen. The yolk precursor, vitellogenin, is synthesized in the liver and is well-known by the acronym VTG, a glycolipophosphoprotein. VTG is transported by the circulatory system to the secondary growth follicles in the ovary. It enters follicle cells and oocytes by endocytosis (Selman and Wallace 1989, Patiño and Sullivan 2002, Lubzens et al. 2010, Reading and Sullivan 2011). Oestrogens produced by the ovary, as a response to the hypophyseal gonadotropin, the follicle stimulating hormone (FSH), stimulate the synthesis and liberation of VTG from the liver. VTG binds to oocyte receptors in coated pits at the oocyte surface, is internalized to the ooplasm and is processed and deposited in yolk gradually during the next steps: Early Secondary Growth (SGe) and late Secondary Growth (SGl) (Figs. 10A–D). During vitellogenesis, oils rich in mono-unsaturated fatty acids, wax and esters are stored in oil droplets in the ooplasm (Patiño and Sullivan 2002, Grier et al. 2009, Hiramatsu et al. 2006). The genes encoding these fish reproductive proteins are conserved widely in the animal kingdom and are products of several hundred million years of evolution (Guraya 1986, Selman and Wallace 1986, Arukwe and Goksoyr 2003, Jalabert 2005, Grier et al. 2016). In most fishes, yolk exists as discrete platelets, having a crystalline structure, or globules that are composed of a membrane-bound fluid (Selman and Wallace 1989). In atherinomorphs, as one diagnostic character that supports their monophyly (Parenti and Grier 2004), yolk fuses progressively, becoming a homogeneous fluid (Figs. 11A, C, D). During the Secondary Growth Stage, in addition to yolk, glycogen granules and oil droplets are also stored forming an essential complex for embryonic nutrition (Figs. 11A–D, 12A). The germinal vesicle moves towards the animal pole and becomes irregular in shape (Figs. 11C, D).

The increases in the number of microvilli during oocyte growth, that penetrate the zona pellucida, amplify significantly the surface of interchange between the oocyte and the follicle cells (Figs. 12B, C). Kunz (2004) revised widely the different morphology and components of the zona pellucida in fishes. At the animal pole of the oocyte, an opening, the micropyle (Fig. 12C), is located through which one sperm enters, contacts the oocyte during fertilization and has a role to prevent polyspermy. The opening

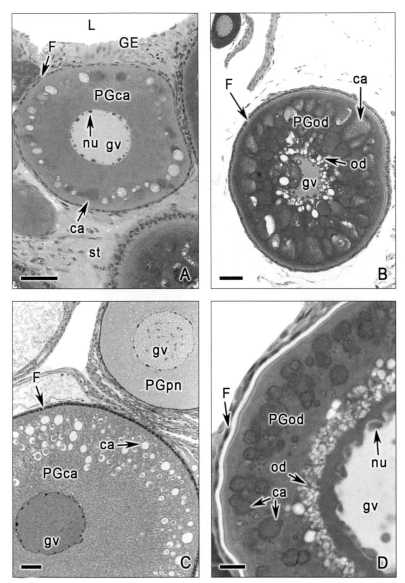

Fig. 9. Oogenesis. Oocytes in Primary Growth Stage. A, B. In Gulf Killifish *Fundulus grandis*. Scale bars = 50 μm.
C. In Rainbow Trout *Oncorhynchus mykiss*. Scale bar = 50 μm. **D.** In Red Drum *Sciaenops ocellatus*. Scale bar = 10 μm.
A, C. Oocytes during the Cortical Alveolar Step of Primary Growth (PGca) have cortical alveoli (ca) at the periphery. Stroma
(st), germinal epithelium (GE) and ovarian lumen (L). **B, D.** Oocytes during Oil Droplets Step of Primary Growth (PGod)
which contain cortical alveoli (ca) and oil droplets (od) surrounding the germinal vesicle. The germinal vesicle (gv) contains
perinucleoli (nu). Squamous follicle cells (F) surround the oocytes.

Color version at the end of the book

diameter of the micropyle differs between species, commonly being 2–3 μm, coinciding with the size
of the spermatozoon. The micropyle consists of a vestibule and a micropylar canal in the zona pellucida
(Kobayashi and Yamamoto 1981, Nagahama 1983, Guraya 1986, Patiño and Takashima 1995). A micropyle
has been reported in species such as: *Fundulus heteroclitus* (Kutchnow and Scott 1977, Dumont and
Brummett 1980, Grier et al. 2009), Chum Salmon, *Oncorhynchus keta* (Kobayashi and Yamamoto 1981),
Anchovy, *Engraulis japonica* (Hirai and Yamamoto 1986), Rosy Barb, *Barbus conchonius* (Amanze and
Iyengar 1990), Swamp Eel, *Synbranchus marmoratus* (Ravaglia and Maggese 2002), *Crenichthys bailey*

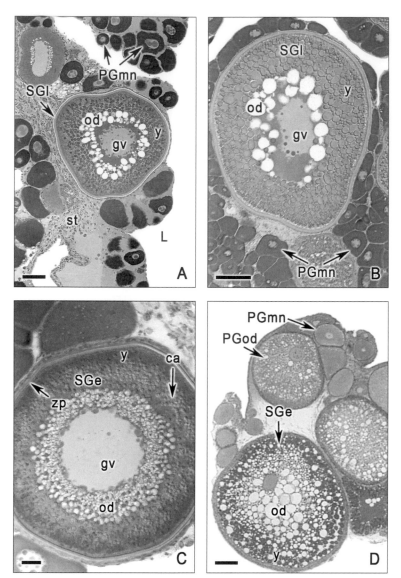

Fig. 10. Oogenesis. Oocytes in Primary and Secondary Growth Stages. A, B. In Gray Angelfish *Pomacanthus arcuatus* and in Red Drum *Sciaenops ocellatus*. Oocytes during late Secondary Growth Step (SGe) have oil droplets (od) around the periphery of the germinal vesicle (gv). Some yolk globules (y) are also seen in the ooplasm. Oocytes during Multiple Nucleoli Step of Primary Growth (PGmn), stroma (st) and lumen (L) are seen. Scale bars = 50 μm, 100 μm. **C, D.** Oocytes during the early Secondary Growth Step (SGe) in *Sciaenops ocellatus,* and in *Poecilia latipinna* with some yolk globules (y) at the periphery, cortical alveoli (ca) extending to the periphery and abundant oil droplets (od) adjacent to the germinal vesicle (gv). The zona pellucida (zp) surround the oocyte. Oocytes during Multiple Nucleoli Step of Primary Growth (PGmn), Oil Droplets (od), Step of Primary Growth (PGod). Scale bars = 10 μm, 100 μm.

Color version at the end of the book

and *Empetrichthys latos* (Uribe et al. 2012), the Gar-fish, *Atractosteus tropicus* (Méndez-Marín et al. 2012) and *Jordanella floridae* (Uribe et al. 2016).

The follicle cells remain as a single layer throughout oogenesis, but may become columnar or pseudostratified in shape during secondary growth (Figs. 12C, 13A, B), thicker than the squamous and cuboidal epithelium described in primary growth oocytes (Fig. 8D). During vitellogenesis, the animal-vegetal polarity is established, initiated during previtellogenesis by the irregular deposition of organelles

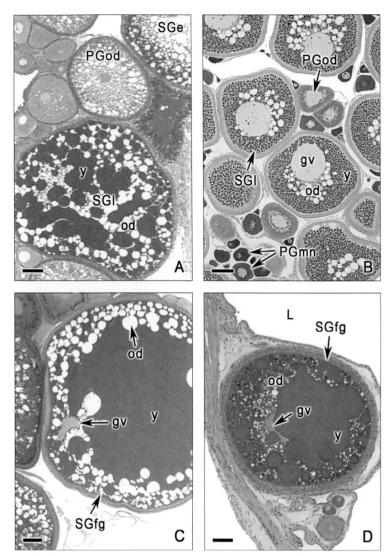

Fig. 11. Oogenesis. Oocytes in Secondary Growth Stage. A. Oocytes during Late Secondary Growth Step (SGl) in *Poecilia latipinna* have coalescing, fluid yolk globules (y), and **B**, in the Spotted Seatrout *Cynoscion nebulosus*, the abundant yolk globules (y) do not fuse during Secondary Growth. Oil droplets (od) are in the ooplasm in vicinity of the germinal vesicle (gv), oocytes during Multiple Nucleoli Step of Primary Growth (PGmn), Oil droplets Step of Primary Growth (PGod) and Early Secondary Growth Step (SGe) are also seen. Scale bars = 100 μm. **C, D.** Oocytes during Full-Grown Step of Secondary Growth (SGfg) in the Gulf killifish *Fundulus grandis* and in the goodeid *Xenotoca eiseni*. The yolk (y) is fluid, as is typical of Secondary Growth in atherinomorphs; there are oil droplets (od) toward the periphery of the ooplasm. The germinal vesicle (gv) is displaced towards the animal pole of the oocytes. Lumen (L). Scale bars = 100 μm.

Color version at the end of the book

and RNAs, and then in this Stage, by the deposition of the yolk. At the end of Secondary Growth Stage, vitellogenesis is completed, and the oocyte is full-grown (Figs. 11C, D). Follicle cells of full-grown oocytes are reduced in height becoming squamous (Fig. 12B) (Nagahama 1983, Wallace and Selman 1990, Patiño and Takashima 1995, Grier et al. 2009).

Follicle cells of developing oocytes of some teleosts form elongate, attachment filaments over all or a portion on the surface of the zona pellucida (Figs. 13C, D). Filaments are described in species such as *Fundulus heteroclitus* (Kuchnow and Scott 1977, Dumont and Brummett 1980, Brummett and Dumont

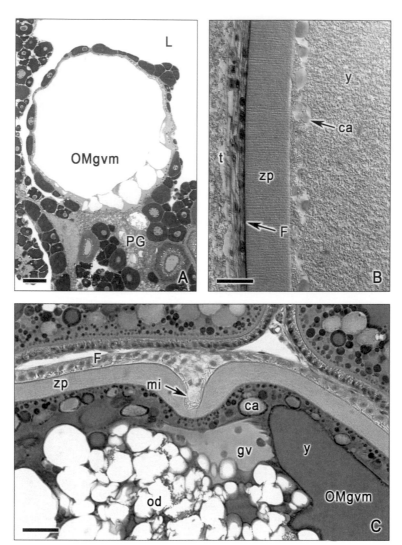

Fig. 12. Oogenesis. Maturation Stage Oocytes. A. Preovulatory oocyte of Red Drum *Sciaenops ocellatus*, after yolk hydration and Germinal Vesicle Migration to the animal pole (OMgvm). Abundant oocytes during Multiple Nucleoli Step of Primary Growth (PG), with basophilic ooplasm. Lumen (L). Scale bar = 100 μm. **B.** Oocyte surface of the Rainbow Trout *Oncorhynchus mykiss*, the ooplasm contain yolk (y), and cortical alveoli (ca) at the periphery. The zona pellucida (zp) has a striated morphology. The follicle cells (F) are squamous. The theca (t) surrounds the follicle. Scale bar = 20 μm. **C.** Animal pole of an oocyte in the Flagfish, *Jordanella floridae* during formation of the fluid yolk (y) and Germinal Vesicle (gv) Migration Step (OMgvm) to the animal pole. The zona pellucida (zp) has a micropyle (mi) adjacent to the germinal vesicle. At the periphery of the oocyte, there are cortical alveoli (ca) and oil droplets (od). The follicular cells (F) are cuboidal. Scale bar = 20 μm.

Color version at the end of the book

1981); *Oryzias latipes* (Iwamatsu 1992); *Crenichthys baileyi* and *Empetrichthys latos* (Uribe et al. 2012); and *Jordanella floridae* (Uribe et al. 2016). Filaments permit the fertilized eggs adhere to vegetation; they may also protect the egg as they trap debris and obscure the embryo from view (Brummett 1966).

The theca that surrounds each follicle, derived from stroma, initially is an incomplete cell layer over the basement membrane around primary growth oocytes, but differentiates into a theca interna and a theca externa (Grier 2000, 2012) as the follicle grows. Blood vessels, spherical or polygonal cells identified with activity of steroid biosynthesis (Lambert 1970, Guraya 1986) are located within the theca interna and abundant collagen fibers in the theca externa (Fig. 13A).

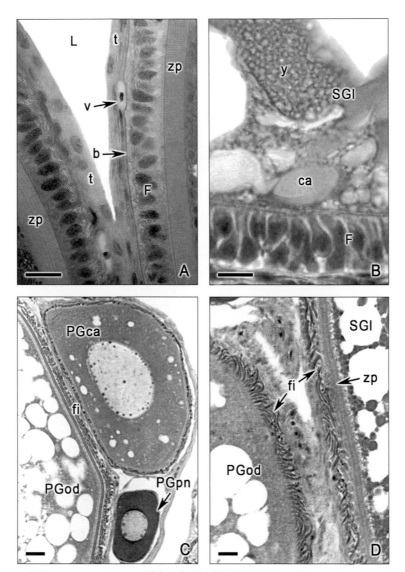

Fig. 13. Oogenesis. Follicle components. A. Follicle cells (F) of Gulf Killifish *Fundulus grandis* form a columnar epithelium around the zona pellucida (zp). The basement membrane (b) separates the follicle cells from the theca (t). The theca contains blood vessels (v). Scale bar = 20 μm. **B.** Follicle cells (F) surrounding an oocyte during the Late Secondary Growth Step (SGl) in the poeciliid *Heterandria formosa* form a pseudostratified epithelium with their nuclei situated at different levels of the cells. Yolk globules (y) are in the ooplasm. Scale bar = 10 μm. **C, D.** Filaments (fi) irregularly situated among the follicle cells in follicles of the oviparous goodeid *Crenichthys baileyi* during the Oil Droplets Step of Primary Growth (PGod), Cortical Alveoli Step of Primary Growth (PGca), and Late Secondary Growth Step (SGl) of oocyte development are also observed. The zona pellucida (zp) surround the oocyte. An oocyte during Perinucleolar Step of Primary Growth (PGpn) is seen. Scale bars = 50 μm, 20 μm.

Oocyte Maturation Stage (OM)

Oocyte maturation consists of a sequence of cellular processes, including the ooplasm and the germinal vesicle. During this Stage, several morphological and physiological changes occur, such as: displacement of the germinal vesicle towards the animal pole, as oil droplets coalesce; breakdown of the germinal vesicle and the resumption of meiosis. In atherinomorphs as poeciliids and goodeids, during oocyte maturation the hydration of yolk occurs (Fig. 12A) and becomes fluid. Some species spawn non-hydrated eggs such as Zebrafish, *Danio rerio* or Medaka, *Oryzias latipes* (Parenti and Grier 2004, Parenti 2005, Lubzens et al. 2010). The formation of fluid yolk is described in oviparous fishes by Jalabert (2005) and Viedma et al. (2011), as a consequence of a deep reorganization of the lipoprotein yolk. Most saltwater fish species, such as *Sciaenops ocellatus*, produce pelagic eggs that are buoyant (Grier 2012). The oocytes of these species possess a single oil globule. The oil globule provides energy for the developing embryo (Patiño and Sullivan 2002). When the germinal vesicle completes its migration, rupture of the nuclear membrane and resumption of meiosis occur. At ovulation, the oocyte is released from the follicle into the ovarian lumen, and it becomes an egg ready for fertilization. During oogenesis, the oocytes remain arrested in diplotene of the first meiotic division until they receive the hormonal signal to reinitiate meiosis before ovulation (Wallace and Selman 1990).

Postovulatory Follicle Complex

After ovulation, the follicle complex collapses, becoming a postovulatory follicle complex (POC) (Figs. 14A–D), composed of both the postovulatory follicle (POF) which consists of the follicle cells that surrounded the oocyte during oocyte growth and a postovulatory theca (Grier 2012). Both components are separated by the same basement membrane observed during the oocyte growth. As used in fishery literature, the POF was composed of both the follicle cells and the theca. But, taking into account the origin of these components, derived from two different tissue compartments, the follicle cells from the epithelial compartment, and the theca derived from the stroma compartment, this stage corresponds to a postovulatory follicle complex (POC). Then, the follicle complex (oocyte, follicle cells, basement membrane and theca) that exists during oocyte growth becomes a POC at ovulation. POCs have been analyzed in their main function of synthesizing steroid hormones. Smith and Haley (1987) described the morphological characteristics of postovulatory follicles of the tilapia, *Oreochromis mossambicus*, with enzyme histochemistry and evidenced their steroid-hormone production. Progressively, the POC becomes resorbed. The signs of resorption are the decrease in size of the cells, pyknosis of the cell nucleus, and an accumulation of lipid droplets in the cytoplasm of luteal cells.

Atresia

Atresia, the process of degeneration and removal of ovarian follicles, is an essential aspect of the ovarian physiology (Figs. 15A–D). This process is a common phenomenon in all vertebrates and occurs at any stage of oogenesis. Lambert (1970) described the changes taking place in the ovary of *Poecilia reticulata* during atresia. He divided atresia into two parts: (1) resorption of the oocyte characterized by fragmentation of the zona pellucida and hypertrophy of the follicle cells which become phagocytes and engulf the oocyte followed by (2) gradual regression and decrease of the cells of the atretic follicle. We illustrate in different species the sequential morphological changes taking place in atretic follicles, summarized by authors as Mendoza (1943) in the goodeid *Skiffia* (=*Neotoca*) *bilineata*, Rajalakshmi (1966) in *Gobius giuris*, Rastogi (1969) in the mud-eel *Amphipnous cuchia* and Uribe et al. (2006) in oocytes of the goodeids *Ilyodon whitei* and *Goodea atripinnis*: (1) the follicle cells lose their regular shape (Fig. 15A); (2) the zona pellucida was thrown into folds and breaks up (Figs. 15B, C); (3) the follicle cells remain in their position and from there they phagocytize the oocyte (Fig. 15B), or the follicle cells become distributed within the ooplasm and phagocytize the oocyte (Fig. 15C); (4) after the removal of the oocyte residue, the follicle collapses and forms a group of cells in the stroma (Fig. 15D); and (5) removal of the degenerating mass of follicle cells from the ovarian stroma.

It is not well known why some oocytes are able to develop until ovulation, whereas others undergo degeneration. It has been suggested that atretic follicles contribute to steroid secretions and to regulate the number of ovulated oocytes (Guraya 1986, Viedma et al. 2011). Tyler and Sumpter (1996) considered that atresia plays an important role in batch fecundity by controlling the number of oocytes that attain maturity, therefore, defining the number of eggs that are produced.

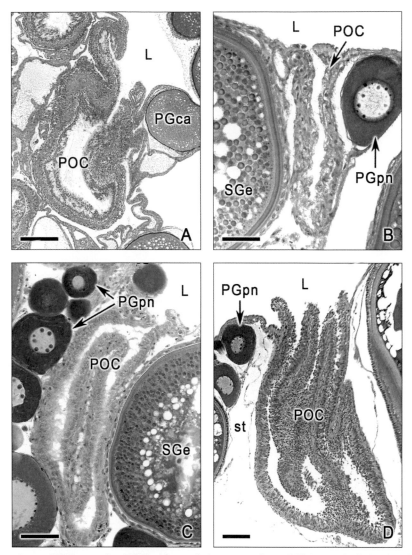

Fig. 14. Postovulatory follicle complex (POC). After ovulation, postovulatory follicle complex is developed. POC opened to the lumen (L). **A.** In Rainbow Trout *Oncorhynchus mykiss*, among oocytes during Cortical Alveolar Step of Primary Growth (PGac). Scale bar = 100 µm. **B, C.** Postovulatory follicle complex (POC) opened to the lumen (L) in Red Drum *Sciaenops ocellatus*, between oocytes during Perinucleolar Step of Primary Growth (PGpn) and Early Secondary Growth Step (SGe). Scale bars = 50 µm. **D.** Postovulatory follicle complex (POC) opened to the ovarian lumen (L) in Gulf Killifish, *Fundulus grandis* between oocytes during Perinucleolar Step of Primary Growth (PGpn). St, stroma. Scale bar = 100 µm.

Color version at the end of the book

Diameter of Eggs

The shape of the eggs is usually spherical, but their diameter varies widely among species. Within a species, changes in egg size are related with diverse factors such as: latitude-dependent temperature and food supply, advancement of the spawning season, and egg batch sequence in multi-batch spawners (Pandian 2011, Ponce de León et al. 2011). With a better food supply, the females produce a larger number of smaller eggs but a small number of larger eggs when receiving low food supply (Hutchings 1991). However, the fishes seem to attain optimum egg sizes, which ensure maximum survival of their progeny (Pandian 2011). Demersal species produce larger eggs than pelagic species. Pelagic species produce smaller but a larger

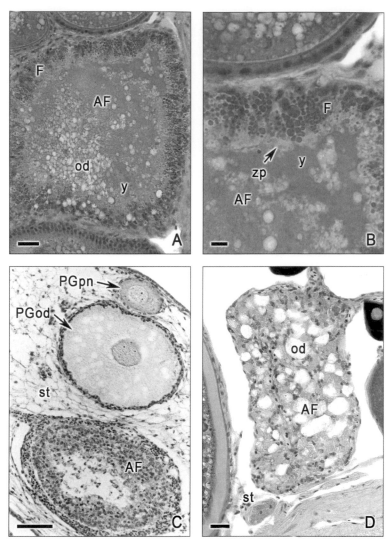

Fig. 15. Atretic follicles. A, B. Early atretic follicles (AF) in the ovary of the poeciliid *Gambusia affinis* indicated by the disorganization of the follicle cells (F), yolk (y) and oil droplets (od). The zona pellucida (zp) is folded. Scale bars = 20 μm, 10 μm. **C.** Mid atretic follicle (AF) in the ovary of the goodeid *Ilyodon whitei*. The oocyte is clearly reduced and the proliferation of the follicle cells is evident. The follicle cells become distributed within the ooplasm and phagocytize the oocyte. Oocytes, during Perinucleolar Step of Primary Growth (PGpn) and Oil Droplets Step of Primary Growth (PGod), are also observed. Stroma (st). Scale bar = 50 μm. **D.** Late atretic follicle (AF) in the ovary of the Gray Angelfish *Pomacanthus arcuatus*, the follicle collapses and forms a group of cells surrounded by the stroma (st). The nuclei of the follicle cells appear as small, black granules and invaded the oocyte. Some oil droplets (od) remain at this time. Yolk no longer exists. Scale bar = 20 μm.

Color version at the end of the book

number of eggs (Kunz 2004, Pandian 2011). Pelagic spawners release a few thousands to millions of eggs. In contrast, the demersal spawners lay a few hundred to thousands eggs only (Pandian 2011). Egg sizes are reported to vary significantly among batches and this variation is usually revealed as decreases in egg sizes as spawning proceeds (Chambers and Leggett 1996, Coward and Bromage 1999, Kennedy et al. 2007). In viviparous species, the nutrition of embryos may be, in addition to the yolk of the egg, supplemented by maternal contributions. Some examples of approximate maximum mean diameter of oocytes are:

In oviparous species: *Epinephelus marginatus* 0.6–0.8 mm (Mandich et al. 2002); *Centropomus undecimalis* 0.6–0.8 mm (Taylor et al. 1998); *Danio rerio* 0.7 mm (Koc et al. 2008); *Sciaenops ocellatus* 1 mm (Grier 2012); *Prochilodus affinis* 1.08 mm (Rizzo et al. 1998); *Perca fluviatilis* L. 1.2 mm (Bonislawska et al. 2001); *Carassius auratus* (Lorenzoni et al. 2009) and *Syngnathus scovelli* (Begovac and Wallace 1988) 1.3 mm; *Tilapia zilli* 1.4 mm (Coward and Bromage 1999); *Fundulus heteroclitus* 1.6 mm (Selman and Wallace 1986); *Fundulus sciadicus* 1.6 mm (Rahel and Thel 2004); *Esox lucius* L. 2 mm (Benzer et al. 2010); *Pleuronectes platessa* 2 mm (Kennedy et al. 2007). Eggs with a large diameter are observed in salmonids and bagrids, such as: *Salmo trutta* 4–6 mm (Bonislawska et al. 2001); *Oncorhynchus mykiss* 5 mm (Bonislawska et al. 2001); *Oncorhynchus keta* 8–9.5 mm (Kunz 2004); *Bagre marinus* 20 mm (Pinheiro et al. 2006, Segura-Berttolini and Mendoza-Carranza 2013).

In viviparous species: *Heterandria formosa* 0.4 mm (Uribe and Grier 2011); *Skiffia bilineata* 0.5 mm (Turner 1933); *Ilyodon whitei* 0.7 mm (Uribe et al. 2005); *Goodea atripinnis, Characodon lateralis* 0.8 mm (Uribe et al. 2005); *Poeciliopsis turneri* 1 mm (Thibault and Schultz 1978); *Ateniobius toweri* 1 mm (Wourms 1981); *Poeciliopsis lucida* 1.4 mm, *Poecilia reticulata* 1.7 mm (Thibault and Schultz 1978); *Poeciliopsis occidentalis* 2 mm (Scrimshaw 1946); *Poeciliopsis monacha* 2 mm (Thibault and Schultz 1978); *Limia vittata* 2 mm (Ponce de León et al. 2011).

The revision of the themes developed in this chapter, based on numerous species, considered the characterization of the ovaries and eggs of teleosts, in the context of the ovarian structure and the process of oogenesis. Of special interest is the origin of the follicle development when oogonia, situated in the germinal epithelium, proliferate and are surrounded by somatic cells, the follicle cells. The analysis of this process is established in the staging proposed by Grier et al. (2009). The addition of basement membrane and theca layers formed the follicle complex. The identification of the activity of the germinal epithelium and Stages of the oocyte development are essential characteristics in the analysis of the seasonal cycles of the teleost ovaries. The great diversity of reproductive strategies in fishes is a challenge for understanding this vital function, such as: the type of spawning patterns that differ among the recruitment and timing of release of gametes; the oviparity and the viviparity that implicate specific modifications for the intraovarian fertilization and gestation; the hermaphroditism that reveals both ovarian and testicular tissues in an ovotestis (Parenti et al. 2015); the change of sex, as protogynous (Lo Nostro et al. 2003) and protandric species; or, the different habitat as marine or freshwater species where the fishes may be located, are some evidences of this significant diversity.

Acknowledgements

We thank Noretta Perry, Adriana García-Alarcón and Gabino De la Rosa-Cruz for valuable assistance with histology; Ana Isabel Bieler Antolin and Gabino De la Rosa-Cruz for the kind, helpful technical assistance with digital photography; and José Antonio Hernández Gómez for the excellent digital preparation of histological figures.

References Cited

Amanze, D. and A. Iyengar. 1990. The micropyle: a sperm guidance system in teleost fertilization. Development 109: 495–500.

Amoroso, E.C. 1960. Viviparity in fishes. Symp. Zool. Soc. Lond. 1: 153–181.

Anderson, E. 1968. Cortical alveoli formation and vitellogenesis during oocyte differentiation in the Pipefish *Syngnathus fuscus* and the Killifish, *Fundulus heteroclitus*. J. Morphol. 125: 23–31.

Arukwe, A. and A. Goksoyr. 2003. Eggshell and egg yolk proteins in fish: hepatic proteins for the next generation: oogenetic, population, and evolutionary implications of endocrine disruption. Comp Hepatol. 2: 4 Online: doi: 10.1186/1476-5926-2-4.

Asturiano, J.F., L.A. Sorbera, J. Ramos, D.E. Kime, M. Carrillo and S. Zanuy. 2002. Group-synchronous ovarian development, ovulation and spermiation in the European Sea Bass (*Dicentrarchus labrax* L.) could be regulated by shifts in gonadal steroidogenesis. Sci. Mar. 66(3): 273–282.

Bailey, R.J. 1933. The ovarian cycle in the viviparous teleost *Xiphophorus helleri*. Biol. Bull. 64: 206–225.

Begovac, P.C. and R.A. Wallace. 1987. Ovary of the pipefish *Syngnathus scovelli*. J. Morphol. 193: 117–133.

Begovac, P.C. and R.A. Wallace. 1988. Stages of oocyte development in the Pipefish *Syngnathus scovelli*. J. Morphol. 197: 353–369.

Benzer, S., A. Gül and M. Yilmaz. 2010. Breeding properties of *Esox lucius* (L., 1758) living in Kapulukaya Dam Lake (Kirikkale, Turkey). Afr. J. Biotechnol. 9(34): 5560–5565.

Bonislawska, M., K. Formicki, A. Korzelecka-Orkisz and A. Winnicki. 2001. Fish egg size variability: biological significance. EJPAU 4(2), #02. Online: http://www.ejpau.media.pl/volume4/issue2/fisheries/art-02.html.

Brummett, A.R. 1966. Observations on the eggs and breeding season of *Fundulus heteroclitus* at Beaufort, North Carolina. Copeia 1966: 612–620.

Brummett, A.R. and J.N. Dumont. 1981. A comparison of chorions from eggs of northern and southern populations of *Fundulus heteroclitus*. Copeia 1981: 607–614.

Brummett, A.R., J.N. Dumont and J.R. Larkin. 1982. The ovary of *Fundulus heteroclitus*. J. Morphol. 173: 1–16.

Burns, J.R. and S.H. Weitzman. 2005. Insemination in ostariophysan fishes. pp. 105–132. *In*: M.C. Uribe and H.J. Grier (eds.). Viviparous Fishes. New Life Publications, Homestead, Florida.

Campuzano-Caballero, J.C. and M.C. Uribe. 2014. Structure of the female gonoduct of the viviparous teleost *Poecilia reticulata* (Poeciliidae) during non-gestation and gestation stages. J. Morphol. 275: 247–257.

Chambers, R.C. and W.C. Leggett. 1996. Maternal influences on variation in egg sizes in temperate marine fishes. Amer. Zool. 36: 180–196.

Coward, K. and N.R. Bromage. 1999. Spawning periodicity, fecundity and egg size in laboratory held stocks of a substrate-spawning tilapiine *Tilapia zilli* (Gervai). Aquaculture 171: 251–267.

Coward, K. and N.R. Bromage. 2000. Reproductive physiology of female tilapia Broodstock. Rev. Fish. Biol. Fisher 10: 1–25.

Dodd, J.M. 1977. The structure of the ovary of nonmammalian vertebrates. pp. 219–163. *In*: S. Zuckerman and B.J. Weir (eds.). The Ovary. Vol. 1, Academic Press, New York, USA.

Dodd, J.M. and J.P. Sumpter. 1984. Fishes. pp. 1–126. *In*: G.E. Lamming (ed.). Marshall's Physiology of Reproduction, Vol. 1. London: Churchill. Livingstone.

Dumont, J.N. and A.R. Brummett. 1980. The vitelline envelope, chorion and micropyle of *Fundulus heteroclitus* eggs. Gamete Res. 3: 24–44.

Erickson, D.L. and E.K. Pikitch. 1993. A histological description of short spine thorny head, *Sebastolobus alascanus*, ovaries: structures associated with the production of gelatinous egg masses. Environ. Biol. Fish 36(3): 273–282.

Freijo, R.O., A.M. García, E.L. Portiansky, C.G. Barbeito, G.J. Macchi and A.O. Díaz. 2009. Morphological and histochemical characteristics of the epithelium of ovarian lamellae of Genypterus blacodes (Schneider, 1801). Fish Physiol. Biochem. 35(3): 359–67.

Grier, H.J. 2000. Ovarian germinal epithelium and folliculogenesis in the common snook, *Centropomus undecimalis* (Teleostei: Centropomidae). J. Morphol. 243: 265–281.

Grier, H.J., M.C. Uribe and L.R. Parenti. 2007. Germinal epithelium, folliculogenesis, and postovulatory follicles in ovaries of rainbow trout, *Oncorhynchus mykiss* (Walbaum, 1792) (Teleostei, Protacanthopterygii, Salmoniformes). J. Morphol. 268: 293–310.

Grier, H.J., M.C. Uribe and R. Patiño. 2009. The ovary, folliculogenesis, and oogenesis in teleosts. pp. 25–84. *In*: B.G.M. Jamieson (ed.). Reproductive Biology and Phylogeny of Fishes (Agnathans and Bony Fishes), Vol. 8A. Enfield, New Hampshire: Science Publishers.

Grier, H.J. 2012. Development of the follicle complex and oocyte staging in red drum, *Sciaenops ocellatus* Linnaeus, 1776 (Perciformes, Sciaenidae). J. Morphol. 273(8): 801–29.

Grier, H.J., M.C. Uribe, F. Lo Nostro, M. Steven and L.R. Parenti. 2016. Conserved form and function of the germinal epithelium through 500 million years of vertebrate evolution. J. Morphol. 277: 1014–1044.

Guraya, S.S. 1986. The cell and molecular biology of fish oogenesis. Karger. Basel, Switzerland, pp. 223.

Hirai, A. and T.S. Yamamoto. 1986. Micropyle in the developing eggs of the Anchovy, *Engraulis japonica*. Jpn. J. Ichthyol. 33(1): 62–66.

Hiramatsu, N., T. Matsubara, T. Fujita, C.V. Sullivan and A. Hara. 2006. Multiple piscine vitellogenins: Biomarkers of fish exposure to estrogenic endocrine disruptors in aquatic environments. Mar. Biol. 149: 35–47.

Hoar, W.S. 1969. Reproduction. pp. 1–72. *In*: W.S. Hoar and D.J. Randall (eds.). Fish Physiology, Vol. 3. Ac Press. New York and London.

Hutchings, J.A. 1991. Fitness consequences of variation in egg size and food abundance in brook trout *Salvelinus fontinalis*. Evolution 45: 1162–1168.

Iwamatsu, T. 1992. Morphology of filaments on the chorion of oocytes and eggs in the Medaka. Zool. Sci. 9: 589–599.

Jalabert, B. 2005. Particularities of reproduction and oogenesis in teleost fish compared to mammals. Reprod. Nutr. Dev. 45: 261–279.

Kendall, W.C. 1921. Peritoneal membranes, ovaries, and oviducts of salmonid fishes and their significance in fish-cultural practices. Bull Bureau Fish 37: 184–208.

Kennedy, J., A.J. Geffen and R.D.M. Nash. 2007. Maternal influences on egg and larval characteristics of plaice (*Pleuronectes platessa* L.). J. Sea Res. 58: 65–77.

Kirschbaum, F. and C. Schugardt. 2002. Reproductive strategies and developmental aspects in mormyrid and gymnotiform fishes. J. Physiol. (Paris) 96(5-6): 557–566.

Kloc, M. and L.D. Etkin. 2005. RNA localization mechanisms in oocytes. J. Cell Sci. 118: 269–282.

Kobayashi, W. and T.S. Yamamoto. 1981. Fine structure of the micropylar apparatus of the chum salmon egg, with a discussion of the mechanism for blocking polyspermy. J. Exp. Zool. 217(2): 265–275.

Koc, N.D., Y. Aytekin and R. Yüce. 2008. Ovary maturation stages and histological investigation of ovary of the Zebrafish (*Danio rerio*). Braz. Archives Biol. Techn. 51(3): 513–522.

Kuchnow, K.P. and J.R. Scott. 1977. Ultrastructure of the chorion and its micropyle apparatus in the mature *Fundulus heteroclitus* (Walbaum) ovum. J. Fish Biol. 10: 197–201.

Kunz, R.Y. 2004. Developmental Biology of Teleost Fishes. Springer, Dordrecht, The Netherlands, pp. 636.

Lambert, J.G.D. 1970. The ovary of the guppy *Poecilia reticulata*. The atretic follicle, a *corpus atreticum* or a *corpus luteum praeovulationis*. Z. Zellforsch. 107: 54–67.

Lombardi, J. 1998. Comparative Vertebrate Reproduction. Kluwer Academic Publications. Boston, Dordrecht, London, pp. 1–469.

Lo Nostro, F., H.J. Grier, L. Andreone and G.A. Guerrero. 2003. Involvement of the gonadal germinal epithelium during sex reversal and seasonal testicular cycling in the protogynous swamp eel, *Synbranchus marmoratus* Bloch 1795 (Teleostei, Synbranchidae). J. Morphol. 257: 107–126.

Lorenzoni, M., L. Ghetti, G. Pedicillo and A. Carosi. 2009. Analysis of the biological features of the goldfish *Carassius auratus auratus* in Lake Trasimeno (Umbria, Italy) with a view to drawing up plans for population control. Folia Zool. 59(2): 142–156.

Lubzens, E., G. Young, J. Bobe and J. Cerda. 2010. Oogenesis in teleosts: How fish eggs are formed. Gen. Comp. Endocrinol. 165: 367–389.

Mandich, A., A. Massari, S. Bottero and G. Marino. 2002. Histological and histochemical study of female germ cell development in the dusky grouper *Epinephelus marginatus* (Lowe, 1834). Eur. J. Histochem. 46: 87–100.

Marlow, F.L. and M.C. Mullins. 2008. Bucky ball functions in Balbiani body assembly and animal-vegetal polarity in the oocyte and follicle cell layer in zebrafish. Dev. Biol. 321(1): 40–50.

Mayer, I., S.E. Shackley and J.S. Ryland. 1988. Aspects of the reproductive biology of the bass, *Dicentrarchus labrax* L. I. A histological and histochemical study of oocyte development. J. Fish Biol. 33: 609–622.

Mazzoni, T.S., H.J. Grier and I. Quagio-Grassiotto. 2010. Germline cysts and the formation of the germinal epithelium during the female gonadal morphogenesis in *Cyprinus carpio* (Teleostei: Ostariophysi: Cypriniformes). Anat. Rec. 293: 1581–1606.

Méndez Marín, O., A. Hernández-Franyutti, C.A. Alvarez-González, W. Contreras Sánchez and M.C. Uribe. 2012. Histología del ciclo reproductor de hembras del pejelagarto *Atractosteus tropicus* (Lepisosteiformes: Lepisosteidae) en Tabasco, Mexico. Rev. Biol. Trop. 60(4): 1857–1871.

Mendoza, G. 1943. The reproductive cycle of the viviparous teleost *Neotoca bilineata*, a member of the family Goodeidae. IV. The germinal tissue. Biol. Bull. 84: 87–97.

Mendoza, G. 1965. The ovary and anal processes of *Characodon eiseni*, a viviparous cyprinodont teleost from Mexico. Biol. Bull. 129: 303–315.

Muñoz, M., M. Casadevall and S. Bonet. 2001. Gonadal structure and gametogenesis of *Aspitrigla obscura* (Pisces, Triglidae). Italian J. Zool. 68(1): 39–46.

Nagahama, Y. 1983. The functional morphology of teleost gonads. pp. 223–275. *In*: W.S. Hoar, D.J. Randall and E.M. Donaldson (eds.). Fish Physiology, Vol. IX. Academic Press, New York.

Pandian, T.J. 2011. Sexuality in Fishes. Enfield, New Hampshire: Science Publishers, Inc., pp. 189.

Parenti, L.R. and H.J. Grier. 2004. Evolution and phylogeny of gonad morphology in bony fishes. Integr. Comp. Biol. 44: 333–348.

Parenti, L.R. 2005. The phylogeny of atherinomorphs: Evolution of a novel fish reproductive system. pp. 13–30. *In*: M.C. Uribe and H.J. Grier (eds.). Viviparous Fishes. Homestead, Florida: New Life Publications.

Parenti, L.R., H.J. Grier and M.C. Uribe. 2015. Reproductive biology of *Chlorophthalmus agassizi* Bonaparte, 1840 (Teleostei: Aulopiformes: Chlorophthalmidae) as revealed through histology of archival museum specimens. Copeia 103(4): 821–837.

Patiño, R. and F. Takashima. 1995. Gonads. pp. 128–153. *In*: F. Takashima and T. Hibiya (eds.). An Atlas of Fish Histology, Normal and Pathological Features. Kodanska Ltd/Gustav Fisher Verlag. Tokyo, Stuttgart, New York.

Patiño, R. and C.V. Sullivan. 2002. Ovarian follicle growth, maturation and ovulation in teleost fish. Fish Physiol. Biochem. 26: 57–70.

Pinheiro, P., M.K. Broadhurst, F.H.V. Hazeri, T. Benzera and S. Hamilton. 2006. Reproduction in *Bagre marinus* (Ariidae) of perrambuco, northeastern Brazil. J. Appl. Ichthyol. 22: 189–192.

Ponce de León, J.L.P., R. Rodríguez, M. Acosta and M.C. Uribe. 2011. Egg size and its relationship with fecundity, newborn length and female size in Cuban poeciliid fishes (Teleostei: Cyprinodontiformes). Ecol. Freshw. Fish 20: 243–250.

Potter, H. and C.R. Kramer. 2000. Ultrastructural observations on sperm storage in the ovary of the platyfish, *Xiphophorus maculatus* (Teleostei: Poeciliidae): the role of the duct epithelium. J. Morphol. 245: 110–129.

Quagio-Grassiotto, I., H.J. Grier, T.S. Mazzoni, R.H. Nóbrega and J.P. Arruda Amorim. 2011. Activity of the ovarian germinal epithelium in the freshwater catfish, *Pimelodus maculatus* (Teleostei: Ostariophysi: Siluriformes): Germline cysts, follicle formation and oocyte development. J. Morphol. 272: 1290–1306.

Rahel, J. and L.A. Thel. 2004. Plains Topminnow (*Fundulus sciadicus*): a technical conservation assessment. Online. USDA Forest Service, Rocky Mountain Region. http://www.fs.fed.us/r2/projects/scp/assessments/plainstopminnow.pdf.

Rajalakshmi, M. 1966. Atresia of oocytes and ruptured follicles in Gobius giuris (Hamilton-Buchanan). Gen Comp. Endocrinol. 6: 378–385.

Rastogi, R.K. 1969. The occurrence and significance of ovular atresia in the freshwater mud-eel, Amphipnous cuchia (Ham), Acta Anat. 73: 148–160.

Ravaglia, M.A. and M.C. Maggese. 2002. Oogenesis in the swamp eel *Synbranchus marmoratus* (Bloch, 1795), (Teleostei, Synbranchidae). Ovarian anatomy, stages of oocyte development and micropyle structure. Biocell (Mendoza) 26: 325–337.

Reading, B.J. and C.V. Sullivan. 2011. Vitellogenesis in fishes. pp. 2272. *In*: A.P. Farrell (ed.). Encyclopedia of Fish Physiology: From Genome to Environment. San Diego, CA: Elsevier. Chapter 257.

Rizzo, E., T.F.C. Moura, Y. Sato and N. Bazzoli. 1998. Oocyte surface in four teleost fish species postspawning and fertilization. Braz. Arch. Biol. Techno. 41(1). doi.org/10.1590/S1516-89131998000100005.

Robertson, J.G. 1953. Sex differentiation in the Pacific salmon *Oncorhynchus keta* (Walbaum). Can. J. Zool. 31: 73–79.

Sales, N.G., S.A. dos Santos, F.P. Arantes, R.E.S. Hojo and J.E. dos Santos. 2013. Ovarian structure and oogenesis of Catfish *Pimelodella vittata* (Lütken, 1874) (Siluriformes, Heptapteridae). Anat. Histol. Embryol. 42(3): 213–219.

Schindler, J.F. and W.C. Hamlett. 1993. Maternal–embryonic relations in viviparous teleosts. J. Exp. Zool. 266: 378–393.

Scrimshaw, N.S. 1946. Egg size in poeciliid fishes. Copeia. (1): 20–23.

Segura-Berttolini, E.C. and M. Mendoza-Carranza. 2013. Importance of male gafftopsail catfish, *Bagre marinus* (Pisces: Ariidae), in the reproductive process. Cienc. Mar. 39(1): 29–39.

Selman, K. and R.A. Wallace. 1986. Gametogenesis in *Fundulus heteroclitus*. Amer. Zool. 26(1): 173–192.

Selman, K., R.A. Wallace and V. Barr. 1988. Oogenesis in *Fundulus heteroclitus*. V. The relationship of yolk vesicles and cortical alveoli. J. Exp. Zool. 246: 42–56.

Selman, K. and R.A. Wallace. 1989. Cellular aspects of oocyte growth in teleosts. Zool. Sci. 6: 211–231.

Sequeira, V., S. Vila, A. Neves, P. Rifes, A.R. Vieira, M. Muñoz and L. Serrano Gordo. 2011. The gelatinous matrix of the teleost *Helicolenus dactylopterus dactylopterus* (Delaroche, 1809) in the context of its reproductive strategy. Marine Biol. Res. 7(5): 478–487.

Smith, C.J. and S.R. Haley. 1987. Evidence of steroidogenesis in postovulatory follicles of the tilapia, *Oreochromis mossambicus*. Cell Tissue Res. 247: 675–687.

Strüssmann, C.A. and M. Nakamura. 2002. Morphology, endocrinology and environmental modulation of gonadal sex differentiation in teleost fishes. Fish Physiol. Biochem. 26: 13–29.

Taylor, R.G., H.J. Grier and J.A. Whittington. 1998. Spawning rhythms of common snook in Florida. J. Fish Biol. 53: 502–520.

Thibault, R.E. and R.J. Schultz. 1978. Reproductive adaptations among viviparous fishes (Cyprinodontiformes: Poeciliidae). Evolution 32: 320–333.

Thiry, M. and P. Poncin. 2005. Morphological changes of the nucleolus during oogenesis in oviparous teleost fish, *Barbus barbus* (L.). J. Struct. Biol. 152(1): 1–13.

Turner, C.L. 1933. Viviparity superimposed upon ovoviparity in the Goodeidae, a family of cyprinodont teleosts fishes of the Mexican Plateau. J. Morphol. 55: 207–251.

Turner, C.L. 1947. Viviparity in teleost fishes. Sci. Monthly 65: 508–518.

Tyler, C.R. and J.P. Sumpter. 1996. Oocyte growth and development in teleosts. Rev. Fish. Biol. Fisher 6: 287–318.

Tyler, C.R., T.G. Pottinger, E. Santos, J.P. Sumpter, S.A. Price, S. Brooks and J.J. Nagler. 1996. Mechanisms controlling egg size and number in the Rainbow Trout, *Oncorhynchus mykiss*. Biol. Reprod. 54: 8–15.

Uribe, M.C., G. De la Rosa-Cruz and A. Garcia-Alarcón. 2005. The ovary of viviparous teleosts. Morphological differences between the ovaries of *Goodea atripinnis* and *Ilyodon whitei*. pp. 217–235. *In*: M.C. Uribe and H.J. Grier (eds.). Viviparous Fishes. New Life Publications, Homestead, FL.

Uribe, M.C., G. De la Rosa-Cruz, A. García-Alarcón, S. Guerrero-Estévez and M. Aguilar-Morales. 2006. Características histológicas de los estadios de atresia de folículos ováricos en dos especies de teleósteos vivíparos: *Ilyodon whitei* (Meek, 1904) y *Goodea atripinnis* (Jordan, 1880) (Goodeidae). Hidrobiológica. 16(1): 255–264.

Uribe, M.C., H.J. Grier, G. De La Rosa-Cruz and A. García-Alarcón. 2009. Modifications in ovarian and testicular morphology associated with viviparity in teleosts. pp. 85–117. *In*: B.G.M. Jamieson (ed.). Reproductive Biology and Phylogeny of Fishes (Agnathans and Bony Fishes), Vol. 8A. Chapter 3. Enfield, New Hampshire: Science Publishers, Inc.

Uribe, M.C. and H.J. Grier. 2011. Oogenesis of microlecithal oocytes in the viviparous teleost *Heterandria formosa*. J. Morphol. 272: 241–257.

Uribe, M.C., H.J. Grier and L.R. Parenti. 2012. Ovarian structure and oogenesis of the oviparous goodeids *Crenichthys baileyi* (Gilbert, 1893) and *Empetrichthys latos* Miller, 1948 (Teleostei, Cyprinodontiformes). J. Morphol. 273: 371–387.

Uribe, M.C., G. Dela Rosa-Cruz and H.J. Grier. 2014. Proliferation of oogonia and folliculogenesis in the viviparous teleost *Ilyodon whitei* (Goodeidae). J. Morphol. 275: 1004–1015.

Uribe, M.C., H.J. Grier, A. García-Alarcón and L.R. Parenti. 2016. Oogenesis: From oogonia to ovulation in the flagfish, *Jordanella floridae* Goode and Bean, 1879 (Teleostei: Cyprinodontidae). J. Morphol. DOI: 10.1002/jmor.20580.

Viedma, R., J. Franco, C. Bedia, G. Guedea Fernández, H Villa Zevallos and H. Barrera Escorcia. 2011. Estructura y ultraestructura del ovario de *Cichlasoma urophthalmus* (Perciformes: Cichlidae). Rev. Biol. Trop. 59(2): 743–750.

Wallace, R.A. and K. Selman. 1981. Cellular and dynamic aspects of oocyte growth in teleosts. Am. Zool. 21: 325–343.

Wallace, R.A. and K. Selman. 1990. Ultrastructural aspects of oogenesis and oocyte growth in fish and amphibians. J. Electron Micr. Tech. 16: 175–201.

Wildner, D.D., H.J. Grier and I. Quagio-Grassiotto. 2013. Female germ cell renewal during the annual reproductive cycle in Ostariophysians fish. Theriogenology 79(4): 709–724.

Wourms, J.P. 1981. Viviparity: The maternal-fetal relationship in fishes. Am. Zool. 21: 473–515.

Wourms, J.P., B.D. Grove and J. Lombardi. 1988. The maternal-embryonic relationship in viviparous fishes. pp. 1–134. *In*: W.S. Hoar and D.J. Randall (eds.). Fish Physiology, Vol. XIB. New York: Ac Press.

Wourms, J.P. and J. Lombardi. 1992. Reflections on the evolution of piscine viviparity. Am. Zool. 32: 276–293.

Yanwirsal, H., P. Bartsch and F. Kirschbaum. 2017. Reproduction and development of the asian bronze featherback *Notopterus notopterus* (Pallas, 1769) (Osteoglossiformes, Notopteridae) in captivity. Zoosyst. Evol. 93(2): 299–324. DOI: 10.3897/zse.93.13341.

Chapter 11

Egg Envelopes

*Krzysztof Formicki** and *Agata Korzelecka-Orkisz*

INTRODUCTION

The egg envelope is the external part of a mature fish egg. The fish egg envelope has no cells or other characteristics of a living matter and is thus of acellular structure. However, it remains in close connection with the living matter—the developing embryo which it provides with permanent contact with the external environment depending on the requirements. The egg envelope is a dynamic structure and its multiple functions resulted, in the course of evolution, in differentiation of its structure. The structure of the egg envelope is adapted to environmental conditions in which the egg develops. It plays a variety of functions which change during the embryonic development (embryogenesis). During egg activation, it attracts spermatozoa and prevents polyspermy, protects the embryo against mechanical damage, later it protects the developing embryo from microorganisms and other small organisms, and ensures gas exchange; it also plays an important role in the regulation of the egg's water balance, makes it possible to excrete metabolic wastes, and in many fish species ensures anchoring in the substratum in a place which is adequate for the embryo's development. The last function can be performed in two ways—through special structures, e.g., attachment filaments on the surface of egg envelope or through special sticky layers. The egg envelope constitutes also a partial barrier for environmental pollutants. The egg envelope of some fish species (*Cyprinus carpio, Tribolodon hakonensis* of the Cyprynidae and *Plecoglossus altivelis* of the Plecoglossidae) are surrounded by an additional layer which forms a chemical barrier of strong bacteriostatic properties due to the presence of hydrogen peroxide. Such a variety of functions are possible thanks to the complex architecture of this external part of the egg (Riehl 1996, Dumont and Brummet 1980, Kudo et al. 1988, Kudo and Inoue 1986, 1988, Riehl and Appelbaum 1991, Depêche and Billard 1994, Paxton and Willoughby 2000, Kunz 2004, Sadowski 2004, Siddique 2016, Pelka et al. 2017, Riehl and Patzner 1998, Dulčić et al. 2008, Żelazowska 2010).

The egg envelopes are produced by ovaries and oviducts and surround the central (live) part of the egg: the genetic material and reserve substances (yolk). The structure of the fish egg envelope depends on the species (Riehl and Schulte 1977, Riehl 1979, Mikodina 1987). The external structure of the egg envelope can be used as a character to identify eggs of teleost fishes (Riehl and Schulte 1978, Riehl 1993, Riehl and Kokoscha 1993). However, depending on ecological factors such characters may vary within some limits, even within the same species (Ivankov and Kurdyayeva 1973, Riehl 1978, Li et al. 2000).

Department of Hydrobiology, Ichthyology and Biotechnology of Reproduction; West Pomeranian University of Technology in Szczecin (Poland).
* Corresponding author: Krzysztof.Formicki@zut.edu.pl

Egg envelopes occur in two forms: as membranes or as shells. The division is based not only on the structural differences, but also on the way they are formed (Hamazaki et al. 1989, Rościszewska 1994, Heiden et al. 2005). The egg envelopes of teleost and chondrichthyan fishes differ. The structure of egg envelopes in Chondrichthies will be discussed below.

Because of the non-uniform terminology and different origin of particular membranes, several attempts were made concerning their classification. Homologous structures have different names in literature, while different structures are given the same names and hence it was suggested to classify the envelopes based on the place of their origin. In 1874, Ludwig proposed one of the most commonly known classifications of egg envelopes, based on the site of origin:

– primary membranes (envelopes)—formed by the oocytes (oocyte origin) within the ovary-zona radiata,
– secondary membranes (envelopes)—formed by the ovarian follicle cells, especially in phytophilous fishes: homogenous membranes which surround the whole zona radiata of the egg or its part, and serve for attaching the egg to the substratum when it becomes sticky upon contact with water,
– tertiary membranes (envelopes)—formed by the cells of oviduct or lower parts of reproductive ducts.

The classification was later adopted by numerous researchers, among others by Ginzburg (1968), Anderson (1967, 1974), Dumont and Brumet (1980), Yamagami et al. (1992).

One of the first researchers to perform histological examination of fish egg envelopes was Henneguy (1889). He studied salmonid eggs and found that the envelope was perforated by radial canals which, according to him, served exchange of substances. He introduced the term zona radiata. There is a great variety in the nomenclature of external membrane of the teleost egg. The terms used are zona radiata, zona pellucida, chorion, radiate membrane, egg membrane, primary membrane, vitelline membrane, vitelline envelope, egg envelope, egg shell, or egg capsule (Yamagami et al. 1992, Bielańska-Osuchowska 1993, Kunz 2004, Mekkawy and Osman 2006).

Most often, particular layers of fish egg envelopes are termed: oolemma—internal layer surrounding the oocyte, zona radiata—middle layer of the egg envelope, and viscous layer—the outermost layer of egg envelope. In oviparous chondrichthyan fishes, the embryos are surrounded by hard capsules which may bear long adhesive fibres to anchor the eggs in the substratum. The oolemma is also called plasma membrane, plasmalemma or vitelline membrane (Depêche and Billard 1994, Kunz 2004, Heiden et al. 2005).

Oolemma

On the outside, the teleost egg cell is surrounded by a very thin membrane—the oolemma, which is the egg cell membrane and is located between the zona radiata, and the layer of cytoplasm surrounding the whole yolk. The external layer of the cytoplasm holds cortical alveoli filled with mucopolysaccharides, which upon contact with water (and possibly penetration by a spermatozoon) extrude their contents through exocytosis into the space between the oolemma and the zona radiata; as a result, the oolemma moves away from the zona radiata, to form a perivitellin space, while the high osmotic pressure causes intensive water absorption from the external environment (Kryżanowski 1960, Yamagami et al. 1992, Scapigliati et al. 1994, Korzelecka et al. 1998, Tański et al. 2000, Korzelecka-Orkisz et al. 2005, 2012). Since the oolemma constitutes a barrier for water and ions, the development of eggs of such fishes as the medaka *Oryzjas latipes* or asp *Aspius aspius* without an external egg envelope—the zona radiata—is possible (Smithberg 1969, Depêche and Billard 1994, Korzelecka-Orkisz et al. 2013).

Zona Radiata

Irrespective of the species, the zona radiata in teleost eggs is built up of two layers: a thin external layer (zona radiata externa–ZRE) and a thicker internal layer (zona radiata interna–ZRI) (Fig. 1) (Balon 1977, Riehl 1991, Riehl and Patzner 1998). Depending on the family, ánd even species of fish, the layers differ in thickness and vary greatly in their structure (Figs. 2 and 3) (Lönning 1972, Yamamoto and Yamagami 1975, Johnson and Werner 1986, Mayer and Shackley 1988, Jasiński 2004). The significance of the consecutive layers of the egg envelope and its composition is not completely understood. It is known, however, that the different layers react to chemical compounds and enzymes in different ways. These differences depend also on the developmental stage of the egg.

According to some authors who studied the structure of the zona radiata, three layers can be distinguished: Z1, Z2, Z3, from the outside to the inside of the egg. In *Xenoophorus captivus*, Z1 is a thin homogeneous layer, Z2 is also thin, opaque and granular, and Z3 is the thickest layer with a lamellate structure (Schindler and Vries 1989). In zebrafish *Danio rerio*, the egg envelope is also composed of three layers penetrated by pore canals (Hisaoka 1958, Laale 1977, Bonsignorio et al. 1996, Rawson et al. 2000); the canals are evenly distributed over the whole surface of the unfertilised egg (Hart and Donovan 1983). In the egg of zebra fish *Danio rerio*, the envelope is also composed of three layers: (1) An external, electron dense, the zona radiata externa containing pore canal plugs; (2) A middle layer, the superficial zona radiata interna, which is of fibrillary structure; (3) An internal layer, the deep zona radiata interna built of 16 electron dense, horizontally arranged lamellae penetrated by patent pore canals (Hart and Donovan 1983).

The egg envelope of the Antarctic fish *Chionodraco hamatus* (Notothenioidei) is organised in a different way compared to non-polar species. It consists of several concentric layers of different thickness. The thickness of the zona radiata of the unfertilised egg is 50 μm. The external surface of the egg envelope is acellular and sponge-like. The external layer is built of lamellae forming parabolic arches, the middle layer has irregular lamellae, and the internal layer, the thickest one, is composed of interconnected lamellae. That layer also contains fibrous material arranged in parabolic arches (Baldacci et al. 2001).

The structure of the zona radiata changes not only during ontogeny, its spatial structure changes during egg activation and at the end of embryogenesis. Bian et al. (2010) analysed the structure of mature, still unfertilised, as well as fertilised and developing eggs in four species of flounders. In the starry flounder *Platichtys stellatus*, the surface of unfertilised eggs is covered with irregular grooves creating an impression of wrinkling. The pores on the egg surface form a hexagonal pattern, and around each pore canal there are six other pore canals of the same diameter. In fertilised eggs, the above-mentioned grooves on the external surface of the egg envelope are less visible, and the surface around the pores is thickened and raised. In unfertilised eggs, the micropyle is cylindrical with a small, shallow vestibule, while the pores and shallow concavities of different size are irregularly distributed. When the embryos are at the stage of blastopore closure, the closed micropyle in the zona radiata appears to be stretched and the rough micropylar canal does not have such distinct spiral-shaped ridges.

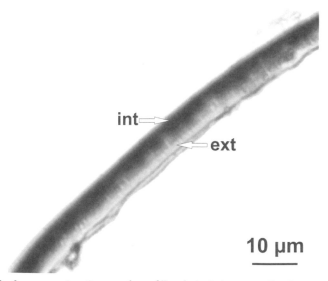

Fig. 1. Light micrograph of a cross-section of egg envelope of *Esox lucius* L. int, zona radiata interna; ext, zona radiata externa.

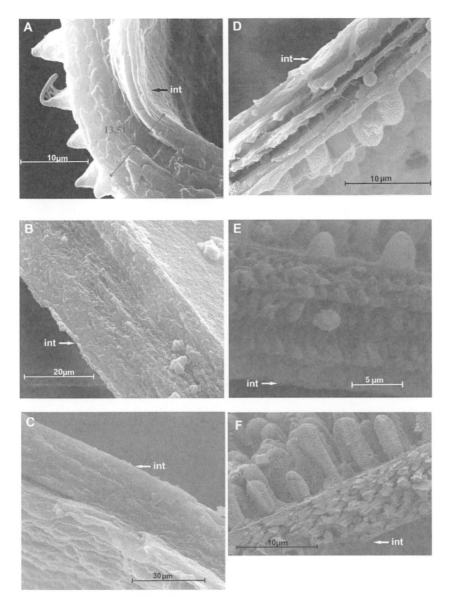

Fig. 2. Cross-section of egg envelopes (SEM) of *Coregonus lavaretus* (L.) (**A**), *Salmo trutta* (**B**), *Salmo salar* (**C**), *Leuciscus idus* (**D**), *Aspius aspius* (**E**), and *Vimba vimba* (**F**). Arrows point to the internal surface (int) of the egg envelope. **D**, **E**, **F**. Sticky plugs attach the eggs to the substrate; in cyprinoid fishes, layers in the egg envelope are loosely arranged.

Thickness of Egg Envelopes

The egg size is correlated with its surface and thus with the surface of the zona radiata, since its dimensions laid down in the ovaries at the moment of its formation determine the volume of the egg interior which should accommodate both the ectoplasm of the future egg cell and its most important component, the nucleus as well as the storage containing yolk and perivitelline space. This space during activation becomes filled with perivitelline fluid formed by hydrophilous colloids from the contents of bursting cortical alveoli and the water "sucked in" by them from the outside (Bogucki 1930, Yamamoto 1961, Afzelius et al. 1968).

The principal role of the egg envelope is protection of the embryo against mechanical forces that act on the egg (Mansour et al. 2009a, b).

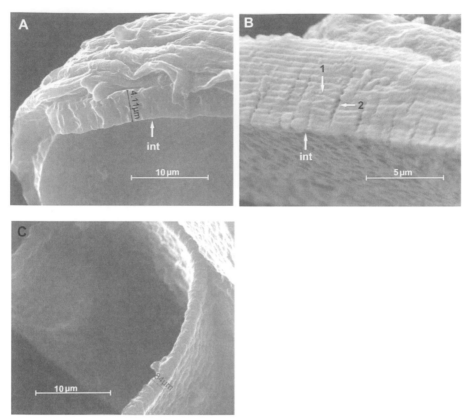

Fig. 3. Cross-section of egg envelopes (SEM). **A.** *Perca fluviatilis.* **B.** Note on the external surface tubular structures in the gelatinous layer (*Esox lucius*). 1—tighly packed lamellae with small channel between the internal layer (int); 2—radial canals. **C.** *Branchydanio rerio* (Hamilton-Buchanan 1822).

The ultrastructure and thickness of egg envelopes depends on the environmental conditions of spawning; this is why they vary so much in their structure (Riehl 1978, Stehr and Hawkes 1979, Li et al. 2000, Andrade et al. 2001). The more exposed the laid egg is to mechanical stress, the thicker the egg envelope. The eggs of teleost fishes which develop in water column have relatively thinner envelopes in relation to the egg diameter, while the eggs developing on the bottom have relatively thicker envelopes. In *Brachydanio rerio*, the envelope thickness is 2.5 µm (Rawson et al. 2000), in *Melanogrammus aeglefinus* 8.5 µm (Morrison et al. 1999), in *Romanichtys valsanicola* 30 µm, in many species laying eggs on gravel (Salmonidae) the thickness may reach 70 µm, while in species spawning on reefs or in surf region of shores (e.g., Agonidae, hook-nose *Agonus cataphractus*) envelope thickness may even go up to 100 µm (Göting 1967). When the eggs develop inside the parent's body, the thickness of their envelopes is significantly smaller—in mouthbrooders the egg envelope thickness ranges from 3 to 5 µm. In teleosts, whose eggs are only slightly burdened, for example, those hatching in mud or in viviparous species, the envelope is especially thin. Among the Goodeidae or viviparous species of Poeciliidae, the envelope thickness is only 0.3–2 µm at an egg diameter of 1.5 mm (Riehl and Greven 1993, Pelka et al. 2017). The egg envelope in zebrafish (*Danio rerio*) is equally thin, 1.5 µm to 2.5 µm thick (Fig. 3C) (Pelka et al. 2017). The envelope of eggs of *Platichthys stellatus*, which develops in the water column, is 2–5 µm thick (egg diameter 0.9 mm), and the egg envelope of the European white-fish *Coregonus*, which develop on sandy bottom, is 29 µm thick (Fig. 2A) (egg diamater 1.15–1.56 mm). Likewise, eggs of phytophilous fishes which are attached to various kinds of substrats, such as vegetation, where they have good conditions for development, have thinner envelopes, for example pike *Esox lucius* (sticky eggs, attached to vegetation, diameter 3 mm) (Fig. 3B) and carp *Cyprinus carpio* (eggs of 1.15–1.56 mm) have envelopes 1.5–8 µm thick. Salmonid eggs are buried in gravel which requires high mechanical resistance: the egg envelope in trout *Salmo trutta* is 50 µm thick (egg diameter 4.5–6 mm) (Fig. 2B), in salmon *Salmo salar* 43 µm (Fig. 2C) (Zotin 1953, Hisaoka 1958, Laale 1977, Vorobieva et al. 1986, Bonsignorio et al. 1996, Rawson et al. 2000). In the Antarctic fish *Chionodraco hamatus* (Nototenioidei), the egg envelope is 50 µm thick, which is associated with the specific conditions of the region (Baldacci et al. 2001).

Experiments on the cod *Gadus morhua* revealed differences between egg envelopes of conspecific spawners. The thickness of egg envelope in the cod from northern Norway was greater than in the southern cod. According to the authors, it might be associated with the ecological adaptation to ensure egg buoyancy in conditions of higher salinity (Kjesbu et al. 1992).

The structure and thickness of the egg envelope change after fertilisation. The egg envelope of the unfertilised egg of pike *E. lucius* is 0.6–0.9 µm thick, and its external surface has a honeycomb structure whose concavities (0.8–1.6 µm in diameter) hold centrally located orifices which lead to canals which penetrate the envelope. The inner part of the egg envelope is thicker, it ranges from 7–9 µm, and is composed of 10–12 layers, and canals are also visible in it. Following fertilisation, the envelope structure undergoes changes—the outer layer retains its honeycomb structure, but it is less distinct. The envelope thickness also changes; the envelope becomes thinner, 6–8 µm (external 3.5–4 µm, internal 0.3–0.5 µm), the canal diameter increases, and the shape changes from circular to oval (Gajdusek and Rubcov 1983) (Figs. 3B, 5F).

Resistance of Egg Envelope

Upon contact with water teleost eggs, independent from fertilisation, absorb water from the external environment into the perivitelline space, resulting in stretching of the zona radiata. The contact with water triggers processes in the egg envelope which cause an increase in the envelope's resistance. During the first hours after fertilisation, the egg envelope straightens and hardens which prevents polyspermy and protects the developing embryo (Pelka et al. 2017). The first reports on "egg hardening" upon contact with water date from the 19th century (Vogt 1842, Ransom 1854). In 1942, Hayes interpreted it as the effect of water on the egg envelope, and called the process "water hardening". The hardening of the zona radiata follows the so called egg activation and contact with water and is independent from fertilisation. It results from releasing of colloid substances into the perivitelline space. The biological active substance shows enzymatic properties and causes a very rapid increase in the resistance of the egg envelope (Kusa 1949, Yamamoto 1957, Zotin 1958, Iwamatsu 1969, Iuchi et al. 1985, 1996, Masuda et al. 1991, Iwamatsu et al. 1995, Robles et al. 2007). The teleost egg envelope is composed of 3–4 proteins. These proteins are probably responsible for the hardening of the envelope after fertilisation (Modig et al. 2006, 2007, Lubzens et al. 2010).

The resistance of egg envelopes can vary depending on environmental conditions in which the fish has been bred for generations. There are certain regularities, namely: the resistance is the greatest in lithophilous species, especially those which lay eggs in special nests dug in gravel, smaller in those which spawn (incubate their eggs) in rubble and the smallest in phytophilous species and those spawning in the pelagial. There is a distinct dependence between the resistance and the thickness of egg envelopes.

Studies on eggs of rainbow trout (*Oncorhynchus mykiss* Walb.) showed that the resistance of egg envelopes immediately after oviposition was less than 20 g, while 24 hours after fertilisation it increased from 2 to 3 kg (Iuchi et al. 1996): according to Dąbrowski and Stanuch (1998) as well as Jasiński (2004), it may even reach 5 kg. The resistance of rainbow trout *Oncorhynchus mykiss* egg envelopes on the 4th day after fertilisation was ca. 2 kg, and in trout *Salmo trutta* ca. 1.2 kg (Formicki 1986). Among the trout *Salmo trutta* eggs, single cases of increase in the resistance of egg envelope to more than 14 kg were observed (Winnicki and Domurat 1964). The resistance of the egg envelope of Atlantic salmon *Salmo salar* L. before fertilisation was on average from 45 to 97 g; 12 hours after fertilisation it ranged from 514 to 770 g, and 60 hours after fertilisation from 888 to 925 g. The resistance of egg envelopes 12 hours after activation of eggs of spawners from the Vistula River was nearly twice as high as that of egg envelopes of spawners from the Wieprza River; 60 hours after activation the difference decreased to

1/4. The resistance of egg envelopes of the Atlantic salmon was also found to change over a few years (Biernaczyk et al. 2012).

The resistance of egg envelopes achieved in the first minutes or hours remains at a very high level during embryonic development and decreases only immediately before hatching (Zotin 1958, Winnicki 1967a, b, Sobociński and Winnicki 1974, Lönning et al. 1984, Iuchi et al. 1996, Sobociński et al. 2005).

External and Internal Surface of the Zona Radiata

The surface of the fish egg is often covered with a characteristic sculpture or tubercular outgrowths (Figs. 4–8). The surface sculpture may change even within species, depending on the geographical region (Riehl and Kokoscha 1993, Długosz 1994, Patzner and Gleichner 1996).

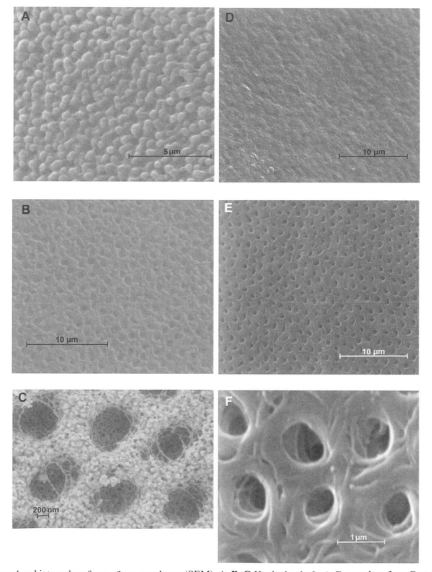

Fig. 4. External and internal surfaces of egg envelopes (SEM). **A, B, C** *Hucho hucho* L. **A**. External surface; **B** and **C** internal surfaces. **D, E, F**. *Salmo trutta*. **D**. External surface. **E** and **F** internal surface. Note the numerous pores (cavities) arranged hexagonally; these cavities may be the radial canals with their openings on the surface (Figure 4C with kind permission from Carl Zeiss Microscopy GmbH and Dr. Martin Dass).

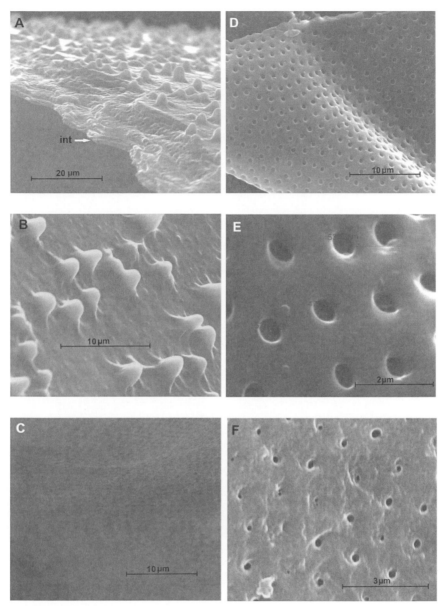

Fig. 5. External and internal surfaces of egg envelopes (SEM). **A**, **B**, **C.** *Coregonus lavaretus* (L.). External surface (**A**, **B**) and internal surface (int) (**C**). **A.** External surface with a visible break of the egg envelope. **B.** Knobs on the external surface of the egg envelope. **D**, **E.** *Danio rerio,* internal surface. **F.** *Esox lucius,* internal surface; note hexagonal pattern of the canals.

The external surface may bear pores—entrances to canals whose number and size do not have to be the same over the whole egg surface. The occurrence of pores is observed, among others, in *Platichthys stellatus* and *Oncorhynchus gorbuscha* (Stehr and Hawkes 1979), *Epinephelus coioides*, *Epinephelus malabaicus* (Li et al. 2000) and in *Serrasalmus spilopleura* (Rizzo et al. 2002). In the case of the three families Characidae, Anostomidae, and Curimatidae, the diameter and density of canals increases with decreasing distance from the micropyle (Rizzo et al. 2002).

Often the envelope of the fish egg has a honeycomb structure, among others, in *Polycentropsis abbreviata* (Britz 1997), *Zingel streber* (Patzner et al. 1994), *Romanichthys valsanicola* (Riehl and Bless 1995), *Gymnocephalus cernuus,* and *Gymnocephalus baloni* (Riehl and Meinel 1994). In the zebrafish,

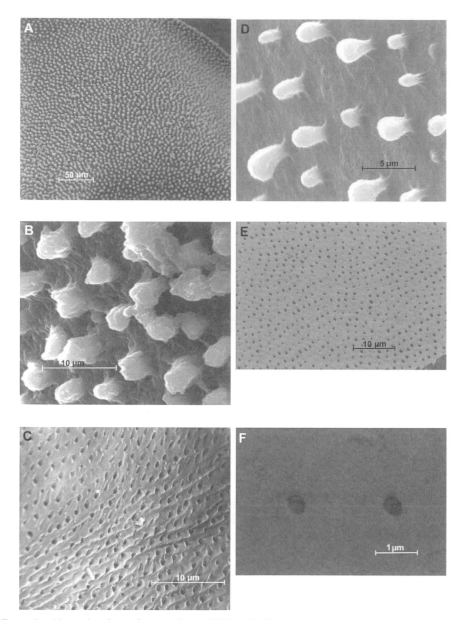

Fig. 6. External and internal surfaces of egg envelopes (SEM). **A, B, C.** *Leuciscus idus.* **A, B** external, **C** internal surface. **D, E, F.** *Vimba vimba.* **D** external, **E** and **F** internal surface. **B, D.** Attachment knobs (projections) of different size and density allow gas exchange.

the pores on the envelope are arranged fairly evenly, and the canal diameter is 0.2 μm in unfertilised eggs (Hart and Donovan 1983) and 0.5–0.7 μm in fertilised eggs at gastrula stage (Rawson et al. 2000).

Another modification is the presence of particular villi on the surface of the zona radiata—they attach the egg to the substratum. Depending on the species, they can have the form of small tubercles, e.g., in *Schizodon knerii* (Rizzo et al. 2002); they are elongated in the region of the animal pole, and become flattened toward the micropylar region (a kind of adhesive pads). Such structures are characteristic of cyprinid eggs, for example in the chub *Leuciscus cephalus*, vimba *Vimba vimba* and roach *Rutilus rutilus* (Riehl and Patzner 1998).

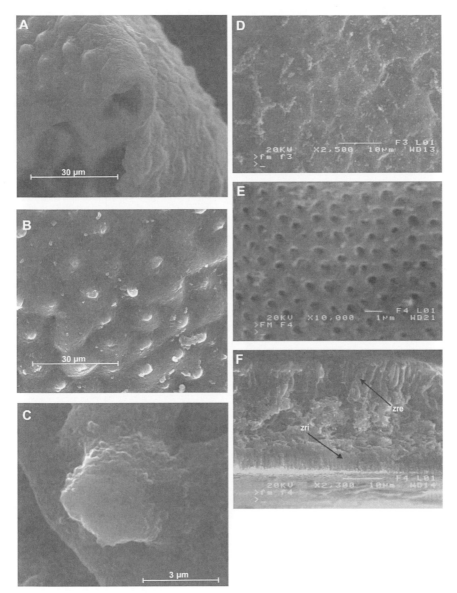

Fig. 7. External and internal surfaces of egg envelopes (SEM). **A, B, C.** *Heteropneustes fossilis.* **A.** Cross-section of egg envelope. **B, C.** External surface; the knobs contribute to the stickiness of the zona radiata externa. The nearby pits facilitate penetration of oxygen through the egg membrane. **D, E, F.** *Ancistrus dolichopterus.* **D.** External surface with honeycomb pattern. **E.** Internal surface. **F.** Cross-section of egg envelope. ZRE, zona radiata externa; ZRI, zona radiata inerna (Figs. **A, C** after Korzelecka-Orkisz et al. 2010; Figs. **D, E, F** after Brysiewicz et al. 2011 with permission from Publishing House West Pomeranian University of Technology in Szczecin).

Fibres which fasten the egg to the sustratum represent another modification of the zona radiata. They were found in eggs of the freshwater families Cichlidae, Gobiidae, Blenniidae, Serrasalmidae, Clarriidae, Pseudochromidae (Riehl and Patzner 1998) as well as in Nandidae and Badidae (Britz 1997) or Belonidae (Korzelecka-Orkisz et al. 2015).

In salmonids, the zona radiata differs from that of the above-mentioned taxa. The external layer is smooth and without any distinct structure, the internal layer is thicker and perforated by canals which are perpendicular to the surface. On cross sections no distinct layers are visible, unlike other families and species (Vorobieva et al. 1986).

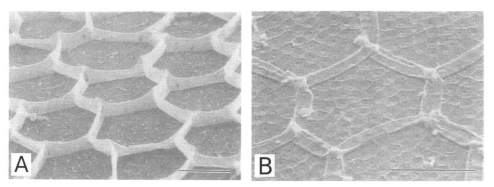

Fig. 8. External and internal surfaces of egg envelopes. **A, B.** *Pleuronichthys cornutus.* **A.** External surface, sculpturing of the egg membrane consists of regular, hexagonally-arranged walls. **C.** Higher magnification of the external surface; note pore canal openings between the walls (Fig. **A, B** after Hirai 1993; with permission from the Ichthyological Society of Japan).

Detailed structural analysis of the zona radiata in the Pleuronectinae showed that the egg envelopes varied both in thickness and structure. In pelagic eggs of *Hippoglossoides dubius* the zona radiata is smooth, as opposed to those of *Pleuronichthys cornutus* in which the surface of the zona radiata is sculptured, composed of regular, hexagonally-arranged units (3 μm high and 10 μm long); inside each hexagonal unit there are much smaller polygonal areas (ca. 1.5 μm in diameter), and in the middle of each polygonal area there is a canal opening (Fig. 8B). The zona radiata is composed of six lamellae of equal thickness. The area around the micropyle is devoid of this characteristic structure (Hirai 1993).

The egg envelope in *Pleuronectes yokohamae* consists of 10 lamellae of different thickness. The external lamellae are 5 times thicker than the internal ones (Hirai 1993).

Depending on additional structures on the surface of the egg envelope, or their absence, playing a role in adapting the egg envelope in the habitat, the following categories can be distinguished:

- Eggs without sticky external envelope and not sticking to the substratum—salmon, *Salmo salar* L., rainbow trout, *Oncorhynchus mykiss* (Walb.), trout, *Salmo trutta* L., Danube salmon, *Hucho hucho* (L.),
- Poorly adhering eggs—chub (*Leuciscus cephalus*), streber (*Zingel streber*),
- Eggs with very sticky envelopes—vimba (*Vimba vimba*), catfish (*Silurus glanis*) (Korzelecka et al. 2010), tench (*Tinca tinca*) (Korzelecka-Orkisz et al. 2009).

Till now, more than a couple of adhesive mechanisms closely associated with the structure of egg envelopes have been described: zona radia externa or muco-follicular epithelium (Terminology according to Riehl 1996).

Adhesive mechanisms associated with the zona radiata externa include the occurrence of:

- An adhesive layer, when the surface of the zona radiata is smooth and devoid of adhesive villi or filaments; examples are eggs of the pike *Esox lucius* and *Esox masquinongy* (Riel and Patzner 1992), of the burbot *Lota lota*, the Danube bleak *Chalcalburnus chalcoides mento* (Riel 1993), or the minnow *Phoxinus phoxinus* (Riel and Schulte 1977).
- A honeycomb surface—honeycomb structures occur as a rule on the whole surface of the egg envelope. The stickiness of the envelope is due to acid mucopolysaccharides on the outermost layer; this kind of adhesive mechanism occurs, for example, in the streber *Zinger streber* (Patzner et al. 1994), the zingel *Zingel zingel*, the Romanian perch *Romanichthys valsanicola* Dumitrescu, Banarescu and Stoica 1957 (Riehl and Bless 1995), the ruffes *Gymnocephalus cernutus* (Riehl and Meinel 1994, Riehl and Werner 1994), and *Pleuronichthys cornutus* (Hirai 1993) (Fig. 8).
- Villi-like processes evenly distributed over the whole surface of the zona radiata, frequently occurring in cyprinids, for example in *Leciscus leuciscus* the villi are short, 5 μm long, spaced every 6 μm (Figs. 6A, B), in *Vimba vimba elongata* the villi are even shorter, 4 μm long (Fig. 6D) (Riehl

et al. 1993), in roach *Rutilus rutilus* they are relatively longer, of 11.5 μm in length, and very closely spaced (Riehl 1996), while in *Alburnoides bipunctatus* the villi vary in length, the longest being 10 μm long (Glechner et al. 1993). Sticky villi of different length and diameter were also observed in the stinging catfish *Heteropneustes fossilis* (Fig. 7) (Korzelecka-Orkisz et al. 2010).

– Attachment filaments are much longer than villi, arranged in bundle-like assemblages, which may be located on the animal (e.g., in goby eggs) or vegetal pole (e.g., in cichlid eggs), depending on which pole serves the attachment of the egg to the substratum. Oval cichlid eggs are laid in cavities and attached to the substratum in such a way that the micropyle is on the opposite side from the animal pole, while eggs of other cichlids are attached with their longer side, with the micropyle facing sideways, on the shorter side. In *Atherina boyeri,* attachment filaments are evenly distributed around the egg envelope (Fig. 9).

In some species, the arrangement of attachment filaments is regular (Terminology according to Riehl 1996):

– Attachment filaments arranged on the disc—the disc is composed of a large number of short attachment filaments arranged around the micropyle in the centre of its area (e.g., piranha) (Wirz-Hlavacek and Riehl 1990).

– Attachment filaments forming a bulge built of numerous, short filaments, for example in a clarid *Clarias gariepinus* (Riehl and Applebaum 1991).

– Jelly layer—in some fish species between the zona radiata and the follicular cells of the ovary, there arises a gelatinous egg envelope. It surrounds the whole egg except the micropylar region. The layer is formed before egg-laying, in the ovary or the oviduct. The jelly layer is mainly built of glycoproteins (Kobayashi 1982, Długosz 1994). After egg-laying, the layer swells under the effect of water and

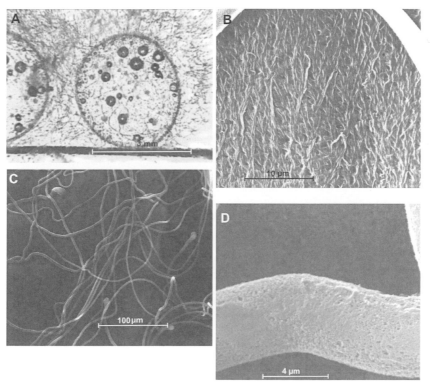

Fig. 9. Egg envelope of *Atherina boyeri,* Risso, 1810. **A.** Developing egg. **B.** External surface with fibrillar net. **C.** Higher magnification of the sticky filaments attaching the egg. **D.** Higher magnification of a single filament. **A**—light microscopy; **B, C, D**—SEM.

the egg surface becomes sticky and forms lace-like tapes fastened to the vegetation (Flügel 1967, Kucharczyk et al. 1997, Riehl and Patzner 1998, Korzelecka et al. 1998, Mansour et al. 2009b); their role is to protect the eggs with the embryos against mechanical damage. The jelly layer also enables the eggs to float in the water column to ensure optimum conditions for gas exchange, and constitutes a reservoir and filter of water absorbed by the egg (Riehl and Patzner 1998, Korzelecka et al. 1998). This enables the eggs to attach to substrata of different physical properties. An example is the perch *Perca fluviatilis*, whose eggs are laid in form of long tapes fastened to the vegetation (Fig. 10A) (Wintrebert 1923, Flügel 1966, Riehl and Patzner 1998, Formicki et al. 2009). Tubular structures were observed just below the surface of the jelly layer (Fig. 3A). Some of them open to the outside (on the surface) as ring-like structures resembling nozzle-like openings (Formicki et al. 2009). These opening are entrances to canals which have their own walls. The canals are located in the jelly layer covering the egg and near the junctions in the jelly layer. It is very likely that the system of tubular structures in the jelly layer plays a skeletal part; it is characterised by elasticity and by high resistance to stretching, and ensures preserving of the original shape of the tape after it has been bent by water currents. The microtubules in the jelly layer of the tape play a part in water and gas exchange on the boundary of the jelly layer and the zona radiata (Formicki et al. 2009) (Figs. 10B–D).

The jelly layer is also present in the stickleback (*Gasterosteus aculeatus*) (Yamamoto 1963, Winnicki et al. 1998) and in *Dendrochirus brachypterus* (Riehl and Patzner 1998). In most sturgeons, adhesiveness is ensured by the outer jelly-like layer, produced by the follicular epithelium (Dettlaff

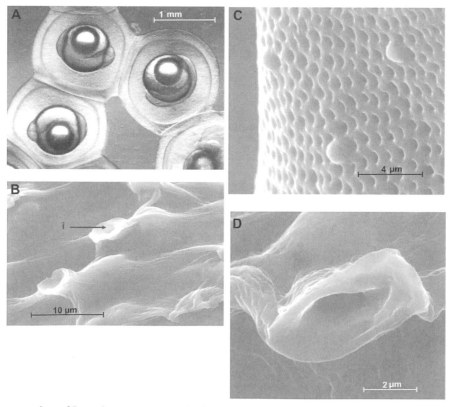

Fig. 10. Egg envelope of *Perca fluviatilis* L. **A.** Developing eggs, each surrounded by a gelatinous layer. **B.** External surface, note the pores of the microtubular network, arrow indicates nozzle-like opening of the microtubular network. **C.** Internal surface. **D.** Higher magnification of the single pore of the microtubular network. A—light microscopy; **B, C, D**—SEM (**A, B, D** after Formicki et al. 2009; with permission from Publishing House West Pomeranian University of Technology in Szczecin).

et al. 1993, Vorob'eva and Markov 1999). The jelly coat varies in thickness depending on the sturgeon species. In *Acipenser transmontanus*, it is 0.3–0.64 µm thick (Clark et al. 1982), in *Acipenser baerii* 0.5–0.9 µm (Dulčić et al. 2008).

- Adhesive mechanisms associated with the muco-follicular epithelium—mucosomes-developed during ontogeny in the follicular epithelium are released during spawning which causes stickiness of eggs in, e.g., the catfish *Silurus glanis* (L.) (Abraham et al. 1993, Korzelecka-Orkisz et al. 2010).

Micropyle

The envelope of the fish egg, and properly the zona radiate, is relatively thick which prevents penetration by a spermtozoon through local lysis of the zona radiata, unlike in other taxa. The envelope of the fish egg has a micropyle located at the animal polc; it constitutes a specific canal for the spermatozoon to enter the egg interior. Teleosts have one micropyle (Zotin 1953, Kuchnow and Scott 1977, Gajdusek and Rubcov 1983, Mikodina 1987, Dzierżynskij et al. 1992, Riehl and Kokoscha 1993, Riehl and Werner 1994, Morrison et al. 1999, Riehl 1999), sturgeons have a few or about a dozen (Ginzburg 1968, Markov 1975, Szagajeva et al. 1993, Vorobieva and Markov 1999).

The micropyle was first described by von Baer in 1835 in the egg envelope of *Cyprinus blicca* (*Blicca bjoerkna*). More than ten years later, micropyles were observed in *Syngnathus acus* and *Silurus glanis*.

The micropyle has the form of a funnel directed to the egg's interior. The micropyle has the following diameters (depending on the species): external diameter is 15–16 µm and the internal is 2 µm in *Oncorhynchus gorbuscha*, 8–10 µm and 4 µm in *Platichthys stellatus* (only one spermatozoon can pass through the narrowest place in the micropyle) (Stehr and Hawkes 1979, Hart and Donovan 1983). After fertilisation, the mycropyle is closed by a plasma plug (Sakai 1961, Renard et al. 1990, Riehl 1996/1997, Riehl and Patzner 1998, Britz and Cambray 2001). Closing of the micropyle may also be an effect of stretching of the egg envelope following water absorption into the perivitellan space; on the one hand it prevents polyspermy, on the other it protects the developing embryo from pathogenes which could penetrate the egg's interior (Yamamoto and Kobayashi 1992).

The structure of the micropyle, like the external structure of the zona radiata, is a taxonomic character and can serve as a basis for species identification of eggs (Figs. 11–13) (Riehl 1993). Before ovulation, the micropyle is plugged by one large micropylar cell which is structurally associated with the follicular tunicle and together with it, drops off the ovulating oocyte. The micropylar cell is large, triangular and contains organelles which are characteristic of secretory cells; this shows that it is not only the plug closing the micropyle, but also a secretory cell (Bieniarz and Epler 1991).

Two terminologies of micropyles were proposed based on the shape and size of the micropylar canal and size and shape of the canal's entrance.

The first term proposed by Riehl and Göting (1974, 1975) as well as Riehl (1980) and followed by Hosaja and Łuczyński (1984) and Mikodina (1987) included three types of micropyles, and somewhat later, Riehl and Schulte (1977, 1978) introduced a division into 4 types since studies on Antarctic fishes revealed the existence of a fourth type of micropyle in their zona radiata.

The terminology is as follows:

- Type I—deep micropyle pit or funnel-like depression on the surface of the egg envelope with a short canal, found in *Gobio gobio*, *Tinca tinca*, *Leuciscus cephalus*, *Lota lota* (Fig. 12E), and *Hucho hucho* (Fig. 12B).
- Type II—shallow pit passing into a long canal, e.g., in *Oncorhynchus keta*, *Salmo salar*, *Salvelinus alpinus*, *Cyprinus carpio*, *Vimba vimba* (Figs. 11E, F), and *Salmo trutta* (Fig. 12D).
- Type III—no micropylar pit, only a canal which can be somewhat wider in its upper part: found in *Coregonus lavaretus* (Fig. 12C), *Esox lucius* (Figs. 11B, C), or *Salvelinus fontinalis*, *Oncorhynchus mykiss*, *Clupea harengus* (Fig. 12F), and *Pleuronectes yokohamae* (Figs. 13A, B).
- Type IV—two funnel-like depressions (external and internal) on the surface of the egg envelope and a short canal—as seen in such Antarctic fishes as *Chinodraco mersi*, *Chionobathiscus dewitti*.

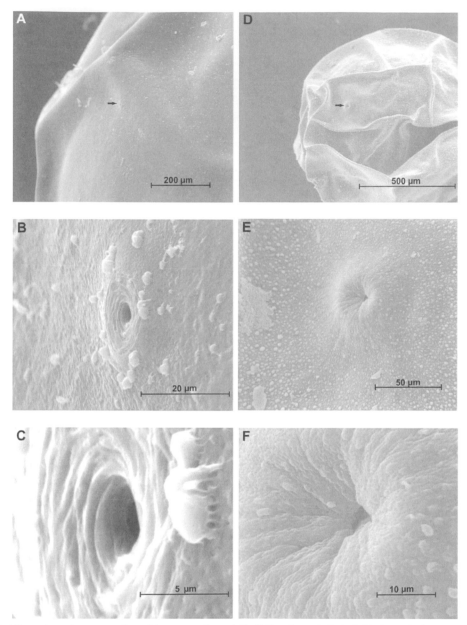

Fig. 11. Micropyle on the outer surface of the egg envelope. **A, B, C.** *Esox lucius* L. **D, E, F.** *Vimba vimba* (L.). **A, D.** Arrows indicate micropyle; wrinkles are the results of SEM preparation. **D.** Helicoid ribs surrounding the micropylar canal. **F.** Ridges and grooves guide to the micropylar canal.

The salmonid micropyle consists of a pit and a canal. According to Groot and Alderdice (1985), in such Salmonidae as the Pacific salmons *Oncorhynchys nerka*, *O. gorbuscha*, *O. keta*, *O. kisutch*, *O. tsawytscha* and the steelhead trout *Salmo gairdnieri*, the micropyle is very similar. Around the micropylar canal, the zona radiata externa is thinner than in the other parts of the egg surface, or it is absent. In *Coregonus nasus* and *C. lavaretus*, in the region of funnel there is no honeycomb pattern, unlike the rest of the egg surface (Riehl 1980).

In rainbow trout *Salmo gairdneri*, the micropyle canal opens to the outside as a round funnel. In that area, the funnel has no plug. The outer opening of the canal is in the centre of the funnel, with a diameter

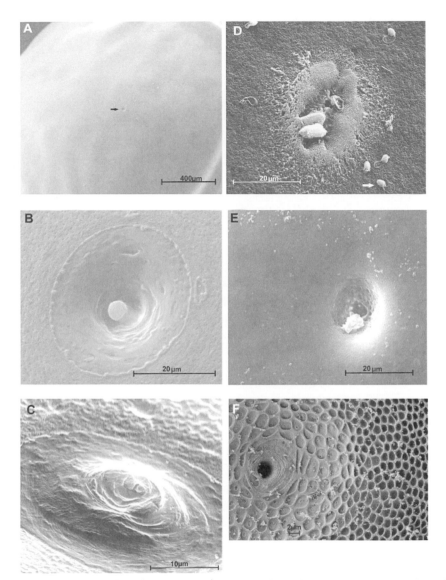

Fig. 12. Micropyle on the outer surface of the egg envelope. **A, B.** *Hucho hucho* L. **A.** Arrow indicates micropyle. **B.** The micropyle consists of a flat micropyle pit and a longer micropyle canal; the canal is plugged with a plug after fertilization. **C.** *Coregonus lavaretus* (L.). Closed micropyle during embryogenesis (2/3 after fertilisation). **D.** *Salmo trutta.* **E.** *Lota lota.* Pore canals in micropyle pit. **F.** (SEM) (**D** after Smaruj 2010, with permission from the author and West Pomeranian University of Technology, Szczecin, Poland; **F** with kind permission from Carl Zeiss Microscopy GmbH and Dr. Martin Dass).

of 3.3 to 4.3 µm. In *Salmo trutta* m. *fario* and *Salmo trutta* m. *lacustris*, the micropylar canal is always located in the funnel's centre, while in *Salmo trutta* the canal is not placed centrally in the funnel, and the funnel itself is oval (Fig. 12D). Among these three fish taxa, the greatest diameter of the micropylar funnel was observed in *S. trutta* m. *lacustris* (12–14 µm), and the smallest in *S. trutta* (7–10 µm). The diameter of the micropylar canal was the smallest in *S. trutta* m. *lacustris* (2.5 µm), and the greatest in *S. trutta* m. *fario* (4 µm) (Riehl 1980).

In *Salvelinus alpinus*, the micropyle has the shape of a shallow pit, 14–16 µm in diameter, and the external, non-centrally located opening of the micropyle canal has a diameter of 1.8–2.7 c. The canal walls bear annular thickenings. In *Salvelinus fontinalis*, the diameter of the micropyle funnel is 7 µm and the diameter of the micropyle canal is 2–2.5 µm (Riehl 1980). In *Coregonus pidschian* (Gmelin 1788),

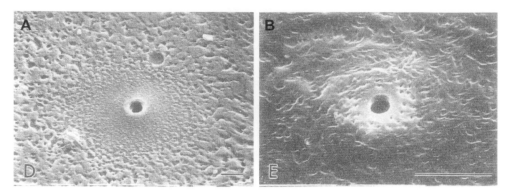

Fig. 13. Outer (**A**) and inner (**B**) surface of the micropyle region of *Pleuronectes yokohamae*. The outer opening of the micropyle is surrounded by pores and shallow cavities of various sizes and the inner opening by pore canals. (SEM) (after Hirai 1993; with permission from the Ichthyological Society of Japan).

Coregonus nasus (Pallas, 1776), and *Coregonus lavaretus* L. 1758, there is a simple orifice to a canal through the zona radiata; there is no pit, and the micropyle canal opens like a funnel (Riehl 1980).

In unfertilised eggs of the rosy barb *Barbus conchonius*, the micropylar region has a non-sticky area which can be termed sperm catchment area of a diameter of ca. 20 μm and an area of 314 μm². The centre of this area holds a funnel-shaped vestibule of 4.5 μm in diameter, and a micropylar entrance; the micropylar pit has a diameter less than 1 μm. The sperm catchment area is composed of 7 to 10 micropylar grooves and ridges, which converge toward the micropylar orifice. Following fertilisation or water activation, the structure of the micropylar region changes such that the grooves and ridges become less visible as a result of stretching of the egg envelopes (Amanze and Iyengar 1990). Similar structures can be observed in *Vimba vimba* (Fig. 11F).

Sturgeon eggs are provided with numerous micropyles, and numerous micropylar openings are located in a small space in the animal pole region of the egg (Linhard and Kudo 1997, Debus et al. 2008). The number of the opening varies not only among species, but also between eggs from various females of the same species, for example *Huso huso* L. can have up to 52 micropyles (Ginsburg 1972), in the white sturgeon *Acipenser transmontanus* Richardson, the mean number is 7 (Cherr and Clark 1982) and in the Russian sturgeon *Acipenser gueldenstaedtii* 30 and more (Dettlaff et al. 1993). In sturgeons, the mechanism preventing polyspermy is such that spermatozoa enter each micropylar canal but only one of them gets in contact with the oolema, causing a very fast cortical reaction which blocks the remaining spermatozoa from oolemma penetration and closes the micropylar openings (Cherr and Clark 1985).

Formation of Egg Envelopes

During early vitellogenesis, an acellular layer called zona radiata is formed between the layer of the follicular cells and the oocyte. The zona radiata externa in teleosts is the first to appear and has the form of a thin homogeneous layer. The zona radiata interna is formed of material synthesised in the oocyte and transported by the Golgi apparatus vesicles toward the oolemma, and then into the space between the oocyte and the follicular cells. The formation of the zona radiata interna causes removal of follicular cells from the oocyte surface. The layer develops till the egg becomes mature (Długosz 1994). Filamentous structures form on the oocyte surface; they then grow to form unordered tangles surrounding the egg. They may later serve as material to form adhesive structures to attach the eggs to the vegetation or other objects.

The formation of the zona radiata in oocytes was traced in many species. The development of oocytes and the zona radiata is often presented in five stages. In the Chinese perch *Siniperca chuatsi*, in the first stage (perinucleolar oocytes), the zona radiata is not formed but the margin of the oocyte begins to be surrounded by a layer of follicular cells. In the next stage (cortical alveolus stage) the zona radiate, 1–2 μm thick, starts being visible around the oocyte; microvilli are also visible. This is the first zone of the zona radiata–Z1. In the stage of previtellogenic oocytes, the envelope is thicker and the projections to the oocyte originating from follicular cells are longer and more numerous. In this stage, subsequent layers of the zona radiata are formed–Z2 and Z3. Z2 appears in transmission electron microscope as a dense granular region. Z3 develops and accumulates as a mesh-like amorphous material. In these layers, pore-canals were observed as projections of follicular cells and microvilli from the oocyte. In that period of development the zona radiata is 2.7 to 10.3 μm thick, of which Z1 is 0.8 to 1 μm thick, Z2 ranges from 0.6 to 0.8 μm, and Z3 from 3 to 5.2 μm. In the fourth stage—in vitellogenic oocytes, the thickness of the zona radiata still increases and ranges from 14.6 μm to 24.1 μm, with the thickness of layer Z3 increasing significantly (23.0–25.2 μm). Z2 and Z3 are compressed. Z3 in this stage loses its mesh-like structure and is composed of homogeneous material with pore-canals. In the fifth stage (mature oocytes), the thickness of the zona radiata ranges from 23.5 μm to 26.3 μm and its structure is similar to that observed in mature (ovulated) oocytes (Jiang et al. 2010).

In the carp *Cyprinus carpio*, as in the Chinese perch, the zona radiate, when 3–6 μm thick becomes visible at the cortical alveolus stage as an acellular layer between the follicular epithelium and the oolemma. Pores become visible on the internal and external side; the diameter of the pores in the external layer is larger—330 nm, and the diameter of the canals in the internal layer is 280 nm. In the next stage, the thickness of the zona radiata increases to ca. 8 μm and filamentous projections are visible on its surface. In the fifth stage, the thickness of the zona radiata decreases to ca. 6 μm (Shabanipour and Hossayni 2010).

In the sea bass *Dicentrarchus labrax* (L. 1758), in the stage called lipidic vitellogenesis, layer Z1 is the first to form as a uniform electron-dense material. The layer Z1 closely adheres to the oocyte plasma membrane, and the perivitelline space is located between Z1 and the follicle cells. With progressing development, Z1 thickens. Also the new layer Z2 appears, which is more electron dense than Z1, resulting in an increase of the total thickness of the egg envelope. With progressing development, sub-layers Z1a and layer Z3 (as the innermost part of the egg envelope) become visible in the ultrastructure of the egg envelope. Z1a is a thin layer between Z1 and Z2. Both Z1a and Z3 are built of fibrous material, and layer Z3 has an ordered structure in the form of undulating lamellae which form spirals and loops. In the stage of proteic vitellogenesis oocytes (late-developing oocytes), 3 layers are visible in a rather thick egg envelope, and in this stage Z3 increases in thickness, becoming the thickest layer. In the stage of mature oocytes, the zona radiata has a more compact structure—layers Z1 and Z2 decrease in thickness, while in Z3 12 electron-dense fibrous lamellae are visible, with material of smaller density in-between.

Changes in Egg Envelope Structure During Hatching

Immediately before hatching the egg envelope resistance decreases under the effect of a substance called hatching enzyme (chorionase), produced by special gland cells which are scattered on the embryo's head, body and yolk sac. The cells contain secretory granules (Łuczyński et al. 1986, 1987, Schoots et al. 1982, Yamagami 1981, 1988, Depeche and Billar 1994). The hatching enzyme of teleosts contains two proteases—high choriolitic enzyme and low choriolitic enzyme. The high choriolitic enzyme causes swelling of the internal layer of zona radiata, the low choriolitic enzyme digests this layer (Yasumasu et al. 1988, 2010, Kawaguchi et al. 2010). The hatching enzyme chorionase digests only the internal layer, the zona radiata interna. In the process of digestion of the zona radiata interna, the number of its lamellae decreases gradually. The swelling of the zona radiata is a side effect of its degradation (Fig. 14). As a rule, the zona radiata externa is not digested by the hatching enzyme, and is only mechanically broken as a result of the embryo's energetic movements (Łuczyński 1985), caused, among others, by its hypoxia (Kawaguchi et al. 2008, 2010); the hypoxia is an effect of the greatest oxygen requirements just before hatching, while oxygen diffusion through egg envelope is decreased as a result of encrusting of its surface (Korwin-Kossakowski 2012).

The time of enzymatic hydrolysis of the internal layer of the zona radiata varies depending on environmental conditions, mainly temperature, and on the species. For example, in natural conditions this time in *Oryzias latipes* is ca. 0.5 h, in coreginidae 1.2–2.0 h, and in rainbow trout *Oncorhynchus mykiss* 2 h (Yamagami 1981, Łuczyński et al. 1986).

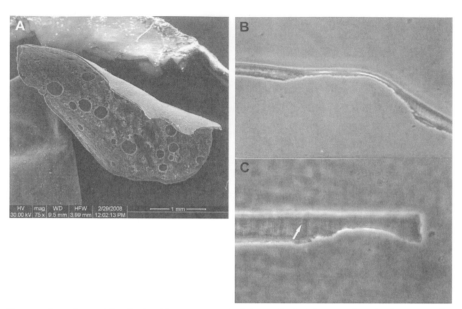

Fig. 14. Egg envelope digested by the hatching enzyme. **A.** *Salmo salar* L. Note openings in the zona radiata interna. **B**, **C**. *Salmo gairdneri.* Cross-section of egg envelope, digested by hatching enzyme in different stages. Arrow, pore canals penetrating egg envelope (SEM) (**B**, **C**. after Winnicki et al. 1970 with permission from Publishing House West Pomeranian University of Technology in Szczecin).

Egg Capsule of Elasmobranch Fishes

In elasmobranchs, the egg capsule is produced by the nidamental gland. Such capsules are formed in oviparous species and many viviparous species. In oviparous fishes, the egg capsule is thick and resistant during the development in the marine environment (Knight et al. 1996). In viviparous fishes, the egg capsule is much thinner (Hamlett and Koob 1999). Thus, there is a dependence between the egg capsule structure and the reproductive strategy (Heiden et al. 2005). The egg capsule of oviparous taxa like *Scyliorhinus* or *Raja* has more fibrous laminae compared to viviparous species. In *Scyliorhinus canicula* the egg capsule consists of 30 laminae (Knight and Feng 1992), and the capsules surrounding the eggs of *Mustelus canis* are composed of 4 laminae (Lombardi and Files 1993). The egg capsules of elasmobranch fishes–viviparous and oviparous—have fibres running in two directions in their acellular laminae and have no visible pores (Hunt 1985, Knight et al. 1996). Such encapsulation does not occur in some rays (Hamlett and Koob 1999). In the bonnethead shark *Sphyrna tiburo*, the egg capsule is ellipsoid and composed of two components. Its thickness is almost even and is 1.14 ± 0.29 μm. The egg capsule is built of three layers of homogeneous fibrous material. It contains branch-like fibres which are oriented in two directions, and the egg capsule is composed of three separate laminae (of which the middle one is the thickest). The middle layer is divided in two sub-layers (Heiden et al. 2005).

In skates the egg capsules have adhesive fibres, which make it possible to attach the eggs to the substratum (Koobe 1997). In oviparous dogfishes, e.g., *Scyliorhinus canicula*, the corners of the capsule bear spiral tendrils, which attach it to the substratum (Knight et al. 1996). In viviparous species, the embryos stay in the mother's reproductive tract; their egg capsules have no such structures and are much thinner, for example in the spiny dogfish *Squalus acanchias*.

The thickness of elasmobranch egg capsules varies rather widely. The egg capsules of the bonnethead shark *Sphyrna tiburo* are relatively thin compared to oviparous species. The thickness of the egg capsules of *Sphyrna tiburo* is only 0.25% of that of the egg capsule of the oviparous *S. canicula* (thickness 0.30 mm), 0.33% of that of *Raja erinacea* (thickness 0.40 mm) and 0.16% of that of *Raja ocellata* (thickness 0.60 mm) (Hornsey 1978, Kormanik et al. 1992, Heiden et al. 2008).

References Cited

Abraham, M., V. Hilge, R. Riehl and Y. Iger. 1993. Mucofollicle cells of the jelly coat in the oocyte envelope of the sheatfish (*Silurus glanis* L.). J. Morphol. 217(1): 37–43. https://doi.org/10.1002/jmor.1052170103.

Afzelius, B.A., L. Nicander and I. Sjödén. 1968. Fine structure of egg envelopes and the activation changes of cortical alveoli in the river lamprey, *Lampetra fluviatilis*. Development 19: 311–318.

Amanze, D. and A. Iyengar. 1990. The micropyle: a sperm guidance system in teleost fertilization. Development 109: 495–500.

Andrade, R.F., N. Bazzoli, E. Rizzo and Y. Sato. 2001. Continuous gametogenesis in the neotropical freshwater teleost, *Bryconops affinis* (Pisces: Characidae). Tissue Cell 33(5): 524–532. https://doi.org/10.1016/j.tice.2018.05.008.

Baer, K.E. von. 1835. Studies on the Histology of Development in Fishes, pp. 52. Leipzig: Friedrich Vogel Verlag.

Baldacci, A., A. Taddei, M. Mazzini, A. Fausto, F. Buonocore and G. Scapigliati. 2001. Ultrastructure and proteins of the egg chorion of the Antarctic fish *Chionodraco hamatus* (Teleostei, Notothenioidei). Polar Biology 24(6): 417–421.

Balon, E.K. 1977. Early ontogeny of *Labeotropheus* Ahl, 1927 (Mbuna, Cichlidae, Lake Malawi), with a discussion on advanced protective style in fish reproduction and development. Environ. Biol. Fish 2: 147–176.

Bian, X., X. Zhang, T. Gao, R. Wan, S. Chen and Y. Sakurai. 2010. Morphology of unfertilized mature and fertilized developing marine pelagic eggs in four types of multiple spawning flounders. Ichthyol. Res. 57: 343–357. DOI: 10.1007/s10228-010-0167-1.

Bielańska-Osuchowska, Z. 1993. Embriologia. PWRiL, Warszawa [in Polish].

Bieniarz, K. and P. Epler. 1991. Rozród ryb. Akademia Rolnicza w Krakowie [in Polish].

Biernaczyk, M., K. Formicki, R. Bartel and Z. Mongiałło. 2012. Characteristics of gametes of the Atlantic salmon (*Salmo salar* L.) restored in northern Poland. J. Appl. Ichthyol. 28 : 66–74.

Bogucki, N. 1930. Recherches sur la permeabilité des membranes et sur la presion osmotique des oeufs des salmonides. Protoplasma 9: 345–369.

Bonsignorio, D., L. Perego, L. Del Giacco and F. Cotelli. 1996. Structure and macromolecular composition of the zebrafish egg chorion. Zygote 4(2): 101–108.

Britz, R. 1997. Egg surface structure and larval cement glands in nandid and badid fishes with remarks on phylogeny and biogeography. American Museum Novitates 3195: 1–17.

Britz, R. and J.A. Cambray. 2001. Structure of egg surfaces and attachment organs in anabantoids. Ichtyol. Explor. Freshwaters 12(3): 267–288.

Cherr, G.N. and W.H. Clark. 1982. Fine structure of the envelope and micropyles in the eggs of the white sturgeon, *Acipenser transmontanus* Richardson: micropyle/chorion/egg envelopes/sturgeon/egg jelly. Dev. Growth Differ. 24(4): 341–352.

Cherr, G.N. and W.H. Clark Jr. 1985. Gamete interaction in the white sturgeon *Acipenser transmontanus*: a morphological and physiological review. Environ. Biol. Fish 14(1): 11–22.

Clark, W.H., G.N. Cherr jr. and W.H. Clark jr. 1982. Fine Structure of the envelope and micropyles in the eggs of the White sturgeon, *Acipenser transmontanus* Richardson Dev. Growth Differ. 24(4): 341–352. https://doi.org/10.1111/j.1440-169X.1982.-00341.x.

Dąbrowski, P. and G. Stanuch. 1998. Wpływ pola magnetycznego na wytrzymałość osłonek jajowych pstrąga tęczowego (*Oncorhynchus mykiss*) i siei (*Coregonus lavaretus*), Akademia Rolnicza w Szczecinie, Praca magisterska.

Debus, L., M. Winkler and R. Billard. 2008. Ultrastructure of the oocyte envelopes of some Eurasian acipenserids. J. Appl. Ichthyol. 24 (Suppl. 1): 57–64. https://doi.org/10.1111/j.1439-0426.2008.01093.x.

Depêche, J. and R. Billard. 1994. Embryology in fish: a review. Société Francaise d'Ichtyologie 8–25.

Dettlaff, A.T., A.S. Ginsburg and O.J. Schmalhausen. 1993. Sturgeon fishes—developmental biology and aquaculture. Springer Verlag, Berlin, Germany.

Długosz, M. 1994. VI. Oogeneza. Oogeneza u ryb kostnoszkieletowych. W: Ultrastruktura i funkcja komórki. [Biliński Sz., Bielańska-Osuchowska Z., Kawiak J., Przełęcka A. (red.)]. PWN, Warszawa. 115–132 [in Polish].

Dulčić, J., L. Grubišić, A. Pallaoro and B. Glamuzina. 2008. Embryonic and larval development of big-scale sand smelt *Atherina boyeri* (Atherinidae). Cybium 32(1): 27–32.

Dumont, J.N. and A.R. Brummett. 1980. The vitelline envelope, chorion, and micropyle of *Fundulus heteroclitus* eggs. Mol. Reprod. Dev. 3(1) 25–44. https://doi.org/10.1002/mrd.1120030105.

Dzierżinskij, K.F., D.A. Pavlov and E.K. Radzichovskaja. 1992. Osobennosti strojenia oboločki jajca belomorskoj zubatki *Anarhichas lupus marisalbi*: obnaruženije neskolkich mikropile. Voprosy Ichtiologii. 1992: 182–186.

Flügel, H. 1966. Elektronenmikroskopische Untersuchung an den Hüllen der Oozyten und Eier des Flussbarches *Perca fluviatilis*. Cell Tissue Res. 77(2): 244–256.

Formicki, K. 1986. The effect of magnetic field on resistance of egg membranes is some salmonid fish. Pol. Arch. Hydrobiol. 33: 105–114.

Formicki, K., I. Smaruj, J. Szulc and A. Winnicki. 2009. Microtubular network of the gelatinous egg envelope within the egg ribbon of European perch, *Perca fluviatilis* L. Acta Icht. Piscat. 39(2): 147–151.

Gajdusek, J. and V. Rubcov. 1983. The micropylar of egg membranes in carp (*Cyprinus carpio*). Folia Zool. 32(3): 217–279.

Ginzburg, A.S. 1968. Fertilization in fishes and the problem of polyspermy. Moscow Academy of Science USSR; Translation: NOOAA and National Science Foundation, New York.

Ginsburg, A.S. 1972. Fertilization in Fishes and the Problem of Polyspermy. p. 290. Jerusalem: Israel Program for Scientific Translations.

Glechner, R., R.A. Patzner and R. Riehl. 1993. The eggs of native fishes. 5. Schneider—*Alburnoides bipunctatus* (Bloch, 1782) (Cyprinidae). Österr. Fisch. 46: 169–172 (in German).

Götting, K.J. 1967. Der Follikel und die peripheren Strukturen der Oocyten der Teleosteer und Amphibien. Z. Zellforsch. Mikrosk. Anatomie. 79(4): 481–491. https://doi.org/10.1007/BF00336308.

Groot, E.P. and F. Alderdice D. 1985. Fine structure of the external egg membrane of five species of Pacific salmon and steelhead trout. Can. J. Zool. 63: 552–566.

Hamazaki, T.S., Y. Nagahama, I. Iuchi and K. Yamagami. 1989. A glycoprotein from the liver constitutes the inner layer of the egg envelope (zona pelicula interna) of fish, *Oryzias latipes*. Dev. Biol. 133: 101–110.

Hamlett, W.C. and T.J. Koob. 1999. Female reproductive system. pp. 398–443. *In*: W.C. Hamlett (ed.). Sharks, Skates, and Rays: The Biology of Elasmobranch Fishes. Johns Hopkins University Press, Baltimore and London.

Hart, N.H. and M. Donovan. 1983. Fine structure of the chorion and site of sperm entry in the egg of *Brachydanio rerio*. J. Exp. Zool. 41: 447–460.

Hayes, F.R. 1942. The hatching mechanism of salmon eggs. J. Exp. Zool. 89: 357–373.

Heiden, T.C.K., A.N. Haines, Ch. Manire, J. Lombardi and T.J. Koob. 2005. Structure and permeability of the egg capsule of the bonnethead shark, *Sphyrna tiburo*. J. Exp. Zool. 303A: 577–589.

Henneguy, L.F. 1889. Recherches sur le développment des poissons osseux. Embryogénie de la truite. Felix Alcan Ed. Paris.

Hirai, A. 1993. Fine structure of the egg membranes in four species of Pleuronectinae. Japan J. Ichthyol. 40(2): 227–235.

Hisaoka, K.K. 1958. Microscopic studies of the teleost chorion. Transactions of the American Microscopical Society 77(3): 240–243.

Hosaja, M. and M. Łuczyński. 1984. Micropyle of three corregoniae species (Teleostei). Z. Angew. Zool. 7: 21–27.

Hunt, S. 1985. The selachian egg case collagen. pp. 409–434. *In*: A. Bairati and R. Garrone (eds.). Biology of Invertebrates and Lower Vertebrate Collagens. Plenum Press, New York.

Ivanikov, V.N. and V.N. Kurdjaeva. 1973. Sistematičeskoje različja i ekologičeskoje značenije strojenija oboloček jajcekletok ryb. 13, 6(83): 1035–1045.

Ivankov, V.N. and V.P. Kurdyayeva. 1973. Systematic dofferences and the ecological importance of the membranes in fish eggs. J. Ichtyol. 13: 864–873.

Iuchi, I., T. Hamazaki and K. Yamagami. 1985. Mode of action of some stimulants of the hatching enzyme secretion in fish embryos. Dev. Growth Differ. 27: 573–581.

Iuchi, I., C.-R. Ha, H. Sugiyama and K. Nomura. 1996. Analysis of chorion hardening of eggs of rainbow trout, *Oncorhynchus mykiss*. Dev. Growth Differ. 38(3): 299–306.

Iwamatsu, T. 1969. Changes of the chorion upon fertilization in the medaka, *Oryzias latipes*. Bull Archiv. Univ. Educat. 18(Nat. Sci.): 43–56.

Iwamatsu, T., Y. Shibata and T. Kanie. 1995. Changes in chorion proteins induced by the exudate released from the egg cortex at the time of fertilization in the teleost, *Oryzias latipes* Dev. Growth Differ. 37(6): 747–759.

Jiang, Y.-Q., T.-T. Zhangb and W.-X. Yangb. 2010. Formation of zona radiata and ultrastructural analysis of egg envelope during oogenesis of Chinese perch *Siniperca chuatsi*. Micron. 41(1): 7–14. https://doi.org/10.1016/j.micron.2009.07.004.

Jasiński, M. 2004. Mechanizm wylęgania się ryb—możliwość zastosowań wybranych farmaceutyków dla modulowania tego procesu. Ph Thesis. Uniwersity of Agriculture in Szczecin.

Johnson, E.Z. and R.G. Werner. 1986. Scanning electron microscopy of the chorion of selected freshwter fishes. J. Fish Biol. 29: 257–265.

Kawaguchi, M., M. Nakagawa, T. Noda, N. Yoshizaki, N. Hiroi, M.I. Iuchi and S. Yasumasu. 2008. Hatching enzyme of the ovoviviparous black rockfish *Sebastes schlegelii*—Environmental adaptation of the hatching enzyme and evolutionary aspects of formation of the pseudogene. FEBS Journal 275(11): 2884–2898.

Kawaguchi, M., S. Yasumasu, A. Shimizu, K. Sano, I. Iuchi and M. Nishida. 2010. Conservation of the egg envelope digestion mechanism of hatching enzyme in euteleostean fishes. FEBS Journal 277(23): 4973–4987. https://doi.org/10.1111/j.1742-4658.2010.07907.x.

Kjesbu, O.S., H. Kryvi, S. Sundbady and P. Solemdal. 1992. Buoyancy variations in eggs of Atlantic cod (*Gadus morhua* L.) in ralation to chorion thickness and egg size; theory and observations. J. Fish Biol. 41: 581–599.

Knight, D.P. and D. Feng. 1992. Formation of the dogfish egg capsule; a co-extruded, multilayer laminate. Biomimetics 1: 151–175.

Knight, D.P., D. Feng and M. Stewart. 1996. Structure and function of the salachian egg case. Biol. Rev. 71: 81–111. https://doi.org/10.1111/j.1469-185X.1996.tb00742.x.

Kobayashi, W. 1982. The fine structure and amino acid composition of the envelopes of the chum salmo egg. J. Fac. Sci. Zool. 23: 1–12.

Koob, T.J. 1997. On the attachment fibers on little skate (*Raja erinacea*) egg capsules Bull. Mt. Desert Isl. Biol. Lab. 36: 114–116.

Korwin-Kossakowski, M. 2012. Fish hatching strategies: A review. Rev. Fish Biol. Fish 22(1): 225–240.

Korzelecka, A., M. Bonisławska and A. Winnicki. 1998. Structure, size and spatial distribution of perch (*Perca fluviatilis* L.) egg components during incubation. EJPAU 1(1): #05. http://www.ejpau.media.pl/volume1/issue1/fisheries/art-05.html.

Korzelecka-Orkisz, A., K. Formicki, A. Winnicki, M. Bonisławska, J. Szulc, M. Biernaczyk, A. Tański and W. Wawrzyniak. 2005. Peculiarities of egg structure and embryonic development of garfish (*Belone belone* (L.)). Electr. J. Ichthyol. 1(2): 1–13.

Korzelecka-Orkisz, A., M. Bonisławska, D. Pawlos, J. Szulc, A. Winnicki and K. Formicki. 2009. Morphophysiological aspects of the embryonic development of tench *Tinca tinca* (L.). EJPAU 12(4): #21. http://www.ejpau.media.pl/-volume12/issue4/art-21.html.

Korzelecka-Orkisz, A., I. Smaruj, D. Pawlos, P. Robakowski, A. Tanski, J. Szulc and K. Formicki. 2010. Embryogenesis of the stinging catfish, *Heteropneustes fossilis* (Actinopterygii: Siluriformes: Heteropneustidae). Acta Icht. Piscat. 4(2): 187–197.

Korzelecka-Orkisz, A., Z. Szalast, D. Pawlos, I. Smaruj, A. Tański, J. Szulc and K. Formicki. 2012. Early ontogenesis of the angelfish, *Pterophyllum scalare* Schultze, 1823 (Cichlidae). Neotropical Ichthyology. 10(3): 567–576.

Korzelecka-Orkisz, A., M. Bonisławska, A. Tański, I. Smaruj, J. Szulc and K. Formicki. 2013. Embryonic development of *Aspius aspius* L. (Actinopterygii: Cypriniformes: Cyprinidae). EJPAU 16(3): #09. http://www.ejpau.media.pl/volume16/issue3/art-09.html.

Kryžanowski, S.G. 1960. O znaczeni zirowych wkluczenij w jajcach ryb. A.N. SSSR. Zool. Żurnał. 39: 111–123.

Kuchnow, K.P. and J.R. Scott. 1977. Ultrastructure of the chorion and its micropyle apparatus in the mature *Fundulus heteroclitus* (Walbaum) ovum. J. Fish Biol. 10: 197–201.

Kudo, S. and M. Inoue. 1986. A bactericidal effect of fertilization envelope extract from fish eggs. Zool. Sci. 3: 323–329.

Kudo, S., A. Sato and M. Inoue. 1988. Chorionic peroxidase activity in the eggs of the fish *Tribolodon hakonensis*. J. Exp. Zool. 245(1): 63–70. https://doi.org/10.1002/jez.1402450110.

Kudo, S. and M. Inoue. 1989. Bacterial action of fertilization envelope extract from eggs of the fish Cyprinus *carpio* and *Plecoglossus altivelis*. J. Exp. Zool. 250(2): 219–228. https://doi.org/10.1002/jez.1402500214.

Kucharczyk, D., M. Luczynski, R. Kujawa and P. Czerkies. 1997. Effect of temperature on embryonic and larval development of bream (*Abramis brama* L.). Aqu. Sc. 59(3): 214–224.

Kunz, Y.W. 2004. Developmental Biology of Teleost Fishes, Dordrecht, The Netherlands: Springer.

Kusa, M. 1949. Hardening of the chorion of salmon egg. Cytologia. 131–137 1-2. DOI: https://doi.org/10.1508/cytologia.15.131.

Laale, H.W. 1977. Culture and preliminary observations of follicular isolates from adult zebra fish, *Brachydanio rerio*. Canad. J. Zool. 55(2): 304–309. https://doi.org/10.1139/z77-041.

Li, Y.H., C.C. Wu and J.S. Yang. 2000. Comparative ultrastructural studies of the zona radiata of marine fish eggs in three genera in Perciformes. J. Fish Biol. 56: 615–621.

Linhart, O. and S. Kudo. 1997. Surface ultrastructure of paddlefish eggs before and after fertilization. J. Fish Biol. 51: 573–582.

Lombardi, J. and T. Files. 1993. Egg capsule structure and permeability in the viviparous shark, *Mustelus canis*. J. exp. Zool. 267(1): 76–85.

Lönning, S. 1972. Comparative electron microscopic studies of teleostean eggs with special reference to the chorion. Sarsia 49: 41–48.

Lönning, S., E. Kjørsvik and J. Davenport. 1984. The hardening process of the chorion of the cod (*Gadus morhua*) and lumpsucker (*Cyclopterus lumpus* L.) egg. J. Fish Biol. 24: 505–522.

Lubzens, E., G. Young, J. Bobe and J. Cerdà. 2010. Oogenesis in teleosts: How fish eggs are formed. Gen. Comp. Endocrinol. 165(3): 367–389.

Ludwig, H. 1874. Über die Eibildung im Thierreiche. Arb. Zool. Zoot. Inst (Würzburg). 1: 287–510.

Łuczyński, M. 1985. Fizjologia ryb. 1. Wykluwanie się ryb. Wydawnictwo ART. Olsztyn [in Polish].

Luczynski, M., M. Hosaja and K. Dąbrowski. 1986. Halching gland cells in Coregonidae embryos. Z. Angew. Zool. 73: 63–73.

Luczynski, M., T. Strzezek and P. Brzuzan. 1987. Secrelion of hatching enzyme and its proteolytic activity in Coregoninae (*Coregonus albula* L. and *C. lavaretus* L.) embryos. Fish Physiol. Biochem. 4(2): 57–62.

Mansour, N., F. Lahnsteiner and R.A. Patzner. 2009a. Physiological and biochemical investigations on egg stickiness in common carp. An. Reprod. Sc. 114(1-3): 256–268.

Mansour, N., F. Lahnsteiner and R.A. Patzner. 2009b. Ovarian fluid plays an essential role in attachment of Eurasian perch, *Perca fluviatilis* eggs. Theriogenology 71(4): 586–593.

Markov, K.P. 1975. Izučenije mikrostruktury oboločki jajc russkogo osetra *Acipenser güldenstädti* Brandt c pomoščju elektronnogo skanirujuščrgo mikroskopa. Voprosy Ichtiologii 15(5): 94.

Masuda, K., I. Luchi and K. Yamagami. 1991. Analysis of hardening of the egg envelope (Chorion) of the fish, *Oryzias latipes* (Egg envelope (chorion)/Egg activation/Chorion hardening/Fish egg/Chorion proteins). Dev. Growth Differ. 33(1): 75–83.

Mayer, S. and E. Shackley. 1988. Aspects of the reproductive biology of the bass *Dicentrarchus labrax*. I. An histological and histochemical study of oocyte development. J. Fish Biol. 33: 609–622.

Mekkawy, I.A.A. and A.G.M. Osman. 2006. Ultrastructural studies of morphological variations of the egg surface and envelopes of the African catfish *Clarias gariepinus* (Burchell, 1822) before and after fertilisation, with a discussion of the fertilisation mechanism. Scientia Marina 70S2: 23–40.

Mikodina, E.V. 1987. O strukturę poverchnosti оболочек ikrinok kostistych ryb. Voprosy Ichtiologii 27(1): 300–306.

Modig, C., T. Modesto, A. Canario, J. Cerdà, J. Von Hofsten and P.-E. Olsson. 2006. Molecular characterization and expression pattern of zona pellucida proteins in gilthead seabream (*Sparus aurata*). Biol. Reprod. 75(5): 717–725.

Modig, C., L. Westerlund and P.E. Olsson. 2007. Oocyte zona pellucida proteins. pp. 113–139. *In*: J. Patrick Babin, J. Cerdà and E. Lubzens (eds.). The Fish Oocyte: From Basic Studies to Biotechnological Applications. Springer Netherlands DOI: 10.1007/978-1-4020-6235-3-5.

Morrison, C., C. Bird, D. O'Niel, C. Leggiadro, D. Marti-Robichaud, M. Rommesns and K. Waiwood. 1999. Structure of the egg envelope of the haddock, *Melanogrammus aeglefinus*, and effects of microbial colonization during icubation. Can. J. Zool. 77: 890–901.

Patzner, R., R. Glechner and R. Riehl. 1994. The eggs of native fishes. 9. Sterber—Zingel streber Siebold, 1863 (Percidae). Österr. Fisch. 47: 122–125.

Patzner, R. and R. Glechner. 1996. Attaching structures in eggs of native fishe. Limnologica. 26(2): 179–182.

Paxton, C.G.M. and L.G. Willoughby. 2000. Resistance of perch eggs to attack by aquatic fungi. J. Fish Biol. 57(3): 562–570. doi:10.1006/jfbi.2000.1332.

Pelka, K.E., K. Henn, A. Keck, B. Sapel and T. Braunbeck. 2017. Size does matter—Determination of the critical molecular size for the uptake of chemicals across the chorion of zebrafish (*Danio rerio*) embryos. Aqu. Toxicol. 185: 1–10.

Ransom, W.H. 1854. On the impregnation of the ovum in the stickleback. Proc. Roy. Soc. London 7: 168–172.

Rawson, D.M., T. Zhang, D. Kalicharan and W.L. Jongebloed. 2000. Field emission scanning electron microscopy and transmission electron microscopy studies of the chorion, plasma membrane and syncytial layers of the gastrula-stage embryo of the zebrafish *Branchydanio rerio*: a consideration of the structural and functional relationships with respect to cryoprotectant penetration. Aquacult. Res. 31: 325–336.

Renard, P., B. Fléchon, R. Billard and R. Christen. 1990. Biochemical and morphological changes in the chorion of the carp (*Cyprinus carpio*) oocyte, following the cortical reaction. J. Appl. Ichtyol. 6: 81–90.

Riehl, R. and K.J. Gotting. 1974. Zu Struktur und Vorkommen der Mikropyle an Eizellen und Eiern von Knochenfischen. Arch. Hydrobiol. 74: 393–402.

Riehl, R. and K.J. Gotting. 1975. Bau und Entwicklung der Mikropylen in den Oocyten einiger Süßwasser-Teleosteer. Zool. Anz. 195: 363–373.

Riehl, R. and E. Schulte. 1977. Vergleichende rasterelektronenmikroskopische Untersuchungen an den Mikropilen ausgewählter Süsswasser-Teleosteer. Arch. Fisch. Wiss. B 28, 2/3: 95–107.

Riehl, R. 1978. Ultrastructure, development and significance of the egg membrane of teleosts. Rivista Italiana di Piscicotura e Ittiopatologia 13: 113–121.

Riehl, R. and E. Schulte. 1978. Bestimmungsschlüssel der wichtigsten deutschen Süsswasser—Teleosteer anhand ihrer Eier. Arch. Hydrobiol. 83(2): 200–212.

Riehl, R. 1979. Ein erweiterber und verbesserter Bestimmungsschlüssel für die Eier deutschen Süsswasser. Teleosteer. Z. angew. Zool. B 66(2): 199–216.

Riehl, R. 1980. Micropyle of some salmonins and coregonins. Env. Biol. Fish 5(1): 59–66.

Riehl, R. 1991. Structure of oocytes and egg envelope in oviparous teleosts—an overview. Acta Biol. Benrodis. 3: 27–65.

Riehl, R. and S. Appelbaum. 1991. A unique adhesion apparatus on the eggs of the catfish *Clarias gariepinus* (Teleostei, Clariidae). Jap. J. Ichthyol. 38: 191–197.

Riehl, R. and R.A. Patzner. 1992. The eggs of native fishes. 3. Pike *Esox lucius* L., 1758. Acta Biol. Benrodis. 4: 135–139 (in German).

Riehl, R. 1993. Surface morphology and micropyle as a tool for identifying fish eggs by scanning electron microscopy. Microsc. Analysis May: 29–31.

Riehl, R. and H. Greven. 1993. Fine structure of egg envelopes in some viviparous goodeid fishes, with comments on the relation of envelope thinness to viviparity. Can. J. Zool. 71(1): 91–97. https://doi.org/10.1139/z93-014.

Riehl, R. and M. Kokoscha. 1993. A unique surface pattern and micropylar apparatus in the eggs of Luciocephalus sp. (Perciformes, Luciocephalidae). J. Fish Biol. 43: 617–620.

Riehl, R., R.A. Patzner and R. Glechner. 1993. The eggs of native fishes. 6. *Vimba vimba elongata* (Valenciennes, 1844) (Cyprinidae). Österr. Fisch. 46: 266–269 (in German).

Riehl, R. and M. Werner. 1994. Die Eier heimischer Fische, 8. Kaulbarsch—*Gymnocephalus cernuus* (Linnaeus, 1758) mit Anmerkungen zum taxonomischen Status von *Gymnocephalus baloni* (Holcik and Hensel, 1974). Fischökologie 7: 25–33.

Riehl, R. and W. Meinel. 1994. The eggs of native fishes. 8. Ruffe—*Gymnocephalus cernuus* (Linnaeus, 1758) with remarks to the taxonomical status of *Gymnocephalus baloni* (Holcik and Hensel, 1974). Fischökologie 7: 25–33.

Riehl, R. and R. Bless. 1995. First report on the egg deposition and egg morphology of the endangered endemic *Romanian perch*. J. Fish Biol. 46: 1086–1090.

Riehl, R. 1996. The ecological significance of the egg envelope in teleosts with special reference to limnic species. Limnologica. 26(2): 183–189.

Riehl, R. 1996/97. Ein ganz besonderes "Loch" in der Eihülle von Knochenfischen—die Mikropyle. Aquatica 2, 12.1996-01.1997: 122–123.

Riehl, R. and R.A. Patzner. 1998. Minireview: The modes of egg attachment in teleost fishes. Ital. J. Zool. 65 suppl.: 415–420.

Riehl, R. 1999. Minireview. The micropyle of teleost fish eggs: morphological and functional aspects. Soc. Fr. Ichtyol. 589–599.

Rizzo, E., Y. Sato, B.P. Barreto and H.P. Godinho. 2002. Adhesiveness and surface patterns of eggs in neotropical freshwater teleosts. J. Fish Biol. 61: 615–632.

Robles, V., E. Cabrita, P. de Paz and Herráez. 2007. Studies on chorion hardening inhibition and dechorionization in turbot embryos. Aquaculture 262(2-4): 535–540.

Rościszewska, E. 1994. VI Oogeneza. Osłony jajowe. In. Ultrastruktura i funkcja komórki. [Biliński Sz., Bielańska-Osuchowska Z., Kawiak J., Przełęcka A. (red.)]. PWN, Warszawa pp. 72–86 [in Polish].

Sadowski, M. 2004. Morfo-funkcjonalne skutki oddziaływań polichlorowanych bifenyli i detergentu w przebiegu embriogenezy ryb i modulujących wpływ pola magnetycznego na te procesy. Praca doktorska. Akademia Rolnicza w Szczecinie [in Polish].

Šagajeva, V.G., M.P. Nikolskaja, N.V. Akimova, K.P. Markov and N.G. Nikolskaja. 1993. Issledovanije rannego ontogeneza volžskich osetrovych (Acipenserdiae) v sviazi c antropogennym vozdejstvijem. Voprosy Ichtiologii 33(2): 230–240.

Sakai, Y.T. 1961. Method for removal of chorion and fertilization of the naked egg in *Oryzias latipes*. Embryologia. 5(4): 357–368.

Scapigliati, G., M. Carcupino, A.R. Taddei and M. Mazzini. 1994. Characterization of the main egg envelope proteins of the sea bass *Dicentrarchus labrax* L. (Teleostea, Serranidae). Mol. Reprod. Dev. 38(1): 48–53.

Schoots, A.F.M., J.J.M. Stikkelbroeck, J.F. Bekhuis and J.M. Denucé. 1982. Hatching in teleostean fishes: Fine structural changes in the egg envelope during enzymatic breakdown *in vivo* and *in vitro*. Journal of Ultrastructure Research 80(2): 185–196.

Shabanipour, N. and S.N. Hossayni. 2010. Histological and ultrastructural study of Zona Radiata in oocyte of common carp *Cyprinus carpio* (Linnaeus 1758). Micron. 41(7): 877–881. https://doi.org/10.1016/j.micron.2010.04.012.

Siddique, M.A.M., M. Psenicka, J. Cosson, B. Dzyuba, M. Rodina, A. Golpour and O. Linhart. 2016. Egg stickiness in artificial reproduction of sturgeon: An overview. Reviews in Aquaculture 8(1): 18–29.

Smaruj, I. 2010. Morphomechanical base of the turgor during fish embryogenesis in magnetic field. Ph Thesis. West Pomeranian University of Technology in Szczecin [in Polish].

Sobociński, A. and A. Winnicki. 1974. Influence of NaCl solution at various contentrations on hardening of egg membranes of trout (*Salmo trutta* L.). Acta Ichthyol. Piscat. 4(2): 11–17.

Sobociński, A., M. Biernaczyk, J. Szulc, K. Formicki and A. Winnicki. 2005. Zmiany wytrzymałości osłonek jajowych troci (*Salmo trutta* m. *trutta* L.) z kilku rzek polskiego wybrzeża w wieloleciu. Wylęgarnia 2005, 14–16 Wrzesień, Ślesin [in Polish].

Stehr, C.M. and J.W. Hawkes. 1979. The comparative ultrastructure of the egg membrane and associated pore structures in the starry flounder, *Platichthys stellatus* (Pallas), and pink salmon, *Oncorhynchus gorbuscha* (Walbaum). Cell and Tissue Research 202(3): 347–356.

Tański, A., A. Korzelecka, M. Bonisławska, A. Winnicki and K. Formicki. 2000. New data on morphomechanical changes during embryogenesis of pike (*Esox lucius* L.). Univ. Agric. Stetin. 214, ser. Piscaria (27): 207–213.

Vogt, C. 1842. Embryologie des Salmones, with atlas, V.I. pp. 328. *In*: L. Agassiz (ed.). Histoire naturelle des poissons d'eau douce de l'Europe centrale, part 2. L. Agassiz (ed.). 1842. Pepitpierre, Institut Lithographique de H. Nicolet, Neuchâtel.

Vorobiev, E.I., W.W. Rubcov and K.P. Markov. 1986. Wlijanije wniesznich faktorov na mikrostrukturu oblocek ikry ryb. Izd. "Nauka" Moskva.

Vorobieva, E.I. and K.P. Markov. 1999. Ultrastrukturnyje osobennosti ikry u predstavitelej Acipenseridae v sviazi c biologiej razmnoženija i filogeniej. Voprosy Ichtiologii 39(2): 197–209.

Winnicki, A. and J. Domurat. 1964. Wytrzymałość osłonek jajowych niektórych ryb rozwijających się w różnych środowiskach. [Durability of egg membranes of selected fish developing in various environments]. Zesz. Nauk. WSR Olsztyn 18, 381: 315–324.

Winnicki, A. 1967a. The plasticity and hardening of the rainbow-trout egg membranes. Bull. Acad. Pol. Sci. 12: 779–783.

Winnicki, A. 1967b. The turgor of salmonid fish eggs during hatching. Bull. Acad. Pol. Sci. 12: 785–787.

Winnicki, A., M. Stańkowska-Radziun and K. Radziun. 1970. Structural and mechanical changes in the egg membranes of *Salmo gairdneri* Rich. during the period of hatching of larval fishes. Acta Icht. Piscat. 1: 7–20.

Winnicki, A., A. Korzelecka and A. Tański. 1998. New data on breeding and early ontogenesis of three-spined stickleback (*Gasterosteus aculeatus* L.). Arch. Ryb. Pol. 6(1): 115–122.

Wintrebert, P. 1923. L'eclosion de la Perche (Perca fluviatilis). Comptes Rendus Hebdomodaires de la Sciente de Biologie. Vol. II.

Wirz-Hlavacek, G. and R. Riehl. 1990. Reproductive behaviour and egg structure of the piranha *Serrasalmus nattereri* (Kner, 1860). Acta Biol. Benrodis. 2: 19–38 (in German).

Yamagami, K. 1981. Mechanisms of hatching in fish: Secretion of hatching enzyme and enzymatic choriolysis. Integ. Comp. Biol. 21(2): 459–471.

Yamagami, K. 1988. 7 Mechanisms of hatching in fish. Fish Physiol. 11(PA): 447–499.

Yamagami, K., T.S. Hamazaki, S. Yasumasut, K. Masuda and I. Luchi. 1992. Molecular and cellular basis of formation, hardening and breakdown of the egg envelope in fish. Int. Rev. Cyt. 136: 51–92. https://doi.org/10.1016/S0074-7696(08)62050-1.

Yamamoto, K. and F. Yamazaki. 1961. Rhythm of development in the oocyte of the gold-fish, *Carassius auratus*. Bulletin of the Faculty of Fisheries Hokkaido University 12(2): 93–110. Http://hdl.handle.net/2115/23128.

Yamamoto, M. and K. Yamagami. 1975. Electron microscopic studies on choriolysis by the hatching enzyme of the teleost, *Oryzias latipes*. Develop. Biol. 43: 313–321.

Yamamoto, T.S. 1958. Biological property of the membrane of the herring egg, with special reference to the role of the micropyle in fertilization. J. Fac. Sci. Hokkaido Univ. Ser. 6, 14(1): 9–16.

Yamamoto, T.S. 1963. Eggs and ovaries of the stickleback, *Pungitius tymensis*, with a note on the formation of a jelly-like substance surrounding the egg (With 4 Plates). Journal of the Faculty of Science Hokkaido University Series VI Zoology 15(2): 190–201. http://hdl.handle.net/2115/27363.

Yamamoto, T.S. and W. Kobayashi. 1992. Closure of the micropyle during embryonic development of some pelagic fish eggs. J. of Fish Biol. 40(2): 225–241.

Zotin, A.I. 1953. Načalnyje stadii processa zatverdevanija obloček jajc lososevych ryb. Dokl. AN SSSR 89: 573–576.

Zotin, A.I. 1958. The mechanism of hardening of the salmonid egg membrane after fertilization or sponatneous activation. J. Embryol. Exp. Morphol. 6: 546–568.

Zotin, A.I. 1961. Fiziologija vodnogo obmiena u zarodyšej ryb i kruglorotych. Izd. AN SSSR, Moskva.

Żelazowska, M. 2010. Formation and structure of egg envelopes in Russian sturgeon *Acipenser gueldenstaedtii* (Acipenseriformes: Acipenseridae). J. Fish Biol. 76(3): 694–706.

Chapter **12**

Testis Structure, Spermatogenesis, and Spermatozoa in Teleost Fishes

Anna Pecio

INTRODUCTION

During more than 310 million years of evolutionary history, including great diversification during the Mesosoic and Cenozoic eras, teleosts adapted to aquatic environments all over the world and evolved a wide range of reproductive strategies. About 97% of the teleosts are oviparous species, having external fertilization with both eggs and sperm being released into the water for fertilization, and about 600 species practice insemination, followed often by internal fertilization and viviparity. Insemination appeared independently in species of about 10 orders belonging to the Osteoglossiformes, Characiformes, Siluriformes, Osmeriformes, Ophidiiformes, Perciformes, Atheriniformes, Beloniformes, Cyprinodontiformes, and Scorpaeniformes. These orders represent both primitive and most advanced taxa (Pecio 2010). In males, various evolutionary modifications of the testis structure are functionally connected to the formation of sperm packets (spermatozeugmata/spermatophores), and changes in spermatozoon ultrastructure (Grier et al. 1978, 1981, Pecio and Rafiński 1994, Downing and Burns 1995, Pecio et al. 2001, Javonillo et al. 2009).

In all teleosts, the renal system is completely separated from the reproductive (genital) system. Testes are located within the dorsal abdomen, below the swim bladder, and attached to the dorsal wall by a mesorchium. Testes are mainly paired organs, yet in a few species (e.g., the poeciliids) they are fused into a single, medial organ (Parenti et al. 2010).

Each testis is covered by a tunica albuginea and consists of seminiferous tubules or lobules, testicular efferent ducts, testicular main ducts, and a spermatic duct, which joins the spermatic duct of the other testis to form an unpaired structure opening at the genital papilla (Grier et al. 1980, Lahnsteiner et al. 1993). The seminiferous tubules/lobules are filled with spermatocysts, in which the developing germ cells undergo spermatogenesis, whereas the testicular efferent ducts, the testicular main duct, and spermatic ducts lined with epithelium seem to serve as the main structures for the storage of spermatozoa. Additionally, the epithelial cells of these ducts, depending on the phase of the reproductive cycle, exhibit either an inactive or active state. The active state in the epithelial cells is characterized by proliferation, synthesis, and secretion of proteins, monosaccharides, steroids, and other lipids, which have important roles in the nutrition of spermatozoa and formation of an ionic gradient in the seminal fluid (Lahnsteiner et al. 1993, 1994, Lahnsteiner and Patzner 2009).

Department of Comparative Anatomy, Institute of Zoology and Biomedical Research, Jagiellonian University, 9 Gronostajowa St., 30-387 Kraków, Poland.
Email: anna.pecio@uj.edu.pl

As in all vertebrates, testes comprise germinal and interstitial compartments (Grier 1993). The germinal compartment has a tripartite structure: germ cells, Sertoli cells, and an acellular basement membrane which separates germinal from interstitial compartments. The germ cells include all cells, ranging developmentally from spermatogonia to spermatozoa, undergoing mitotic and meiotic divisions, whereas the Sertoli cells are somatic cells, which support germ cell survival, development and physiological functioning. The interstitial compartment is formed by Leydig cells, smooth muscle cells and elements of connective tissue such as collagen fibers and fibroblasts and they contain nerve fibers and blood vessels.

The unit of the spermatogenic function in teleosts is the spermatocyst: an association of Sertoli cells with a single germ cells' clone through the different stages of spermatogenesis (Grier 1993). The spermatocyst starts to form when Sertoli cells enclose a single spermatogonium. Then, the cyst is enlarged due to the division of Sertoli cells and germ cells (Schultz et al. 2005). All germ cells within the cyst are derived from a single spermatogonium and develop synchronously up to the end of transformation of haploid spermatids into spermatozoa, after which the cyst ruptures and the spermatozoa or sperm packets are freed into the lumen of the tubules or lobules.

The size of the testes in teleosts is correlated with the progress of spermatogenesis. In many teleost species the process of spermatogenesis is continuous, but most species have an annual reproductive cycle with cyclic gonadal development. Grier and Uribe (2009) updated earlier terminology based on histological examination of the testis structure in the common snook, *Centropomis undecimalis* (Teleostei: Perciformes; Taylor et al. 1998, Grier and Taylor 1998). They distinguished five phases during the annual reproductive cycle, according to changes in the continuity and discontinuity of the germinal epithelium (GE): (1) Regressed, (2) Early GE Development, (3) Mid GE Development, (4) Late GE Development, and (5) Regression.

Testis Structure in Teleosts

The morphological organization of the teleost testis, based on histological examination of representative species from different groups, reveals that more basal taxa have an anastomosing tubular testis whereas derived taxa—the neoteleosts—have a lobular testis. Grier (1993) defines tubule as a structure with a lumen used for conveying sperm (milt) and having looped germinal compartments that do not terminate at the testis periphery but have a dual drainage of sperm into duct system, whereas a lobule is a structure of rounded or globular form with germinal compartments terminating at the periphery of the testis.

The lobular testis in neoteleosts, according to the distribution of spermatogonia, are classified in two types: (a) unrestricted spermatogonial testis type, where spermatogonia are situated along the lobules, and (b) restricted spermatogonial testis type, where spermatogonia are situated only at the distal end of the lobule just beneath the tunica albuginea (Grier 1993, Parenti and Grier 2004, Grier and Uribe 2009, Uribe et al. 2014).

Anastomosing Tubular Testis Type

This type is characterized by germinal compartments that form loops at the testis periphery and double back to the efferent ducts, forming a highly branched and anastomosing network similar to the tubules present in the mammalian testis. This testis type was described in many species belonging to phylogenetically primitive taxa of Teleostei, namely representatives of the families Elopidae (i.e., *Megalops atlanticus*), Salmonidae (i.e., *Oncorhynchus mykiss*), Clupeidae (i.e., *Opisthonema oglinum* and *Dorosoma petenense*) (Grier and Uribe 2009).

In this chapter, this testis type is exemplified by histological examination of externally fertilizing species like the rainbow trout, *Oncorhynchus mykiss* (Salmonidae) (Fig. 1), and the freshwater butterfly fish *Pantodon buchholzi* (Pantodontidae) (Fig. 2) of unknown fertilization method, but possessing modified testicular efferent ducts leading into the testicular gland (Dymek and Pecio, unpublished data). Another modification of the testis is exemplified by some of the inseminating species of characids (e.g., *Mimagoniates barberi*, *Tyttocharax cochui*, *Scopeocharax* sp. and *Xenurobrycon macropus*) (Figs. 3 and 4).

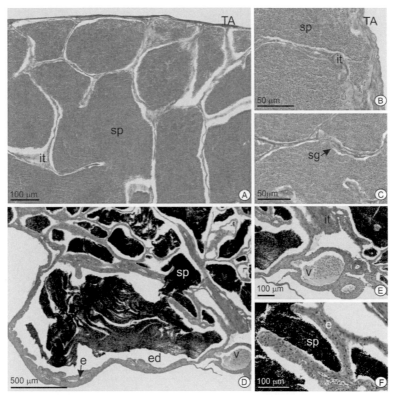

Fig. 1. Anastomosing tubular testis in *Oncorhynchus mykiss* (Salmonidae). Paraffin sections: **A–C.** Mallory staining. **D–F.** H&E. **A.** The part of the periphery during the pre-spawning period shows the anastomosing tubules filled with stored spermatozoa (sp). Connective tissue covering the testis as the tunica albuginea (TA) and the component of the interstitium (it) stain blue. **B.** The periphery of the testis with tunica albuginea (TA) penetrating the inner of the testis as the somatic component of the interstitium (it); tubules filled with spermatozoa (sp). **C.** The intratubular area of the tubules is lined by developing primary spermatogonia (sg). **D.** The anastomosing tubules filled with stored spermatozoa (sp) coalesce into the efferent duct (ed) lined with epithelium (e). **E.** The distal part of the testis with blood vessels (V) and abundant interstitium (it). **F.** Spermatozoa (sp) in the distal tubules lined with epithelium (e).

Color version at the end of the book

In *O. mykiss* the lumen of the anastomosing tubules is filled by spermatozoa (Fig. 1A). The testis is covered by a fibrous capsule, the tunica albuginea, consisting of connective tissue and penetrating the inner space as loose connective tissue between tubules (Fig. 1B). The germinal compartment is mainly formed by the Sertoli cells and primary spermatogonia are distributed alongside the tubules (Fig. 1C). The efferent ducts and collecting duct are located laterally to the testis (Fig. 1D). The lumen of efferent ducts and main ducts is filled with spermatozoa and the germinal compartment is formed by columnar epithelial cells, homologous to the Sertoli cells (Figs. 1E and F).

The tubules in the testes of most externally fertilizing species show a similar structure alongside the entire testis. But in species practicing insemination, which evolved independently many times, this reproductive strategy is coupled with different modifications that facilitate the transfer of spermatozoa to the reproductive tract of the female (Pecio and Rafiński 1994, Burns and Weitzman 2002, Javonillo et al. 2007, 2009). The morphological adaptations in particular comprise modifications of the testis structure and the formation of sperm bundles. The mode of fertilization in the freshwater butterflyfish *Pantodon buchholzi*, belonging to the ancient teleost group Osteoglossomorpha, is unknown but the unique spermatozoon with a highly modified and elongated midpiece, and the presence of sperm bundles, may suggest insemination (Deurs and Lastein 1973). The testis tubules in *P. buchholzi* are filled with spermatocysts having germ cells with progressive stages of spermatogenesis situated around the lumen filled with a secretion

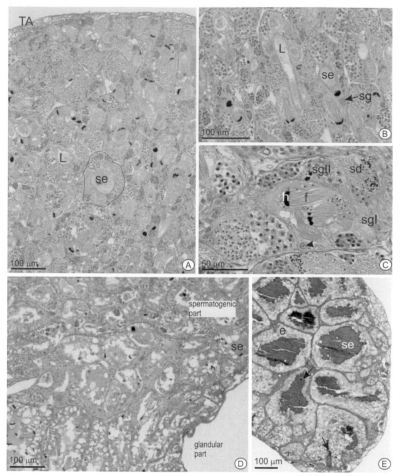

Fig. 2. Anastomosing tubular testis with continuous spermatogenesis in *Pantodon buchholzi* (Pantodontidae). Paraffin sections. **A–C.** H&E staining. **D–E.** Mallory staining. **A.** The panoramic view shows tubules with spermatocysts (area within contour) containing developing germ cells in spermatogenesis. The lumen (L) of the tubules is located centrally and filled with secretion (se). TA, *tunica albuginea*. **B.** Cross and longitudinally sectioned tubules show that single spermatogonia (sg) and spermatocysts are situated peripherally; the lumen (L) with secretion (se) is filled of the freed sperm bundles. **C.** The spermatocysts containing germ cells in different stages of spermatogenesis are located peripherally, whereas in the lumen sperm bundles embedded in the secretion are visible. f, flagella; h, heads of spermatozoa; sgI, primary spermatogonia; sgII, secondary spermatogonia; sd, spermatids. **D.** The transition zone between the spermatogenic part of the testis and glandular part. Se, secretion; area within contour show the outline of tubule. **E.** Cross sectioned tubules lined by epithelial cells (e) producing secretion (se). Sperm bundles (black arrow) embedded in secretion.

(Figs. 2A and B). Late spermatids, each possessing an elongated nucleus with highly condensed chromatin and fully formed flagella, show parallel arrangement and bipolar organization within the cyst (Fig. 2C). The spermatozoa released into the lumen of the tubule remain in parallel arrangement and are embedded in the secretion (Fig. 2D). The anterior part of the testis in *P. buchholzi* is the location of spermatogenesis but the much narrower posterior part is transformed into a gland (Fig. 2D). The tubules are lined by a glandular epithelium producing the secretion in which sperm bundles are present (Fig. 2F).

Inseminating characids of the subfamilies Glandulocaudinae and Stevardiinae, as well some *incertae sedis* species in the family Characidae, have testes divided into two parts: a spermatogenic and an aspermatogenic part (Pecio and Rafiński 1994, Burns et al. 2005, Javonillo et al. 2007, 2009). In *Mimagoniates barberi*, a representative of the Glandulocaudinae, the anterior spermatogenic part contains

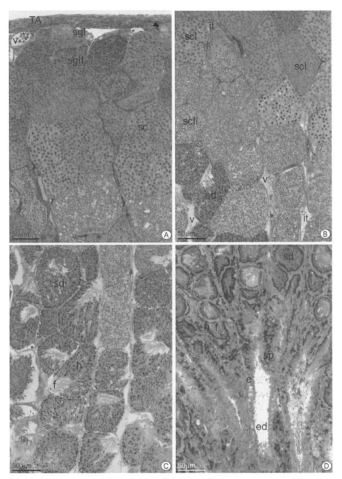

Fig. 3. Restricted spermatogonial testis type in *Anableps anableps*. Semithin section stained with methylene blue **A.** Periphery of the testis shows distal part of the lobules with primary spermatogonia (sg I) situated just beneath the tunica albuginea (TA). The cysts with successive stages of spermatogenesis are arranged nearly linearly in lobules, which are separated by interstitial tissue (it). SgII, secondary spermatogonia; scI, primary spermatocytes; scII, secondary spermatocytes; V, blood vessels. **B.** Spermatocysts with developing spermatids (sd) in various stages of chromatin condensation and different arrangement within the cysts. F, flagella; h, heads of spermatids. **C.** Spermatocysts containing spermatids during nuclear rotation showing the bipolar organization with head and flagella located at the opposite site pole of the cyst. **D.** Spermatids after nuclear rotation show radial arrangement with the head embedded in the Sertoli cells and flagella occupying the center of the cyst. The cysts located near the main duct (ed) release the free spermatozoa (sp) into the lumen of the duct. E, epithelium; f, flagella; h, heads of spermatids; sd, spermatids.

spermatogonia situated alongside the tubules together with groups of clonal germ cells developing synchronously within the spermatocysts (Fig. 3A). While the spermatogenic part is the place where the sperm are produced, the posterior aspermatogenic part is a storage organ for spermatozoa as well as spermatozeugmata, where they are formed (Figs. 3B and C). In the most caudal efferent ducts and collecting spermatic duct, only spermatozeugmata are present (Fig. 3D). The spermatozeugmata are polymorphic: the smaller are one-sided with parallel arrangement of spermatozoa and contain dozens of spermatozoa, whereas the bigger spindle shaped ones possess flagella on both sides, and contain hundreds of spermatozoa aligned in parallel (Figs. 4A and B).

In representative species of the Stevardiinae, the spermatozeugmata are formed in the spermatogenic part of the testis. At the end of spermiogenesis, fully formed spermatozeugmata are released into the lumen of the tubule (Figs. 4B and C). All Stevardiinae examined to date have one-sided spermatozeugmata,

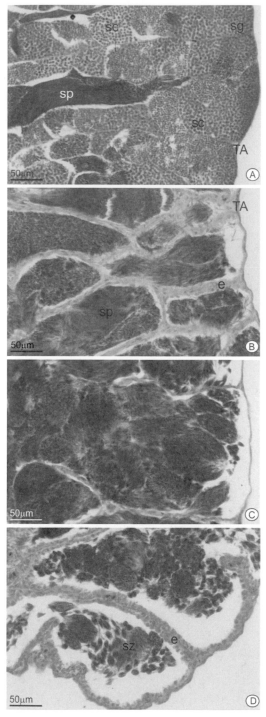

Fig. 4. Anastomosing tubular testis type with modification of the testis structure in *Mimagoniates barberi*. Paraffin sections staining with H&E. **A.** The anterior spermatogenic part of the testis containing lobules filled with cysts in different stages of spermatogenesis and freed spermatozoa (sp). Spermatogonia (sg) are situated alongside the tubules lined by epithelial cells (e). TA, *tunica albuginea*. **B.** The posterior aspermatogenic part is filled with free and coalescing spermatozoa (sp). e, epithelium; TA, *tunica albuginea*. **C.** The cortical fragment of the aspermatogenic part of the distal part of the testis filled with spermatozeugmata (sz). **D.** The distal part with two types of spermatozeugmata (sz): small–one sided and bigger–spindle shaped ones. e, epithelium.

but contain very different numbers of spermatozoa, from a dozen in *Tyttocharax cochui* to thousands in *Xenurobrycon macropus* (Figs. 4D and E).

Unrestricted Spermatogonial Testis Type

This type is found in neoteleosts, but not in Atherinomorpha. The lobular testis is characterized by the germinal compartment terminating beneath the tunica albuginea as solid rounded projections. An example of a species with the unrestricted spermatogonial testis type is *Gymnocephalus cernuus* (Percidae) (Fig. 5). The lobules are filled with spermatocysts having germ cells in various stages of spermato- and spermiogenesis (Fig. 5A). Primary spermatogonia, associated with Sertoli cells, occur alongside the lobules closer to the basal lamina (Fig. 5B). During spermiation, the spermatozoa are released into the lobular lumen.

Restricted Spermatogonial Testis Type

This type is found in all the Atherinomorpha, which supports the monophyletic character of this group (Parenti and Grier 2004). In this testis type, lobules terminate blindly at the periphery of the testis and spermatogonia are restricted to the most distal part. Lobules are separated by interstitial tissue with vessels. Here, this testis type is exemplified by the foureyed fish, *Anableps anableps* (Fig. 6A). The primary spermatogonia are situated just beneath the tunica albuginea, and are visible singly or in clusters

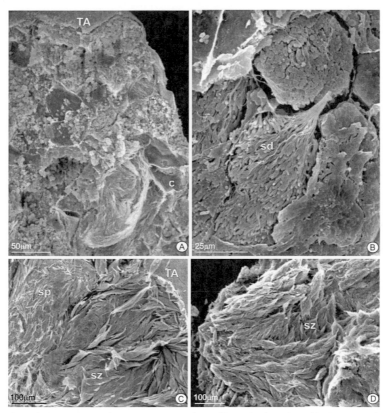

Fig. 5. Anastomosing tubular testis type in some representatives of inseminating characids. SEM. **A.** The panoramic view of the spermatogenic part in *Corynopoma rissei* shows cysts filled with germ cells in different stages of spermatogenesis as well as free spermatozoa in the lumen of the tubules. C, spermatocysts; TA, tunica albuginea. **B.** Cysts with spermatids (sd) in different stage of nucleus elongation from round to elongated with bipolar organization within cyst. **C.** The cortical region of the aspermatogenic part in *Mimagoniates barberi* shows the coalescing spermatozoa (sp) and spermatozeugmata (sz). **D.** Different types of spermatozeugmata in *M. rheocharis*: one sided and spindle shaped ones.

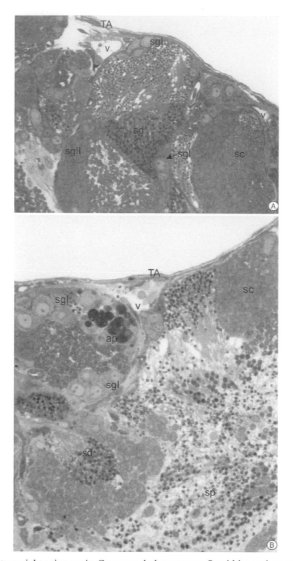

Fig. 6. Unrestricted spermatogonial testis type in *Gymnocephalus cernuus*. Semithin sections stained with methylene blue.
A. Lobules contain primary spermatogonia (sgI) alongside the wall terminating at the periphery beneath the tunica albuginea
(TA). Secondary spermatogonia (sgII) and spermatocytes (sc) fulfill tightly cyst, whereas the cysts with spermatids (sd) shows
the lumen between them. V, capillary. **B.** Peripheral lobules with single primary spermatogonia (sgI) surrounded by Sertoli
cells (arrow) and cysts with developing germ cells as well as with apoptosis (ap). Released spermatozoa (sp) in the lumen of
lobules. sc, spermatocytes; sd, spermatids; TA, tunica albuginea.

in association with processes of the Sertoli cells. The cysts with secondary spermatogonia and subsequent
stages of spermatogenesis are located more centrally in a nearly linear arrangement. The lobules lack
lumens because the spermatocysts (actually Sertoli cells' processes) fill the entire width of the lobules
(Figs. 6B, C). During spermatogenesis, the cysts with later stages of spermatogenesis migrate alongside
the lobules towards the efferent ducts. The cysts with spermatogonia and spermatocytes exhibit an even
distribution of cells in the whole volume of the cyst, whereas the cysts during spermiogenesis show a
different organization of the spermatids. Round spermatids are distributed evenly in the spermatocyst,
whereas the spermatids with flagella tend to have bipolar organization (Fig. 6C). The spermatids with
elongated nuclei and condensed chromatin become oriented radially with their heads embedded in peripheral
Sertoli cells' cytoplasm and flagella occupying the central region of the cysts (Fig. 6D). At the end of

spermiogenesis, the cysts open up to release free spermatozoa into the efferent ducts which lead into the main, centrally located testis duct. In *A. anableps* free spermatozoa are seen in the testis duct, whereas in the viviparous species of atherinomorphs, e.g., *Xenotoca eiseni* (Goodeidae) and *Poecilia latipinna* (Poeciliidae), sperm are packed into spermatozeugmata (Grier and Uribe 2009).

Spermatogenesis

Independent of testis type, spermatogenesis is a process comprising of morphological changes from spermatogonia via spermatocytes to spermatids, developing finally into species specific spermatozoa. This process of differentiation was studied by several authors describing particular stages in different species (Pudney 1995, Grier and Uribe 2009, Schultz et al. 2010, Uribe et al. 2014).

Spermatogonia

Primary Spermatogonia

Primary spermatogonia (= spermatogonia A), considered as stem cells, reside in a niche created by a supporting cell type, the Sertoli cells, and the surrounding extracellular matrix (Li and Xi 2005). These cells belong to the largest germ cells among all germ cell types, receiving in diameter 12–16 μm. They possess light granular cytoplasm and a centrally located spherical nucleus containing 1–2 nucleoli (Fig. 7A). Primary spermatogonia are mainly single cells or arranged in small groups. After a few mitotic divisions they give rise to secondary spermatogonia (= spermatogonia B).

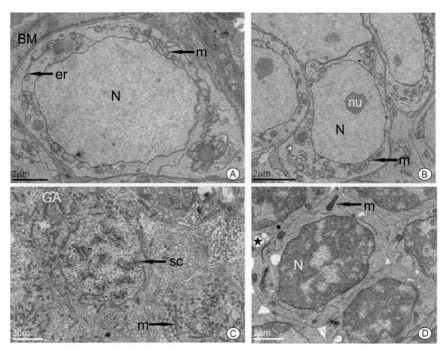

Fig. 7. Developing germ cells: from primary spermatogonium to spermatids. TEM. **A–B.** *Cobitis taenia.* **A.** Primary spermatogonium with prominent nucleus (N) surrounded by a basemant membrane (BM). Mitochondria (m) and endoplasmic reticulum (er) are randomly distributed in the cytoplasm. **B.** Secondary spermatogonia with nuclei (N) having a well-defined nucleolus (nu). Note numerous mitochondria (m) in the cytoplasm. **C–D.** *P. buchholzi.* **C.** Primary spermatocytes with clumped chromatin and synaptonemal complex (sc) within the nucleus. The wide cytoplasmic zone around the nucleus contains numerous mitochondria (m) and a prominent Golgi apparatus (GA). **D.** Early spermatids at the beginning of flagellum formation connected by cytoplasmic bridges. Within cysts appears a free space (star) due to decreasing volume of spermatids. m, mitochondria; N, nucleus.

Secondary Spermatogonia

In contrast to the primary spermatogonia, the secondary spermatogonia (= spermatogonial B) are completely enclosed by Sertoli cell processes which form the wall of the spermatocyst. The secondary spermatogonia are smaller than the primary ones, having 9–12 μm in diameter and they are interconnected within the cysts by cytoplasmic bridges (Fig. 7B). They divide mitotically more rapidly than spermatogonia A and after 6 divisions in rainbow trout (Loir 1999) or 14 in the guppy (Billard 1969) they enter into meiosis becoming primary spermatocytes.

Spermatocytes

Primary Spermatocytes

Primary Spermatocytes, similar in shape and size to secondary spermatogonia, enter the meiosis which in males consists of two consecutive cell divisions: the result of the first division are two secondary spermatocytes, and the second division results in four haploid spermatids. The spermatocysts containing the primary spermatocytes of the meiotic prophase I at the pachytene stage with a synaptonemal complex are relatively abundant because this step has the longest duration, whereas primary spermatocytes in leptotene and zygotene stage are rather rare, and the rarest are those in the diplotene stage because this stage has the shortest duration of all different stages (Fig. 7C). When prophase I is completed, spermatocytes enter the metaphase I, anaphase I, and telophase I resulting in two secondary spermatocytes.

Secondary Spermatocytes

Secondary Spermatocytes are cells with diameters of 4–7 μm, are spherical in shape and possess a spherical nucleus with filamentous chromosomes. They divide rapidly resulting in a short metaphase II, anaphase II, and telophase II, giving rise to two spermatids of each secondary spermatocyte.

Spermatids

Early spermatids are cells with diameters of 2–4 μm, spherical in shape and they possess a nucleus (Fig. 7D). They undergo spermiogenesis, which is the process of significant morphological transformation of spermatids into species-specific, highly specialized and motile spermatozoa, possessing a head with highly condensed chromatin, mitochondria in the midpiece and the flagellum. In teleosts, spermiogenesis has been studied in many species and revealed diverse processes present during the transformation of spermatids into spermatozoa (Billard 1969, Burns et al. 2009, Grier and Uribe 2009, Mattei 1970, Uribe et al. 2014, Pecio and Rafiński 2001, Pecio et al. 2007, Pecio 2009, 2010, Sprando et al. 1988). The early spermatids are spherical and possess a nucleus with heterogeneous chromatin surrounded by a wide zone of cytoplasm with several mitochondria and both centrioles in perpendicular arrangement located lateral to the nucleus (Fig. 8A). When the distal centriole begins to extend into the flagellar axoneme, the nucleus of the spermatids starts to rotate (Fig. 8B), which causes the flagellar axis to become perpendicular to the base of the nucleus and centrioles are located in the more or less deeply nuclear fossa (Fig. 8C). Spermatozoa of this type are termed Type I spermatozoa, whereas in Type II spermatozoa—the flagellum-remains parallel to the base of nucleus because the rotation does not take place (Mattei 1970). Spermiogenesis resulting in Type I sperm with a small rounded or ovoid nucleus, approximately 2–3 μm in diameter, is more typical for teleosts belonging to the orders of Ostariophysi, whereas Type II sperm is found mostly in species of the Perciformes. In 2000, for the first time, Type III spermatozoa were described in *Suribim lima* (Siluriformes: Pimelodidae) by Quagio-Grassiotto and Carvalho and this was confirmed later in Heptapteridae and other species of Pimelodidae (Quagio-Grassiotto et al. 2005, Quagio-Grassiotto and Oliveira 2008). In this type of spermiogenesis, the nuclear rotation does not occur because centrioles migrate to the cellular membrane and the distal centriole starts to extend into the flagellum. The flagellar axis in the spermatozoon remains perpendicular to that of the nucleus. The cytoplasmic canal is more or less absent. If it is present, it forms

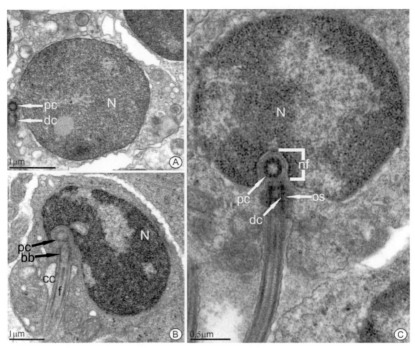

Fig. 8. Spermatids during formation of the flagellum and nuclear rotation in *Phenacogrammus interruptus*. TEM. **A.** An early spermatid with nucleus (N) having finely granular chromatin and laterally situated proximal (pc) and distal centriole (dc). **B.** A spermatid during flagellum extension from the basal body (bb) and rotation of the nucleus (N) above the proximal centriole (pc). The proximal part of a newly formed flagellum (f) is located in the cytoplasmic canal (cc). **C.** A spermatid after nuclear rotation with extended flagellum in one axis with a nucleus, a centriolar complex and a midpiece. Proximal centriole (pc) is situated in the nuclear fossa (nf). Distal centriole (dc) is surrounded by the osmiophilic material (os). After Pecio (2009). Folia biologica (Krakow). 57: 13–21.

due to displacement of cytoplasm, rather than due to centriolar movements as is in the case of Types I and II spermiogenesis (Quagio-Grassiotto and Oliveira 2008).

Simultaneously with the nuclear morphological changes, the process of chromatin condensation occurs which may involve the gradual enlargement of chromatin granules in the entire volume of the nucleus observed, e.g., in some characids (Pecio et al. 2007). The condensation of chromatin may occur only at one pole of the nucleus as in *Lepomis macrochirus* (Centrarchidae) (Sprando et al. 1988), *Gymnocephalus cernuus* (Percidae) (Figs. 9A, B; own observation), *Panthodon buchholzi* (Pantodontidae) (Figs. 9C–E; own observation) or only from the outer side of nucleus exemplified in *Anableps anableps* (Anablepidae) (Figs. 9F–H). In *L. macrochirus* the condensation of chromatin is connected with nuclear envelope breakdown and reformation only at the border of the condensed chromatin, causing an 80% reduction of the nuclear volume of the spermatozoon as compared with the early spermatid. Such significant reduction of the nuclear volume due to increasing degree of chromatin condensation is coupled with the elimination of the watery component of the nucleo- and cytoplasm and an excess of the nuclear membrane. These components are eliminated from spermatids into the lumen of the spermatocysts as residual bodies, which are later phagocytized by Sertoli cells. The variation in chromatin condensation suggests the presence of different types of protamines associated with DNA (Saperas et al. 1993).

Spermatozoon Ultrastructure in Teleost Fishes

The earliest observation and descriptions of sperm morphology of teleosts come from the 19th century (Kunz 2004). The early investigators as Remak (1854) referred spermatozoa "as pin-shaped with small spherical head and a very fine threadlike tail", whereas Wagner (1854) and Miescher (1872–78) were

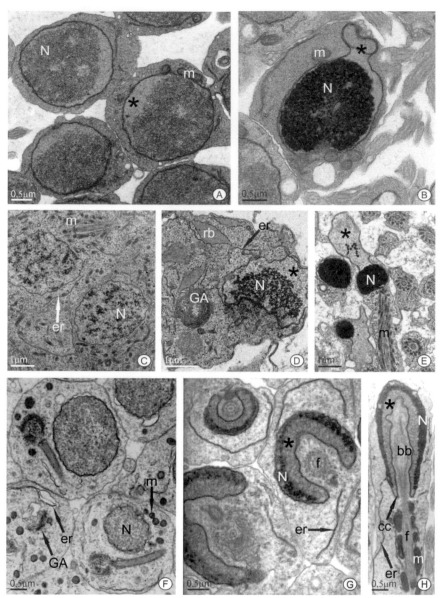

Fig. 9. Different types of chromatin condensation in the developing spermatids (TEM). **A–B.** *Gymnocephalus cernuus.*
A. The beginning of chromatin condensation takes place in the part of the nucleus (N) volume in early spermatids; note future excess of nucleoplasm (star). **B.** Condensed chromatin in late spermatid; an excess of cytoplasm (star) will be eliminated as residual bodies. M, mitochondria. **C–E.** *Pantodon buchholzi.* **C.** Early spermatids with nuclei (N) having euchromatin and patches of fine granular heterochromatin and a wide zone of cytoplasm with mitochondria (m) and endoplasmic reticulum (er). **D.** A spermatid with nucleus (N) having most part filled with finely granular heterochromatin and peripherally located an excess of nucleoplasm (star). Wide zone of cytoplasm contains prominent Golgi apparatus (GA) and spindle shaped residual bodies (rb) as well endoplasmic reticulum (er). **E.** Late spermatid with nucleus having condensed chromatin and discarding nucleoplasm (star). M, mitochondria. **F–H.** *Anableps anableps.* **F.** Early spermatids having fine granular chromatin in the entire volume of the nucleus (N). Wide zone of the cytoplasm with newly formed flagella running in the cytoplasmic canal, prominent Golgi apparatus (GA) and randomly distributed spherical mitochondria (m). er, endoplasmic reticulum. **G.** Cross sectioned spermatids before nuclear rotation showing condensed heterochromatin at the periphery of nucleus (N); cytoplasm around nucleus with endoplasmic reticulum (er). star, nucleoplasm. **H.** Frontally sectioned spermatid after rotation of the nucleus (N) with peripherally condensed chromatin. Deep nuclear fossa contains proximal centriole and elongated basal body (bb). The proximal part of the flagellum (f) is surrounded by a cytoplasmic canal (cc) and a midpiece with elongated mitochondria (m). star, uncondensed chromatin; er, endoplasmic reticulum.

the first who observed, besides head and tail, a middle piece adjacent to the head of the spermatozoa of the perch *Perca fluviatilis*, pike *Esox lucius* and salmon *Salmo salar*. These observations were confirmed later by Kupffer (1878) and His (1873), who described the spermatozoon of the herring, *Clupea harengus* and the salmon, *S. salar* and revealed in the middle portion 4–5 intensively stained "tail globules" later named mitochondria (Retzius 1905, 1910, Ballowitz 1915, 1916). Additionally, Kupffer (1873) was the first researcher, who gave the measurements for the salmon sperm: length of head 0.0025 mm, breath of head 0.0020 mm and length of tail 0.062–0.075 mm.

More details of sperm ultrastructure and great diversity of spermatozoa among fishes were revealed when the electron microscopical techniques started to be more available among researchers in the 7th decade of the 20th century (Billard 1970, Baccetti 1970, Mattei 1970). These TEM observations have shown that spermatozoa of many species are quite different from those previously described (Remak or Miescher) and this caused, in the next decades of the twenty century, great interest in studies of sperm ultrastructure of species belonging to the Characiformes, Cypriniformes, Siluriformes, Perciformes, Salmoniformes, and Tetraodontiformes (Jamieson 1991, Mattei 1991).

Until now, sperm ultrastructure has only been investigated in about 1–1.5% of all the species from 100 families of teleosts, and gave detailed evidence of the presence of a head, a midpiece and a flagellum (Lahnsteiner and Patzner 2008). The only common structural features in the spermatozoon of teleosts are that they are lacking an acrosome, despite the presence of a micropyle in the egg chorion. Details of spermatozoa of a limited number of teleosts demonstrated a great diversity reflected in the shape and size of the head (mainly the nucleus), size, location and number of mitochondria, number of flagella or lacking flagella, mutual arrangement of centrioles and intercentriolar apparatus as well presence of additional structures such as pseudoflagella, striated rootlet, accessory microtubules, lateral fins, cytoplasmic sleeve, fenestrated membrane, dense fibres, etc. (Baccetti 1991, Jamieson 1991, Mattei 1991). Baccetti (1991), Jamieson (1991, 2009) and Mattei (1991) have described in many reviews the ultrastructure of spermatozoa as a valuable taxonomic and phylogenetic criterion, especially at the family-specific level. In 2008, Lahnsteiner and Patzner have described the differences in sperm structure at the subfamily and species-specific level in families of the orders Perciformes, Salmoniformes, and Scorpeniformes according to data from families with different reproductive strategies and from different habitats.

In 1991, Jamieson had introduced the classification of teleost's spermatozoa into two main groups: aquasperm, which refers to the free swimming phase in water before fertilization and introsperm, referring to inseminating or internally fertilizing sperm. This classification will be considered in this survey.

Aquasperm

Teleostean aquasperm mainly possesses a small rounded or slightly ovoid head (= nucleus), approximately 2–3 μm in diameter, two centrioles, proximal and distal ones, consisting mainly of nine triplet microtubules located at or in the basal fossa of the nucleus (= nuclear fossa), a midpiece with one to a few simple cristae, rounded mitochondria situated at the base of the nucleus, as shown in Fig. 10. The centrioles have different mutual arrangement varying from perpendicular to oblique, but the distal centriole in the flagellate sperm always arises into the flagellum. In the case, when the centrioles are parallel, it happens that both form the basal body for two flagella. The axonemal pattern of microtubules in most cases is $9 \times 2 + 2 \times 1$. The proximal part of the flagellum is surrounded by the cytoplasmic collar which can contain mitochondria in the anterior half, but in the posterior half it consists mainly of two apposed plasma membranes with small amount of cytoplasm. The periaxonemal space between flagellum and cytoplasmic sleeve is termed cytoplasmic canal.

The length of the flagellum varies in different species from 25 to 70 μm. In some externally fertilizing species, they may possess additional structures. For example in characids, in *Phenacogrammus interruptus* (Alestidae), the initial part of the flagellum is additionally associated with membranous compartments (Fig. 11A), whereas in *Chilodus punctatus* (Chilodontidae) the flagellum possess lateral fins, which are the extensions of cytoplasmic membrane in the plane of the central singlets (Fig. 11B) (Pecio 2003, 2009). Lateral fins are unique structures among Characiformes, Siluriformes and Cypriniformes (Ostariophysi);

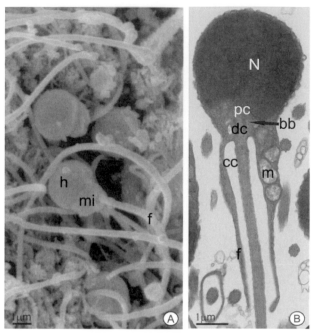

Fig. 10. Aquasperm. SEM and TEM. **A.** A spermatozoon of *Eupallasella percnurus* with spherical head (h), basally located midpiece (mi) and single fagellum (f) (own observation). **B.** Longitudinally sectioned spermatozoon of *Hemigrammus erythrozonus* (Characidae). Nucleus with highly condensed chromatin and basally located centrioles in parallel arrangement. Shallow nuclear invagination contains only the anterior part of the proximal centriole (pc) and a basal body (bb). Midpiece with round mitochondria (m) surrounds the anterior part of the flagellum (f) running in the elongate cytoplasmic canal (cc). After Pecio et al. (2007). Neotr. Ichthyol. 5: 457–470.

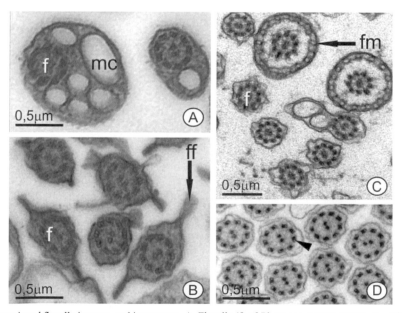

Fig. 11. Cross sectioned flagella in aqua- and introsperm. **A.** Flagella (f) of *Phenacogrammus interruptus* with classic 9 + 2 axoneme and membranous compartments (mc) in the initial part. **B.** Flagella (f) of *Chilodus punctatus* with lateral fins (lf) extending in the plane of flagellar singlets. **C.** Flagella (f) of *Panthodon bucholzi* surrounded beneath the midpiece region by a fenestrated membrane (fm) and a free portion distally. **D.** Free flagella of *Anableps anableps* with intratubular differentiation in the microtubule A of all doublets (arrowheads).

however, forming the additional surface to move in water, they are widespread in many species belonging to distantly related teleosts as in the Esociformes, Aulopiformes and Perciformes (Jamieson 1991, Matos et al. 2002).

According to the presence or absence of a flagellum, the aquasperm are aflagellated, uniflagellated or biflagellated, but the structure of the spermatozoa could be simple or complex.

Aflagellate Spermatozoa

Aflagellate spermatozoa evolved only in osteoglossomorphs and were described only in the five species from different genera of the family Mormyridae: *Brienomyrus niger, Marcusenius* (= *Gnathonemus*) *senegalansis, Hyperopisus bebe, Mormyrus rume* and *Petrocephalus bovei* and in the monospecific *Gymnarchus niloticus* (Gymnarchidae) (Mattei et al. 1967, 1972, Mattei 1991). The sperm of all the above mentioned species does not look like a spermatic cell because they contain uncondensed or slightly condensed chromatin in the nucleus surrounded by abundant vacuolated cytoplasm containing mitochondria and centrioles and other species-specific structures (Fig. 12). Among mormyrids, only the spermatozoa of *Petrocephalus* exhibit a polarization with one pole occupied by the nucleus and the opposite one by the abundant vacuolated cytoplasm containing mitochondria and centrioles between them (Fig. 12A (Mattei et al. 1972)). The spermatozoon of *G. niloticus* has an uncondensed polymorphic nucleus and its double nuclear membrane possesses pores. The other features, namely, the thick cell membrane lined internally by thick tubular fibrils, globular mitochondria situated in the cytoplasm containing lipid masses as well as myelin vesicles are very peculiar and are present only in this species. Abundant microtubules present in cytoplasm of all above described species suggest adaptation to amoeboid movement during fertilization (Kirschbaum and Schugardt 2002).

Uniflagellate Spermatozoa

This type of spermatozoon is the most common in teleosts, and with some exceptions (e.g., spermatozoa of Elopomorpha, see below) it exhibits a simple structure with a spherical nucleus, having, at the base or laterally, invagination called nuclear fossa. Both centrioles, the proximal and the distal ones, could be located inside the nuclear fossa in various mutual arrangements, or outside it, which happens in the case when the flagellum is tangential to the nucleus. The midpiece with mitochondria is located usually at the base of the nucleus. The flagellum arises from the distal centriole forming the basal body. The sperm of Elopomorpha, as shown in Fig. 13(A–C), possesses several unique features being spermatozoal synapomorphies for the orders of Elopiformes, Anguilliformes, and Albuliformes (Jamieson and Mattei 2009). All these unique features were described in the sperm of *Anguilla anguilla* (Gwo et al. 1992), *Albula vulpes* (Mattei and Mattei 1973, 1974), *Lycodontis afer* (Mattei 1970, Mattei and Mattei 1974, Mattei 1991). The sperm has an

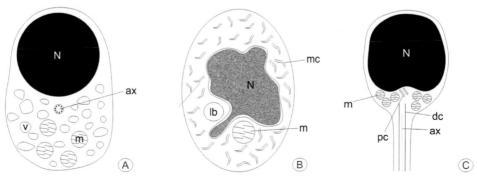

Fig. 12. Diagrammatic structure of spermatozoa in representatives of the Osteoglossomorpha. **A.** *Petrocephalus bovei.* After a micrograph of Mattei et al. (1972). J. Microsc. (Paris) 15: 67–78. **B.** *Gymnarchus niloticus.* After a micrograph of Mattei et al. (1967). C.R. Hebd. Seanc. Acad. Sci. D 265: 2010–2012. **C.** *Notopterus afer.* After a micrograph of Mattei et al. (1972). J. Microsc. (Paris). ax, axoneme; dc, distal centriole; lb, lipid body; m, mitochondria; mc, microtubules; N, nucleus; pc, proximal centriole; v, vacuoles.

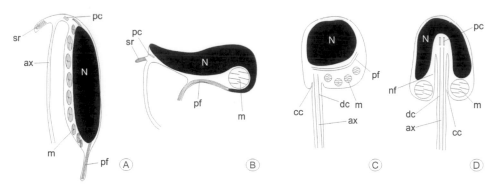

Fig. 13. Diagrammatic structure of spermatozoa in representatives of Elopomorpha. **A.** *Anguilla anguilla*. After Gibbons et al. (1983) J. Submicrosc. Cytol. 15: 15–20. Figure 10. **B.** *Albula vulpes*. After Mattei and Mattei (1973). Zeitschr. Zellforsch. Mikrosk. Anat. 142: 171–192. **C.** *Lycodontis afer*. After Mattei (1970). *In*: Comparative Spermatology (ed. B. Baccetti). pp. 59–69. **D.** *Anchoa guineensis*. After Mattei et al. (1981). J. Ultrastr. Res. 74: 307–312. ax, axoneme; cc, cytoplasmic canal; dc, distal centriole; f, flagellum; m, mitochondria; N, nucleus; nf, nuclear fossa; pc, proximal centriole; ps, pseudoflagellum; sr, striated rootlet.

elongate nucleus, crescentic in shape and with a hook-shaped anterior end, where the only mitochondrion is situated. The centrioles are located at the posterior end. The proximal centriole is extended and dissociated into two bundles of microtubules consisting of five and four triplets, running from the centriolar region towards the anterior tip of the elongated nucleus and forming the pseudoflagellum. Another feature is the striated rootlet, a strong projection supported by a central axis present at the posterior tip near the anterior part of the flagellum. The flagellum arises from the distal centriole and the elopomorph axoneme is composed of nine peripheral doublets and lacks central singlets.

The spermatozoa of Clupeiformes were investigated in species of three families: the Clupeidae (e.g., *Clupea harrengus*, *Sardinops melanosticus*), Pristigasteridae (e.g., *Ilisha africana*), and Engraulidae (e.g., *Anchoa* (= *Engraulis*) *guineensis*) (Mattei et al. 1981, Mattei 1991, Hara et al. 1994, Gwo et al. 2006). Clupeoid sperm is characterized by an oliviform head, a very deep penetration of the nucleus by basal fossa (Fig. 13D). The midpiece contains only a single, large, C-shaped mitochondrion, which is situated at the base of the nucleus and surrounds the base of the flagellum. The cytoplasmic canal is not present. The flagellum has a 9 + 2 microtubular pattern and the A microtubules of axonemal doublets 1,3 and 5–7 are characterized by intratubular differentiation (ITD). In some species: e.g., *Spratelloides gracilis*, no proximal centriole and ITD has been identified. In all investigated species, the flagellar fins were absent.

The uniflagellate, anacrosomal aquasperm among the Ostariophysi were investigated in many species being representatives of the orders Gymnotiformes, Cypriniformes, Characiformes, and Siluriformes, but only in one species of Gonorchynchiformes, *Chanos chanos*, which differs from the other ostariophysan aquasperm having a single mitochondrion (Fig. 14) (Burns et al. 2009).

The ostariophysan aquasperm can be characterized by a nearly spherical head, measuring about 2 μm in diameter, a midpiece with several mitochondria, presence of two centrioles and a flagellum of moderate length as shown in Fig. 15 in *Phenacogrammus interruptus* (Characiformes: Alestidae). Nevertheless, the spermatozoa described in this taxa display a great variability relating to the shape of the nucleus varying from spherical to elliptical or slightly elongate showing the highly condensed chromatin with scattered pale lacunae or a granular chromatin, presence of different number of mitochondria (mostly 1 to 10) and their different arrangement related to the length of the cytoplasmic sleeve, which is often different due to mutual arrangement of the centriole position and their position within a basal nuclear fossa. The midpiece of several species of cypriniforms contains glycogen granules and vesicles. The length of the flagellum varies from 25 μm to 60 μm and the flagellar fins are generally (with some exceptions) absent, what is considered as the ostariophysan synapomorhpy (Jamieson 1991). The flagella show a classic 9 + 2 axonemes and all microtubules are electron-lucent. All the above mentioned features of the sperm ultrastructure are exemplified here by the spermatozoon of the cypriniform species, *Leuciscus cephalus* (Cyprinidae) (Baccetti et al. 1984), which in many characters is similar to the characiform sperm, with the exceptions of

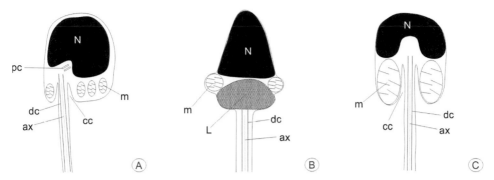

Fig. 14. Diagrammatic structure of spermatozoa in the representatives of Ostariophysi. **A.** *Leuciscus cephalus.* From Jamieson 1991, after Baccetti et al. (1984). Gamete Research 10: 373–396. **B.** *Citharinus* sp. After Mattei et al. (1995). J. Submicrosc. Cytol. Pathol. 27: 189–191. **C.** *Chanos chanos.* After Gwo et al. (1995). Submicrosc. Cytol. Pathol. 27: 94–104. ax, axoneme; cc, cytoplasmic canal; dc, distal centriole; f, flagellum; L, lattice tubule; m, mitochondria; N, nucleus; nf, nuclear fossa; pc, proximal centriole.

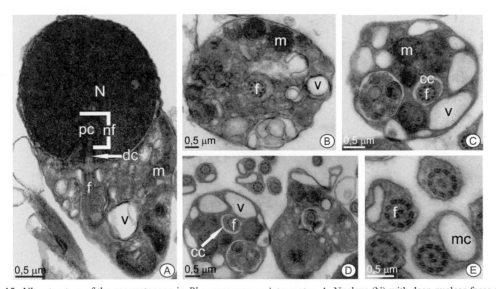

Fig. 15. Ultrastructure of the spermatozoon in *Phenacogrammus interruptus*. **A.** Nucleus (N) with deep nuclear fossa (nf) at the base containing proximal (pc) and distal centrioles (dc). Midpiece with mitochondria (m) and vesicles (v) embrace the initial part of the flagellum (f). **B.** Cross sectioned anterior part of midpiece with mitochondria (m) and vesicles (v) located more peripheral and the flagellum (f) in the center. **C.** Cross sectioned posterior part of the midpiece with mitochondria (m) and vesicles (v) surrounding the flagellum (f) in the cytoplasmic canal (cc). **D** and **E.** Free flagellum behind the cytoplasmic canal (cc) with membranous compartments (mc).

the spermatozoon of *Citharinus* sp. (Citharinidae) (Mattei et al. 1995), showing some peculiarities in the midpiece (Fig. 14B) as well the spermatozoon of *Chanos chanos* (Chanidae) (Fig. 14C) (Gwo et al. 1995) resembling the clupeoid spermatozoa. The spermatozoon of *L. cephalus* is characterized by a spheroidal nucleus with nuclear fossa containing the proximal centriole. This centriole lies in the sagittal plane of the sperm, and is always eccentric. The mitochondria are distributed asymmetrically due to the asymmetry of centrioles; most of them are located adjacent to the nucleus and only one in the opposite area. Between the plasma membrane and the axoneme in the anterior part of the flagellum membranous vesicles are visible. The axonemal axis is tangential to the nucleus. The spermatozoon of *Citharinus* sp. has a slightly elongate head containing a conical nucleus with highly condensed flocculent chromatin measuring 2 µm in length and 1.5 µm in width (Fig. 14B). The nucleus possess a shallow nuclear fossa, where is located the proximal portion of the distal centriole, whereas the proximal centriole, nearly parallel to the distal one, is located at the periphery of the cell. The midpiece is divided in two portions; the anterior one, just below the nucleus containing several mitochondria and the posterior one, where is situated the peculiar "tubular-vesicular system" (= "lattice tubule"), formed from vesicles during spermiogenesis. The cytoplasmic canal surrounds only the initial part of the flagellum lacking fins. The aquasperm of the milkfish, *Chanos chanos* exibits some characters present in clupeoid sperm, namely the single C-shaped mitochondrion located at the posterior part of the ovoid nucleus with deep basal fossa. However, in the deep fossa are located two, (not only the distal one) centrioles in perpendicular arrangement. The axoneme has the 9 + 2 construction typical for most teleosts, but the microtubules in doublets 1, 5, and 8 contain dense material. These features may support the close clupeiform-ostariophysan relationship. Most ostariophysan species of externally fertilizing species produce Type I aquasperm, and all the above-described spermatozoa belong to this type, which means that the flagellar axis is perpendicular to the base of the nucleus, caused by rotation of the nucleus during spermiogenesis.

The sperm ultrastructure of Esociformes and Salmoniformes resembles in the form of the spherical head the spermatozoa of Clupeidae and Cyprinidae, but the flagellum has a cytoplasmic expansion (= flagellar fins) present on one or both sides (Fig. 16). For example, the sperm of the pike, *Esox lucius* (Esocidae) has a rounded nucleus 2 µm long and wide with a maximum diameter of 1.8 µm, slightly flattened in one plane, leading to a bilateral asymmetry (Billard 1970, Stein 1981). The midpiece is asymmetrical due to an uneven distribution of separate mitochondria in the posterior region of the nucleus. The centrioles are situated in the nuclear fossa, and the flagellum is inserted slightly lateral with the proximal portion located in the cytoplasmic canal. The flagellar fin is present only on one side. In *Oncorchynchus mykiss* and *Salmo trutta* (Salmonidae), the nucleus is elongate and ovoid and possesses a moderately deep basal nuclear fossa, which houses two centrioles in perpendicular arrangement (Billard 1970, Stein 1981). The base of the flagellum is surrounded by a cytoplasmic canal and a short mitochondrial collar. The flagellum possesses two lateral fins shortly behind the midpiece; however, Lahnsteiner and Patzner (2008) observed in about 12% of flagellar cross sections a flagellum exhibiting three fins.

Some spermatozoa belonging to the uniflagellate aquasperm type show an external symmetry, but possess an asymmetrical nucleus such as in the spermatozoon of *Trachinocephalus myops* (Synodontidae) (Mattei 1970, 1991), which has a conical tip and one mitochondrion, located on one side on the base of flagellum, whereas the opposite is occupied by a part of the nucleus (Mattei 1970, 1991).

The spermatozoa of species of the Perciformes, a dominating group of marine species, develops during spermiogenesis mainly without rotation (= Type II aquasperm); thus, the flagellum remains tangential relative to the nucleus. The nuclear fossa develops after migration of centrioles in the direction of the nucleus, but they remain located outside of the nuclear depression and usually are in perpendicular

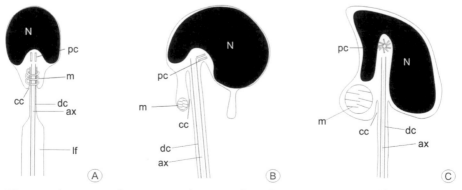

Fig. 16. Diagrammatic structure of spermatozoa of representatives of Esociformes and Aulopiformes. **A.** *Oncorhynchus tchawytscha.* From Jamieson 1991, after a micrograph from Zirkin (1975). J. Ultrastuc. Res. 50: 174–184. **B.** *Esox lucius.* From Jamieson after Stein (1981). Zeitschr. Angew. Zool. 68: 183–198. **C.** *Trachinocephalus myops.* From Jamieson after Mattei (1970). *In*: Comparative Spermatology (ed. B. Baccetti). pp. 59–69. cc, cytoplasmic canal; dc, distal centriole; f, flagellum; lf, lateral fins; m, mitochondria; N, nucleus; nf, nuclear fossa; pc, proximal centriole.

arrangement as in *Thalassoma bifasciatum* (Labridae) (Lahnsteiner and Patzner 1997). Mitochondria are situated at the base of the flagellum, which has a classical axoneme arrangement 9 + 2 and possess short flagellar fins formed by membrane extensions at the singlets plane. The perciform sperm here is exemplified by the ultrastructural examination in *Gymnocephalus cernuus* (Percidae) (Fig. 17).

In Blennidae, in which the spermatozoa have a simple structure (Lahnsteiner and Patzner 1990a, b), four types of spermatozoa (Fig. 18) have been distinguished. In *Salaria pavo* (Type A), the head and midpiece are integrated in one region, but the mitochondria are situated in an invagination of the nucleus at its basal part. The centrioles are located in a shallow fossa at the basal part of the nucleus. In Type B, represented here by *Aidablennius sphynx*, the head and midpiece with mitochondria form two spherical regions. Type C is here exemplified by the spermatozoon of *Lipophrys adriaticus*, which possesses six mitochondria located in an invagination of the nucleus, forming an integrated region with a midpiece. Similar to *Rhabdoblennius ellipes*, or *Entomacrodus striatus*, representing Type D, the midpiece is not distinguished from the head because the mitochondria are located in the invagination at the apical part of the nucleus.

A very unique structure represents the spermatozoon of *Macquaria ambigua* (Percichthyidae) (Jamieson 1991) (Fig. 19). It has an elliptical nucleus with weekly developed nuclear fossa. Several mitochondria may sometimes fuse and form an incomplete ring. The basal body (= distal centriole), in perpendicular arrangement to the proximal centriole, is extended to 9 + 2 axoneme of the flagellum running tangential to the nucleus. The cytoplasmic canal is long and most of it is bounded by a thin cytoplasmic sleeve which only anteriorly expands towards the mitochondrion-containing midpiece. Behind the cytoplasmic sleeve the flagellum extends into two fins, which disappear terminally (Jamieson 1991).

Biflagellate Aquasperm

Biflagellarity is known in 32 species in 17 families of the Euteleostomi. In all the biflagellate spermatozoa, both centrioles have a parallel orientation and in young spermatids each produces an intact 9 + 2 flagellar

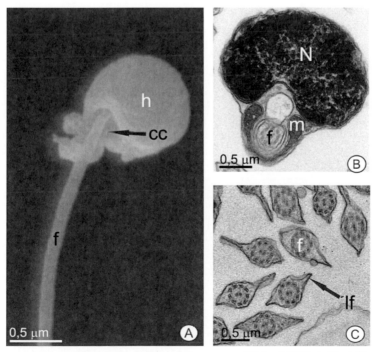

Fig. 17. Ultrastructure of the spermatozoon in *Gymnocephalus cernuus*. SEM and TEM. **A.** Head (h) with midpiece and flagellum (f) extending from the cytoplasmic canal (cc). **B.** Transverse section through the basal part of the nucleus and C-shaped mitochondrion embracing the flagellum (f) situated lateral to the nucleus (N). **C.** Flagella (f) with lateral fins (lf).

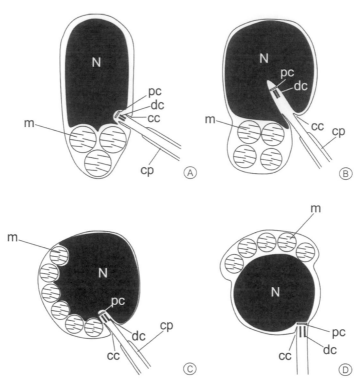

Fig. 18. Diagrammatic model of spermatozoa in representatives of the Blennidae. **A.** *Salaria pavo*. **B.** *Aidablennius sphynx*. **C.** *Lipophrys adriaticus*. **D.** *Entomacrodus striatus*. After Lahnsteiner and Patzner (2008). *In*: Fish Spermatology. S.M.H. Sadi, J.J. Cosson, K. Coward and G. Rafiee (eds.). cc, cytoplasmic canal; cp, cytoplasmic protrusions; dc, distal centriole; f, flagellum; m, mitochondria; N, nucleus; nf, nuclear fossa; pc, proximal centriole.

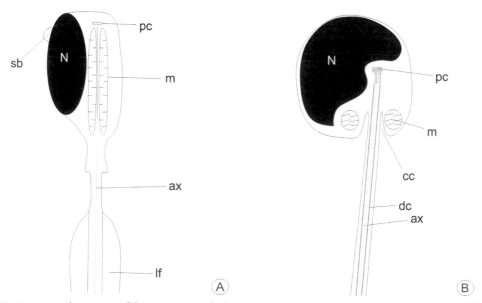

Fig. 19. Diagrammatic structure of the spermatozoon in Percomorpha. **A.** *Macquaria ambigua*. From Jamieson 1991, after Marschal. **B.** *Thalasoma pavo*. After Lahnsteiner and Patzner. J. Submicrosc. Cytol. Pathol. 29: 477–485. Ax, axoneme; cc, cytoplasmic canal; cp, cytoplasmic protrusions; dc, distal centriole; lf, lateral fins; m, mitochondria; N, nucleus; nf, nuclear fossa; pc, proximal centriole; sb, spherical body.

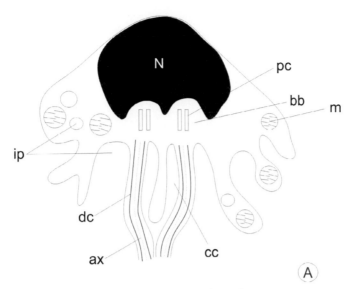

Fig. 20. Diagrammatic structure of a biflagellate spermatozoon in *Ictalurus punctatus*. From Jamieson 1991. After a micrograph by Poirer and Nicholson (1982). J. Ultrastruc. Res. 80: 104–110. Bb, basal body; cc, cytoplasmic canal; f, flagellum; ip, inpocketings; m, mitochondria; N, nucleus; nf, nuclear fossa.

axoneme. In Siluriformes, the biflagellate spermatozoon was described by Poirer and Nicholson (1982) in *Ictalurus punctatus* (Ictaluridae) (Fig. 20). The total length of the head with the midpiece is about 4 μm and both flagella achieve 95 μm in length. The nucleus has granular chromatin and at the base it possesses moderate nuclear fossa, in which lie both centrioles. The extending flagella surround independently cytoplasmic canals in their anterior part. The midpiece, which spreads slightly anteriorly to the nucleus, contains several mitochondria, glycogen granules and double-walled vacuoles. Additionally, the midpiece has numerous microtubules, which radiate throughout the length and width of it.

In Batrachoidiformes, the biflagellate spermatozoa are known in species belonging to the family Batrachiididae, in *Opsanus tau* (Hoffman 1963, Casas et al. 1981) and *Halobatrachus didactylus* (Mattei 1991) exhibiting a simple structure and in *Porichthys notatus* (Stanley 1965), where the spermatozoon has a more complex structure (Fig. 20). The *O. tau* spermatozoon has a spheroidal nucleus with a high degree of chromatin compression. The midpiece contains several round mitochondria. The nucleus has, at the base, two separate slight indentations in which mutually parallel centrioles are located. The sperm of *H. didactylus* has a more elongate nucleus and only two bilateral mitochondria, whereas the spermatozoon of *P. notatus*, in contrast to *Opsanus* and *Halobatrachus*, possesses an elongate spiral nucleus, as well elongate mitochondria lying close to the basal portion of the flagella to form a distinct midpiece. In all these spermatozoa, the parallel centrioles are connected by a fan-like array of filaments to the nuclear membrane. The biflagellate sperm in *Lepadogaster lepadogaster* (Gobiesciformes: Gobiescidae) is very similar to that in *P. notatus*, especially concerning the shape of the helical elongate nucleus and the presence of 6–10 elongate mitochondria (Fig. 21) (Mattei and Mattei 1978). The midpiece consists of two distinct regions. The anterior part is shorter, about 3 μm long, and contains mitochondria situated parallel to each other and to the flagella from which they are separated by two cytoplasmic canals. The posterior part is longer, about 10 μm long, and contains two flagella enclosed in a common sheath, consisting of the wall of the cytoplasmic canal. The flagella arise from parallel arranged centrioles, and have classic 9 + 2 axonemes without flagellar fins. The total length of the spermatozoon in *L. lepadogaster* is about 90 μm.

The biflagellarity is very common in species of the Perciformes and evolved independently in several families: in the Apogonidae as well as among species of the Siluriformes belonging to the Nematogenyidae, Cetopsidae, Ictaluridae, Malapteruridae, Doradidae, and Plotosidae (Jamieson 2009).

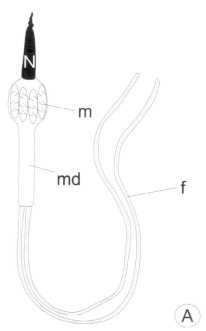

Fig. 21. The diagrammatic structure of the spermtozoon in *Lepadogaster lepadogaster.* From Jamieson 1991 after Mattei and Mattei (1978). Biologie Cellulaire 32: 267–274. f, flagellum; m, mitochondria; N, nucleus.

Introsperm

The teleostean introsperm in most cases comprises highly modified cells with nuclei ranging from slightly to extremely elongated, an enlarged midpiece containing extensive mitochondria as well as glycogen deposits, microtubules alongside the nucleus and intratubular differentiation in one of the axonemal microtubules. All these features are considered adaptations facilitating the forward movement of the spermatozoa through viscous secretion and narrow pathways in the ovary between the oocytes. Besides the elongation of the sperm nucleus, many spermatozoa have flattened nuclei which may aid the clumping and/or adhesion during the formation of sperm into high density bundles called spermatozeugmata or spermatophores. However, in a few cases among internally fertilizing species, especially in those where internal fertilization evolved recently, simple spermatozoa (= simple secondary introsperm) are known, and in one case of introsperm, in *Lepidogalaxias salamandroides*, a spermatozoon with an acrosome was described (Fig. 22) (Leung 1988).

Modifications in the spermatozoon ultrastructure involving nucleus and midpiece elongation evolved independently as many times among teleosts as insemination had evolved.

To illustrate the various introsperm modifications, different examples of spermatozoa from distantly related group will be described: in *Pantodon buchholzi* (Osteoglossiformes: Pantodontidae) (Deurs and Lastein 1973), *Mimagoniates barberi* (Characiformes: Characidae: Glandulocaudinae) (Pecio and Rafiński 1994, 1999, Burns et al. 1998), *Scopeocharax rhinodus* (Characiformes: Characidae: Stevardiinae) (Pecio et al. 2005) and *Anableps anableps* (Cyprinodontiformes: Anablepidae) (Pecio et al. 2010); however, the insemination has evolved also in species belonging to the Osmeriformes, Ophidiiformes, Perciformes, Atheriniformes, Beloniformes, and Scorpaeniformes (Pecio 2010).

The spermatozoon in *P. buchholzi*, shown in Fig. 23, reaching the length of over 80 μm, is the longest and the most derived introsperm having many features not observed in any other teleost (Deurs and Lastein 1973; own observations). The nucleus, sharply pointed anteriorly, is about 7 μm long and 0.6 μm wide, but the midpiece reaches up to 45 μm length and consists of nine helical mitochondrial derivatives surrounding the anterior part of the flagellum; between the axoneme and mitochondria nine helical dense fibres are present. Beneath the mitochondrial derivatives, the axoneme is surrounded by the fenestrated

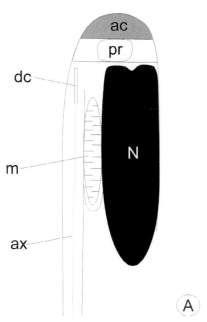

Fig. 22. Diagrammatic structure of the spermatozoon of *Lepidogalaxias salamandroides.* From Jamieson 1991, after Leung (1988). Gamete Research 19: 41–19. ac, acrosome; ax, axoneme; dc, distal centriole; m, mitochondria; N, nucleus; pr, perforatorium.

sheath, about 6 μm long. The proximal centriole was described only in some spermatozoa as lying parallel to the distal one. The distal centriole forms the basal body and is separated from the basal part of nucleus by the lamellate body. The flagellum has a classical 9 + 2 arrangement of microtubules.

The spermatozoon of *M. barberi* shown on Fig. 24 is very unique being flail-shaped with an elongated and highly flattened nucleus, especially at its anterior and posterior portion. Additionally, the centrioles and mitochondria lying laterally to the nucleus are situated on the opposite poles; the centrioles are at its anterior portion, whereas mitochondria are located at the posterior portion. The nucleus is about 13 μm long and is situated alongside the free flagellum extending from the basal body. The short cytoplasmic canal is present only at the very anterior part of the flagellum. Long cristate mitochondria are situated not only alongside the posterior portion of the nucleus, but also extend posterior to the nucleus. Numerous microtubules situated just beneath the plasmalemma are running along the nucleus and mitochondrial region on one side only.

The spermatozoon of *M. barberi* possesses a head about 20 μm long and the length of flagellum reaches 60 μm. Some degree of nucleus elongation is typical for nearly all the many inseminating characids.

The head of another inseminating characid, *Scopeocharax rhinodus*, is very similar in shape, has a great elongation of the nucleus and mitochondria, the presence of a short cytoplasmic canal, and reaches the length of about 17 μm. However, some characters are typical for the species-specific *S. rhinodus* spermatozoon. Namely the centrioles, in nearly parallel arrangement form the tip of the spermatozoon, the mitochondria are found only alongside the nucleus and never appear posterior to the nucleus, and the accessory microtubules are found on both sides of the nucleus (Fig. 25).

The flagellum running parallel to the nucleus is the characteristic feature of many inseminating characids such as *Corynopoma rissei* (Stevardiinae, Stevardiini) (Pecio et al. 2007), *Pseudocorynopoma doriae* (Stevardiinae, Hysteronotini) (Burns et al. 1998), *Brittanichthys axelrodi* (insertae sedis) (Javonillo et al. 2007), and *Bryconadenos tanaothoros* (insertae sedis) (Weitzman et al. 2005), but in the latter four species the flagellum is contained within a cytoplasmic collar which is attached alongside the entire length of nuclear and mitochondrial regions and continues posteriorly as in a thin sleeve. The flagellar axoneme has the typical 9 + 2 arrangement, possessing occlusions on all A-tubules of the peripheral doublets in the region outside the cytoplasmic canal. All the above mentioned characid introsperms develop during

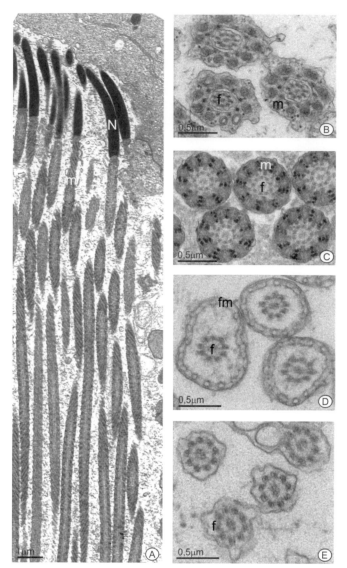

Fig. 23. The spermatozoon of *Pantodon buchholzi* (own data). **A.** Longitudinal sectioned spermatozoa with elongate nucleus (N) and midpiece. **B–C.** Cross sections through the midpiece with flagella (f) surrounded by dense fibers and helically arranged mitochondria (m). **D.** Cross sectioned flagella (f) situated in a fenestrated membrane (fm). **E.** Free flagella (f) behind the level of the fenestrated sheath.

spermiogenesis in which the process of nuclear rotation does not take place (Type II spermiogenesis), although spermiogenesis with rotation (Type I spermiogenesis) is typical for the characid aquasperm as for many other inseminating and internally fertilizing species of the Cyprinodontiformes (e.g., *A. anableps, Poecilia reticulata*) or Beloniformes (e.g., *Hemirhamphodon pogonognathus* (Hemirhamphidae))) (Fig. 25) (Jamieson 1989).

The spermatozoon of *A. anableps* is approximately 40 μm long from the tip of nucleus to the end of flagellum (Fig. 25). The nucleus is only 4 μm long and is conical in shape. The nuclear fossa is very deep and contains both centrioles in perpendicular arrangement; the distal centriole forms an elongate basal body covered by amorphous electron dense substance. The midpiece is comprised of a long cytoplasmic collar containing 12–15 C-shaped elongated mitochondria. The cytoplasmic collar surrounds the proximal part of the flagellum, having 9 + 2 axoneme with A-tubules in all doublets filled by electron dense material.

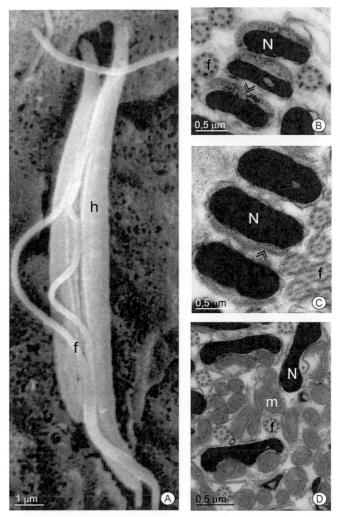

Fig. 24. The spermatozoa of inseminating characids. SEM and TEM. **A.** Two spermatozoa of *Glandulocauda melanogenys* having very elongated heads and a laterally located flagellum (f) alongside the entire nucleus (N). **B–D.** Cross sectioned spermatozoa of *Mimagoniates barberi* from the anterior part (B) through the middle part (C) up to the mitochondrial tip (D). The flattened nucleus is associated with few microtubules (arrowheads) on one side.

Biflagellarity evolved also in internally fertilizing species belonging to the families Zoarcidae (Perciformes) (Jamieson 2009). The biflagellate spermatozoa in *Zoarces elongatus* (Koya et al. 1993) and *Macrozoarces americanus* (Yao et al. 1995) have a discoidal head, six mitochondria of various sizes in *Z. elongatus* and nine in *M. americanus*. Two parallel flagella, extending from parallel arranged centrioles at the base of the head, are associated with each other by an enclosing cytoplasmic membrane at the end of the midpiece but they are posteriorly separated. The length of these spermatozoa is about 80 μm.

Paraspermatozoa

Species of the Hemitripteridae and some Cottidae are unusual in having two sperm types: euspermatozoa and paraspermatozoa. In *Blepsias cirrhosus* (Hemitripteridae), the euspermatozoa possess a disc-like head, about 2 μm long and 1.6 μm wide and 1 μm thick (Hayakawa and Munehara 2004). The nucleus, which is flattened in lateral and transverse views, possesses condensed chromatin. The proximal centrioles are

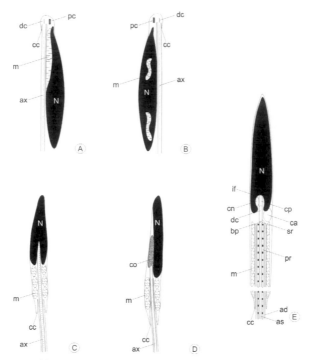

Fig. 25. Diagrammatic structure of introsperm. **A.** *Scopeocharax rhinodus.* **B.** *Tyttocharax tambopatensis.* After Pecio et al. (2005). J. Morphol. 263: 216–226. **C.** *Poecilia reticulata.* From Jamieson 1991, pp. 204–205. **D.** *Cymatogaster aggregata.* From Jamieson 1991 after a micrograph by Gardiner (1978) J. Fish Biol. 13: 435–438. **E.** *Hemirhamphodon pogonognathus.* After Jamieson (1989). Gamete Research 24: 247–259. Ad, axonemal dublets; as, axonemal singlets; co, electron dense cone; bp, basal plate; ca, centriolar adjunct; cn, lateral centriolar connection; cp, centriolar plug; if, implantation fossa; m, mitochondria; N, nucleus; pr, radial periaxonemal rods; sr, satellite ray.

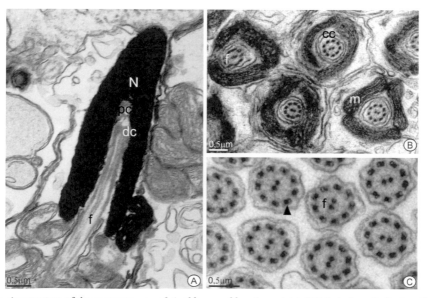

Fig. 26. The ultrastructure of the spermatozoon of *Anableps anableps* (own data). TEM. **A.** Longitudinal section through the elongated nucleus (N) with highly condensed chromatin and deep nuclear fossa containing both centrioles, the proximal (pc) and distal centriole (dc), which forms the basal body. **B.** Cross sectioned midpieces with mitochondria (m) and flagella situated in the cytoplasmic canal (cc). **C.** Cross sectioned flagella with 9 + 2 axonemal pattern but with electron dense A-microtubules (arrowheads).

located laterally at the anterior part of the nucleus in slight indentation. The midpiece contains several mitochondria and surrounds the proximal portion of the flagellum from which it is separated by a deep cytoplasmic canal. The total length of the spermatozoon is about 35 µm. The paraspermatozoa develop from aberrant spermatids which differ from the normal euspermatids in the abundance of cytoplasm and is often binuclear. The paraspermatozoa at the completion of development are spherical cells of about 5 µm in diameter, and have two shorter flagella than the euspermatozoa. They are considered to be hyperpyrenic.

Acknowledgments

I would like to thank two people of the staff of the Department of Comparative Anatomy for their contribution to the preparation of figures; to Dr. Dagmara Podkowa for Figs. 1–7 and to Anna Dymek for Figs. 8–26 and all diagrammatic models of the different spermatozoa. Without the help of these two people, my chapter would have never been finished.

References Cited

Arratia, G. 2000. Phylogenetic relationships of teleostei: past and present. Estudios Oceanológicos 19: 19–51.

Baccetti, B. 1970. Comparative Spermatology. Academia Nationale dei Lincei. Rome. Academic Press. New York. London.

Baccetti, B., A.G. Burrini, G. Callaini, G. Gibertini, M. Mazzini and S. Zerunian. 1984. Fish germinal cells 1. Comparative spermatology of 7 cyprinid species. Gamete Res. 10: 383–396.

Baccetti, B. 1991. Comparative Spermatology 20 years after. pp. 1112. Raven Press New York.

Ballowitz, E. 1915. Ueber die Samenkörner der Forellen. Arch. Zellforsch. 14: 185–192.

Ballowitz, E. 1916. Ueber die körnige Zusammensetzung des Verbindungsstückes der Samenkörper der Knochenfische. Arch. Zellforsch. 14: 355–358.

Billard, R. 1969. La spermatogenèse de *Poecilia reticulata*. I. Estimation du nombre de generations goniales et rendement de la spermatogenèse. Ann. Biol. Anim. Bioch. Biophys. 9: 251–271.

Billard, R. 1970. Ultrastructure comparée de spermatozoïdes des quelques poisson Téléostéens. pp. 71–79. *In*: B. Baccetti (ed.). Comparative Spermatology. Academic Press, New York, USA.

Burns, J.R., S.H. Weitzman, H.J. Grier and N.A. Menezes. 1995. Internal fertilization, testis and sperm morphology in glandulocaudine fishes (Teleostei: Characidae: Glandulocaudinae). J. Morphol. 224: 131–145.

Burns, J.R., S.H. Weitzman and L.R. Malabarba. 1997. Insemination in eight species of cheirodontine fishes (Teleostei: Characidae: Cheirodontinae). Copeia 1997: 433–438.

Burns, J.R., S.H. Weitzman, K.R. Lange and L.R. Malabarba. 1998. Sperm ultrastructure in characid fishes (Teleostei: Ostariophysi). pp. 35–244. *In*: L.R. Malabarba, R.E. Reis, R.P. Vari, Z.M.S. Lucena and C.A.S. Lucena (eds.). Phylogeny and Classification of Neotropical Fishes. Edipucrs, Porto Alegre, Brazil.

Burns, J.R., A.D. Menezes, S.H. Weitzman and L.R. Malabarba. 2002. Sperm and spermatozeugma ultrastructure in the inseminating catfish, *Trachelyopterus lucenai* (Ostariophysi: Siluriformes: Auchenipteridae). Copeia 2002: 173–179.

Burns, J.R. and S.H. Weitzman. 2005. Insemination in ostariophysan fishes. pp. 107–134. *In*: H.J. Grier and M.C. Uribe (eds.). Viviparous Fishes. New Life Publications, Homestead Florida.

Burns, J.R., I. Quagio-Grassiotto and B.G.M. Jamieson. 2009. Ultrastructure of spermatozoa: Ostariophysi. pp. 287–388. *In*: B.G.M. Jamieson (ed.). The Reproductive Biology and Phylogeny in Fishes. Vol. I. Science Publishers, Enfield (NH).

Casas, M.T., S. Munoz-Guerra and J.A. Subirana. 1981. Preliminary report on the ultrastructure of chromatin in the histone containing spermatozoa of a teleost fish. Biol. Cell (Paris) 40: 87–92.

Deurs, B.V. and U. Lastein. 1973. Ultrastructure of the spermatozoa of the teleost *Pantodon buchholzi* Peters, with particular reference to the midpiece. J. Ultrastruc. Res. 42: 517–533.

Downing, A.L. and J.R. Burns. 1995. Testis morphology and spermatozeugma formation in three genera of viviparous halfbeaks: *Nomorhamphus*, *Dermogenys* and *Hemirhamphodon* (Teleostei: Hemiramphidae). J. Morphol. 225: 329–343.

De Pinna, M.C.C. 1996. Teleostean monophyly. pp. 193–207. *In*: M.L.J. Stiassny, L.R. Parenti and D.G. Johnson (eds.). Interrelationships of Fishes. Acadamic Press, San Diego.

Gibbons, B.H., I.R. Gibbons and B. Baccetti. 1983. Structure and motility of the 9 + 0 flagellum of eel spermatozoa. J. Submicrosc. Cytol. 15: 15–20.

Grier, H.J., J.M. Fitsimons and J.R. Linton. 1978. Structure and ultrastructure of the testis and sperm formation in goodeid teleosts. J. Morphol. 156: 419–438.

Grier, H.J. 1981. Cellular organization of the testis and spermatogenesis in fishes. Amer. Zool. 21: 345–357.

Grier, H.J. 1984. Testis structure and formation of spermatophores in the atherinomorph teleost *Horaichthys setnai*. Copeia 1984: 833–839.

Grier, H.J. and B.B. Collette. 1987. Unique spermatozeugmata in the testis of halfbeaks of the genus *Zenarchopterus* (Teleostei: Hemiramphidae). Copeia 1987: 300–311.

Grier, H.J. 1993. Comparative organization of the Sertoli cells including the Sertoli cell barrier. pp. 703–739. *In*: L.D. Russell and M.D. Griswold (eds.). The Sertoli Cell. Cache River Press Clearwater FL.

Grier, H.J. and R.G. Taylor. 1998. Testicular maturation and regression in the common snook. J. Fish Biol. 53: 521–542.

Grier, H.J. and M.C. Uribe. 2009. The testis and spermatogenesis in teleosts. pp. 119–142. *In*: B.G.M. Jamieson (ed.). Reproductive Biology and Phylogeny of Fishes (Agnathans and Bony Fishes). Phylogeny, Reproductive System, Viviparity, Spermatozoa. Vol. 8A of Series: Reproductive Biology and Phylogeny. Science Publishers, Enfield (NH).

Gwo, J.C., H.H. Gwo and S.L. Chang. 1992. The spermatozoon of the Japanese eel, *Anguilla anguilla* (Teleostei, Anguilliformes, Anguillidae). J. Submicrosc. Cytol. Pathol. 24: 571–74.

Gwo, J.C., Y.S. Kao, X.W. Lin, S.L. Chang and M.S. Su. 1995. The ultrastructure of the milkfish, *Chanos chanos* (Forrskal), spermatozoon (Teleostei, Gonorchynciformes, Chanidae). J. Submicrosc. Cytol. Pathol. 27: 99–104.

Gwo, J.C., C.Y. Lin, W.L. Yang and Y.C. Chou. 2006. Ultrastructure of the sperm of blue sprat, *Spratelloides gracilis*; Teleostei, Clupeiformes, Clupeidae. Tissue Cell 38: 285–91.

Hara, M., S. Ishijima and M. Okijama. 1994. Ultrastructure and motility of spermatozoa of the Japanese sardine, *Sardinops melanosticus*. Jap. J. Ichthyol. 41: 322–25.

Hara, M. and M. Okiyama. 1998. An ultrastructural review of the spermatozoa of Japanese fishes. Bull. Ocean Res. Inst. University of Tokyo 33: 1–138.

Hayakawa, Y. and H. Munehara. 2004. Ultrastructural observations of euspermatozoa and paraspermatozoa in a copulatory cottoid fish *Blepsias cirrhosus*. J. Fish Biol. 64: 1530–1559.

His, W.1873. Untersuchungen über das Ei und die Eientwicklung bei Knochenfischen. F.C.W. Vogel. Leipzig.

Hofman, R.A. 1963. Gonads, spermatic ducts, and spermatogenesis in the reproductive system male toadfish, *Opsanus tau*. Chespeak Sci. 4: 21–29.

Javonillo, R., J.R. Burns and S.H. Weitzman. 2007. Reproductive morphology of *Brittanichthys axelrodi* (Teleostei: Characidae), a miniature inseminating fish from South America. J. Morphol. 268: 23–32.

Javonillo, R., J.R. Burns and S.H. Weitzman. 2009. Sperm modifications related to insemination, with examples from the Ostariophysi. pp. 723–763. *In*: B.G.M. Jamieson (ed.). Reproductive Biology and Phylogeny of Fishes (Agnathans and Bony Fishes). Phylogeny, Reproductive System, Viviparity, Spermatozoa. Vol. 8A of Series: Reproductive Biology and Phylogeny. Science Publishers, Enfield (NH).

Jamieson, B.G.M. 1989. Complex spermatozoon of the live-bearing half-beak *Hemirhamphodon pogonognathus* Bleeker Ultrastructural description Euteleostei Atherinomorpha Beloniformes. Gamete Res. 24: 247–260.

Jamieson, B.G.M. 1991. Fish Evolution and Systematics: Evidence from Spermatozoa. Cambridge Univesity Press, Cambridge.

Jamieson, B.G.M. 2009. Ultrastructure of spermatozoa: Acanthopterygii continued: Percomorpha. pp. 503–684. *In*: B.G.M. Jamieson (ed.). Reproductive Biology and Phylogeny of Fishes (Agnathans and Bony Fishes). Phylogeny, Reproductive System, Viviparity, Spermatozoa. Vol. 8A of Series: Reproductive Biology and Phylogeny. Science Publishers, Enfield (NH).

Jamieson, B.G.M. and X. Mattei. 2009. Ultrastructure of spermatozoa: Elopomorpha and Clupeomorpha. pp. 256–285. *In*: B.G.M. Jamieson (ed.). Reproductive Biology and Phylogeny of Fishes (Agnathans and Bony Fishes). Phylogeny, Reproductive System, Viviparity, Spermatozoa. Vol. 8A of Series: Reproductive Biology and Phylogeny. Science Publishers, Enfield (NH).

Kirschbaum, F. and C. Schugardt. 2002. Reproductive strategies and developmental aspects in mormyrid and gymnotiform fishes. J. Physiol. Paris 96: 557–566.

Koya, Y., S. Ohara, T. Ikeuchi, S. Adachi, T. Matsubara and K. Yamauchi. 1993. Testicular development and sperm morphology in the viviparous teleost, Zoarces elongatus. Bull. Hokkaido Natl. Fish. Res. Inst. 57: 21–31.

Kunz, Y.W. 2004. Developmental Biology of Teleost Fishes. Springer. Netherlands.

Kupffer, C.V. 1878. Die Entwicklung des Herings im Ei. Jahresbericht der Commission zur wissenschaftlichen Untersuchung der deutschen Meere in Kiel für die Jahre 1874–76; 4-6: 175–224.

Lahnsteiner, F. and R.A. Patzner. 1990a. Spermiogenesis and structure of mature spermatozoa in blennid fishes (Pisces, Blennidae). J. Submicrosc. Pathol. Cytol. 22: 565–576.

Lahnsteiner, F. and R.A. Patzner. 1990b. The mode of male germ cell renewal and ultrastructure of early spermatogenesis in *Salaria* (= *Blennius pavo*) (Teleostei: Blennidae). Zool. Anz. 224: 129–139.

Lahnsteiner, F., R.A. Patzner and T. Weismann. 1993. The efferent duct system of the male gonads of the European pike (*Esox lucius*): Testicular efferent ducts, testicular main ducts and spermatic ducts. J. Submicr. Cytol. Pathol. 25: 487–498.

Lahnsteiner, F., R.A. Patzner and T. Weismann. 1994. The testicular main ducts and spermatic ducts in cyprinid fishes. Morphology, fine structure and histochemistry. J. Fish. Biol. 44: 937–951.

Lahnsteiner, F. and R.A. Patzner. 1997. Fine structure of spermatozoa of four littoral teleosts, *Symphodus ocellatus*, *Coris julis*, *Thalassoma pavo* and *Chromis chromis*. J. Submicrosc. Cytol. Pathol. 29: 477–485.

Lahnsteiner, F. and R.A. Patzner. 2008. Sperm morphology and ultrastructure in fish. pp. 1–61. *In*: S.M.H. Sadi, J.J. Cosson, K. Coward and G. Rafiee (eds.). Fish Spermatology. Alpha Science International Ltd. Oxford, U.K.

Lahnsteiner, F. and R.A. Patzner. 2009. Male reproductive system: spermatic duct and accessory organs of the testis. pp. 141–186. *In*: B.G.M. Jamieson (ed.). Reproductive Biology and Phylogeny of Fishes (Agnathans and Bony Fishes). Phylogeny, Reproductive System, Viviparity, Spermatozoa. Vol. 8A of Series: Reproductive Biology and Phylogeny. Science Publishers, Enfield (NH).

Leung, L.K.P. 1988. The ultrastructure of the spermatozoon of *Lepidogalaxias salamandroides* and its phylogenetic significance. Gamete Res. 19: 41–50.

Li, L. and T. Xi. 2005. Stem cell niche: structure and function. Ann. Rev. Cell Dev. Biol. 21: 605–631.

Loir, M. 1999. Spermatogonia of rainbow trout: II. *In vitro* study of the influence of pituitary hormones, growth factors and steroids on mitotic activity. Mol. Reprod. Develop. 53: 434–442.

Matos, E., M.N.S. Santos and C. Azevedo. 2002. Biflagellate spermatozoon structure of the hermaphrodite fish *Satanoperca jurupari* (Heckel, 1840) (Teleostei, Cichlidae) from Amazon River. Braz. J. Biol. 62: 847–852.

Mattei, C. and X. Mattei. 1973. La spermiogenése d'*Albula vulpes* (L. 1758) (Poissin Albulidae). Zeitschr. Zellforsch. Mikrosk. Anat. 142: 171–192.

Mattei, C. and X. Mattei. 1974. Spermatogenesis and spermatozoa of the elopomorpha (teleost fish). pp. 211–21. *In*: B.A. Afzelius (ed.). The Functional Anatomy of the Spermatozoa, Pergamon Press, Oxford.

Mattei, C. and X. Mattei. 1978. La spermiogenèse d'un poisson téléostéen (*Lepadogaster lepadogaster*). II. Le spermatozoide. Biol. Cell. 32: 267–274.

Mattei, C., X. Mattei, B. Marchand and R. Billard. 1981. Réinvestigation de la structure des flagellas spermatique: cas particulier des spermatozoïdes à mitochondrie annulaire. J. Ultrastruc. Res. 74: 307–312.

Mattei, X., C. Boisson, C. Mattei and C. Reitzer. 1967. Spermatozoïdes afflageles chez un poisson: *Gymnarchus niloticus* (Télcostéen, Gymnarchidae). Comptes Renduz Hebdomadaires des Seances de l'Academie dse Sciences D 265: 2010–2012.

Mattei, X. 1970. Spermiogenése comparée des poissons. pp. 57–69. *In*: B. Baccetti (ed.). Comparative Spermatology. Academic Press. New York.

Mattei, X., C. Mattei, C. Reitzer and J.-L. Chevalier. 1972. Ultrastructure des spermatozoides aflagelles des Mormyres (Poisson, Téleostéen). J. Microsc. (Paris) 15: 67–78.

Mattei, X. 1991. Spermatozoon ultrastructure and its implications in fishes. Can. J. Zool. 69: 3038–3055.

Mattei, X., B. Marchand and O.T. Thiaw. 1995. Unusual midpiece in the spermatozoon of a teleost fish *Citharinus* sp. J. Submicrosc. Cytol. Pathol. 27: 189–191.

Miescher, F. 1878. Die Spermatozoen einiger Wirbelthiere. Ein Beitrag zur Histochemie (after lectures given in April 1872 and November 1873). Verhandlungen der Naturforschenden Gesellschaft in Basel 1878, VI: 138–208.

Parenti, L.R. and H.J. Grier. 2004. Evolution and phylogeny of gonad morphology in bony fishes. Integr. Comp. Biol. 44: 333–348.

Parenti, L.R., F.L. Lo Nostro and H.J. Grier. 2010. Reproductive histology of *Tomeurus gracilis* Eigenman 1909 (Teleostei: Atherinomorpha: Poecillidae) with comments on evolution of viviparity in atherinomorph fishes. J. Morphol. 271: 1399–1406.

Pecio, A. and J. Rafiński. 1994. Structure of the testis, spermatozoa and spermatozeugmata of *Mimagoniates barberi* Regan, 1907 (Teleostei: Characidae), an internally fertilizing, oviparous fish. Acta Zool. (Stockholm) 75: 179–185.

Pccio, A. and J. Rafiński. 1999. Spermiogenesis in *Mimagoniates barberi* (Teleostei: Ostariophysi: Characidae), an oviparous, internally fertilizing fish. Acta Zool. (Stockholm) 80: 35–45.

Pecio, A. and J. Rafiński. 2001. Spermatozeugmata formation in *Mimagoniates barberi* (Teleostei: Characidae). J. Morphol. 248: 270.

Pecio, A., F. Lahnsteiner and J. Rafiński. 2001. Ultrastructure of the epithelial cells in the aspermatogenic part of the testis in *Mimagoniates barberi* (Teleostei: Characidae: Glandulocaudinae) and the role of their secretions in spermatozeugmata formation. Ann. Anat. 183: 427–435.

Pecio, A., J.R. Burns and S.H. Weitzman. 2005. Sperm and spermatozeugma ultrastructure in the inseminating species *Tyttocharax cochui*, *T. tambopatensis* and *Scopaeocharax rhinodus* (Pisces: Teleostei: Characidae: Glandulocaudinae: Xenurobryconini). J. Morphol. 263: 216–226.

Pecio, A., J.R. Burns and S.H. Weitzman. 2007. Comparison of spermiogenesis in the externally fertilizing *Hemigrammus erythrozonus*, Durbin 1909 and the inseminating *Corynopoma riisei*, Gill 1858 (Teleostei: Characiformes: Characidae). Neotr. Ichthyol. 4: 457–470.

Pecio, A. 2009. Ultrastructural examination of spermiogenesis and spermatozoon ultrastructure in Congo tetra *Phenacogrammus interruptus* Boulenger, 1899 (Ostariophysi: Characiformes: Alestidae). Folia Biol. (Kraków) 57: 13–21.

Pecio, A. 2010. Modifications coupled with insemination in male reproductive system in the representatives of subfamily Glandulocaudinae and Stevardiinae (Teleostei: Characiformes: Characidae), pp. 105. WUJ. Kraków (in polish).

Pecio, A., J.R. Burns and H.J. Grier. 2010. Testis structure, spermiogenesis and spermatozoon ultrastructure in a four-eyed fish, *Anableps anableps* L. 1758 (Teleostei: Atherinomorpha: Cyprinodontiformes: Anablepidae). *In*: H.J. Grier and M.C. Uribe (eds.). Viviparous Fishes. II Book. New Life Publications.

Poirer, G.R. and N. Nicholson. 1982. Fine structure of the testicular spermatozoa from the Channel Catfish, Ictalurus punctatus. J. Ultrastruc. Res. 36: 455–465.

Pudney, J. 1995. Spermatogenesis in nonmammalian vertebrates. Microsc. Res. Tech. 32: 459–497.

Retzius, M.G. 1905. Die Spermien der Leptokardier, Teleostier and Ganoiden. Biologische Untersuchungen. N.F. Stockholm n.s. 12: 103–115.

Retzius, M.G. 1910. Weitere Beiträge zur Kentnis der Spermien mit besonderer Berüksichtigung der Kernsubstanz. Biologische Untersuchungen. Stockholm n.s. 15: 63–82.

Remak, R. 1854. Ueber Eihüllen und Spermatozoen. Müller's Archiv. pp. 252–56 (Hoffmann 1881. P.22 Zeichnung Tafel II).

Saperas, N., C. Ribes, C. Buesa, F. Garcia-Hegart and M. Chiva. 1093. Differences in chromatin condensation during spermiogenesis in two species of fish with distinct protamines. J. Exp. Zool. 265: 185–194.

Schulz, R.W., S. Menting, J. Bogerd, L.R. Franca, D.A.R. Vilela and H.P. Godinho. 2005. Sertoli cell proliferation in the adult testis—evidence from two fish species belonging to different orders. Biol. Reprod. 73: 891–898.

Schultz, R.W., L.R. de Franca, J.J. Lareyre, F. LeGac, H. Chiarini-Garcia, R.H. Nobrega and T. Miura. 2010. Spermatogenesis in fish. Gen. Comp. Endocrinol. 165: 390–411.

Sprando, R.L, R.C. Heidinger and L.D. Russell. 1988. Spermiogenesis in the bluegill (*Lepomis macrochirus*): a study of cytoplasmic events including cell volume changes and cytoplasmic elimination. J. Morphol. 198: 165–177.

Stanley, H.P. 1965. Electron microscopic observations on the biflagellate spermatids of the teleost fish, Porichthys notatus. Anatomical Record 151: 477.

Stein, H. 1981. Licht- und elektronenoptische Untersuchungen an den Spermatozoen verschiedener Süsswasserknochenfische (Teleostei). Zeit. Angew. Zool. 68: 183–98.

Taylor, R.G., H.J. Grier and A.J. Whittington. 1998. Spawning rhythms of common snook in Florida. J. Fish Biol. 53: 502–520.

Uribe, M.C., H.J. Grier and V. Mejia-Roa. 2014. Comparative testicular structure and spermatogenesis in bony fishes. Spermatogenesis 4(3): e983400. doi 10.4161/21565562.2014.983400.

Weitzman, S.H., S.H. Menezes, H.-G. Evers and J.R. Burns. 2005. Putative relationships among inseminating and externally fertilizing characids, with a description of a new genus and species of Brazilian inseminating fosh bearing an anal-fin gland males (Characiformes: Characidae) Neotr. Ichthyol. 3: 329–360.

Wourms, J.P. and J. Lombardi. 1992. Reflections on the evolution of piscine viviparity. Amer. Zool. 32: 276–293.

Yao, Z., C.J. Emerson and L.W. Crim. 1995. Ultrastructure of the spermatozoa and eggs of the ocean pout (*Macrozoarces americanus* L.), an internally fertilizing marine fish. Mol. Reprod. Devel. 42: 58–64.

Zirkin, B.R. 1975. The ultrastructure of nuclear differentiation during spermiogenesis in the salmon. J. Ultrastuc. Res. 50: 174–184.

Chapter 13

Cardiovascular System and Blood

José M. Icardo

INTRODUCTION

The heart is the main organ of the circulatory system. Classical descriptions indicate that most fish hearts follow the typical piscine pattern of organization, with four chambers arranged in series: the sinus venosus, atrium, ventricle and outflow tract (OFT) (Santer 1985, Satchell 1991, Farrell and Jones 1992, Burggren et al. 1997, Farrell 2007, 2011). However, this description is incomplete since the OFT consists of two components: the conus arteriosus and the bulbus arteriosus (Icardo 2012, 2017). In addition, the atrioventricular (AV) segment appears to be interposed between the atrium and the ventricle (Icardo and Colvee 2011). Developmental analyses indicate that the hearts from ancient (Fig. 1A) (Guerrero et al. 2004, Icardo et al. 2004) and modern fishes (Fig. 1B) (Icardo 2006) show, like the developing heart of higher vertebrates, all of these six components. With the exception of Cyclostomes (hagfishes, lampreys) (Kardong 2006, Farrell 2007, Icardo et al. 2016a), the adult fish heart shows all the components, albeit with a variable degree of morphological expression. Noticeably, the AV region, the bulbus in ancient fishes and the conus in modern teleosts are mostly reduced to connecting segments interposed between large chambers. This is evident both in external views (Fig. 1C) and in histological sections (Figs. 2A–B).

This chapter reviews the histological and structural characteristics of the fish heart. Of note, the teleosts constitute the most evolved fish group and have experienced the widest radiation in vertebrate evolution. Because of this, they constitute the backbone of this chapter. Data from other fish groups are also presented, and comparative analyses are made when considered appropriate. It should be underscored that, given the high diversity of fish, any descriptive data may not fit one particular species or group of species. Also, the general characteristics of species pertaining to small groups may have been overlooked. However, care has been taken to assure that the descriptions included here cover the general histological characteristics of the hearts of most fish groups. The different parts of the heart are analyzed following the direction of blood, from the caudal to the cranial end. The general characteristics of the arterial vessels, and an overview of the blood cells, have also been included.

Department of Anatomy and Cell Biology, University of Cantabria, 39011-Santander, Spain.
Email: icardojm@unican.es

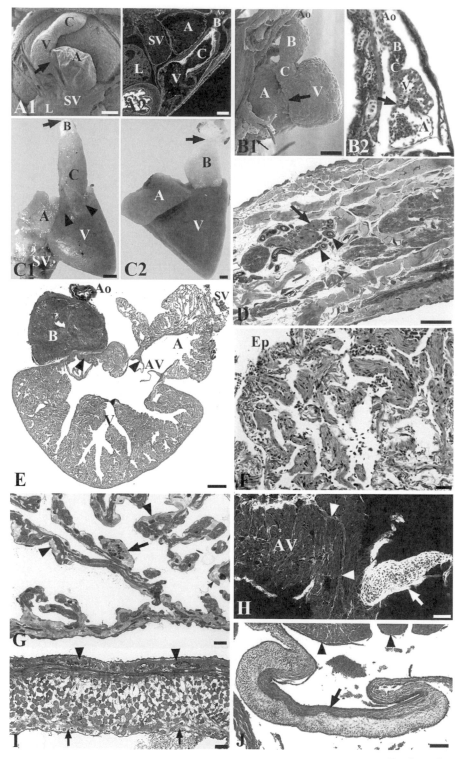

Fig. 1 contd. ...

Sinus Venosus and Atrium

The sinus venosus is the most caudal portion of the heart (Fig. 1C). The wall of the sinus venosus is normally thin and has a fibrous appearance in most species (Fig. 1C1, Fig. 2B). However, its structure varies widely among the different fish groups. In teleosts, the sinus venosus wall may be mostly made up of connective tissue, may contain a few myocardial bundles, or may be mostly made up of myocardium or of smooth muscle cells (Yamauchi 1980, Santer 1985, Farrell and Jones 1992). This variability is shared by other fish groups. For instance, the sinus venosus of hagfishes contains both myocardial and connective tissue (Icardo et al. 2016b), that of elasmobranchs is rich in myocardium (Yamauchi 1980) and that of lungfishes contains sparse myocardial bundles only (Fig. 2C). This area is also densely innervated and, in many species, contains nerve fibres pertaining to the vagus system and ganglion cells (Fig. 1D) (Santer 1985, Haverinen and Vornanen 2007, Zaccone et al. 2009, 2010, Newton et al. 2014). The sinus venosus also contains the heart pacemaker, which appears as a specialized ring of tissue located at the sinoatrial region in most species (Vornanen et al. 2010, Jensen 2014). The sinus venosus conveys the blood into the atrium from which it is separated by the sinus valve.

...Fig. 1 contd.

Fig. 1. Development and anatomy of the heart of various fish groups. **A.** *Acipenser naccarii* (Adriatic sturgeon). **A1.** 3 days post-hatching (dph). All heart segments except the bulbus are apparent. Arrow, atrioventricular segment. SEM. **A2.** 9 dph. Wheat germ agglutinin (WGA) lectin staining. The bulbus has been incorporated at the cranial end of the heart. The subendocardial conal tissue is intensely positive to WGA. Scale bars: 100 μm (After Icardo et al. 2004, Anat. Embryol. 208: 439–449). **B1.** *Sparus auratus* (Gilthead seabream). 16 dpf. Right lateral view. All of the six heart components are present. Thick arrow indicates atrioventricular segment, thin arrow the sinus venosus. SEM. **B2.** *Danio rerio* (Zebrafish). 5 dpf. Haematoxylin-eosin. The conus arteriosus appears as a distinct heart segment interposed between bulbus and ventricle. The sinus venosus has been left out of the section plane. Scale bars: 25 μm (After Icardo 2006, Anat. Rec. 288A: 900–908). **C.** Adult fish heart. Right-sided view. In the two panels, arrow indicates upper limit of bulbus. **C1.** *Acipenser naccarii*. The heart outflow tract is formed by a long, cylindrical conus and a short bulbus. An arterial coronary trunk runs along the conal surface. Arrowheads, boundary between conus and ventricle. **C2.** *Sparus auratus*. The outflow tract is dominated by the presence of a large, robust bulbus. The conus is very short and remains out of sight. Scale bars: 0.1 cm. **D.** *Acipenser naccarii*. Toluidine blue. Sinus venosus wall. Myocardial bundles alternate with collagen bundles. Small vessels supply the myocardium. The centre of the figure is occupied by myelinated nerve fibres surrounded by Schwann cells (arrowheads). Arrow, large ganglion cells. Scale bar: 50 μm. **E.** *Balistes carolinensis* (Trigger fish). Haematoxylin-eosin. Panoramic view. The atrium, the sac-like ventricle and the bulbus arteriosus are large chambers. A small portion of the sinus venosus appears caudal to the atrium. The atrium shows a main lumen and a delicate trabecular network (pectinate muscles). The ventricle is completely trabeculated. Trabecular sheets face the main ventricular lumen. The size of the intertrabecular spaces decreases towards the ventricular periphery. The conus and the atrioventricular segment are formed by compact myocardium (arrowheads) that supports the corresponding heart valves. The bulbus has no myocardium. Cranially, the bulbus opens into the ventral aorta. Scale bar: 0,1 cm. **F.** *Myxine glutinosa* (Atlantic hagfish). Hematoxylin-eosin. Atrium. Detail of atrial trabeculae. Myocardial cells have different orientations. Myocardial nuclei appear as dark dots. Scale bar: 50 μm. **G.** *Acipenser naccarii*. Toluidine blue. Atrial wall. A nerve bundle (arrow) containing myelinated nerve fibres courses along the myocardial muscle. The nerve bundle runs under the endocardium. The external wall of the atrium shows discontinuous myocardium and large amounts of collagen. Abundant collagen (arrowheads) is also observed under the endocardium. Scale bar: 20 μm. **H.** *Sparus auratus*. Lectin staining. Atrioventricular segment and valve. Intense lectin-binding occurs in the extracellular components and in the collagenous fibrosa (arrow) of the valve leaflets. The lectin also marks the collagenous component of the AV muscle ring and defines the presence of a vessel (arrowheads) traversing the compact myocardium. Scale bar: 100 μm. **I.** *Mullus surmuletus* (Striped mullet). Toluidine blue. Detail of AV valve leaflet. Rounded interstitial cells appear separated by large spaces. The atrial fibrosa (arrowheads) contains abundant collagen and small cell groups. Collagen is also deposited on the ventricular side of the leaflet (arrows). Scale bar: 20 μm. **J.** *Thunnus alalunga* (Albacore). Haematoxylin-eosin. Atrioventricular leaflet. The thick atrial fibrosa (arrow) is formed by several parallel layers of collagen. Note high cell density in the rest of the leaflet. Collagen is also present under the subendocardium facing the leaflet (arrowheads). Scale bar: 100 μm. A, atrium; Ao, ventral aorta; AV, atrioventricular; B, bulbus arteriosus; C, conus arteriosus; Ep, epicardium; L, liver; SV, sinus venosus; V, ventricle.

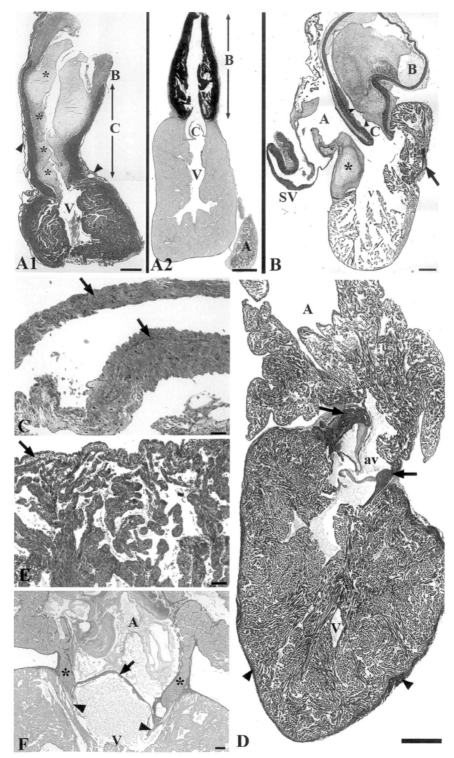

Fig. 2 contd. ...

The atrium is a single chamber that shows considerable variability in size and shape between species (Fig. 1C). As an exception, an incomplete septum partially divides the atrium in the heart of lungfishes. In teleosts, the wall of the atrium is formed of an external rim of myocardium and of a complex network of trabeculae (pectinate muscles). Two arcuate systems of delicate trabeculae, fanning out from the atrioventricular aperture, have been described in several teleosts (Fig. 1E) (Santer 1985). However, this architectural arrangement is not always easy to see (Fig. 2D). Thin atrial trabeculae are also observed in hagfishes (Fig. 1F), lungfishes (Fig. 2B), Polypteriformes and holosteans (Icardo 2017). However, thicker pectinate muscles occur in other primitive species such as elasmobranchs (Fig. 2E) and sturgeons. In many species, such as in sturgeons (Fig. 1G), nerve bundles course along the trabeculae. In addition, elasmobranchs, sturgeons and tuna show intra-atrial vessels in areas where the pectinate muscles appear more densely packed (Tota 1989, Icardo 2017).

Atrioventricular Segment and Atrioventricular Valves

The AV region constitutes a distinct morphological segment of the fish heart. In most teleosts, the AV segment is a ring of compact, vascularized myocardium (Figs. 1E, 2D) (Icardo and Colvee 2011). The compactness of this myocardium contrasts with the spongy appearance of the atrium and ventricle in many teleost species (Fig. 1E). The AV muscle is also easily distinguishable in those species exhibiting a ventricular compacta (Fig. 2D). The AV muscle ring is surrounded by a collagenous ring that isolates the AV muscle from the adjacent regions (Fig. 2F). However, isolation is not complete since the AV muscle remains connected to the atrial and ventricular myocardium (Fig. 2F). This muscle continuity has been

---

...Fig. 2 contd.

Fig. 2. Anatomy and histology of the heart of various fish groups. **A1.** *Raja clavata* (Thornback ray). Martin's trichrome. This heart shows a saccular ventricle with compact and spongy components, a long muscular conus arteriosus and a short, collagen-rich bulbus. The upper limit of the bulbus is not clear in this image. Bulky conal valves (asterisks) arranged in rows are formed by loose connective tissue and show unequal development. Note coronary vessels in the subepicardium (arrowheads). **A2.** *Trigla lucerna* (Yellow gounard). Orcein staining. This heart shows an elongated, completely trabeculated ventricle, a short conus that supports two thin conal valves and a long, elastin-rich bulbus. Scale bars: 0.1 cm (after Icardo 2017, In Cardiovascular Design, Control and Function, Fish Physiology Series, Cardiovascular Physiology, Part A, eds. A.K. Gamperl, T. Gillis, C.J. Brauner and A.P. Farrell). **B.** *Protopterus annectens* (West African lungfish). Martin's trichrome. Panoramic view of the entire heart. The sinus venosus and atrium constitute the most caudal components of the heart. A large piece of cartilage, the AV plug (asterisk), is interposed between the atrium and the entirely trabeculated ventricle. The outflow tract comprises a proximal conus and a distal, arterial-like bulbus. The conus is formed by compact, vascularized myocardium and shows two rows of vestigial conal valves (arrowheads). The conal myocardium extends cranially to wrap the entire bulbar wall. Two ridges, the conal and bulbar folds, formed by loose connective tissue, protrude into the lumen and follow the irregular course of the outflow tract. Arrow indicates the gubernaculum cordis, a thick tendon that attaches the ventral heart surface to the pericardial wall. Scale bar: 0.1 cm. **C.** *Protopterus annectens*. Martin's trichrome. Detail of sinus venosus. The thick collagenous wall encloses a few muscle cells (arrows) that do not form a continuous layer. Scale bar: 100 µm. **D.** *Echiichthys vipera* (Lesser weever). Martin's trichrome. The atrium contains delicate trabeculae that appear more densely packed near the atrioventricular orifice. The AV valves are surrounded by a thick ring of compact, vascularized myocardium (arrows). The AV muscle ring is partially isolated from the atrial and ventricular myocardium by connective tissue rich in collagen. The AV valve leaflets are also rich in collagen. The architecture of the ventricle is dominated by the presence of myocardial trabeculae. However, it also has a thin compacta (arrowheads). Scale bar: 100 µm. **E.** *Myliobatis aquila* (Eagle ray). Martin's trichrome. Atrium. Thick pectinate muscles form the atrial wall. The subepicardium shows a thick collagenous layer (arrow). Scale bar: 100 µm. **F.** *Spondyliosoma cantharus* (Black bream). Sirius red. The AV muscle ring (asterisks) is surrounded by a collagenous ring (here in red). Collagen is also present under the endocardium of the atrial trabeculae, in the subepicardium, and in the fibrosa (arrow) of the AV valve leaflets. Arrowheads: connections between the ventricular trabeculae and the AV muscle. Scale bar: 200 µm. A, atrium; B, bulbus arteriosus; C, conus arteriosus; SV, sinus venosus; V, ventricle.

Color version at the end of the book

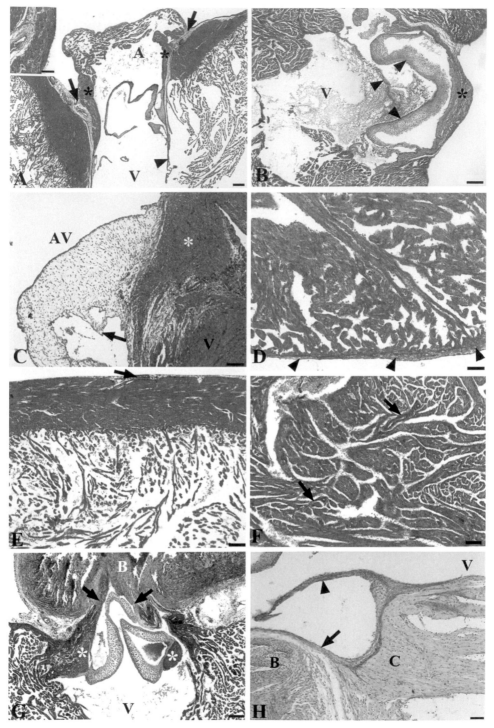

Fig. 3 contd. ...

considered to constitute a preferential way for the transmission of electrical impulses from the heart pacemaker (Sedmera et al. 2003, Icardo and Colvee 2011).

Vascular supply for the AV myocardium comes from the subepicardial coronaries that run in the AV sulcus (Fig. 2F). In many cases, the coronaries are easily distinguishable both in hearts with completely trabeculated ventricles (Fig. 1H) and in those having a ventricular compacta (Fig. 3A).

In a few cases, however, coronaries cannot be demonstrated. In these cases, the AV myocardium appears to be supplied directly from the blood through endocardial sinusoids, or through the development of vascular sinuses continuous with the atrial endocardium (Icardo and Colvee 2011).

With the exception of lungfishes (see below), the AV muscle ring gives support to the AV valves. In general, the AV valves of teleosts are formed by two leaflets (Figs. 1E, 2D). The leaflets show an atrial fibrosa and a dense core formed by interstitial cells (Figs. 1I–J, 3B) and by abundant extracellular material, presumably rich in glycosaminoglycan/proteoglycan complexes (Fig. 3B). The leaflets contain cells of variable size and shape and exhibit different cell densities. An example is provided in Fig. 1I. The thickness and organization of the atrial fibrosa is also variable. It may be thin and formed by densely packed collagen (Fig. 3B), may contain cells embedded in a collagenous matrix (Fig. 1I), or may be much thicker with the collagen organized into several layers (Fig. 1J). Thus, the anatomical and structural configuration of the AV segment is close to being species-specific. Papillary muscles and chordae tendineae are lacking in all the teleost species. However, the observation of ventricular trabeculae attached to the AV muscle ring is a common finding (Figs. 2F, 3A).

The anatomical characteristics of the AV segment may be different in more primitive fish groups. In hagfishes, the AV segment is elongated and funnel-like and contains just a few scattered bundles of myocardial tissue (Icardo et al. 2016b). The heart of lungfishes lacks a distinct AV segment and AV valves. Instead, the transition between the atrium and ventricle is dominated by the presence of a large piece of

---

...Fig. 3 contd.

Fig. 3. Anatomy and histology of the heart of various fish groups. **A.** *Oncorhynchus mykiss* (Rainbow trout). Martin's trichrome. The compact AV myocardium (asterisks) contains numerous vessels. A collagenous ring (arrows), rich in vessels, surrounds the muscle ring. Numerous vessels are also in the ventricular compacta. A few ventricular trabeculae (arrowhead) attach to the AV muscle. Inset: Detail of the AV muscle ring. Numerous vessels are surrounded by collagen. Scale bars: 200 μm; inset, 100 μm. **B.** *Coris julis* (Rainbow wrasse). Martin's trichrome. The atrial fibrosa (arrowheads) of the AV valve leaflets is formed by dense collagen. Cell density in the leaflet core decreases at the areas of valve attachment to the compact AV myocardium (asteriks). Scale bar: 100 μm. **C.** *Raja clavata* (Thornback ray). Martin's trichrome. The thick AV ring (asterisk) is separated from the adjacent ventricular muscle by a ring of connective tissue rich in collagen and vessels. The AV valve leaflet is thick and bulky and appears formed by loose connective tissue. Note the presence of chordae tendineae (arrow). Scale bar: 100 μm. **D.** *Spondyliosoma cantharus* (Black bream). Martin's trichrome. Completely trabeculated ventricle. Detail of ventricular periphery. The external myocardial layer appears thickened due to muscle contraction. The subepicardium is rich in collagen and vessels (arrowheads). Collagen fibres are also present in the trabecular subendocardium. Scale bar: 50 μm. **E.** *Oncorhynchus mykiss*. Martin's trichrome. The external ventricular layer is formed by a thick, vascularized compacta. Myocardial cells organize into layers with different orientations. Subepicardial coronaries (arrow) supply the compacta. The boundary between the compacta and the spongiosa is well defined. Note perpendicular or oblique attachment of the trabeculae onto the compacta. Scale bar: 100 μm. **F.** *Thunnus alalunga* (Albacore). Haematoxylin-eosin. Spongy layer of ventricle. Numerous vessels (arrows) course among the trabeculae. Scale bar: 100 μm. **G.** *Serranus cabrilla* (Comber). Martin's trichrome. The conus arteriosus (asterisks) is a ring of compact, vascularized myocardium interposed between the ventricle and the bulbus arteriosus. The conal myocardium supports the conal valves. Note the spongy ventricle and the presence of vessels in the subepicardium. Arrows indicate the fibrous cylinder connecting the ventricle, the conus and the bulbus. Scale bar: 100 μm. **H.** *Oncorhynchus mykiss*. Haematein trichrome. Detail of conus arteriosus. Longitudinal section. Note the presence of collagen and vessels. The valve leaflet (arrowhead) is rich in collagen. The fibrous wall of the valve sinus (arrow) connects the bulbus with the ventricular subendocardium. Scale bar: 50 μm. A, atrium; AV, atrioventricular; B, bulbus arteriosus; C, conus arteriosus; V, ventricle.

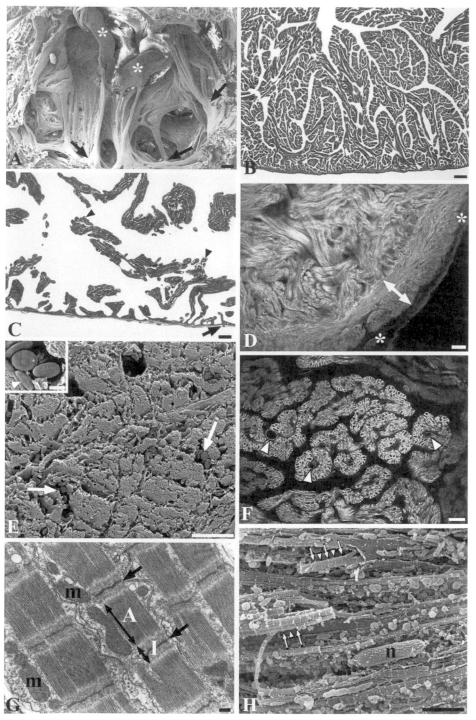

Fig. 4 contd. ...

cartilage, the AV plug (Fig. 2B) (Bugge 1961, Icardo et al. 2005a), that likely controls the size of AV aperture during the cardiac cycle. In elasmobranchs (Fig. 3C) and sturgeons, the AV region is formed by a ring of compact, vascularized myocardium surrounded by a connective ring rich in collagen and vessels. These vessels supply the AV myocardium. However, the muscle ring is very thick when compared to that of teleosts. The AV region also contains compact, vascularized myocardium in the holostean group (Icardo 2017).

In primitive fishes, the AV valves present several differences with respect to those of the teleosts. First, the valve tissue may show several indentations (commissures) that divide the tissue into several leaflets (Fig. 4A).

Second, the leaflets are thick and bulky and contain numerous cells enmeshed in loose connective tissue (Figs. 2A1, 3C). A discrete fibrosa can be observed on the atrial side of the leaflets. Third, the AV valves of elasmobranchs, sturgeons (Fig. 4A) and holosteans show numerous chordae tendineae. Chordae appear to be exclusively formed by wavy collagen bundles. They may attach to the valve marginal border, as in holosteans (Icardo 2017), or to the margins and to the ventricular side of the leaflets, as in sturgeons (Fig. 4A) and elasmobranchs (Fig. 3C) (Hamlett et al. 1996). The chordae emerge directly from the ventricular wall and true papillary muscles appear to be absent in all the cases (however, see Hamlett et al. 1996). Noticeably, the AV valve of hagfishes has two leaflets formed by a few interstitial cells and a dense collagenous core lined by endocardium (Icardo et al. 2016b), thus being closer to that of several teleosts.

Ventricle

The fish heart ventricle is a single, muscular chamber responsible for generating the heart output. As an exception, the ventricle of the lungfishes is partially divided by an incomplete muscular septum (Bugge 1961, Icardo et al. 2005b) that contributes to separate the oxygenated from the deoxygenated blood streams that flow through the heart (Szidon et al. 1969, Burggren et al. 1997).

...Fig. 4 contd.

Fig. 4. Anatomy and fine structure of the heart of various fish groups. **A.** *Acipenser naccarii* (Adriatic sturgeon). Partial view of the AV valve showing two leaflets (asterisks) separated by commissures. The chordae tendineae arise directly from the ventricular trabeculae (arrows). No papillary muscles are apparent. The corrugated appearance of the chordae is due to the wavy collagen bundles that constitute their structural core. Scale bar: 200 μm. **B.** *Notothenia coriiceps* (Black rockcod). Haematoxylin-eosin. Completely trabeculated ventricle. The ventricular periphery is formed by single trabeculae and the corresponding lacunary spaces. Scale bar: 200 μm. **C.** *Periophthalmodon schlosseri* (Giant mudskipper). Toluidine blue. The external myocardium is formed by a very thin muscle layer. Trabeculae in the spongiosa are also thin. Note myofibril banding. Endocardial cells (arrowheads) are prominent and show small granules and empty vacuoles. Arrow, subepicardial capillary. Scale bar: 20 μm. **D.** *Acipenser naccarii*. FITC-conjugated phalloidin staining. Thick section of the ventricular chamber. Double-headed arrow indicates compacta thickness. Note opposite orientation of cell layers. The spongiosa is formed by thick trabeculae. Asterisks indicate subepicardial space. Scale bar: 100 μm. **E.** *Myliobatis aquila* (Eagle ray). SEM. Ventricular compacta. Muscle bundles are separated by a collagenous meshwork that contains capillaries filled with erythrocytes. Inset: Detail of erythrocytes. Note the centrally located nucleus. Arrowhead indicates sectioned nucleus. Scale bars: 50 μm; inset: 3 μm. **F.** *Acipenser naccarii*. FITC-conjugated phalloidin staining. Thick section of the ventricular chamber. Detail of trabeculae. Arrowheads indicate capillaries supplying the trabecular myocardium. Scale bar: 100 μm. **G.** *Anguilla anguilla* (European eel). TEM. Ventricle. Myofibrils in register. Each sarcomere is delimited by the dark, Z-discs (arrows). The A-band (large double-headed arrow) mostly contains thick (myosin) filaments. The I-bands (small double-headed arrow) mostly contain thin (actin) filaments. Actin and myosin slide past each other during muscle contraction. The H-zone is barely visible, indicating muscle contraction. Mitochondria (m) show regular cristae. Scale bar: 300 nm. **H.** *Trachurus trachurus* (Atlantic horse mackerel). SEM. Ventricular myocardium. The cell cytoplasm contains elongated, branching myofibrils. Z-discs (arrows) delimit the sarcomeric units. M-bands are indicated by arrowheads. Numerous rounded or oval mitochondria align along the myofibril axis. n, cell nucleus. Scale bar: 5 μm.

Several attempts have been made to classify the fish heart ventricle into distinct categories according to the external shape, inner architecture and functional capabilities (Santer 1985, Tota et al. 1983, Tota 1989, Farrell and Jones 1992, Farrell 2007, Farrell et al. 2012). However, the heart ventricle shows considerable species variability making categorization very difficult in many cases. In teleosts, the ventricle adopts one of three main shapes: pyramidal (Fig. 1C2), elongated (Fig. 2A2) or sac-like (Fig. 1E) (Santer 1985, Farrell and Jones 1992, Icardo 2012). Internally, the majority (between 50% and 80%—Santer and Greer Walker 1980, Farrell et al. 2012) of the teleost species has a completely trabeculated, spongy ventricle (Figs. 1E, 2A2). Trabeculae are grouped into sheets that face the main ventricular lumen. In the ventricular periphery, trabeculae may form arch systems (Fig. 4B) with muscle cells oriented in several directions (Munshi et al. 2001, Icardo et al. 2005c, Icardo 2012). The most external ventricular layer is one- or two-cell thick (Figs. 4B, 3D), or it may barely cover the outer surface of the ventricle (Fig. 4C). Nonetheless, care has to be taken when analyzing the characteristics of this layer in thick sections since fixation of the heart in contraction may modify the appearance of the external myocardium to give the impression of compactness (compare Figs. 3D and 4B).

The ventricle of a smaller number (between 20% and 50%—Santer and Greer Walker 1980, Farrell et al. 2012) of teleost species has a different ventricular architecture, showing both an inner spongiosa and an external compacta (Figs. 2D, 3A). The thickness of the compacta varies widely between species, ranging from being two- to three-cell thick to occupy a large proportion of the ventricular wall (Fig. 3E). The ratio compacta/spongiosa appears to be species-specific but it is not constant through life since it has been shown to vary with smoltification, changing seasons (Farrell and Jones 1992) and growth (Farrell et al. 1988, Cerra et al. 2004, Gamperl and Farrell 2004, Farrell et al. 2007). Also, the compacta thickness is not uniform across the ventricle. Myocardial cells in the compacta are arranged into layers that show different orientations (Sanchez-Quintana and Hurle 1987, Icardo 2012). Subepicardial vessels invade the compacta (Fig. 3E), that is always vascularized. By contrast, the spongiosa has no vessels except in the heart of tuna where the spongy layer shows a rich vascular supply (Fig. 3F) (Tota 1983, 1989, Icardo 2012).

The existence in the ventricular wall of two muscle compartments, compact and spongy, is also a common feature of the ventricle of the more primitive elasmobranchs (Fig. 2A1) and sturgeons (Fig. 4D). In these fish groups, both the compacta (Fig. 4E) and the spongiosa (Fig. 4F) are heavily vascularized. By contrast, hagfishes, lungfishes (Fig. 2B) and holosteans (except *Amia calva*, Icardo 2017) have completely trabeculated ventricles. Hence, ventricular vessels are restricted to the subepicardium.

It should be underscored that "compact" and "spongy" are largely intuitive terms that may be used to define adequately the ventricular structure in a large number of species. However, given the enormous fish diversification and the variability in heart morphology, definitional and categorization problems have often aroused. This has been addressed in recent works (Farrell and Smith 2017, Icardo 2017).

Myocardial cells throughout the heart show the ultrastructural features typical of muscle cells (Fig. 4G). They contain myofibrils in register that exhibit Z-discs, "M", "A" and "I" bands, and numerous mitochondria aligned with the myofibrils (Figs. 4G–H). Myocardial cells are joined by tight and gap junctions. However, the complexity of the junctional complexes is lower than that in higher vertebrates. In addition, myofibrils are shorter. On the other hand, the myofibrillar content appears to vary widely. In general, cardiomyocytes from active species appear to be filled with myofibrils (Figs. 4G–H) whereas myocardial cells from less active species, or from species having lower blood pressure, present a reduced myofibrillar content (Farrell 2007). This occurs, for instance, in Antarctic teleosts, lungfishes and hagfishes (Icardo 2017).

The Outflow Tract (OFT)

The fish OFT is the portion of the heart located between the ventricle and the ventral aorta. It is formed by two segments: a proximal, muscular, conus arteriosus, and a distal, arterial-like, bulbus arteriosus. The primitive Cyclostomes constitute the exception to this rule since, in these hearts, the aorta arises directly from the base of the ventricle (Icardo et al. 2016a).

In modern teleosts, the conus arteriosus appears as a short, discrete segment interposed between the ventricle and the bulbus arteriosus (Fig. 1E) (Schib et al. 2002, Icardo et al. 2003, Grimes et al. 2006,

Icardo 2006, Genten et al. 2009). The conus is formed in most cases by compact, vascularized myocardium (Figs. 3G, 5A).

It is easily distinguishable from the spongy myocardium in completely trabeculated ventricles (Figs. 3G, 5A). However, it may be more difficult to distinguish in hearts with a ventricular compacta (Figs. 3H, 5B). In general, the conus contains more collagen (Figs. 3H, 5B), elastin (Fig. 5A) and laminin than the ventricular muscle (Schib et al. 2002, Garofalo et al. 2012). With a few exceptions, the conus contains vessels even when the neighbouring myocardium is not vascularized (Figs. 3G, 5A). In basal teleosts, the conus arteriosus is relatively longer and thicker, being also formed by compact, vascularized myocardium (Lorenzale et al. 2017).

The teleost conus arteriosus supports the outflow, conal valves. In modern teleosts, the conus is endowed with a single row of pocket-like valves (Figs. 3G, 5A). Normally, there are left and right valves. Each valve consists of the valve leaflet and the corresponding sinus. The leaflets show a stout proximal body anchored to the conus (Figs. 3G, 5A), and a flap-like distal region that enters the bulbus (Figs. 3H, 5B) and attaches to the inner bulbar surface. The leaflets present a variable number of cells and a thick luminal fibrosa (Figs. 3G, 5A). In general, the histological characteristics of the conus and those of the atrioventricular segment are very similar for any given species (Icardo 2012).

In teleosts, the structural connection between the ventricle, the conus and the bulbus is completed by the presence of a fibrous cylinder that extends between the ventricular subendocardium and the proximal portion of the bulbus, covering the inner side of the conal myocardium (Fig. 5A). This cylinder forms the wall of the valve sinus, provides attachment for the valve leaflets and constitutes the fibrous component of the interleaflet areas (Greer-Walker et al. 1985, Icardo 2006). The fibrous cylinder contains abundant collagen (Figs. 3G–H) and elastin (Fig. 5A), and numerous myofibroblasts (Icardo et al. 2003).

In primitive fish, with the exception of the Cyclostomata, the OFT is dominated by the presence of a robust conus arteriosus (Fig. 1C1). While the relative length of the conus is variable, it appears always organized into concentric tissue layers. The conus endocardium underlines a sleeve of connective tissue, the subendocardium, rich in collagen (Fig. 6A) and elastin (Fig. 5C). In sturgeons, the cranial portion of the subendocardium is richer in elastin and myofibroblasts than the caudal portion. By contrast, the caudal portion is richer in collagen and fibroblasts (Icardo et al. 2002a). Differences in extracellular matrix composition along the cranio-caudal axis of the conus subendocardium have also been reported in other basal species (Grimes et al. 2010). In addition, the amount of elastin appears to be much higher in sturgeons than in elasmobranchs. External to the subendocardium, the conal wall contains a sleeve of compact, vascularized myocardium (Figs. 2A1, 2B, 6A) (Santer 1985, Grimes and Kirby 2009, Jones and Braun 2011). Within the conal myocardium, myocardial cells arrange into layers that show different orientations (Figs. 5C, 6A). The conal myocardium is supplied by coronary trunks that branch from the hypobranchial arteries and run along the outflow tract wall (Figs. 1C1, 2A). Small arterial branches invade the myocardium and distribute between the muscle bundles (Fig. 5D). The outer layer of the conus is formed by the subepicardial tissue and the visceral epicardium. The subepicardium is a loose connective tissue that contains interstitial cells, vessels and nerves. However, it may be a much more complex tissue. For instance, the sturgeon subepicardium contains nodular structures (Fig. 5C) that resemble lymphohemopoietic, thymus-like tissue (Icardo et al. 2002b).

Similar to what occurs in modern teleosts, the conus arteriosus of primitive fish supports the conal valves. However, these valves show quite a different organization and structure. First, they are organized into several (between two and eight) transverse rows (Fig. 2A1). Within each row, valves may show different sizes and shapes. In addition, a particular valve row may be absent (as in sturgeons, Fig. 5C), or the valves in one row may be of disproportionate size spanning most of the length of the conus (Parsons 1930, White 1936). This occurs, at least, in several elasmobranchs (Fig. 2A1) and in the holostean *Amia calva* (Icardo 2017). On the other hand, the conal valves are thick and bulky (Figs. 2A1, 6B), show chordae tendineae that arise from the inner side of the leaflets (Fig. 6B), and are formed by numerous interstitial cells enmeshed in a loose connective tissue (Fig. 5E). As in teleosts, most cranial valves surpass the upper boundary of the conus (Fig. 2A1). An intermediate evolutionary stage between ancient fish and modern teleosts appears to be present in the conus of the basal teleost silver arowana. The conus of this species exhibits two bulky leaflets organized into a single valve row (Lorenzale et al. 2017). In general,

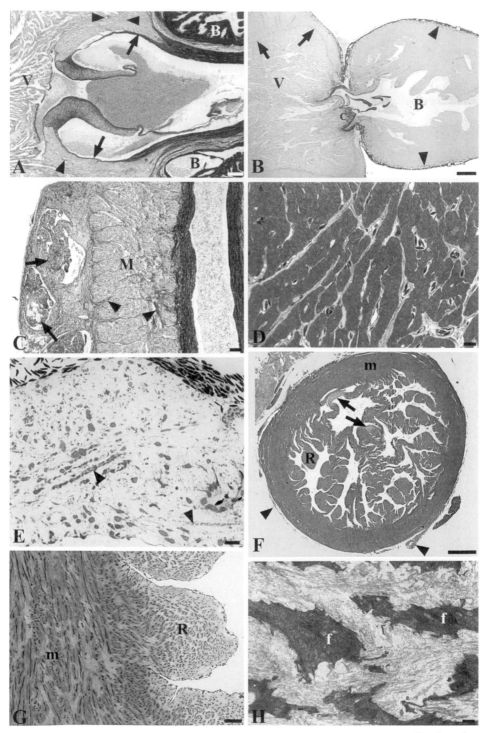

Fig. 5 contd. ...

for any given species, the structure of the conal and atrioventricular valve leaflets appears to be very similar (compare Figs. 3C and 6B). Noticeably, the conal valves of the African lungfish (*Protopterus*) are vestigial (Fig. 2B), being reduced to transverse ridges arranged into two or three rows. By contrast, the Australian lungfish *Neoceratodus forsteri* appears to have three transverse tiers of bulky conal valves (Icardo 2017).

The bulbus arteriosus constitutes the distal portion of the fish OFT. This segment has reached a great development in modern teleosts, dominating the morphology of the OFT (Figs. 1C2, 1E, 2A2) (Santer 1985, Grimes and Kirby 2009, Jones and Braun 2011, Icardo 2012). The external shape of the teleost bulbus is variable, ranging from round, to pear-shaped (Fig. 1C2) to elongated (Fig. 2A2). All the bulbi possess a series of internal columns or ridges (Fig. 5F) that span the entire bulbar length, being attenuated toward the ventral aorta. On the whole, the bulbus is organized into layers: the endocardium, the ridges, and the middle and the subepicardial layers (Figs. 5F, 6C). The visceral epicardium constitutes the external bulbar envelope and faces the pericardial cavity. The subepicardium is a loose, connective tissue layer (Figs. 5F, 6C) that contains interstitial cells, nerves and vessels (Icardo et al. 2000, Icardo 2012). Nonetheless, the structure of the teleost subepicardium may be more complex (Icardo et al. 2000). The middle layer of the bulbus contains, in most teleosts, smooth muscle cells that are arranged in several directions (Fig. 5G) and are surrounded by elastin material (Fig. 5H). The subendocardial ridges contain cell clusters surrounded by amorphous and fibrillar material (Fig. 5G). The bulbar endocardium is a complex layer formed by cells that exhibit different morphologies and cytoplasmic content (Icardo et al. 2000). Consequently, endocardial cells have been implicated in a number of functional activities such as secretion, waste removal and autocrine-paracrine control of heart function (Icardo 2007, Imbrogno and Cerra 2017). The function of the teleost bulbus is to dampen the peaks of blood pressure produced during systolic contraction of the heart (Priede 1976, Satchell 1991, Farrell and Jones 1992, Braun et al. 2003a, b). To this end, the bulbar wall is provided with precise amounts of extracellular matrix components. The entire bulbar wall stains intensely with orcein (Fig. 2A2), indicating the abundance of elastin. The bulbus also contains large amounts of amorphous

---

...Fig. 5 contd.

Fig. 5. Anatomy and histology of the heart of various fish groups. **A.** *Pagellus erythrinus* (Pandora). Orcein. The conal myocardium contains vessels positive for orcein (arrowheads). The conal valves show an elastin scaffold most pronounced in the inner side of leaflets and in the ventricular fibrosa. The wall of the valve sinuses (fibrous cylinder, arrows) is also rich in elastin. Scale bar: 100 μm. **B.** *Scomber japonicus* (Chub mackerel). Sirius red. Longitudinal section. In the ventricle, collagen accumulates in the subepicardium, in the ventricular compacta and along the compacta/spongiosa boundary (arrows). The conal valves are also rich in collagen. In the bulbus, collagen accumulates in the subepicardium. Discrete amounts of collagen appear also under the endocardium and in the most external part (arrowheads) of the middle layer. Scale bar: 500 μm. **C.** *Acipenser naccarii* (Adriatic sturgeon). Orcein. Conus arteriosus. Longitudinal section. Under the myocardium (M), the subendocardium contains elastic fibres oriented longitudinally. Some elastin material (arrowheads) accompanies the connective tissue and vessels that distribute between the myocardial bundles. The subepicardium contains nodes of thymus-like tissue (arrows). This segment of the conal wall is devoid of valves. Scale bar: 200 μm. **D.** *Acipenser naccarii*. Toluidine blue. Transverse section of the conus. Myocardial bundles are separated by connective tissue. Note abundant vascular supply. Scale bar: 20 μm. **E.** *Acipenser naccarii*. Toluidine blue. Conal valve leaflet. Interstitial cells appear loosely distributed throughout the extracellular matrix. Collagen fibres (arrowheads) appear oriented in many directions. A structured fibrosa is lacking. Scale bar: 20 μm. **F.** *Trachurus trachurus* (Atlantic horse mackerel). Haematoxylin-eosin. Bulbus arteriosus. Cross-section. The ridges (R) protrude into the lumen. The middle layer (m) and the subepicardium (arrowheads) are indicated. Arrows indicate bulbar insertion of conal valves. Scale bar: 500 μm. **G.** *Anguilla anguilla* (European eel). Toluidine blue. Detail of the bulbus arteriosus wall. The middle layer (m) contains smooth muscle cells oriented in several directions. The endocardial ridges (R) contain cell clusters surrounded by dense matrix. The endocardium is formed by flattened cells. Scale bar: 50 μm. **H.** *Serranus cabrilla* (Comber). TEM. Bulbus arteriosus. The middle layer is formed by smooth muscle cells surrounded by elastin material. The cell cytoplasm contains numerous microfilament bundles (f). Many extracellular fibres attach to the cell surface. Scale bar: 1 μm. B, bulbus arteriosus; V, ventricle.

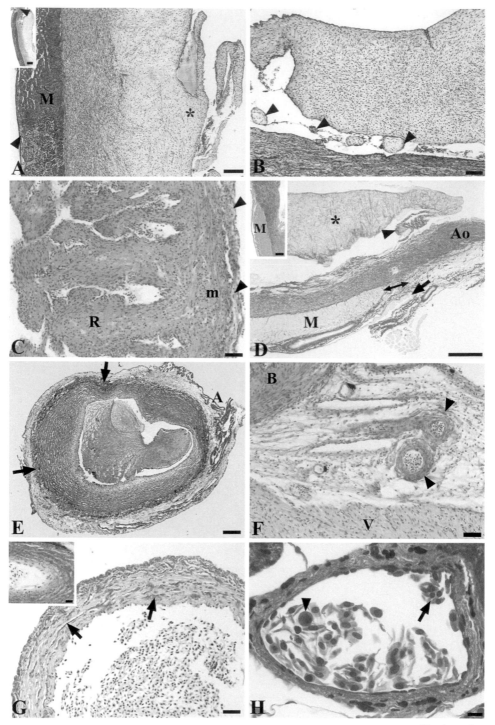

Fig. 6 contd. ...

material. Within the bulbus, collagen is mostly restricted to the subepicardium (Fig. 5B). However, small amounts of collagen appear in the external portion of the middle layer in continuity with the subepicardial collagen (Figs. 5B, 6C). While this constitutes a general description (Icardo et al. 2000, Icardo 2012), the situation is different in many species. For instance, elastin fibres could not be demonstrated in Antarctic teleosts (Icardo et al. 1999a, b). On the other hand, very active Atlantic teleosts like tuna show invasion of the middle layer by connective bundles carrying vessels and nerves (Icardo 2013). Thus, the fine structure of the teleost bulbus is almost species-specific (Icardo et al. 2000, Icardo 2012).

With the exception of the Cyclostomata, a segment that appears to be homologous to the teleost bulbus arteriosus is present in the most distal portion of the OFT in primitive fishes (Icardo et al. 2002a, Durán et al. 2008, Grimes et al. 2010, Durán et al. 2014). The bulbus is a discrete segment in elasmobranchs (Fig. 6D) but becomes more apparent in chondrosteans (Fig. 1C1), holosteans (Icardo 2017) and basal teleosts (Lorenzale et al. 2017). It must be underscored that the bulbus of ancient fish presents many morpho-functional differences with the bulbus of modern teleosts (Icardo 2017). Moreover, the homology of the bulbus across fish groups has been challenged in a recent paper (Moriyama et al. 2016). In sturgeons, the structure of the bulbus resembles that of the ventral aorta, showing ordered layers of collagen and elastin (Fig. 7A).

Indeed, the sturgeon bulbus was initially described as the intrapericardial segment of the ventral aorta (Icardo et al. 2002b). In addition, it should be mostly inextensible making the existence of any dampening function unlikely. Curiously, the arterial-like structure of the sturgeon bulbus is lost at the transition level

...Fig. 6 contd.

Fig. 6. Anatomy and histology of the heart of various fish groups. **A.** *Myliobatis aquila* (Eagle ray). Martin's trichrome. Conus arteriosus. Longitudinal section. The subepicardial tissue (arrowhead) covers the myocardium (M). A sleeve of dense connective tissue, rich in collagen, underlies the myocardium. By contrast, the thick conal valves (asterisk) are formed by loose connective tissue. Inset: Panoramic of the distal portion of the outflow tract. The dense collagenous layer underlying the conus contrasts with the loose connective tissue of the distal, long valve. Arrowhead indicates the very short bulbus. Scale bars: 200 μm; inset: 500 μm. **B.** *Raja clavata* (Thornback ray). Martin's trichrome. Longitudinal section. Conal valve. The leaflet tissue contains many interstitial cells. Some collagen accumulates at the luminal and parietal sides of the leaflet but a well-developed fibrosa is lacking. Arrowheads indicate sections of chordae tendineae. Scale bar: 100 μm. **C.** *Oncorhynchus mykiss* (Rainbow trout). Hematein trichrome. Bulbus arteriosus. Cross-section. The subepicardium (arrowheads) is rich in collagen (blue). The middle layer (m) contains cells oriented in several directions. Discrete amounts of collagen are present in this layer. Flattened endocardial cells cover the ridges (R). Scale bar: 50 μm. **D.** *Myliobatis aquila*. Sirius red. Distal OFT. Longitudinal section. A short segment, arterial in structure (double-headed arrow), appears between the upper limit of the conal myocardium (M) and the insertion of the pericardium (arrow). The distal conal valve (asterisk) extends into the ventral aorta surpassing the distal OFT limit. Arrowhead indicates chordae tendineae. Inset: *Raja clavata* (Thornback ray). Orcein. Distal OFT. Longitudinal section. The oriented elastin fibres in the aorta continue into the bulbus under the conal myocardium (M). Ao, ventral aorta. Scale bar: 500 μm; inset: 200 μm. **E.** *Lepidosiren paradoxa* (South African lungfish). Martin's trichrome. Bulbus arteriosus. Cross-section. A layer of myocardium (arrows) wraps the arterial-like tissue of the bulbus. This myocardium shows irregular thickness and discontinuities. Under the myocardium, the bulbus shows ordered layers of collagen. The bulbar lumen is mostly occupied by the bulbar ridges. In lungfishes, the atrium embraces the outflow tract. The outermost myocardium corresponds to the atrial wall (A). Scale bar: 500 μm. **F.** *Oncorhynchus mykiss*. Haematein trichrome. Subepicardial tissue at the bulboventricular region. The arterial wall contains collagen organized into circumferential layers (arrowheads). Scale bar: 50 μm. **G.** *Myxine glutinosa* (Atlantic hagfish). Haematein trichrome. Ventral aorta. Cross-section. The middle layer of the aorta contains small amounts of irregularly distributed collagen (arrows). Inset: *Acipenser naccarii*. Haematein trichrome. Coronary trunk in the conus arteriosus. The arterial wall contains several ordered layers of collagen. Scale bar: 50 μm; inset: 20 μm. **H.** *Protopterus annectens* (West African lungfish). Martin's trichrome. Pulmonary artery. Erythrocytes are large, rounded or oval cells with a central nucleus. Thrombocytes (arrow) are smaller and show ellipsoidal shape and a smaller cytoplasm/nucleus ratio. A large, rounded granulocyte (arrowhead) displays a dentated nucleus and a cytoplasm coloured in red. Dark cells located in the arterial wall are melanocytes. Scale bar: 20 μm. B, bulbus arteriosus; V, ventricle.

Color version at the end of the book

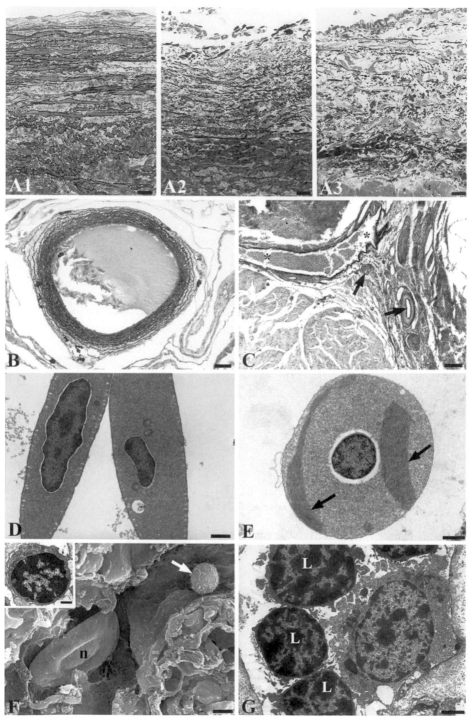

Fig. 7 contd. ...

with the conal myocardium (Fig. 7A) (Icardo et al. 2002b). Then, smooth muscle cells are replaced by myofibroblasts and fibroblasts under the myocardium. These changes in histological organization also occur in other fish groups such as in holosteans (Icardo 2017) and in basal (Lorenzale et al. 2017) and modern teleosts (Icardo et al. 2003) (However, see Grimes et al. 2010, Duran et al. 2014). The myocardium (Icardo et al. 2002b) and/or the extracellular matrix (Moriyama et al. 2016) may generate signals that regulate subendocardial cell phenotypes and restrict full smooth muscle cell differentiation under the myocardium. Nonetheless, the OFT of lungfish stands as a clear exception to this rule (Icardo et al. 2005b). The bulbus of lungfishes is endowed with a thin layer of myocardium that overlies the arterial-like organization of the bulbar wall (Fig. 6E). The myocardium of the bulbus is continuous with the conal myocardium (Fig. 2B), becoming thinner and discontinuous toward the distal portion of the OFT (Fig. 6E). The OFT of lungfishes shows another special feature. It is endowed with a couple of ridges, the spiral and bulbar folds, which run along the outflow tract (Fig. 2B). The two folds are of unequal length, spiral around each other, and form an incomplete division of the OFT (Buggc 1961, Icardo et al. 2005b). They are formed by loose connective tissue that contains collagen, elastin and fibroblast-like cells (Fig. 6E).

Arteries and Blood

The arterial wall of vertebrates is typically described as being composed of a tunica intima formed by the endothelium, a tunica media that contains smooth muscle cells and elastin lamellae, and an adventitia rich in collagen and fibroblasts (Bloom and Fawcett 1994). While the general organization of the arterial wall has been said to be similar in fishes and other vertebrates (Genten et al. 2009), the architecture and distribution of the structural components vary widely across fish groups. For instance, the large arteries of teleosts and many primitive fishes show several layers of circumferentially oriented elastin that surround the arterial endothelium (Fig. 7B). However, the appearance of the elastin sheets is quite irregular. This probably indicates an irregular distribution of elastin fibres, or poor elaboration of elastin sheets. In other large arteries, such as in the ventral aorta of sturgeons, elastin appears in the form of fibres that wrap the smooth muscle cells entirely (Fig. 7A1). By contrast, small arteries in teleosts only show a thin layer of elastin located under the endothelium (Fig. 7C). This layer resembles the internal elastic lamina of higher

---

...Fig. 7 contd.

Fig. 7. Anatomy and histology of blood vessels. **A.** *Acipenser naccarii* (Adriatic sturgeon). Toluidine blue. Histology of the ventral aorta and distal OFT. Longitudinal sections of the same specimen. **A1.** Ventral aorta. Smooth muscle cells are organized into layers separated by collagen bundles. Elastin (black fibres) wraps individual muscle cells. **A2.** Bulbus arteriosus. Elastin fibers are very numerous but they do not wrap cells. The ordered arrangement is lost and cell size is different. **A3.** Cranial end of conus arteriosus. Cells are loosely distributed and show irregular shapes. The contact between elastin fibres and cells is maintained. Scale bar: 20 μm. **B.** *Protopterus annectens* (West African lungfish). Orcein. Pulmonary artery. Cross-section. Elastin is arranged into layers in large arteries. However, elastin layers are irregular and incomplete. Large melanocytes (black spots) are very abundant in many lungfish tissues and organs. Scale bar: 50 μm. **C.** *Thunnus alalunga* (Albacore). Orcein. Subepicardial heart tissue. In small arteries (arrows), elastin localizes mostly to the subendothelium. Very little elastin is observed in the arterial middle layer. Elastin is also present in the thin wall of a large vein that has been opened longitudinally (asterisks). Scale bar: 100 μm. **D.** *Myxine glutinosa* (Atlantic hagfish). TEM. Peripheral blood. Erythrocytes show dense heterochromatin, homogeneous cytoplasm and peripheral pinocytotic vacuoles. Scale bar: 1 μm. **E.** *Gadus morhua* (Atlantic cod). TEM. Peripheral blood. Erythrocytes show thick bundles of cytoplasmic filaments (arrows). Scale bar: 1 μm. **F.** *Neoceratodus forsteri* (Australian lungfish). SEM. Peripheral blood. Within a vein, a large erythrocyte has a centrally located nucleus (n). A much smaller lymphocyte (arrow) shows numerous cell surface processes. Note differences in size. Inset: *Protopterus annectens.* TEM. Small lymphocyte in the spleen. Note the smooth surface. Scale bars: 5 μm; inset: 1 μm. **G.** *Protopterus annectens.* TEM. White pulp of spleen. Several small lymphocytes (L) appear contacting the surface of a macrophage. The macrophage cytoplasm contains numerous inclusions. Scale bar: 3 μm.

vertebrates. Of note, the ventral aorta of hagfishes virtually lacks elastic fibers but shows large amounts of extracellular filaments that may provide the aorta with elastic capabilities (Icardo et al. 2016a). The amount of arterial collagen in fish is also variable. In teleosts, small arteries show a few discrete layers of circumferentially arranged collagen (Fig. 6F). However, the amount of collagen increases in large arteries. On the other hand, very little is known of the collagenous composition of the arterial wall in primitive fishes. For instance, collagen patches distribute irregularly across the wall of the ventral aorta in hagfishes (Fig. 6G) (Icardo et al. 2016a). However, the branches of the hypobranchial artery in sturgeons show several circumferentially-oriented collagen layers (Inset, Fig. 6G). In general, the collagenous component appears to be more abundant in the arteries of fish than elastin.

The circulating blood contains erythrocytes, several types of leukocytes and thrombocytes. Unlike the mammalian red blood cells, fish erythrocytes are nucleated (Fig. 6H). They usually are oval or disc-shaped and show a nucleus with heterochromatin and a homogeneous cytoplasm with a few organelles such as small mitochondria (Fig. 7D). Erythrocytes may also show a large number of pinocytotic vacuoles (Fig. 7D). Characteristically, the cytoplasm of the cod erythrocytes contains large bundles of filamentous material (Fig. 7E). The size of the erythrocytes in relation to other blood cells is variable, but they appear to be exceptionally large in lungfishes (Fig. 7F). Thromobocytes in fish are nucleated cells, usually have oval or spindle shape, contain peripheral vesicles of pinocyotosis and numerous microtubules and may present phagocytic activity (Genten et al. 2009). Thus, they often look like erythrocytes and are easily confounded. However, their size is usually smaller than that of erythrocytes (Fig. 6H).

The term leukocyte comprises a collective of granulocytes (cells with granules) and non-granular cells. Among the latter, lymphocytes constitute the most abundant type (Zapata and Cooper 1990, Bloom and Fawcett 1994). Lymphocytes are small cells (Fig. 7F) that show a rounded or oval nucleus and a variable amount of cytoplasm that contains a very small Golgi complex, a pair of centrioles, a few mitochondria and numerous free ribosomes (Inset, Fig. 7F). The lymphocyte surface may show numerous projections (Fig. 7F) or may be mostly smooth (Inset, Fig. 7F). Three morphological types of lymphocytes—small, medium and large—have been described based on the relative amount of cytoplasm. Lymphocytes can be identified in blood smears because of their basophilic cytoplasm, but they are contained in very high numbers in lymphoid organs. In the spleen, for instance, lymphocytes form clusters or may be associated with other cell types such as macrophages (Fig. 7G). From the functional point of view, two broad lymphocyte categories, B and T, can be discerned. The two types are morphologically undistinguishable. Programmed B-cells, when exposed to the specific antigen, enlarge and are stimulated to divide. During division, B-cells mutate and transform into cell-producing antibodies or plasma cells (Zapata and Cooper 1990, Bloom and Fawcett 1994). Plasma cells are rare in blood but they can be found in many tissues where they can be recognized by the characteristic wheel-like arrangement of chromatin in histological sections (Bloom and Fawcett 1994). Under the transmission electron microscope (Fig. 8A), plasma cells show numerous cisternae of the rough endoplasmic reticulum filled with amorphous material. The cisternae may be flat and form parallel arrays that occupy much of the cytoplasm, or they may be greatly dilated and show a more irregular distribution (Fig. 8A).

Monocytes constitute another type of non-granular leukocyte (Zapata and Cooper 1990, Bloom and Fawcett 1994). They are rarely found in blood where they can be confounded with large lymphocytes. Once formed in the bone marrow, monocytes leave quickly the circulation and invade the different body organs to transform into tissue macrophages. Macrophages are normal tissue residents that have a remarkable capability for phagocytosis. They are able to phagocytose bacteria, dead cells, cellular debris and particulate material, storing the ingested material into vacuoles that show variable size and content (Fig. 7G). In lymphoid organs, like the spleen of lungfishes, macrophages participate in the metabolism of hemosiderin and may appear surrounded by lymphocytes (Fig. 7G) forming "rosettes" (Icardo et al. 2012). Thus, they function as cell-presenting antigens and become involved in the production of the immune responses. Macrophages filled with hemosiderin and melanin may cluster into melanomacrophage centres for further pigment recycling (Zapata and Amemiya 2000) and/or development of immune responses (Vigliano et al. 2006, Icardo et al. 2012).

Granular leukocytes are classified into three different types (eosinophils, basophils and neutrophils) on the basis of the staining properties of the cytoplasmic granules after Romanovsky stains (Bloom and

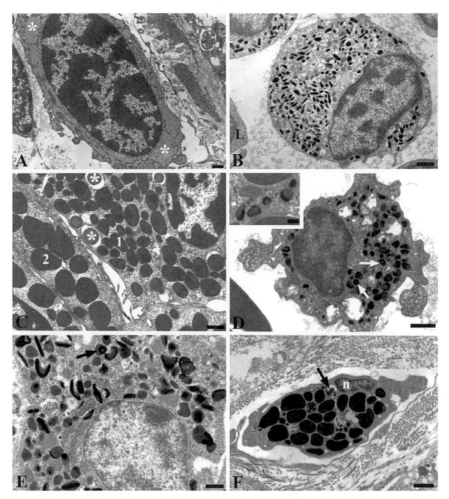

Fig. 8. Blood cells. **A.** *Protopterus annectens* (West African lungfish). TEM. Plasma cell in the spleen. Amorphous material presumably containing Ig proteins fills the dilated cisternae (asterisks) of the rough endoplasmic reticulum. Scale bar: 1 μm. **B.** *Eptatretus cirrhatus* (Pacific hagfish). Peripheral blood. Two granulocytes exhibit dark, membrane-bound, rod-like inclusions of uniform content. A similar cell type has been classified as neutrophil in other primitive fishes. L, small lymphocyte. Scale bar: 1 μm. **C.** *Protopterus annectens*. TEM. Red pulp of spleen. This micrograph illustrates two different types of granulocytes. Type-1 (1) shows membrane-bound granules that contain a dense globular core surrounded by less dense material. Several granules (asterisks) are partially emptied. Similar cells have been classified as eosinophils in *Chimera monstrosa*, but they also resemble neutrophils. Type-2 granulocytes (2) show larger granules with a dense, homogeneous content. Mast cells and basophils show similar morphological characteristics. Scale bar: 1 μm. **D.** *Erpetoichthys calabaricus* (Reedfish). TEM. Peripheral blood. Granulocyte. Granules contain two components: a clear, lucent core and a peripheral, dense component. The dense component may occupy the entire periphery, the granules adopting the appearance of a hollow cylinder (arrows). Inset: *Polypterus senegalus* (Gray bichir). TEM. Detail of similar granules in the leukocyte of another polypteriform. Scale bar: 1 μm; inset, 250 nm. **E.** *Arapaima gigas* (Pirarucu). TEM. Gas bladder parenchyma. Detail of eosinophil leukocyte. The granules are rounded or oval and contain two components of different electron density. The denser component is disk-like and often occupies the equatorial plane of the granule. Arrow indicates complete circles. Scale bar: 500 nm. **F.** *Arapaima gigas*. TEM. Pharyngeal subepithelium. The presence of large, dense granules of uniform content is characteristic of resident mast cells in many species. Of note, similar description also applies to basophil leukocytes. Arrow indicates lysosomal degranulation. N, nucleus. Scale bar: 1 μm.

Fawcett 1994). While this distinction is very clear in mammals, proper identification of fish granulocytes remains conflictive (Zapata and Cooper 1990, Esteban et al. 2000, Zapata and Amemiya 2000, Genten et al. 2009). The granulocyte properties, the staining characteristics and the granule ultrastructure may vary within the same fish group. Furthermore, information on the granule content is scarce and fragmentary

(Genten et al. 2009). On the other hand, related species may have different types, both in number and ultrastructure, of the same basic granulocyte. This occurs, for instance, in lungfishes (Hine et al. 1990, Bielek and Strauss 1993, Icardo et al. 2014). Alternatively, a single type of granulocyte may contain up to three types of structurally different granules (Fang et al. 2014). Combined light- and transmission electron-microscope studies have helped to identify the three basic granulocyte types in a short number of fish species (Zapata and Amemiya 2000, Genten et al. 2009). However, this information may not directly be extrapolated to related species, less so to species from different groups. In fact, the presence of eosinophils and basophils appears to be species specific (see Fang et al. 2014). The following micrographs show several examples of granulocytes in ancient fishes: a single type of granulocyte is found in the blood and tissues of hagfishes (Fig. 8B) (Icardo et al. 2016b), two granulocyte types occur in the African lungfish *Protopterus annectens* (Fig. 8C) (Icardo et al. 2012, 2014), and a single granulocyte type has been observed in the blood of the polypteriforms (Fig. 8D). In the basal osteoglossiform *Arapaima gigas*, a single type of eosinophil leukocyte is found in the circulating blood but it appears to express a more mature phenotype in the tissue parenchyma (Fig. 8E). The granules contain dense, disk-like crystalline inclusions embedded in a less dense matrix. These granules resemble those observed in the blood of the Polypteriformes (Fig. 8D). Noticeably, they are also similar to those found in rodent eosinophils (Bloom and Fawcett 1994). Other blood-borne cells do not express granules until they reach connective tissues. For instance, the pharyngeal capillaries in *Arapaima gigas* contain abundant agranular monocytes while the pharyngeal subepithelium is rich in mast cells (Fig. 8F). Mast cells and basophil leukocytes may be very similar morphologically and are often confounded. Distinction between the two cell types is easy in mammals since basophils do not enter the tissues. In fish, however, the situation is not entirely clear. In the blood of modern teleosts, heterophilic/neutrophilic and acidophilic/eosinophilic granulocytes are generally present, although many variations have been reported (see Esteban et al. 2000). It must also be stressed that the structural appearance of the granules and, hence, granulocyte identification, may depend on the site of cell extraction. While a specific type of granulocyte may not be present in blood (Esteban et al. 2000), other granulocytes appear to mature after they extravasate and become tissue residents (Messenguer et al. 1993). The exact significance of these changes and the variability between related species are far from being understood.

References Cited

Bielek, E. and B. Strauss. 1993. Ultrastructure of the granulocytes of the South American lungfish, *Lepidosiren paradoxa*: Morphogenesis and comparison to other leukocytes. J. Morphol. 218: 29–41.

Bloom, W. and D.W. Fawcett. 1994. A Textbook of Histology, 12th edition. Chapman & Hall, New York.

Braun, M.H., R.W. Brill, J.M. Gosline and D.R. Jones. 2003a. Form and function of the bulbus arteriosus in yellowfin tuna (*Thunnus albacares*), bigeye tuna (*Thunnus obesus*) and blue marlin (*Makaira nigricans*): static properties. J. Exp. Biol. 206: 3311–3326.

Braun, M.H., R.W. Brill, J.M. Gosline and D.R. Jones. 2003b. Form and function of the bulbus arteriosus in yellowfin tuna (*Thunnus albacares*): dynamic properties. J. Exp. Biol. 206: 3327–3335.

Bugge, J. 1961. The heart of the African lungfish, *Protopterus*. Vidensk Meddr. Dansk. natuhr. Foren. 123: 193–210.

Burggren, W.W., A. Farrell and H. Lillywhite. 1997. Vertebrate cardiovascular systems. pp. 215–308. *In*: W.H. Dantzler (ed.). Handbook of Physiology, Sect. 13, Comparative Physiology, Vol. 1. Oxford University Press, New York.

Cerra, M.C., S. Imbrogno, D. Amelio, F. Garofalo, E. Colvee, B. Tota and J.M. Icardo. 2004. Cardiac morphodynamic remodelling in the growing eel (*Anguilla anguilla* L.). J. Exp. Biol. 207: 2867–2875.

Durán, A.C., B. Fernández, A.C. Grimes, C. Rodríguez, J.M. Arqué and V. Sans-Coma. 2008. Condrichthyans have a bulbus arteriosus at the arterial pole of the heart: Morphological and evolutionary implications. J. Anat. 21: 597–606.

Durán, A.C., I. Reyes-Moya, B. Fernández, C. Rodríguez, V. Sans-Coma and A.C. Grimes. 2014. The anatomical components of the cardiac outflow tract of the gray bichir, *Polypterus senegalus*: their evolutionary significance. Zoology 117: 370–376.

Esteban, M.A., J. Muñoz and J. Messeguer. 2000. Blood cells of the sea bass (*Dicentrarchus labrax* L.). Flow cytometry and microscopic studies. Anat. Rec. 258: 80–89.

Fang, J., K. Chen, H.M. Cui, X. Peng, T. Li and Z.C. Zuo. 2014. Morphological and cytochemical studies of peripheral blood cells of *Schizothorax prenanti*. Anat. Histol. Embryol. 43: 386–394.

Farrell, A.P., A.M. Hammons, M.S. Graham and G.F. Tibbits. 1988. Cardiac growth in rainbow trout, *Salmo gairdneri*. Can. J. Zool. 66: 2368–2373.

Farrell, A.P. and D.R. Jones. 1992. The heart. pp. 1–88. *In*: W.S. Hoar, D.J. Randall and A.P. Farrell (eds.). Fish Physiology, Vol. XII, Part A, The Cardiovascular System Academic Press, New York.

Farrell, A.P. 2007. Cardiovascular system in primitive fishes. pp. 53–120. *In*: D.J. McKenzie, A.P. Farrell and C.J. Brauner (eds.). Primitive Fishes. Elsevier, Amsterdam.

Farrell, A.P., D.L. Simonot, R.S. Seymour and T.D. Clark. 2007. A novel technique for estimating the compact myocardium in fishes reveals surprising results for an athletic air-breathing fish, the Pacific tarpon. J. Fish Biol. 71: 389–398.

Farrell, A.P. 2011. Accessory hearts in fishes. pp. 1073–1076. *In*: A.P. Farrell (ed.). Encyclopedia of Fish Physiology. From Genome to Environment. Vol. 2, Circulation, Design and Physiology of the Heart. Academic Press, New York.

Farrell, A.P., N.D. Farrell, H. Jourdan and G.K. Cox. 2012. A perspective on the evolution of the coronary circulation in fishes and the transition to terrestrial life. pp. 75–102. *In*: D. Sedmera and T. Wang (eds.). Ontogeny and Phylogeny of the Vertebrate Heart. Springer, New York.

Farrell, A.P. and F. Smith. 2017. Cardiac form, function and physiology. pp. 155–264. *In*: A.K. Gamperl, T. Gillis, C.J. Brauner and A.P. Farrell (eds.). Fish Physiology: The Cardiovascular System, Vol. 36A: Cardiovascular Design, Control and Function. Academic Press, New York.

Gamperl, A.K. and A.P. Farrell. 2004. Cardiac plasticity in fishes: environmental influences and intraspecific differences. J. Exp. Biol. 207: 2539–3550.

Garofalo, F., S. Imbrogno, B. Tota and D. Amelio. 2012. Morpho-functional characterization of the goldfish (*Carassius auratus* L.) heart. Comp. Biochem. Physiol. A 163: 215–222.

Genten, F., E. Terwinghe and A. Danguy. 2009. Atlas of Fish Histology. Science Publishers, Enfield.

Greer Walker, M., M. Santer, M. Benjamin and D. Norman. 1985. Heart structure of some deepsea fish (Teleostei: Macrouridae). J. Zool. London (A) 205: 75–89.

Grimes, A.C., H.A. Stadt, I.T. Sheperd and M.L. Kirby. 2006. Solving an enigma: Arterial pole development in the zebrafish heart. Dev. Biol. 290: 265–276.

Grimes, A.C. and M.L. Kirby. 2009. The outflow tract of the heart in fishes: anatomy, genes and evolution. J. Fish Biol. 74: 963–1036.

Grimes, A.C., A.C. Durán, V. Sans-Coma, D. Hami, M.M. Santoro and M. Torres. 2010. Phylogeny informs ontogeny: A proposed common theme in the arterial pole of the vertebrate heart. Evol. Dev. 12: 552–567.

Guerrero, A., J.M. Icardo, A.C. Durán, A. Gallego, A. Domezain, E. Colvee and V. Sans-Coma. 2004. Differentiation of the cardiac outflow tract components in alevins of the sturgeon *Acipenser naccarii* (Osteichthyes, Acipenseriformes). Implications for heart evolution. J. Morphol. 260: 172–183.

Hamlett, W.C., F.J. Schwartz, R. Schmeinda and E. Cuevas. 1996. Anatomy, histology, and development of the cardiac valvular system in elasmobranches. J. Exp. Zool. 275: 83–94.

Haverinen, J. and M. Vornanen. 2007. Temperature acclimation modifies sinoatrial pacemaker mechanism of the rainbow trout heart. Am. J. Physiol. 292: R1023–R1032.

Hine, P.M., R.J.G. Lester and J.M. Wain. 1990. Observations on the blood of the Australian lungfish, *Neoceratodus forsteri* klefft. I. Ultrastructure of granulocytes, monocytes and thrombocytes. Aust. J. Zool. 38: 131–144.

Icardo, J.M., E. Colvee, M.C. Cerra and B. Tota. 1999a. Bulbus arteriosus of Antarctic teleosts. I. The white-blooded *Chionodraco hamatus*. Anat. Rec. 254: 396–407.

Icardo, J.M., E. Colvee, M.C. Cerra and B. Tota. 1999b. Bulbus arteriosus of Antarctic teleosts. II. The red-blooded *Trematomus bernacchii*. Anat. Rec. 256: 116–126.

Icardo, J.M., E. Colvee, M.C. Cerra and B. Tota. 2000. The bulbus arteriosus of stenothermal and temperate teleosts: a morphological approach. J. Fish Biol. 57(suppl A): 121–135.

Icardo, J.M., E. Colvee, M.C. Cerra and B. Tota. 2002a. Structure of the conus arteriosus of the sturgeon (*Acipenser naccarii*) heart. I. The conus valves and the subendocardium. Anat. Rec. 267: 17–27.

Icardo, J.M., E. Colvee, M.C. Cerra and B. Tota. 2002b. The structure of the conus arteriosus of the sturgeon (*Acipenser naccarii*) heart. II. The myocardium, the subepicardium and the conus-aorta transition. Anat. Rec. 268: 388–398.

Icardo, J.M., J.L. Schib, J.L. Ojeda, A.C. Durán, A. Guerrero, E. Colvee, D. Amelio and V. Sans-Coma. 2003. The conus valves of the adult gilthead seabream (*Sparus auratus*). J. Anat. 202: 537–550.

Icardo, J.M., A. Guerrero, A.C. Durán, A. Domezain, E. Colvee and V. Sans-Coma. 2004. The development of the sturgeon heart. Anat. Embryol. 208: 439–449.

Icardo, J.M., J.L. Ojeda, E. Colvee, B. Tota, W.P. Wong and Y.K. Ip. 2005a. The heart inflow tract of the African lungfish *Protopterus dolloi*. J. Morphol. 263: 30–38.

Icardo, J.M., E. Brunelli, I. Perrotta, E. Colvée, W.P. Wong and Y.K. Ip. 2005b. Ventricle and outflow tract of the African lungfish *Protopterus dolloi*. J. Morphol. 265: 43–51.

Icardo, J.M., S. Imbrogno, A. Gattuso, E. Colvee and B. Tota. 2005c. The heart of Sparus auratus: a reappraisal of cardiac functional morphology in teleosts. J. Exp. Zool. 303A: 665–675.

Icardo, J.M. 2006. Conus arteriosus of the teleost heart: dismissed, but not missed. Anat. Rec. A 288: 900–908.

Icardo, J.M. 2007. The fish endocardium. A review on the teleost heart. pp. 79–84. *In*: W.C. Aird (ed.). Endothelial Biomedicine. Cambridge University Press, Cambridge.

Icardo, J.M. and E. Colvee. 2011. The atrioventricular region of the teleost heart. A distinct heart segment. Anat. Rec. 294: 236–242.

Icardo, J.M. 2012. The teleost heart: A morphological approach. pp. 35–53. *In*: D. Sedmera and T. Wang (eds.). Ontogeny and Phylogeny of the Vertebrate Heart. Springer, New York.

Icardo, J.M., W.P. Wong, E. Colvee, A.M. Loong and Y.K. Ip. 2012. The spleen of the African lungfish *Protopterus annectens*: freshwater and aestivation. Cell Tissue Res. 350: 143–156.

Icardo, J.M. 2013. Collagen and elastin histochemistry of the teleost bulbus arteriosus: false positives. Acta Histochem. 115: 185–189.

Icardo, J.M., W.P. Wong, E. Colvee, A.M. Loong and Y.K. Ip. 2014. Lympho-granulocytic tissue associated with the wall of the spiral valve in the African lungfish *Protopterus annectens*. Cell Tissue Res. 355: 397–407.

Icardo, J.M., E. Colvee, S. Schorno, E.R. Lauriano, D.S. Fudge, C.N. Glover and G. Zaccone. 2016a. Morphological analysis of the hagfish heart. I. The ventricle, the arterial connection and the ventral aorta. J. Morphol. 277: 326–340.

Icardo, J.M., E. Colvee, S. Schorno, E.R. Lauriano, D.S. Fudge, C.N. Glover and G. Zaccone. 2016b. Morphological analysis of the hagfish heart. II. The venous pole and the pericardium. J. Morphol. 277: 853–865.

Icardo, J.M. 2017. Heart morphology and anatomy. pp. 1–54. *In*: A.K. Gamperl, T. Gillis, C.J. Brauner and A.P. Farrell (eds.). Fish Physiology: The Cardiovascular System, Vol. 36A: Cardiovascular Design, Control and Function. Academic Press, New York.

Imbrogno, S. and M.C. Cerra. 2017. Hormonal and autocoid control of cardiac function. pp. 265–315. *In*: A.K. Gamperl, T. Gillis, C.J. Brauner and A.P. Farrell (eds.). Fish Physiology: The Cardiovascular System, Vol. 36A: Cardiovascular Design, Control and Function. Academic Press, New York.

Jensen, B., B.J.D. Boukens, T. Wang, A.F.M. Moorman and V.M. Christoffels. 2014. Evolution of the sinus venosus from fish to human. J. Cardiovasc. Dev. Dis. 1: 14–28.

Jones, D.R. and M.H. Braun. 2011. The outflow tract from the heart. pp. 1015–1029. *In*: A.P. Farrell (ed.). Encyclopedia of Fish Physiology. From Genome to Environment. Vol. 2, Circulation, Design and Physiology of the Heart. Academic Press, New York.

Kardong, K.V. 2006. Vertebrates: Comparative Anatomy, Function, Evolution. 4th edition. McGraw-Hill, New York.

Lorenzale, M., M.A. Lopez-Unzu, M.C. Fernandez, A.C. Duran, B. Fernandez, M.T. Soto-Navarrete and V. Sans-Coma. 2017. Anatomical, histochemical and immunohistochemical characterisation of the cardiac outflow tract of the silver arowana, *Osteoglossum bicirrhosum* (Teleostei: Osteoglossiformes). Zoology 120: 15–23.

Messeguer, J., M.A. Esteban, J. Muñoz and A. López-Ruiz. 1993. Ultrastructure of the peritoneal exudate cells of the seawater teleosts, seabream (*Sparus aurata* L.) and seabass (*Dicentrarchus labrax* L.). Cell Tissue Res. 273: 301–307.

Moriyama, Y., F. Ito, H. Takeda, T. Yano, M. Okabe, S. Kuraku, F.W. Keeley and K. Koshiba-Takeuchi. 2016. Evolution of the fish heart by sub/neofunctionalization of an elastin gene. Nature Comm. 7: 10397.

Munshi, J.S.D., K.R. Olson, P.K. Roy and U. Ghosh. 2001. Scanning electron microscopy of the heart of the climbing perch. J. Fish Biol. 59: 1170–1180.

Newton, C.M., M.R. Stoyek, R.P. Croll and F.M. Smith. 2014. Regional innervation of the heart in the goldfish, *Carassius auratus*: A confocal microscopy study. J. Comp. Neurol. 522: 456:478.

Parsons, C.W. 1930. The conus arteriosus in fishes. Quart. J. Microsc. Sci. 73: 145–176.

Priede, I.G. 1976. Functional morphology of the bulbus arteriosus of rainbow trout (*Salmo gairdneri* Richardson). J. Fish Biol. 9: 209–216.

Sánchez-Quintana, D. and J.M. Hurle. 1987. Ventricular myocardial architecture in marine fishes. Anat. Rec. 217: 263–273.

Santer, R.M. and M. Greer Walker. 1980. Morphological studies on the ventricle of teleost and elasmobranch hearts. J. Zool. London 190: 259–2372.

Santer, R.M. 1985. Morphology and innervation of the fish heart. Adv. Anat. Embryol. Cell Biol. 89: 1–102.

Satchell, G.H. 1991. Physiology and Form of Fish Circulation. Cambridge University Press, Cambridge.

Schib, J.L., J.M. Icardo, A.C. Durán, A. Guerrero, D. López, E. Colvee, A.V. de Andrés and V. Sans-Coma. 2002. The conus arterious of the adult gilthead seabream (*Sparus auratus*). J. Anat. 201: 395–404.

Sedmera, D., M. Reckova, A. De Almeida, M. Sedmerova, M. Biermann, J. Volejnik, A. Sarre, E. Raddatz, R.A. McCarthy, R.G. Gourdie and R.P. Thompson. 2003. Functional and morphological evidence for a ventricular conduction system in zebrafish and *Xenopus* hearts. Am. J. Physiol. 284: H1152–H1160.

Szidon, J.P., S. Lahiri, M. Lev and A.P. Fishman. 1969. Heart and circulation of the African lungfish. Circ. Res. 25: 23–38.

Tota, B. 1983. Vascular and metabolic zonation in the ventricular myocardium of mammals and fishes. Comp. Biochem. Physiol. 76A: 423–437.

Tota, B., V. Cimini, G. Salvatore and G. Zummo. 1983. Comparative study of the arterial and lacunary systems of the ventricular myocardium of elasmobranch and teleost fishes. Am. J. Anat. 167: 15–32.

Tota, B. 1989. Myoarchitecture and vascularization of the elasmobranch heart ventricle. J. Exp. Zool. (Suppl.) 2: 122–135.

Vigliano, F.A., R. Bermúdez, M.I. Quiroga and J.M. Nieto. 2006. Evidence for melano-macrophage centres of teleost as evolutionary precursors of germinal centers of higher vertebrates: An immunohistochemical study. Fish Shellfish Immunol. 21: 467–471.

Vornanen, M., M. Halinen and J. Haverinrn. 2010. Sinoatrial tissue of crucian carp heart has only negative contractile responses to autonomic agonists. BMC Physiol. 10: 10.

White, E.G. 1936. The heart valves of the elasmobranch fishes. Am. Mus. Novit. 838: 1–21.

Yamauchi, A. 1980. Fine structure of the fish heart. pp. 119–148. *In*: G. Bourne (cd.). Heart and Heart-like Organs, Vol. 1. Academic Press, New York.

Zaccone, G., A. Mauceri, M. Maisano and S. Fasulo. 2009. Innervation of lung and heart in the ray-finned fish, bichirs. Acta Histochem. 111: 217–229.

Zaccone, G., A. Mauceri, M. Maisano, A. Gianneto, V. Parrino and S. Fasulo. 2010. Postganglionic nerve cell bodies and neurotransmitter localization in the teleost heart. Acta Histochem. 112: 328–336.

Zapata, A.G. and E.L. Cooper. 1990. The Immune System: Comparative Histophysiology. Wiley, Chichester.

Zapata, A.G. and C.T. Amemiya. 2000. Phylogeny of lower vertebrates and their immunological structures. pp. 67–107. *In*: L. Du Pasquier and G.W. Litman (eds.). Origin and Evolution of the Vertebrate Immune System. Springer, Berlin.

Chapter **14**

Immune System of Fish[#]

Teresa Wlasow* and Małgorzata Jankun

INTRODUCTION

Two main types of immunity, non-specific (innate, inborn, genetically determined, inherited) and specific (acquired, adaptive), are distinguished in the defense system of vertebrates. However, the elements of both types of resistance work together, and non-specific immunity is important in activating specific defense processes. The significant role of non-specific immunity is one of the main properties of the immune system of fish (Kum and Sekkin 2011).

The fish's non-specific defense system was created for the rapid defense of the organism. This component of the immune system of fish is active within minutes or hours, while the protection based upon the action of specific immunity requires weeks or even months (Köllner and Kotterba 2002, Lamers 1985, Van Muiswinkel and Vervoorn-Van Der Wal 2006). When defining the strategy of the defense process of fish, three defensive lines were distinguished (Lamers 1985). The first line of defense is a relatively solid protection against the penetration of foreign elements into the fish body. The physical barriers like the skin, scales, epithelia and the mucus layer in various parts of the body, e.g., the skin, gills, gill cavity, gastrointestinal tract, are important in controlling and preventing fungal, bacterial or parasitic infections. This is due to the presence of cellular and humoral factors in the mucus. The maintenance of uninjured epithelia is extremely important in fish due to their role in defense and osmoregulation. The healing of the damaged protective barrier in fish is very rapid. In young fish, the repair of the respiratory epithelium which has been damaged by irritant external factors can even lead to proliferative gill disease (PGD), as occurring during the rearing of coregonids larvae. Slow growth of larvae and delays in thymic development during PGD were observed (Własow 1993).

On the first defensive line, there are factors non specific to pathogens but characteristic for fish species or genetic lines of fishes, such as transferrins and lectins, lytic enzymes (chitinase, lyzosyme). Chitinase formed in leukocytes is a weapon directed against the surface structures of fungi, pathogenic crustaceans and nematodes. Lyzosyme produced by phagocytes and occurring not only in the mucus, but also in the serum and eggs of fish is a lytic enzyme directed against bacteria, parasites and viruses. The second line of the fish defense system is launched after an organism has been attacked by pathogenic agents. The response is characterized by low specificity. No production of immunological memory occurs. The elements of this

University of Warmia and Mazury in Olsztyn, Faculty of Environmental Sciences, Department of Ichthyology, Oczapowskiego 5, 10-719 Olsztyn, Poland.
Email: mjpw@uwm.edu.pl
* Corresponding author: tewlasow@uwm.edu.pl
[#] Project financially supported by Minister of Science and Higher Education in the range of the program entitled "Regional Initiative of Excellence". Project No. 010/RID/2018/19, amount of funding 12.000.000 PLN.

line include serum C-reactive protein (CRP) which reacts to the presence of endotoxins of various bacteria, inflammatory factors and damage of own tissues. Another element, interferon, arises in cells attacked by various viruses and has no specific character either. The second line of defense consists of the so-called unspecific phagocytosis, in which non-specific cells such as granulocytes, macrophages, monocytes, and thrombocytes participate. Phagocytic cells of fish are dispersed in the gills and organs of the gill cavity, and in lymphoid organs like the kidney, spleen, and thymus (Lamers 1985, Kum and Sekkin 2011).

Specific defensive reactions that are directed against particular antigens do not appear until the third defensive line. These processes include lymphoid cells, humoral agents and other cells working with lymphoid cells in the immune response (Lamers 1985). The specific reactions are associated with lymphoid cells (B and T lymphocytes), production of specific immunoglobulins/antibodies directed against a definite antigen of the enemy, and the appearance of an immunological memory.

In contrast to higher vertebrates, the immune system of fish is dependent on the conditions of the surrounding aquatic environment. Among the parameters of the environment, temperature is the most important for the functioning of the fish body (Avtalion et al. 1973, Le Morvan et al. 1998). Snieszko (1970), in his work devoted to the immunization of fish, emphasized that an optimal immunological response takes place at an optimal temperature for a species. Generally, this response in thermophilic fish from aquaculture is faster compared to coldwater fish. The temperature range for mammals is narrow (from 36.5°C to 37.5°C) compared to the wide range found in fish (from –2°C to 35°C). Low temperatures do not affect defense reactions of homeothermic mammals, but in jawed fishes, especially in poikilothermic fish, they may induce immunosuppressive effects (Tort et al. 2003).

On the other hand, there are many substances and factors that cause immunosuppression in the body of a fish. The effects of some of these can be illustrated using histological methods (Chilmonczyk 1982, Sayed and Younes 2017, Saxena and Saxena 2007, Własow et al. 2003, 2004). Under certain conditions, many factors can also cause stimulation/increase in activity of defensive reactions in fish. These immunostimulants, such as probiotics, prebiotics, and plant extracts, are of great practical importance in aquaculture (Biller-Takahashi and Urbinati 2014, Cerezuela et al. 2016, Kum and Sekkin 2011, Uribe et al. 2011).

Antibody response of fish is weaker compared to that found in higher vertebrates, and arises with a certain delay. In addition, the rate of reaction can vary depending on the place in the body of a fish. For example, antibodies against *Vibrio vulnificus* in vaccinated young European eel *Anguilla anguilla* from intense brackish water systems (26°C) appeared in the mucus earlier than in serum (Esteve-Gassent et al. 2003). The memory response of mammals is stronger in contrast to the poorly expressed reaction of jawed fish. Mammals have five isotypes of immunoglobulins (IgM, IgA, IgD, IgE, and IgG). Teleostei have IgM and IgD, whereas in Chondrichthyes there are IgM, IgX/IgR and IgW.

Immunological molecules, i.e., the major histocompatibility complex (MHCII), cluster of differentiation 8 (CD8), IgM, IgT and acute phase protein serum amyloid A (SAA), have been shown in many organs and tissues of fishes. SAA, produced mainly in the liver, has many functions including facilitation of macrophage phagocytosis. Immunological molecules have been visualized, using special histological techniques, in many organs and tissues of rainbow trout *Oncorhynchus mykiss* embryos and larvae (Heinecke et al. 2014). Immunohistochemical staining of SAA showed a strong positive reaction in pseudobranches of rainbow trout larvae. Large and round immunoreactive cells were visible (Heinecke et al. 2014). The structure and role of the pseudobranch in fish is poorly known (see also chapter Endocrine Organs). During the ontogenic development of larvae, the structure of this organ changes. In asp *Leuciscus aspius* larvae, the pseudobranchium showed the lamella-free type just after hatching, but 14 days post hatching (dph) lamellae were fused and filaments stayed free (Furgała-Selezniow et al. 2016a). Trouts have both lamellae and filaments fused (84 dph). The pseudobranch in fish is the production site of SAA, an important immunological molecule.

Lymphoid Organs of Fishes

The thymus and bone marrow belong to primary lymphoid organs in the defense system of mammals, whereas the spleen, lymph nodes and Peyer's patches are considered as secondary lymphoid organs. In addition, other lymphoid tissues referred to as tertiary lymphoid organs (TLOs) may be distinguished in mammals under certain conditions, such as cancer, chronic inflammation, autoimmune diseases or microbial infections (Ruddle 2016). TLOs are lymphoid organs without a capsule and they are placed inside other organs after birth. TLOs are considered more as functional units—generators of effector cells rather than histological, cellular structures. There is no information on the presence of TLOs in fish. Fish do not have lymph nodes or Peyer's patches. The recently discovered interbranchial lymphoid tissue is an important organ in the defense system of fish.

The stage in the development of lymphoid centers in an organism determines the defensive capabilities of both fish and higher vertebrates (Gui et al. 2012, Manning et al. 1982, Meštanová and Varga 2016, O'Neill 1989, Wise and Hruska 1988). The thymus gland, which is the central lymphoid organ in vertebrates, plays an especially important role. This organ is responsible for the formation of the young organism's defense system, the development of peripheral lymphoid tissue and the maturation of thymus-dependent T-cells. In longnose gar, *Lepisosteus osseus* (infraclass Holostei), there was either no thymus present or else the thymus underwent an extremely rapid involution (Vasiliev and Polevshchikov 2014).

The thymus is found in almost all fish except Agnatha (lamprey and hagfishes). Among this class, hagfishes (e.g., *Myxine*) have leucocytic infiltration in the pharyngeal velar muscle and lampreys (e.g., *Petromyzon*, *Lampetra*) have associations of lymphoid cells in the branchial area that, according to Fänge (1982), can be considered to be a primitive thymus. The thymus of Holocephali (e.g., *Chimera monstrosa*) is located in the upper part of the pharynx. The gland is divided into a cortex and a medulla. Numerous lymphocytes, especially small cells with a narrow cytoplasmic hemisphere, predominate in histological sections, while irregular macrophages are less numerous. Large lymphocytes have pseudopodia. On histological sections, epithelial reticulocytes have an elongated shape and sometimes reach a length of 100 μm (Mattison and Fänge 1986). They are in very close contact with lymphocytes. The thymus of Holocephali undergoes slight involution, while some representatives of Elasmobranchs (e.g., the dogfish *Scyliorhynchus canicula*) undergo thymic involution in adulthood. On the other hand, in primitive sharks like *Heterodontus*, there is no involution of the thymus associated with age (Fänge 1987).

During the ontogenetic development of many fishes, such as the carp *Cyprinus carpio* or the rainbow trout *Oncorhynchus mykiss*, this gland appears most often before other lymphocytic centers; hence it is referred to as the primary lymphoid organ (Botham and Manning 1981, Grace and Manning 1980). The thymus is also the central and first lymphatic organ throughout the life of Coregonidae larvae. Its development was faster than that of other lymphoid centers (pronephros, mesonephros, spleen) and blood cells (Własow 1993). On the other hand, the spleen and the pronephros appear earlier than the thymus in some fish, for example in the marbled rockfish *Sebasticus marmoratus* and the Antarctic spiny plunderfish *Harpagifer antarcticus* (Nakanishi 1991, O'Neill 1989). But even then, the thymus plays a fundamental role in the development of the fish's defense system and it continues to perform this function up to the period of maturity (Nakanishi 1991).

The site where the thymus is present in teleosts is the gill cavity (Figs. 1A, B). But the thymus of the angler fish *Lophius piscatorius* which lives close to the sea floor is located behind the gill cavity (Fänge 1982).

The thymus is usually an organ composed of two parts, one gland in each of the gill chamber, but the clingfish *Sicyases sanguineus* has one pair of a thymus in each chamber. Three parts of the thymus gland in one of the gill cavities are seen in a histological picture of the thymus in rainbow trout larvae (Heinecke 2014). An additional thymus gland located close to the operculum was found in *Coregonus* larvae two weeks after recovering from the proliferative gill disease (Własow 1993).

The thymus in many fish species can be exposed to the environment of the gill cavity and covered only by a narrow layer of the pharyngeal epithelium (Figs. 1C, D), whereas in the angler fish, the thymus is protected by a layer of loose connective tissue and secured by being placed deeper than in other fish (Fänge 1982). From the inside, the thymus is separated by an outer membrane, which is made up of epithelial cells. The capsule that surrounds the thymus invaginates into the thymic stroma, producing trabeculae and opens the way to the vascularisation of the organ (Bowden et al. 2005, Vasiliev and Polevshchikov 2014).

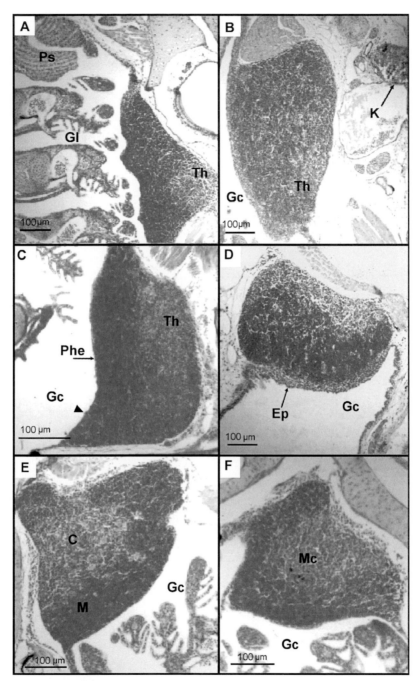

Fig. 1. Anatomy and histology of the thymus of various fish species. **A.** Location of organs in the gill cavity of the asp *Leuciscus aspius* showing thymus, pseudobranch and gills (HE staining). **B.** Cross section of the thymus of the European catfish *Silurus glanis* (HE staining). **C.** Thymus of burbot, *Lota lota* larvae separated from the gill cavity by a single layer of the pharyngeal epithelium. Secretory cells located between epithelial cells at the time of secretion (arrow head) (HE staining). **D.** Several layers of epithelium in the thymus of European catfish larvae (HE staining). **E.** Cross-section of the thymus of *Leuciscus aspius* larva. Cortex and medulla are visible (HE staining). **F.** Cross-section of the thymus of asp larva. Melanomacrophages are visible in the cortex and in the medulla (HE staining). C, cortex; Gc, gill cavity; Gl, gills; K, kidney; M, medulla; MC, melanomacrophages. Phe, pharyngeal epithelium; Ps, pseudobranch; Th, thymus.

The histological structure of the thymus varies depending not only on the position in a body, but also on the individual's age, sexual dimorphism, genetic differences, environmental conditions, and the nutritional status (Gui et al. 2012). All cellular elements, both in the thymus and in other lymphoid organs, are in a network of reticular cells whose role is to support various cells. The thymus as a gland originates in vertebrates from the pharyngeal pouches and is usually the first organ during the ontological development that is infiltrated by lymphoid cells (Chilmonczyk 1992, Press and Evensen 1999). As early as the onset of its development, the thymus shows two types of morphological elements in its structure, which are thymocytes of the hematopoietic origin and epithelial cells. Thymocytes occur in the thymus before the appearance of blood vessels in this organ. Epithelial cells in the thymus are of the non-hematopoietic origin and form a niche of thymic stroma (thymic epithelial space), which is necessary for the development and maturation of thymocytes. Among the classified types of epithelial cells in the thymus of fish sub-populations, the following are mentioned: capsular cells, cortical and medullary reticular cells, perivascular endothelial cells, intermediate cells, nurse-like cells and Hassall-like corpuscles (Bowden et al. 2005, Castillo et al. 1990, Mohammad et al. 2007). Nurse-like cells (or intermediate epithelial cells, nurse cells) occur in the cortex, at the border of the cortex and medulla (Romano et al. 1999). These large (23–35 µm) and irregularly shaped cells have one or several nuclei. The task of these cells is to create an environment for the maturation of T-lymphocytes (Vasiliev and Polevshchikov 2014). Hassall-like corpuscles (Hassall's bodies) are oval shaped epithelial structures (size of 50 µm) and occur both in the cortex and medulla in the thymus of fish (Bowden et al. 2005). Numerous vacuoles are found in their cytoplasm. They participate in the purification of thymus from dead cells. Their task is also to take part in the differentiation of T-lymphocytes.

In the thymus, there are secretory cells with rodlet-like inclusions, the so-called rodlet cells. They are similar to goblet cells but their role has been a subject of discussions. Among the non-lymphoid cells present in the thymus of fish, the following are also mentioned: pale reticular epithelial cells, dark reticular epithelial cells, subcapsular epithelial cells, perivascular cells, intermediate epithelial cells, myoid cells, secretory cells and macrophages. Of the total of eleven types of non-lymphoid thymic cells mentioned by Vasiliev and Polevshchikov (2014), only rodlet cells were not found in mammals. Pale reticular epithelial cells (PRECs or reticular epithelial cells RECs) occur in the cortex and sometimes in the outer part of the cortex. These cells (size from 8 to 12 µm) are irregular in shape and contain a pale, round nucleus with 1–3 nucleoli. The function of these cells is to create a suitable microenvironment for the development of thymocytes and the creation of a niche of thymic stroma in the cortical zone. Dark reticular epithelial cells (DREC or medullary epithelial cells) occur in the medulla and at the border of the cortex and medulla. These cells are star-shaped, somewhat irregular in shape, with long processes, and contain an irregularly shaped nucleus which has large amounts of heterochromatin. They are usually distorted by numerous lymphocytes that surround DREC. In the Australian lungfish *Neoceratodus forsteri*, dark reticular epithelial cells were rare and present only within the thymic cortex (Mohammad et al. 2007). The role of these cells is to partcipate in the formation of the medullary zone of the thymus and a microenvironment for lymphoid cells.

Subcapsular epithelial cells are elongated in shape and contain a round nucleus. They appear under the capsule that surrounds the thymus. They are responsible for the regulation of the thymic epithelium function and migration of mature thymocytes. Subcapsular epithelial cells are associated with the production of thymosin, a thymic hormone, which enables the maturation and proliferation of subcapsular T-cells. Perivascular cells (or limiting cells) occur as a single layer of cells surrounding the blood vessels in the thymic cortex, especially in the zone under the capsule. These cells have the shape of pyramids and possess long cytoplasmic processes, and the irregularly shaped cell nuclei are divided, appearing bright with some condensed heterochromatin near the nuclear membrane. Perivascular cells contribute to the formation of linear collagen fibers around the vascular space in the thymus, making them stronger.

Intermediate epithelial cells (IEC) are elongated and have an oval nucleus. They are frequently observed in the cortex, while being rare in the medullar region of the thymus. These cells usually appear as clusters and have the tendency to form desmosomes. Unlike pale and dark reticular epithelial cells, IEC aim to aggregate and create cell-to-cell links through desmosome connections (Mohammad et al. 2007).

Myoid cells in the thymus are large (30 μm), ovate cells, with a light nucleus, which occurs in the medulla. The adhesion of thymocytes to myoid cells was observed but cell-to-cell connections were not revealed in the electron micrographs of the thymus of the Australian lungfish (Mohammad et al. 2007). Secretory cells are oval or irregularly shaped, and have a bright nucleus placed in the center of the cell. These cells are found in a different site than secreted rodlet cells are. They appear in the subcapsular cortex zone or near the blood vessels, along the connective tissue. Their function is not precisely defined, but they produce cytokines and regulatory factors (Vasiliev and Polevshchikov 2014).

The macrophages in the thymus of fish are similar to those found in mammals. They reach 12–17 μm in size, and have a large, bright nucleus with one to three nucleoli, eccentrically arranged in the cell. The function of these cells is to perform phagocytosis and antigen presentation. There are two populations of macrophages: non-activated and activated. The latter is characterized by the presence of heterogeneous material, electron-dense bodies and phagocytosed thymocytes. Macrophages are present in the cortex and in the medulla (Figs. 1E, F), but macrophages capable to phagocytise apoptotic lymphocytes are located mainly in the medulary zone.

The spleen (Fig. 2A) is an organ in which mainly the hematopoietic tissue is located and whose task is to produce erythrocytes, granulocytes and thrombocytes. There is also lymphoid tissue in this organ but, unlike in mammals, the fish do not have germinal centers and clear white pulp in the spleen.

The structural and functional proportions of these tissues may differ in fish over the ontogenetic development. During the growth and development of lymphoid organs in rainbow trout, the erythropoetic function predominates in the earlier months, but is gradually diminishing with time. In the twelfth month of life, the population of lymphoid cells is visible and equal to the share of red blood cells. The proportion of lymphoid tissue and granulocytes in the spleen of rainbow trout compared to the kidney is clearly smaller (Tatner and Manning 1983). The growth of lymphoid tissue was observed in female Ohrid trout, *Salmo letnica*, during the reproductive cycle (between previtellogenesis and spawning stage); after spawning there was a downward trend (Rebok et al. 2011).

The spleen as a whole organ is enclosed in a capsule made up of connective tissue, cubical epithelial peritoneal epithelium and macrophages. In the fish spleen, capillaries and ellipsoids can be distinguished (Fig. 2B). The latter usually have thick walls formed of endothelial cells, reticulosic fibers and macrophages; ellipsoids are found in all fish.

Melano-macrophages in the carp spleen were noticed inside the connective tissue capsule in close proximity to lymphoid and ellipsoid cells (Lamers and De Haas 1985). The spleen contains two components of the parenchyma, i.e., the white pulp and the red pulp, which are clearly visible in mammals, and each of these parts is surrounded by bound net fibers. The isolation of the white pulp in the spleen of the fish is not always clear, instead, the lymphoid tissue is dispersed and present, for example, as clusters around the blood vessels. The spleen parenchyma in sturgeons contains both the white and red pulp. In hybrid sturgeon (beluga *Huso huso* x sterlet *Acipenser ruthenus*) and in white Pacific sturgeon, *A. transmontanus*, the white pulp of the spleen occurs in the form of vesicles containing both lymphocytes and eosinophilic granulocytes and macrophages (Fänge 1986). The spleen of elasmobranchs is different in shape; it can be divided into two or more parts, up to a hundred ones. The structure of elements that make up the shark's spleen depends on the size of a fish, for example in small sharks this organ is supplied by one artery and vein, and in larger sharks this system is sufficient for a centimeter of the length of the organ (Fänge 1987). Spleen arteries are divided into smaller vessels and have their ending as ellipsoid arterioles (ellipsoids) open to the red pulp. The white pulp is formed by the periarterial lymphoid sheath (PALS), which consists of numerous lymphocytes inside the reticulum and, additionally, from numerous plasma cells associated with the production of immunoglobulins. Ellipsoids are composed of concentrically arranged lamellae from connective tissue cells. In some representatives of the rays, lymphoid foci develop after a few weeks of life in the spleen (Fänge 1987).

The kidney (Fig. 2C) of Teleostei is an organ fulfilling many functions and containing various elements: hematopoietic, reticuloendothelial, endocrine and excretory elements. The anterior part of the kidney is principally composed of hematopoietic elements, while the posterior part is an excretory kidney.

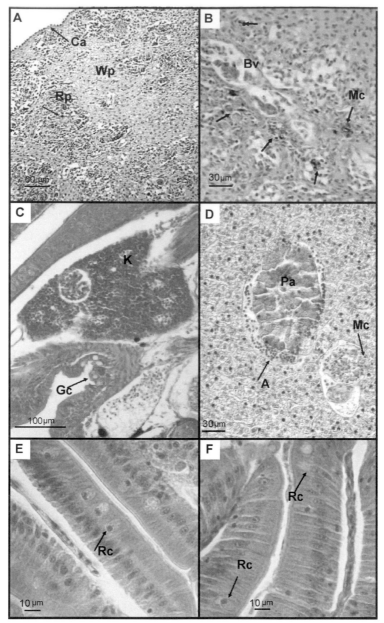

Fig. 2. Histology of various organs involved in immune reaction. **A.** Light micrograph of the spleen of the Ohrid trout illustrating components of the parenchyma, red and white pulp enclosed by a thin capsule (HE staining) (after Rebok et al. 2011; with permission). **B.** Section through the spleen of Ohrid trout. Melamocraphages center around the blood vessels (HE staining) (after Rebok et al. 2011, with permission). **C.** Cross section through the mid-kidney of the burbot, *Lota lota* larva (17 days post hatching). The location of the kidney is shown. Secretory cells (arrow), between pharyngeal epithelium, haemopietic tissue and one round glomerulus are visible (HE staining). **D.** Cross-section through the liver and the pancreas of carp after poisoning with pyrethroids. Leukocytic infiltration (arrow) around the damaged pancreas (HE staining). **E.** Cross-section through the intestine of asp larva. The rodlet cells are next to each other. Their nuclei are at the end of the cell more distant from the surface of the intestinal lumen (HE staining). **F.** Cross-section through the intestine of asp larva (19 days post hatching). The rodlet cells are dispersed (HE staining). A, leukocytic infiltration around the pancreas; Bv, blood vessels; Ca, capsule; Gc, gill cavity; Mc, melamocraphages centers; K, kidney; Mc, melanomacrophages; Pa, pancreas; Rc, rodlet cells; Rp, red pulp; Wp, white pulp.

Color version at the end of the book

In the immune system of fish, the kidney plays the role of a secondary lymphoid organ associated with the production of antibodies, and it also performs this function in a sense as a primary organ, owing to the tissue being abundant in stem cells (Lamers and De Haas 1985). The kidney in cyprinids is divided into two parts: the anterior part of the kidney (pronephros, head kidney) and the trunk kidney (opistonephros, excretory kidney). The carp's head kidney is surrounded by a delicate capsule of connective tissue and is further enhanced by reticular cells (see also chapter Kidney). The parenchyma of this part of the organ contains both myeloid and lymphoidal tissue. The lymphoid tissue is dispersed but its aggregates occur between the small sinuses and along venous and arterial blood vessels as the white pulp. In the non-lymphoid tissue (the red pulp), there are cells representing particular stages of erythropoiesis and granulopoiesis. Clear clusters of melano-macrophages occur in various parts of the kidneys, but most often around larger blood vessels and blood sinuses (Lamers and De Haas 1985). In the trunk kidney of carp, in the space between the tubules, there is myeloid and lymphoid parenchyma, which is supported by a network formed from reticular cells alternating with blood vessels with quite thin walls. Clusters of melanomacrophages are not too numerous and located between the renal channels; the accumulation of lymphoid cells is visible in the histological sections.

In the kidney of sturgeons, both excretory and lymphoid parts are present. The latter fulfills the function of hemopoietic tissue and is primarily associated with the anterior segment of the sturgeon kidney. In this part, large (13–20 μm) blast cells are found, including numerous mitotic stages. There are also large cells similar to plasma cells with the highly stained basophilic cytoplasm and with an eccentrically located nucleus; in addition to the above, lymphocytes, granulocytes and macrophages occur (Fänge 1986).

Specific Lymphoid and Lymphomyeloid Tissues

There is a great diversity of lymphoid and lymphomyeloid organs in fish; in addition to the thymus, kidney and spleen, they include the Leydig organ, a tissue associated with the esophagus, the epigonal organ, linked to the gonads, pericardial tissue, another tissue surrounding the heart, and finally the orbital and the preorbital tissues, that are the tissues located in the region of the eyes of fish (Fänge 1982, 1986, Meštanová and Varga 2016, Oguri 1983). Representatives of Lampreys (Petromyzontiformes, Agnatha) have lymphoid tissue associated with the esophagus, which sometimes occurs as lymphocyte clusters (Amemiya et al. 2007, Ardavin and Zapata 1988, Page and Rowley 1982).

The Leydig organ (Leydig's organ, oesophageal lymphomyeloid tissue) is a hematopoietic organ located in the submucosa of the esophagus of elasmobranchs, but it does not appear in all of the representatives; in some there is only a residual form of this organ (Fänge 1987, Honma et al. 1984, Oguri 1983). The Leydig organ is a component of the defense system in the Elasmobranchii, which contains numerous lymphocytes and granulocytes. During the ontogenetic development of *Scyliorhinus canicula*, immunoglobulin-positive cells appear in the Leydig organ, two months later in the liver and kidney (Lloyd-Evans 1993).

The Leydig organ usually consists of two dorsal and abdominal parts; in the histological image there are numerous patches that reach few arteries and capillaries. These lobes are separated by the endothelium lining of irregular venous spaces. There are huge amounts of leukocytes in the network of the reticulum inside the lobes, as well as numerous promyelocytes, blastic cells and lymphocytes; the latter occur in the form of loosely formed vesicles in dogfish (Fänge 1987). In *Etmopterus spinax* (Elasmobranchii), the widespread occurrence of cells corresponding to mammalian plasma cells was revealed during the observation under the light and electron microscope (Mattison and Fänge 1986).

In many species, both the Leydig organ and the epigonal organ were found. Not all representatives of the Elasmobranchii have an epigonal organ. It is believed that there is some relationship between the occurrences of both parts of these organs; namely when one of them is absent or small, the other one tends to be larger (Fänge 1987). The epigonal organ (Figs. 3A, B, C, and D) is the site of granulocyte maturation; granulocytes occur at various stages of development and after reaching maturity they enter blood vessels (Bircan-Yildirim et al. 2011).

Lymphocytes present in the organ sometimes form clusters; there are small and large round lymphocytes (about 7 to 14 μm), sometimes with irregular cytoplasmic projections, but plasmocytes have not been

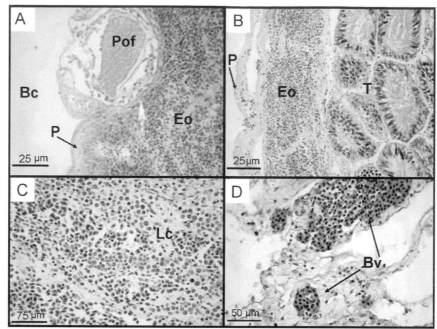

Fig. 3. Histology of the epigonal organ (Eo) of the guitarfish *Rhinobatos rhinobatos* (HE staining). **A.** Cross section through the Eo of the female. Bc, body cavity; P, peritoneum; Pof, post ovulatory follicles (after Bircan-Yildirim et al. 2011 with permission). **B.** Cross section through the Eo of a male. P, peritoneum; T, testes (after Bircan-Yildirim et al. 2011 with permission). **C.** Cross section through the Eo showing lymphomyeloid cells (Lc) (after Bircan-Yildirim et al. 2011 with permission). **D.** Cross section through the Eo, note numerous blood vessels (Bv) (after Bircan-Yildirim et al. 2011 with permission).

found. The organ is equipped with numerous vascular sinuses (Fänge 1982); in the common guitarfish *Rhinobatos rhinobatos* there are numerous blood vessels (Bircan-Yildirim et al. 2011). Depending on the structure and presence of the gonads, this bilateral organ may occur as a single or a geminate one, and can be built differently; the organ surrounds the testes, ovaries and the liver.

The histological structure is similar to that of the Leydig's organ (Fänge 1987). The granulocyte rich tissue of both the Leydig's organ and the epigonal organ in elamobranchs is characterized by a high activity of bacteriolytic enzyme, lysozyme, chitinase (Fänge et al. 1980, Fänge 1987). In sharks (*Ginglymostoma cirratum*), the epigonal organ is the site of B-lymphocyte formation (Rumfelt et al. 2002).

The pericardial tissue has a structure resembling lymph nodes, it surrounds the heart. The task of this organ is to sustain the interaction of the vascular endothelium and lymphoid cells in sturgeons. The lymphoid cells predominate in the pericardial tissue of white sturgeons and hybrid sturgeons (Fänge 1986). The organ divides into smaller parts, separated from each other by a layer of connective tissue. They contain a number of irregular cells in their center that are surrounded by sinuses filled with blood or lymph. The following cells are present in the tissue around the heart: lymphocytes, reticulocytes and granulocytes; macrophages have been detected less often. The whole assemblage of cells is covered by the endothelium of venous sinuses; electron microscopical observations revealed the migration of lymphocytes through the spaces between endothelial cells (Fänge 1986).

The orbital and preorbital tissue is a specific lymphomyeloid tissue described in some fish (Holocephali) that plays the role of bone marrow (Mattison and Fänge 1986). The tissue is located around the eyes and in the periorbital canal of the cranium of the rabbit fish *Chimaera monstrosa*. Ultrastructural studies revealed details of the structure, composed mainly of granulocytes (eosinophilic and heterophilic), which constituted about 80% of all cells of this tissue. Plasma cells occur much less frequently and not in all locations of the orbital tissue; they constitute about 5% of the cellular composition. These cells with dimensions up to 15 µm have certain features indicating that they are a site of protein production. The orbital tissue also contains lymphocytes, macrophages and blast forms of cells, although few in number.

Mucosa-Associated Lymphoid Tissue

Mucosa-associated lymphoid tissue (MALT) is a collection of lymphoid aggregates and dispersed components in various places in the body of animals (Brandtzaeg et al. 2008). Depending on the place of occurrence in Teleostei, the MALT is divided into the following tissues: gut-associated lymphoid tissue (GALT), skin-associated lymphoid tissue (SALT), the gill-associated lymphoid tissue (GIALT) and NALT, nasopharynx-associated lymphoid tissue, connected with the olfactory organ (Salinas 2015, Salinas et al. 2011). There are three characteristic features of the MALT in fish: (1) lymphoid cells dispersed, without organized structures, (2) dominance of specialized mucosa-IgT immunoglobulins and IgT + B cells, (3) a variety of symbiotic bacteria that are overlaid by mucosal Igs of fish (Tacchi et al. 2014).

The GIALT contains both gill filaments and interbranchial lymphoid tissue in the gills. The GIALT consists of scattered elements and just one that is focused, i.e., the interbranchial lymphoid tissue (ILT), which is a unique vertebrate organ found only in fish. This newly discovered tissue, due to its location in the gill cavity, has easy contact with the external environment and antigens can be found therein. This tissue is an important organ in the defense system of fish. The interbranchial lymphoid tissue, whose presence has been confirmed in salmonids so far, is quantitatively a very important site of T-lymphocyte aggregation, strategically located to facilitate the antigen encounter (Aas et al. 2017, Austbø et al. 2014, Dalum et al. 2015, Haugarvoll et al. 2008, Koppang et al. 2010). The interbranchial lymphoid tissue has been defined on histological cross-sections of the gills of rainbow trout and sexually mature Atlantic salmon (Koppang et al. 2010). An interbranchial septum in the gills of the salmonids is attached to the bones of the gill arch. It stretches around 1/3 of the length of the primary gill lamellae, opening to the lumen of the gill cavity. In this site, exposed to the aquatic environment, there is interbrachial lymphoid tissue in the form of leukocyte cell aggregates, predominantly lymphocytes. This tissue in young fish is separated from the external environment only by a narrow epithelial capsule, but with time, as the fish age, its thickness clearly increases; there are also goblet cells but Ig+ cells are rare in ILT. According to Koppang et al. (2010), epithelial leukocytes in the epithelial mesh are T-lymphocytes and interbrachial lymphoid tissue is a secondary lymphoid organ without the property of a primary lymphoid organ. In adult Atlantic salmon, ILT is nearly 13 times as large as the thymus and does not fulfil the function of a primary lymphoid organ (Dalum et al. 2015). This organ is involved in peripheral immune tolerance, which is particularly important due to its location in the distal part of the interbranchial septum and the possibility of exposure to harmless antigens present in water (Aas et al. 2017). The structure of interbranchial lymphoid tissue may undergo variously directed changes under the influence of pathogenic factors. A reduction in ILT size and a slight inflammatory response in Atlantic salmon were found to have been caused by the infection of ISAV responsible for infectious salmon anemia ISA (Austbø et al. 2014). On the other hand, the marine amoeba *Neoparamoeba perurans*, a pathogen responsible for the amoebic gill disease (AGD) causing 50 percentage of mortality among salmon in cage culture in a saltwater environment, triggers the ILT enlargement and multiplication of lymphocytes. These reactions are transient and their severity is related to the presence of the pathogenic factor not only in the gills but also in the ILT (Norte dos Santos et al. 2014, Wright et al. 2015).

The nasopharynx associated lymphoid tissue (NALT) has been detected in fish only recently, although its presence was known before in birds and mammals as an important element of the first line of defense (Salinas 2015, Tacchi et al. 2014). The NALT and nasal immunity are recognized as "an ancient arm" of terrestrial and aquatic vertebrates from different environments (Salinas 2015, Tacchi et al. 2014). The NALT discovered in rainbow trout has a structure basically similar to other mucosa-associated lymphoidal organs of Teleostei. Numerous myeloid and lymphoid cells are scattered in olfactory rosettes of the rainbow trout, Australian eel *Anguilla australis*, goldfish *Carassius auratus* and New Zealand groper *Polypryon oxygeneios*. Lymphoid cells occur in the NALT of rainbow trout in a dispersed form. The olfactory organ in fish can participate in specific and unspecific defensive reactions. IgT+ and IgM+ cells are present in the epithelial layer of the olfactory organ of the rainbow trout (Tacchi et al. 2014).

Polymeric IgT occurs as the basic immunoglobulin in the gut-associated lymphoid tissue of fish, while in mammals this role is played by IgA; these immunoglobulins coat intestinal commensal organisms. In reactions directed against intestinal parasites, fish and mammals produce specific IgT or IgA, respectively (Sunyer 2013). Secretory immunoglobulin transport to the intestinal lumen takes place owing to the polymeric immunoglobulin receptors in both fish and mammals. Lymphoid tissue associated with the digestive tract occurs in Teleostei in the form of dispersed cells (lymphocytes, plasmocytes, granulocytes and macrophages) or in the form of small clusters of both present in the lamina propria and epithelium. In the spotted snakehead *Channa punctatus*, lymphoid cell aggregates have not been found, while leucocytes, plasma cells, mast cells have been detected as being dispersed along the entire intestine, and intraepithelial leucocytes as well as granular cells were present in the submucosa and intralaminal leucocytes in the lamina propria of mucosa (Venkatesh et al. 2014). In an adult Mozambique tilapia *Oreochromis mossambicus*, grouped cells were present in lamina propria but were rare; the highest intensity of GALT takes place in the intestine (Doggett and Harris 1991). In the Australian lungfish *Neoceratodus forsteri* (Sarcopterygii, Dipnoi), the defense system associated with GALT is more active in sexually immature fish than in mature individuals (Hassanpour and Joss 2009). The sites where lymphoid tissue occurs in the intestine of this fish are: a pyloric fold between the foregut and small intestine, aggregations of cells around the posterior part of the spleen and structural elements similar to nodes in the epithelium of mucosal tissue (Hassanpour and Joss 2009).

The lymphoid tissues of sturgeons are very diverse and located in various places in their body. The tissue associated with the gastrointestinal tract of sturgeons (white Pacific sturgeon *A. transmontanus* and sturgeon hybrids) contains numerous lymphocytes and clusters of these cells are mainly located in the spiral valve (Fänge 1986). This organ, recently identified, is also a site where the GALT appears in the form of vesicular clusters of lymphoid cells in sharks, while in other elasmobranchs (e.g., *Torpedo*) this form of the GALT is in the duodenum (Fänge 1987).

Skin-associated lymphoid tissue SALT (or SIS—skin immune system, or cutaneous immune system) in vertebrates is found only in fish and amphibians, although proper mucosal lymphoid tissue is found in fish (Salinas et al. 2011). The fish skin is a large organ (especially in larvae in relation to body volume) that performs multiple functions (e.g., a protective barrier and a site of defense reactions, respiration, osmoregulation, locomotor, sensory and secretory functions). Esteban (2012) reviewed 367 publications related to fish skin defense reactions, fish skin genetics and mucosal immunology. The importance of the tasks undertaken in the publications in practice, including the application of new methods of preventing fish diseases and new diets for farmed fish, was taken into account. The skin in fish, unlike in other vertebrates, is characterized by the lack of keratinization and the presence of living epithelial cells. They are similar to epithelial cells of the gill and gill chamber cells exposed to an aquatic environment, although the structure of the SALT is closer to the gut-associated lymphoid tissue than to the gill-associated lymphoid tissue because the GALT also includes a variety of microbiota that are surrounded by secreted immunoglobulins, and specific IgT proteins appear in both tissues as a reaction against pathogenic agents (Salinas 2015, Xu et al. 2013). There is a variety of the skin structure in various fish species and, similarly, there are differences in the composition of secreted mucus or the presence of molecules (Salinas et al. 2011). The epidermis of healthy fish contains moderately numerous secretory cells (club cells, sacciform cells, goblet cells, Malpighian cells) between epithelial cells; during stress, the secretory cells can be more numerous. Secretory cells in the skin are responsible for the production of mucosal secretions; some of them (Malpighian and goblet cells) can probably also perform other functions, e.g., phagocytosis. In an *in vitro* test, motile malpighian cells derived from the scales of Atlantic salmon have demonstrated the ability to absorb foreign material (Åsbakk and Roy 1998).

Antibodies present in the mucus covering the fish skin are important in the fight against bacterial infection; bacterial diseases of fish in aquaculture are among the most common cases and constitute 54.9% of prevalence of all diseases (McLoughlin and Graham 2007). The positive effect of the immersion vaccination in the European eel against *Vibrio vulnificus* was both a systemic (serum) and local (mucus) reaction. However, antibodies in the mucus play an important role at the beginning of an infection, protecting the body against bacterial colonization (Esteve-Gassent et al. 2003).

Cells of the Immune System

Among these cells there are the morphotic elements of the blood, discussed in one of the chapters, cells appearing in various fish tissues under the influence of various pathogenic agents or as an effect of environmental stress (Lieschke and Trede 2009, Teh et al. 1997, Triebskorn et al. 2008, Sayed and Younes 2017), after poisoning with pyrethroids (Fig. 2D).

Some of them were initially included among organisms pathogenic to fish (e.g., *Rhabdospora thelohani* counted as Sporozoa), and only later they were recognized as a fish cell (rodlet cells) participating in the fish defense system (Reite and Evensen 2006).

Rodlet cells observed in many species of freshwater and marine fish were not immediately recognized as cells participating in the defense system (Manera and Dezfuli 2004, Reite 2005, Reite and Evensen 2006). These intriguing cells were thought to be parasites with an undefined systematic position (Morrison and Odense 1978, Schmachtenberg 2007). Rodlet cells, named *Rhabdospora thelohani*, were discovered by Waluga et al. (1986) in herring, sprat, cod and flounder from the Baltic Sea. Although the main pathogenic factor in the examined fish was poliorganotropic *Pleistophora* (Microsporidia), rodlet cells occurred frequently in gill epithelium, submucosa of the intestine and kidney, and were very numerous, especially in herring with degeneration and necrosis of epithelial tissues. Rodlet cells are oval in shape, with a length close to or greater than the length of red blood cells in fish. They are elongated and positioned perpendicular to the surface of the epithelium (Figs. 2E, F). In the intestine of burbot *Lota lota* larvae (22 days post hatching), the rodlet cells were very rare (Furgała-Selezniow et al. 2016b). In the fish thymus, they are associated with the capsule of this organ (Vasiliev and Polevshchikov 2014). These cells have a nucleus located at one of the poles of the cell, and rodlets that are quite stable after firing out remain unchanged while the cell dies (Schmachtenberg 2007).

The melanomacrophage aggregates or melanomacrophage centers, MMC, or pigmented cells have been observed in fish, amphibians and reptiles, and they do not always include cells with macrophage properties (Agius and Roberts 2003). They occur in various organs (often in the spleen, kidneys and liver of fish); in the tissues of fish they appear after the first nutrition. As fish age, the number of MMCs increases, a development also observed during starvation. The structure of melanomacrophage centers varies between fish species. MMCs have the form of nodules separated by a capsule formed by layers of flattened cells from the surrounding tissue. The capsule, especially in small MMCs, is not always clearly visible. The most common pigments of macrophages are lipofuscin, melanin and haemosiderin, which are visible due to special the Pearl staining method (Borucinska et al. 2009). Melanomacrophage centers of bony fishes are a place where a trapped antigen can interact with the immune system, so small lymphocytes, the cells for which this antigen is presented, are found inside these centers next to the macrophages.

Antibiotic-resistant intracellular bacteria, including those responsible for bacterial kidney disease BKD, can be accumulated inside MMCs; under certain conditions this accumulation may lead to the development of the chronic wasting disease (Agius and Roberts 2003). According to Sitja-Bobadilla et al. (2016), intracellular parasites (Microsporidia and Apicomplexa) induce hardly any host immune responses. In the case of such parasites, macrophages can be used to spread infection in the body of fish like a Trojan horse. Clustered and single MMCs can accumulate around the site of various bacteria in the organs, or close to the larvae of parasites inside the cysts (Borucinska et al. 2009).

We have observed plasmodium with immature spores of *Myxidium* (Myxosporea) (Fig. 4A) in the kidney of roach *Rutilus rutilus* and the formation of melanomacrophage centers and lymphocyte infiltration (Fig. 4B) in the liver as a response to *Myxidium rhodei* infection in fish (Dzika et al. 2006).

Numerous large macrophages with necrotic, eosinophilic cellular fragments were observed in the gills and liver of the wild common carp in Canada during Koi herpesvirus infection (Garver et al. 2010). Melanomacrophage centers are also a non-specific tool used to monitor and assess the impact of various pollutants in an aquatic environment. Melanomacrophage centers as immunohistological biomarkers in the

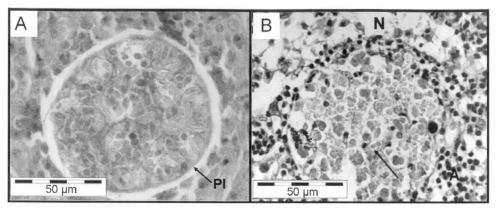

Fig. 4. Infection of the roach *Rutilus rutilus* with the parasite *Myxidium rhodei* (HE staining). **A.** Plasmodium (Pl) with *Myxidium rhodei* spores in the kidney. Plasmodium surrounded by epitheloid cells (after Dzika et al. 2006 with permission). **B.** Cross section through the liver. Numerous necrotic hepatocytes and single damaged spores of *Myxidium* (arrow) are visible. An early macrophages center (asterics) and lymphocytic infiltration (A) were a common reaction. Note necrotic hepatocytes (N) (after Dzika et al. 2006 with permission).

liver, kidney and spleen of the African catfish *Clarias gariepinus* were used for demonstrating the influence of silver nanoparticles (Sayed and Younes 2017). The high accumulation of cadmium and copper in fish (sneep *Chondrostoma nasus*, European chub *Leuciscus cephalus*) from a polluted river has been associated with reactions of the defense system in the form of inflammatory foci in the gills and as inflammation, lymphocyte infiltration, macrophage clusters in histological liver preparations (Triebskorn et al. 2008).

References Cited

Aas, I.B., L. Austbø, K. Falk, I. Hordvik and E.O. Koppang. 2017. The interbranchial lymphoid tissue likely contributes to immune tolerance and defense in the gills of Atlantic salmon. Dev. Comp. Immunol. 76: 247–254.

Agius, C. and R.J. Roberts. 2003. Melano-macrophage centres and their role in fish pathology. J. Fish Dis. 26: 499–509.

Amemiya, C.T., N.R. Saha and A. Zapata. 2007. Evolution and development of immunological structures in the lamprey. Curr. Opin. Immunol. 19: 535–541.

Ardavin, C.F. and A. Zapata. 1988. The pharyngeal lymphoid tissue of lampreys. A morpho-functional equivalent of the vertebrate thymus? Thymus 11: 59–65.

Åsbakk, K. and A.D. Roy. 1998. Atlantic salmon (*Salmo salar* L.) epidermal Malpighian cells—motile cells clearing away latex beads *in vitro*. J. Mar. Biol. 6(1): 30–34.

Austbø, L., I.B. Aas, M. Konig, S.C. Weli, M. Syed, K. Falk and E.O. Koppang. 2014. Transcriptional response of immune genes in gills and the interbranchial lymphoid tissue of Atlantic salmon challenged with infectious salmon anaemia virus. Dev. and Comp. Immunol. 45: 107–114.

Avtalion, R.R., A. Wojdani, Z. Malik, R. Shahrabani and M. Duczyminer. 1973. Influence of environmental temperature on the immune response in fish. Curr. Top. Microbiol. Immunol. 61: 1–35.

Biller-Takahashi, J.D. and E.C. Urbinati. 2014. Fish immunology. The modification and manipulation of the innate immune system: Brazilian studies. An. Acad. Bras. Cienc. 86: 1483–1495.

Bircan-Yildirim, Y., Ş. Çek, N. Başusta and E. Atik. 2011. Histology and morphology of the epigonal organ with special referance to the lymphomyeloid system in *Rhinobatos rhinobatos*. Turk. J. Fish. Aquat. Sci. 11: 351–358.

Borucinska, J.D., K. Kotran, M. Shackett and T. Barker. 2009. Melanomacrophages in three species of free-ranging sharks from the northwestern Atlantic, the blue shark *Prionacae glauca* (L.), the shortfin mako, *Isurus oxyrhinchus* Rafinesque, and the thresher, *Alopias vulpinus* (Bonnaterre). J. Fish Dis. 32: 883–891.

Botham, J.W. and M.J. Manning. 1981. The histogenesis of the lymphoid organs in the carp *Cyprinus carpio* L. and the ontogenetic development of allograft reactivity. J. Fish Biol. 19: 403–414.

Bowden, T.J., P. Cook and J.H.W.M. Rombout. 2005. Development and function of the thymus in teleosts. Fish Shellfish Immunol. 19: 413–427.

Brandtzaeg, P., H. Kiyono, R. Pabst and M.W. Russell. 2008. Terminology: Nomenclature of mucosa-associated lymphoid tissue. Mucosal. Immunol. 1: 31–37.

Castillo, A., B.E. Razquin, P. Lopez-Fierro, F. Alvarez, A. Zapata and A.J. Villena. 1990. Enzyme- and immuno-histochemical study of the thymic stroma in the rainbow trout, *Salmo gairdneri*, Richardson. Thymus 15: 153–166.

Cerezuela, R., F.A. Guardiola, A. Cuesta and M.Á. Esteban. 2016. Enrichment of gilthead seabream (*Sparus aurata* L.) diet with palm fruit extracts and probiotics: Effects on skin mucosal immunity. Fish Shellfish Immunol. 49: 100–110.

Chilmonczyk, S. 1982. Rainbow trout lymphoid organs: cellular effects of corticosteroids and anti-thymocyte serum. Dev. Comp. Immunol. 6: 271–280.

Chilmonczyk, S. 1992. The thymus in fish: Development and possible function in the immune response. Annu. Rev. Fish Dis. 2: 181–200.

Dalum, A.S., L. Austbø, H. Bjørgen, K. Skjødt, I. Hordvik, T. Hansen, P.G. Fjelldal, C.M. Press, D.J. Griffiths and E.O. Koppang. 2015. The interbranchial lymphoid tissue of Atlantic salmon (*Salmo salar* L.) extends as a diffuse mucosal lymphoid tissue throughout the trailing edge of the gill filament: distribution of the ILT. J. Morphol. 276: 1075–1088.

Doggett, T.A. and J.E. Harris. 1991. Morphology of the gut associated lymphoid tissue of *Oreochromis mossambicus* and its role in antigen absorption. Fish Shellfish Immunol. 3: 213–222.

Dzika, E., T. Wlasow and R.W. Hoffmann. 2006. *Myxidium rhodei* Léger, 1905 (Myxozoa: Myxosporea) infection in roach from four lakes of northern Poland. Bull. Eur. Assoc. Fish Pathol. 26: 119–124.

Esteban, M.Á. 2012. An overview of the immunological defenses in fish skin. ISRN Immunol. 2012: 1–29. Article ID: 853470.

Esteve-Gassent, M.D., M.E. Nielsen and C. Amaro. 2003. The kinetics of antibody production in mucus and serum of European eel (*Anguilla anguilla* L.) after vaccination against *Vibrio vulnificus*: development of a new method for antibody quantification in skin mucus. Fish Shellfish Immunol. 15: 51–61.

Fänge, R., G. Lundblad, K. Slettengren and J. Lind. 1980. Glycosidases in lymphomyeloid (hematopoietic) tissues of elasmobranch fish. Comp. Biochem. Physiol. B: Comp. Biochem. 67: 527–532.

Fänge, R. 1982. A comparative study of lymphomyeloid tissue in fish. Dev. Comp. Immunol. Suppl. 2: 23–33.

Fänge, R. 1986. Lymphoid organ in sturgeons (Acipenseridae). Vet. Immunol. Immunopathol. 12: 153–161.

Fänge, R. 1987. Lymphomyeloid system and blood cell morphology in Elasmobranchns. Arch. Biol. (Bruxelles) 98: 187–208.

Furgała–Selezniow, G., T. Własow, A.M. Wiśniewska, R. Kujawa, A. Skrzypczak, P. Woźnicki and M. Jankun. 2016a. Early development of the asp, *Leuciscus aspius* pseudobranch—the histological study. Turkish J. Fish Aquat. Sci. 16: 723–28.

Furgała–Selezniow, G., M. Jankun, R. Kujawa, J. Nowosad, M. Białas, D. Kucharczyk and A. Skrzypczak. 2016b. Histological aspects of the early development of the digestive system of burbot *Lota lota* L. (Lotidae, Gadiformes). Folia Biol-Kraków 64: 11–21.

Garver, K.A., L. Al-Hussinee, L.M. Hawley, T. Schroeder, S. Edes, V. LePage, E. Contador, S. Russell, S. Lord, R.M.W. Stevenson, B. Souter, E. Wright and J.S. Lumsden. 2010. Mass mortality associated with koi herpes virus in wild common carp in Canada. J. Wildl. Dis. 46: 1242–1251.

Grace, M.F. and M.J. Manning. 1980. Histogenesis of the lymphoid organs in rainbow trout *Salmo gairdneri* Rich. 1836. Dev. Comp. Immunol. 4: 255–265.

Gui, J., L.M. Mustachio, D.M. Su and R.W. Craig. 2012. Thymus size and age-related thymic involution: early programming, sexual dimorphism, progenitors and stroma. Aging Dis. 3: 280–290.

Hassanpour, M. and J. Joss. 2009. Anatomy and histology of the spiral valve intestine in juvenile Australian lungfish, *Neoceratodus forsteri*. Open Zool. J. 2: 62–85.

Haugarvoll, E., I. Bjerkas, B.F. Nowak, I. Hordvik and E.O. Koppang. 2008. Identification and characterization of a novel intraepithelial lymphoid tissue in the gills of Atlantic salmon. J. Anat. 213: 202–209.

Heinecke, R.D., J.K. Chettri and K. Buchmann. 2014. Adaptive and innate immune molecules in developing rainbow trout, *Oncorhynchus mykiss* eggs and larvae: Expression of genes and occurrence of effector molecules. Fish Shellfish Immunol. 38: 25–33.

Honma, Y., K. Okabe and A. Chiba. 1984. Comparative histology of the Leydig and epigonal organs in some elasmobranchs. Jpn. J. Ichthyol. 31: 47–54.

Koppang, E.O., U. Fischer, L. Moore, M.A. Tranulis, J.M. Dijkstra, B. Köllner, L. Aune, E. Jirillo and I. Hordvik. 2010. Salmonid T cells assemble in the thymus, spleen and in novel interbranchial lymphoid tissue. J. Anat. 217: 728–739.

Köllner, B. and G. Kotterba. 2002. Temperature dependent activation of leucocyte populations of rainbow trout, Oncorhynchus mykiss, after intraperitoneal immunization with *Aeromonas salmonicida*. Fish Shellfish Immunol. 12(1): 35–48.

Kum, C. and S. Sekkin. 2011. The immune system drugs in fish: immune function, immunoassay, drugs. pp. 169–216. *In:* A. Faruk (ed.). Recent Advances in Fish Farms, InTech Europe Press, Rijeka, Croatia.

Lamers, C.H.J. 1985. The Reaction of the Immune System of Fish to Vaccination. Ph.D. Thesis, Wageningen University, Wageningen, The Netherlands.

Lamers, C.H.J. and M.J.H. De Haas. 1985. Antigen localization in the lymphoid organs of carp (*Cyprinus carpio*). Cell Tissue Res. 242: 491–498.

Le Morvan, C., D. Troutaud and P. Deschaux. 1998. Differential effects of temperature on specific and nonspecific immune defences in fish. J. Exp. Biol. 201: 165–168.

Lieschke, G.J. and N.S. Trede. 2009. Fish immunology. Curr. Biol. 16: 678–682.

Lloyd-Evans, P. 1993. Development of the lymphomyeloid system in the dogfish, *Scyliorhinus canicula*. Dev. Comp. Immunol. 17(6): 501–514.

Manera, M. and B.S. Dezfuli. 2004. Rodlet cells in teleosts: a new insight into their nature and functions. J. Fish Biol. 65: 597–619.

Manning, M.J., M.F. Grace and C.J. Secombes. 1982. Ontogenetic aspects of tolerance and immunity in carp and rainbow trout: studies on the role of the thymus. Dev. Comp. Immunol. Suppl. 2: 75–82.

Mattison, A. and R. Fänge. 1986. The cellular structure of lymphomyeloid tissues in *Chimaera monstrosa* (Pisces, Holocephali). Biol. Bul. 171: 660–671.

McLoughlin, M.F. and D.A. Graham. 2007. Alphavirus infections in salmonids-a review. J. Fish Dis. 30: 511–531.

Meštanová, V. and I. Varga. 2016. Morphological view on the evolution of the immunity and lymphoid organs of vertebrates, focused on thymus. Biologia. 71: 1080–1097.

Mohammad, M.G., S. Chilmonczyk, D. Birch, S. Aladaileh, D. Raftos and J. Joss. 2007 Anatomy and cytology of the thymus in juvenile Australian lungfish, *Neoceratodus forsteri*. J. Anat. 211: 784–797.

Morrison, C.M. and P.H. Odense. 1978. Distribution and morphology of rodlet cells in fish. J. Fish Res. Board Can. 35, 3: 101–116.

Nakanishi, T. 1991. Ontogeny of the immune system in *Sebastiscus marmoratus*: histogenesis of the lymphoid organs and effects of thymectomy. Environ. Biol. Fishes 30: 135–145.

Norte dos Santos, C.C., M.B. Adams, M.J. Leef and B.F. Nowak. 2014. Changes in the interbranchial lymphoid tissue of Atlantic salmon (*Salmo salar*) affected by amoebic gill disease. Fish Shellfish Immunol. 41: 600–607.

Oguri, M. 1983. On the Leydig organ in the esophagus of some elasmobranchs. Bull. Japan Soc. Sci. Fish 49: 989–991.

O'Neill, J.G. 1989. Ontogeny of the lymphoid organs in an Antarctic teleost, *Harpagifer antarcticus* (Notothenioidei: Perciformes). Dev. Comp. Immunol. 13: 25–33.

Page, M. and A.F. Rowley. 1982. A morphological study of pharyngeal lymphoid accumulations in larval lampreys. Dev. Comp. Immunol. 2(Suppl.): 35–40.

Press, C.M.L. and Ø. Evensen. 1999. The morphology of the immune system in teleost fish. Fish Shellfish Immunol. 9: 309–318.

Rebok, K., M. Jordanova and I. Tavciovska-Vasileva. 2011. The spleen histology in the female Ohrid trout, *Salmo letnica* (Kar.) (Teleostei, Salmonidae) during its reproductive cycle. Arch. Biol. Sci Belgrade. 63: 1023–1030.

Reite, O.B. 2005. The rodlet cells of teleostean fish: their potential role in host defence in relation to the role of mast cells/ eosinophilic granule cells. Fish Shellfish Immunol. 19: 253–267.

Reite, O.B. and Ø. Evensen. 2006. Inflammatory cells of teleostean fish: A review focusing on mast cells/eosinophilic granule cells and rodlet cells. Fish Shellfish Immunol. 20: 192–208.

Romano, N., M. Fanelli, G.M. Del Papa, G. Scapigliati and L. Mastrolia. 1999. Histological and cytological studies on the developing thymus of sharpsnout seabream, *Diplodus puntazzo*. J. Anat. 194: 39–50.

Ruddle, N.H. 2016. High endothelial venules and lymphatic vessels in tertiary lymphoid organs: characteristics, functions, and regulation. Front. Immunol. 491: 1–7.

Rumfelt, L.L., E.C. McKinney, E. Taylor and M.F. Flajnik. 2002. The development of primary and secondary lymphoid tissues in the nurse shark *Ginglymostoma cirratum*: B-cell zones precede dendritic cell immigration and T-cell zone formation during ontogeny of the spleen. Scand. J. Immunol. 56: 130–148.

Salinas, I., Y.A. Zhang and J.O. Sunyer. 2011. Mucosal immunoglobulins and B cells of teleost fish. Dev. Comp. Immunol. 35: 1346–1365.

Salinas, I. 2015. The mucosal immune system of teleost fish. Biology 4: 525–539.

Sayed, A.H. and H.A.M. Younes. 2017. Melanomacrophage centers in *Clarias gariepinus* as an immunological biomarker for toxicity of silver nanoparticles. JMAU 5: 97–104.

Saxena, M. and H. Saxena. 2007. Histopathological changes in lymphoid organs of fish after exposure to water polluted with heavy metals. Internet J. Veterin. Med. 5: 1–8.

Schmachtenberg, O. 2007. Epithelial sentinels or protozoan parasites? Studies on isolated rodlet cells on the 100th anniversary of an enigma. Rev. Chil. Hist. Nat. 80: 55–62.

Sitja-Bobadilla, A., I. Estensoro and J. Perez-Sanchez. 2016. Immunity to gastrointestinal microparasites of fish. Dev. Comp. Immunol. 64: 187–201.

Snieszko, S.F. 1970. Immunization of fishes: a review. J. Wildl. Dis. 6: 24–30.

Sunyer, J.O. 2013. Fishing for mammalian paradigms in the teleost immune system. Nat. Immunol. 14: 320–326.

Tacchi, L., R. Musharrafieh, E.T. Larragoite, K. Crossey, E.B. Erhardt, S.A.M. Martin, S.E. LaPatra and I. Salinas. 2014. Nasal immunity is an ancient arm of the mucosal immune system of vertebrates. Nat. Commun. 5. 5205: 1–11.

Tatner, M.F. and M. Manning. 1983. Growth of the lymphoid organs in Rainbow trout, *Salmo gairdneri* from one to fifteen months of age. J. Zool. Lond. 199: 503–520.

Teh, S.J., S.M. Adams and D.E. Hinton. 1997. Histopathologic biomarkers in feral freshwater fish populations exposed to different types of contaminant stress. Aquat. Toxicol. 37: 51–70.

Tort, L., J.C. Balasch and S. Mackenzie. 2003. Fish immune system. A crossroads between innate and adaptive responses. Immunología. 622: 277–286.

Triebskorn, R., I. Telcean, H. Casper, A. Farkas, C. Sandu, G. Stan, O. Colărescu, T. Dori and H.R. Köhler. 2008. Monitoring pollution in River Mureş, Romania, part II: metal accumulation and histopathology in fish. Environ. Monit. Assess. 141: 177–188.

Uribe, C., H. Folch, R. Enriquez and G. Moran. 2011. Innate and adaptive immunity in teleost fish: a review. Vet. Med. (Praha) 56: 486–503.

Van Muiswinkel, W.B. and B. Vervoorn-Van Der Wal. 2006. The immune system of fish. pp. 678–701. *In*: P.T.K. Woo (ed.). Fish Diseases and Disorders. CAB International. Guelph, Canada.

Vasiliev, K.A. and A.V. Polevshchikov. 2014. Fish thymic non-lymphoid cells and the problem of the blood-thymus barrier. Russ. J. Mar. Biol. 40: 323–332.

Venkatesh, P., S.P. Jeyapriya, N. Suresh and T. Vivekananthan. 2014. Report on gut associated lymphoid tissue (GALT) in freshwater fish *Channa punctatus* (Bloch). Int. J. Pure Appl. Zool. 2: 95–99.

Waluga, D., T. Własow, E. Dyner and A. Świątecki. 1986. Studies on the etiepathogenesis of fish diseases in the Baltic Sea. Acta Ichthyol. Pisc. 16: 53–72.

Wise, T.H. and R.L. Hruska. 1986. The thymus: old gland, new persperctives. Domest. Anim. Endocrinol. 5: 109–128.

Własow, T. 1993. Thymic development in coregonid larvae with reference to the influence of pathogens and environmental conditions. Acta Acad. Agricult. Tech. Olst. Water Conservation and Inland Fisheries. Suppl. A 19(432): 47 pp. ISBN 1-5747-087-X (In Polish).

Własow, T., P. Gomułka, M. Łuczyński and A. Szczerbowski. 2003. Effect of hydrogen peroxide exposure to ide *Leuciscus idus* L. Bull. VURH Vodnany 39: 124–128 (In Czech).

Własow, T., P. Gomułka and M. Obara. 2004. Effects of hydrogen peroxide exposure to pike Esox lucius L. Bull. VURH Vodnany 40: 118–124 (In Czech).

Wright, D.W., B. Nowak, F. Oppedal, A. Bridle and T. Dempster. 2015. Depth distribution of the amoebic gill disease agent, *Neoparamoeba perurans*, in salmon sea-cages. Aquacult. Environ. Interact. 7: 67–74.

Xu, Z., D. Parraa, D. Gómeza, I. Salinas, Y.A. Zhang, L. von Gersdorff Jørgensend, R.D. Heinecke, K. Buchmann, S. LaPatra and J.O. Sunyer. 2013. Teleost skin, an ancient mucosal surface that elicits gut-like immune responses. PNAS 110: 13097–13102.

Chapter **15**

Gills
Respiration and Ionic-Osmoregulation

Marisa Narciso Fernandes

INTRODUCTION

Inhabiting a wide diversity of habitats, from almost distilled water to sea water and very saline lakes in which large temperature differences may occur, different fish species face great variabilities in dissolved O_2 availability and ion concentration. Therefore, they show a remarkable ability to maintain the internal O_2 needs for aerobic metabolism and ionic and osmotic levels independent of their habitat. Water-breathing fishes take all O_2 needs from the water and, depending on the dissolved O_2 levels available in the water, physiological (ventilation and cardiovascular mechanisms) and biochemical changes occur to optimize the O_2 uptake (Enders et al. 2016). Nevertheless, some fish need to emerge out of the water and take O_2 from atmospheric air using air-breathing organs in order to complement (facultative air-breathers) or take all (obligatory air-breathers) they need (see Graham 1997, Coolidge et al. 2007, Cruz et al. 2013, Silva et al. 2014).

Freshwater teleosts balance the osmotic gain of water by producing a large volume of diluted urine and the ion diffusion loss into the water is counterbalanced by actively absorbing NaCl through the gills and reabsorbing salt through renal tubes. Conversely, marine teleost that have osmotic loss of water and diffusional gain of ions from the environment maintain ionic and osmotic balance by drinking water, producing a low volume of urine and actively excreting salt through the gills. Marine elasmobranchs are isosmotic to the environment by retaining urea in the blood, but they need to regulate ion gain by diffusion from the environment, while freshwater elasmobranchs have similar ionic and osmotic problems as the freshwater teleost: they have to regulate diffusional gain of water and ion loss. Estuarine fish have to regulate water and ions in an environment in which salinity has remarkable variations over a 24 hr period. Furthermore, anadromous and catadromous fish have to modify the ionic and osmotic regulation to adjust ion absorption/excretion depending on environmental salinity as they live part of their life in freshwater and part in the sea. Physiological processes related to ion uptake and excretion can be found in extensive and comprehensive reviews (Zadunaisky 1984, Evans et al. 2005, Mancera and McCormick 2007, Kaneko et al. 2008, Hiroi and McCormick 2012, McCormick 2013, Wright and Wood 2015). It

Department of Physiological Sciences, Federal University of São Carlos, Via Washington Luiz, km 235, 13565-965 São Carlos, SP Brazil.
Email: dmnf@ufscar.br

is important to emphasize that the physiological alterations to maintain ionic and osmotic homeostasy imply morphological adjustments of gill cells (Perry and Laurent 1993, Moron et al. 2003, Furukawa et al. 2011, Nilsson et al. 2012).

The gills are the main organs for respiration, ionic regulation and acid-base equilibrium. Concerning respiration in air-breathing fish, the gills, in general, have a similar structure as those of water-breathing fish but the gill surface area reflects the fish dependence of O_2 from water or atmospheric air, being extremely reduced in Dipnoi (Graham 1997, Bassi et al. 2005, Moraes et al. 2005, Fernandes et al. 2012, Ramos et al. 2013). In relation to ionic and osmoregulation of the extracellular fluid, besides the gills, the effector organs include the kidney, intestinal tract and body surface. In the gills, the epithelial cells for ion transport actively absorb Na^+, Cl^- and Ca^{+2} from the water, and pump Na^+ and Cl^- outwards in marine fish (Laurent 1984, Evans et al. 2005). These ion transport cells also participate in acid-base equilibrium by eliminating H^+ and HCO_3^- equivalents into the water to maintain blood pH (Heisler 1984, Goss and Wood 1990, 1992, 1994). The skin has large quantities of mucus, which reduces water permeability in freshwater fish (Baldisserotto 2003) but this can contribute to absorption in the swamp eel, *Synbranchus marmoratus*, a facultative air breathing fish (Stiffler et al. 1986). The gut, in freshwater fish, participates in ionic regulation by absorbing ions contained in ingested food. However, in marine fish, the gut has greater importance in water absorption to maintain the osmotic balance while concomitantly it also absorbs ions present in water and food (Grosell 2006, 2007, Genz 2011). Kidneys eliminate the excess of water in freshwater fish and excrete the excess of Ca^{+2}, Mg^{+2}, and S in marine fish, helping the ionic and acid-basic balance (Edwards and Marshall 2013, Hyodo et al. 2014). The urinary bladder also participates in Na^+ and Cl^- absorption (Curtis and Wood 1991) (see also chapter Kidney). In marine elasmobranchs, the rectal gland, at the end of intestinal tube, participates in salt excretion, while in freshwater elasmobranchs which have similar ionic and osmotic problems as the freshwater teleost, this gland is vestigial (Burger 1965, Thorson et al. 1978).

This chapter will be restricted to the structure of gills considering the gas exchange (O_2/CO_2) and the cells related to ionic transport in the epithelium of the gills, with emphasis to freshwater fish. Ion transport cells are generally denominated ionocytes and are characterized by numerous mitochondria that generate energy for the active ionic transport process, specific ion membrane transporters and high density of ATPase units in the apical and basolateral membrane. In the gills, they were first referred to as chloride cells, as they were formerly described in marine fish gills (*Anguilla anguilla*) as "chloride secreting cells" (Keys and Willmer 1932) and, subsequently, generally referred to as chloride cells (Copeland 1948). Actually, the term "chloride cells" have been preferentially used to refer to the ionocytes from the gills of seawater or brackish water fish that specifically secrete Cl^- (Dymowska et al. 2012). In freshwater fish, the ionocytes are generally called mitochondria-rich cells due to high density of mitochondria. In this chapter, I will use "ionocyte" in place of "mitochondrion-rich cell" or "chloride cell".

The Gill Structure

Besides the external respiratory function of O_2/CO_2 gas exchange, the gill is the main site for the active uptake and excretion of ions, respectively, in freshwater and seawater fish. The gills also take part in the acid-base regulation and ammonia and other N catabolites excretion (Evans et al. 2005). Performing all these functions, the gill exhibits a high plasticity, showing morphological adjustments to external and internal media to maintain the respiration-osmoregulation compromise. However, in some environmental conditions, the optimization of one function may occur to the detriment of the other (Perry and Laurent 1993).

The teleost gills consist of four gill arches in which rakers are distributed on the anterior face, oriented to the mouth cavity, and two rows of primary lamellae or filaments (holobranches) are inserted on the posterior region and are accommodated in the opercular cavity (Fig. 1A). Each primary lamella has secondary lamellae, also called lamellae, above and below its horizontal axis (Fig. 1B) which provide a large gill surface area contacting the water. The secondary lamellae are the respiratory region of gills, i.e., they are the structure in which the gas exchange (O_2/CO_2) takes place. The structure of the primary

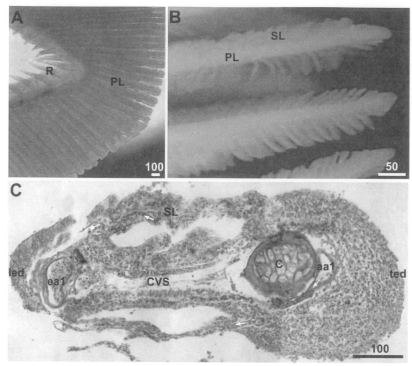

Fig. 1. Gill general structure. **A.** Gill arch showing the rakers (R) and primary lamellae (PL) inserted in the gill arch of *Pimelodus maculatus*. **B.** High magnification of the primary lamellae (PL) showing the secondary lamellae (SL) above and below the horizontal axis. **C.** Cross section of gill primary lamella of *Hoplias malabaricus* showing its structure: cartilage rod (C), afferent primary artery (aa1), efferent primary artery (ea1), central venous sinus (CVS), secondary lamella (SL), leading edge (led), trailing edge (ted). Bouin, Mallory. Scale bars in µm (Photo: M.G. Paulino, S.E. Moron).

lamellae consist of two blood circuits, the arterio-arterial circulation and arterial-venous circulation, a cartilage rod that gives support to the primary lamella and extends alongside its entire length and the central venous sinus which consists of small blood vessels and other structures localized between the afferent and efferent primary arteries (Fig. 1C). The arterio-arterial circulation is related to gas exchange and it is constituted of the afferent primary artery which branches off in afferent secondary arterioles in each secondary lamella. Blood flows through the spaces formed by the flange of pillar cell in the secondary lamella in which the gas exchange takes place, and thereafter, form the efferent secondary arteriole and the efferent primary artery. The proximal and marginal channels in the secondary lamella are formed by endothelial cells and the flange of pillar cells. The proximal channel is buried into the primary epithelium and the marginal channel is located in the secondary lamella border, both are formed by endothelial cells and pillar cells. The arterial-venous circulation is related to the blood supply to primary lamellar structures and ionic transport; it consists of small arteries originating from the afferent and efferent primary arteries, the arterio-venous afferent and efferent anastomosis form small veins, and then the primary lamella vein. The veins of the primary lamella form the branchial vein. The organization of the abductor muscles and the adductor muscles that move the primary lamella during the respiratory cycle varies among species depending on the interbranchial septum length; in species of the Loricariidae, the abductor muscles are restricted to the distal end of interbranquial septum (Fernandes et al. 1995b). Detailed descriptions of gill structure are given by Hughes 1984, Laurent 1984, Fernandes et al. 1995b, Olson 2002, Olson et al. 2003, Kaneko et al. 2008, Hwang et al. 2011, Marshall 2011.

During respiration, water flows from the oral cavity to the opercular cavities through the gill arches and between the lamellae from the leading to trailing edge of primary lamella in counter-current with blood flow inside the lamella from the afferent secondary artery, located in the trailing edge of the primary lamella, to the efferent secondary artery, located in the leading edge of the filament favoring the gas exchange. In general, active fish have a greater number of secondary lamellae per mm of primary lamella while sluggish fish have larger and spaced secondary lamellae; high tolerant-hypoxia water-breathing fish also have a high number of secondary lamellae per mm of primary lamella and a large surface area (Hughes 1984, Fernandes et al. 1994, Mazon et al. 1998, Karakatsouli et al. 2006, Crampton et al. 2008).

Air breathing fish may exhibit a highly modified gill structure depending on the degree of dependence of water or air for respiration or an accessory organ for air respiration, but most of them have a reduced number of primary lamellae, short and large secondary lamellae and consequently, a low respiratory surface area. In *Peryophthalmodon*, the secondary lamellae are adapted to air respiration, they are thick and possess interlamellar fusions that help to support the lamella in air; however, such adaptation increases the diffusion distance and minimize the water flow (Graham et al. 2007). *Clarias* has modified gills located in the upper side of the second and fourth gill arches as a suprabranchial chamber containing an arborescent air-breathing organ and gill fans for ventilation and air retention (Graham 1997). Larvae of *Arapaima gigas* possess a gill structure similar to that of most teleost fish but their morphology changes as the fish grows (Brauner et al. 2011, Ramos et al. 2013). The environmental physical and chemical characteristics of water such as dissolved O_2, salinity, temperature and pH as well as the presence of xenobiotics have a great influence on the gill surface area and water-blood O_2 diffusion distance. Gill plasticity is discussed at the end of this chapter.

In elasmobranchs, the gills have a similar organization as the teleosts, except for the gill arch numbers, organization of the skeletal elements supporting the gill elements and the interbranchial septa length. Elasmobranchs have five arches in each side of the orobranchial cavity, the first arch has only the posterior hemibranch, and the remaining four arches (2, 3, 4 and 5 gill arches) have the anterior and posterior hemibranches (holobranches). The long interbranchial septum connects the entire length of the primary lamellae from anterior and posterior hemibranches ultrapassing the tip of the primary lamellae on the lateral animal body and forming the gill slits. Each primary lamella supports secondary lamellae above and below the horizontal axis (for detailed elasmobranch gill structure, see Wegner 2016). Gill surface area in elasmobranchs, similar to teleost fish, is related to the species' oxygen requirements for aerobic metabolism and the environmental oxygen levels (Duncan et al. 2011, Wooton et al. 2015). Large gill surface areas were described in sharks from the families Lamnidae and Alopiidae, which are active predators (Bernal et al. 2012) and stingray embryos from the *Potamotrygon* genus (Duncan et al. 2015).

The Gill Epithelium

The primary lamellar epithelium is multilayered and consists mainly of pavement cells (PVC), mucous cells (MC), ionocytes, rodlet, neuroepithelial, and undifferentiated cells. The secondary lamellar epithelium consists of two cell layers: the PVC cell layer contacting water and an undifferentiated cell layer on the basement membrane covering the pillar cells (Fig. 2A). PVC constitutes 90% of cells in the outermost cell layer of the gill epithelium; the width of these cells is greater than its height giving them a flat and squamous appearance (Fig. 2A). In air-breathing fish, the PVC in the secondary lamellar epithelium may have a cuboidal shape, which increases the water-blood diffusion distance and ionocytes may be distributed among the PVC in the secondary lamellae (Fig. 2B). The PVC apical surface of the primary lamellar epithelium varies among species. In most species, the PVC is characterized by convoluted microridges, and a transition on the apical architecture from long microridges to a smooth apical surface may be identified in some fish species towards the secondary lamellar epithelium (Moron and Fernandes 1996). In some species, short microvilli or a smooth surface characterizes these cells (Fernandes et al. 1998).

Ionocytes are distributed among PVC in the primary lamellar epithelium close to the secondary lamella and in the interlamellar epithelium. However, depending on the environment, they are also found in the secondary lamellar epithelium (Figs. 2A, B, C) (details of these cells are given below). MC is round or ovoid (goblet) in shape, buried into the epithelial cell layers of the gill epithelium with small contact at the epithelial surface contacting water and it has a basal flattened nucleus and a cytoplasm full of mucus granules. It is distributed mainly in the leading and trailing edges of the primary lamella (Figs. 2D, E); they

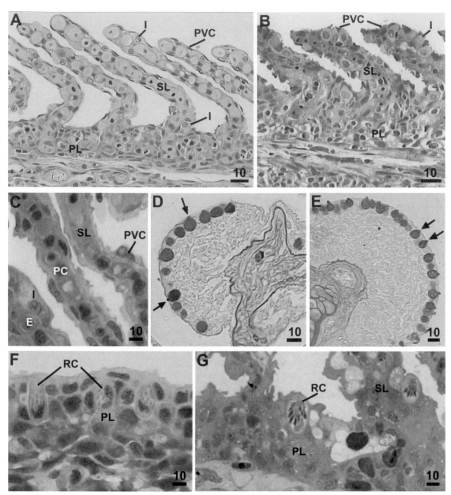

Fig. 2. Primary and secondary lamellae of curimbatá, *Prochilodus lineatus*, a water-breathing fish (**A**) and jeju, *Hoplerythrinus uniteniatus*, a facultative air-breathing fish (**B**). Note the pavement cells (PVC) in the primary and secondary lamellae in (**A**) and the cuboidal appearance of the outermost cell layer of secondary lamellae in (**B**). Ionocytes are present in the primary and secondary lamellae. Glutaraldehyde, Toluidine blue. **C.** High magnification of secondary lamellae of *Prochilodus lineatus* showing erythrocyte, pavement cell, ionocyte, pillar cells and basal membrane (arrow). Glutaraldehyde, Toluidine blue.; **D and E.** Mucous cells (arrows) distribution in the leading (**D**) and trailing (**E**) edge of the primary lamella of the mandi, *Pimelodus maculatus*. Bouin, PAS and Alcian blue; **F and G.** Rodlet cells (RC) in the primary lamellar epithelium of the lambari, *Astyanax fasciatus* (**F**) and lamellar epithelium of traira, *Hoplias malabaricus* (**G**). Glutaraldehyde, Toluidine blue. E, erythrocyte; I, ionocyte; PVC, pavement cell; PC, pillar cell; PL, primary lamella; SL, secondary lamella; RC, rodlet cell. Scale bars in μm (Photo: M.G. Paulino, S.E. Moron).

Color version at the end of the book

are rarely found in the interlamellar region of the primary lamellar epithelium (interlamellar epithelium) and in the secondary lamellar epithelium. These cells produce glycoproteins that have different chemical properties; the MC type can be identified by histochemical staining (periodic acid-Schiff and Alcian blue staining) and the frequency of each MC type may change with the environmental water quality, including salinity (Calabrò et al. 2005, Moron et al. 2009, Paulino et al. 2012a, b). The mucus layer on the epithelial surface, together with the glicocalix on the PVC, constitutes an interface between the gill and water, but this is not considered to be a barrier for water and ion diffusion. However, the glicocalix on the PVC and mucus contains surface negative charges that trap positive ions, forming an ionic gradient close to the cell surface that favor ion uptake, which may reduce ion loss (Handy et al. 1989).

Rodlet cells (RC) are secretory cells whose function is not yet well defined. This cell was first described by Thèlohan in 1892 as a parasite due to the thick capsule around the cell circumference (Mayberry et al. 1979); however, they are now considered to be an endogenous secretory cell (Mazon et al. 2007) that can appear in different organs, including the gills, with a potential anti-inflammatory/anti-infection function (Siderits and Bielek 2009, Matisz et al. 2010, Schultz et al. 2014, Dezfuli et al. 2015). Some studies have also examined the response of RCs in fish exposed to chemical substances, such as metals and herbicides (Hawkins et al. 1984, Dezfuli et al. 2003, Araujo and Borges 2015). RC has an oval-elongated shape, basal nucleus and cytoplasm, contains vacuoles, small mitochondria, ribosomes and endoplasmatic reticulum. This cell is characterized by a fibrous border under the plasma membrane and various dense secretory sacs with a high density rod in its center (Figs. 2F, G). In the gills, immature RCs are buried in the primary lamellar epithelium and mature RCs share the epithelial surface contacting water with PVC, MC and the ionocytes.

Neuroepithelial cells (NEC) are round in shape, have a central nucleus and are located in the trailing edge of the filamental epithelium on the basement membrane (serosal side of the filamental epithelium), either isolated or in clusters. Transmission electron microscopy revealed electron-dense vesicles in the cytoplasm (Dunel-Erb et al. 1982). The fluorescence emitted by these cells after a histochemical reaction with formaldehyde (Falck et al. 1962) indicated the presence of amines, the 5-hydroxytryptamine (5-HT), present in the respiratory tract of mammalian and other vertebrates. Physiological studies and a positive response to the anti-serotonin antibody (5-HT) and an antibody specific for a synaptic vesicle transmembrane protein (SV2) suggested that these cells may respond to aquatic hypoxia, or internally, detect blood chemistry changes (Perry and Reid 2002, Perry et al. 2009, Regan et al. 2011). Undifferentiated cells are localized in the inner cell layers, close to and directly on the basement membrane.

The ionocytes promote active ion transport in the fish gills and other organs involved in ion regulation and have the same general characteristics. Basically, the gill ionocytes are round or oval in shape, have a round nucleus usually in the central or basal cell region; they are easily identified in routine histological sections due to these morphological characteristics and the pale cytoplasm stained with eosin and toluidine blue (Figs. 2A, B). Transmission electron microscopy shows that, in teleost fish, these cells have a basolateral membrane forming numerous folds and an extended intracellular tubular system continuous with the basolateral membrane (Figs. 3A, B), while in elasmobranchs, the tubular system is not highly developed, but the basolateral membrane is deeply folded (Fig. 3D). The tubular system and the basolateral membrane are rich in Na^+/K^+-ATPase units (Figs. 3C, E). At light microscopy, the immunostaining Na^+/K^+-ATPase technique using mouse monoclonal antibody to Na^+/K^+-ATPase α-subunit (IgGα5, Developmental Studies Hybridoma Bank, The University of Iowa, USA) indicates the presence of such ATPase in the tubular system (Figs. 3C), which was confirmed by the immunogold technique of Na^+/K^+-ATPase, also using IgGα5 at TEM (Dang et al. 2000).

The main transporters involved in the Na^+ uptake and acid excretion (epithelial Na^+ channel, ENaC or NaC), Na^+/H^+ exchange (NHE), Na^+/Cl^- co-transporter (NCC), electrogenic Na^+/HCO_3^- co-transporter (eNBC or NBC), in the Cl^- uptake and base secretion, NCC, anion exchange (AE) and protein family SLC26a members, as well as in the Ca^{2+} uptake such as epithelial Ca^{2+} channel (ECaC or CaC) and Na^+/Ca^{2+} exchanger (NCX), are relatively similar in different species (Goss et al. 2005, Lin et al. 2006a, b, Marshall 2011, Hiroi and McCormick 2012). The enzymes of the basolateral plasma membrane, Na^+/K^+-ATPase (NKA) and Ca^{2+}-ATPase (PMCA) as well as the vacuolar-type H^+-ATPase (HA), actively participate in such ion transport by generating a driving force for ion uptake or excretion (Hirose 2003, Tresguerres 2016). Cytosolic carbonic anhydrase provides the ions H^+ and HCO_3^- for Na^+/H^+ and Cl^-/HCO_3^- exchange at the cellular apical surface to maintain ionic homeostasis (Hwang and Lee 2007, Lionetto et al. 2016).

Morphological ionocyte subtypes have been identified according to cell localization, electron density of the cytoplasm and apical surface membrane (Pisam et al. 1987, 2000, Pisam and Rambourg 1991, Goss et al. 1992, 1994, Laurent et al. 2006, Inokuchi et al. 2009, Fernandes et al. 2013, Carmo et al. 2018a), while functional ionocyte subtypes have been identified according to the presence of the different transporters in the cell membrane (Hiroi et al. 2008, Christensen et al. 2012, Wang et al. 2009, Katoh and Kaneko 2002, 2003), with both receiving different nomenclatures. A detailed comparison of

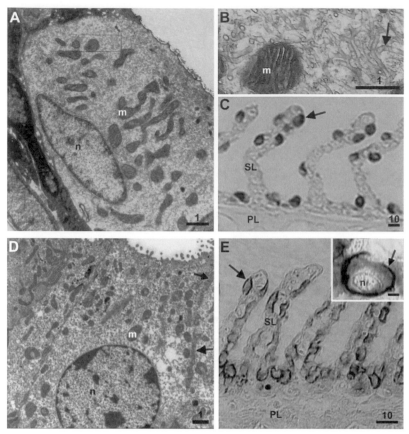

Fig. 3. Ionocytes. **A.** TEM micrographs of ionocytes of the loricariid fish, *Hypostomus Plecostomus* showing tubular system and mitochondria. **B.** High magnification of tubular system (arrow). **C.** Na$^+$/K$^+$-ATPase immunostaining (black stained cytoplasm, arrows) of ionocytes of *Prochilodus lineatus*; **D.** TEM micrographs of ionocytes of the freshwater stingray *Potamotrygon motoro* showing numerous folds of the cellular basolateral membrane (arrow); **E.** Na$^+$/K$^+$-ATPase immunostaining (basolateral membrane stained black, arrows) ionocytes of the stingray *Potamotrygon orbignyi*. Enlarged ionocyte in the upper corner, note intense immunostaining of basolateral membrane (arrow). m, mitochondria; n, nucleus; PL, primary lamella; SL, secondary lamella. Scale bars in µm (Photo: S.E. Moron, W.P. Duncan).

the variable nomenclature facing the recent advances on the molecular mechanisms of ion and acid–base regulation is discussed by Dymowska and colleagues (2012), who focus mainly on the ionocyte subtypes in rainbow trout (*Oncorhynchus mykiss*), killifish (*Fundulus heteroclitus*), tilapia (*Oreochromis species*), and zebrafish (*Danio rerio*) gills.

Marine Teleosts

In the marine teleost, the secondary lamella of gills is thin and the ionocytes are distributed in the primary lamellar epithelium close to the onset of secondary lamella and in the interlamellar epithelium. They have the concave apical surface in contact with water, forming an apical crypt and indenting their lateral surface with an adjacent accessory cell (AC) forming the ionocyte-AC complex. At light microscopic level, it is not easy to identify the ionocyte and the AC individually (Figs. 4A, B). Transmission electron microscopy revealed that AC has extended cytoplasmic processes embedded into the apical ionocyte cytoplasm, forming a mosaic (Laurent 1984). The cellular junctions between ionocyte and AC is characterized by short zonula occludens, usually called "leaky" junctions, which favor Na$^+$ extrusion, via a paracellular route, and accompanying the Cl$^-$ extrusion, via a transcellular route (Sardet et al. 1979, Hwang and Lee 2007,

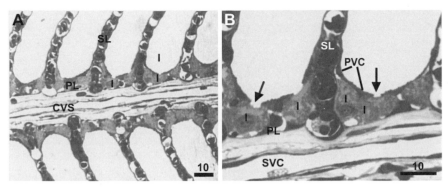

Fig. 4. Ionocytes of puffer fish *Sphoeroides spengleri*. **A.** Primary lamella and secondary lamellae. Note the thin water-blood distance in the secondary lamella and the absence of ionocytes in it. **B.** High magnification of ionocytes. Note the apical crypt (arrows). CVS, central venous sinus; I, ionocyte; PL, primary lamella; PVC, pavement cell; SL, secondary lamella. Alfac, toluidine blue. Scale bars in µm (Photo: M.M. Sakuragui).

Cozzi et al. 2015). The cellular adhesion between ionocytes and the PVC and PVC-PVC is characterized by long zonula ocludens referred to as "tight" junctions (Sardet et al. 1979, Chasiotis et al. 2012).

Freshwater Teleosts

In the freshwater teleost, the gill ionocytes have the same internal morphology as those of marine fish, but there is no AC associated with them. The primary lamellar epithelium consists of several cell layers, and the ionocyte constitutes a part of the outermost cell layer, although due to their large size, they penetrate into the inner cell layers (Figs. 2A, B). Ionocyte distribution occurs mainly in the primary lamellar epithelium (leading and trailing edges), close to and at the onset of the secondary lamella, as well as in the interlamellar epithelium (Figs. 3C and 5A). Their frequency is low in the borders of the primary lamellar epithelium in which MC are predominant. Multicellular ionocyte complexes are observed in some species, such as carp (*Cyprinus carpio*), tilapia (*Oreochromis mossambicus*), and ayu (*Plecoglossus altivelis*) (Hwang 1988). Depending on environmental conditions, ionocytes can also be found in the secondary lamellar epithelium. Hardness and ion concentration in water affect ion regulation and influence the ionocyte distribution throughout the gill epithelium (Goss et al. 1992, Laurent et al. 1992, Mitroviky and Perry 2009, Furukawa et al. 2011).

Fish living in hard and ion-rich water have ionocytes mainly distributed in the primary lamellar epithelium and interlamellar epithelium, while those living in soft and ion-poor water, have ionocytes distributed throughout the secondary lamellar epithelium (Perry and Laurent 1989, Moron et al. 2003). For example, the Brazilian continental waters are extremely soft and ion-poor, except for some saline lagoons in the Pantanal and coastal regions. Most Brazilian fish species have numerous ionocytes distributed throughout the secondary lamellar epithelium which increases the water-blood distance for gas exchange (Fernandes and Perna-Martins 2001, Alberto et al. 2005, Camargo et al. 2009, Fernandes et al. 2013, Costa et al. 2017a, b). Nevertheless, differences in the ionocyte distribution among species living in the same environment seem to be related to species strategy and efficiency of ionic-osmoregulation. In the Erythrinidae family, the traira (*Hoplias malabaricus*), a water-breathing species that occurs in diverse habitats from flowing clear water streams to slow turbid and hypoxic waters (Kenny 1995) and the jeju (*Hoplerythrinus unitaeniatus*), an air-breathing species which use the gas bladder for air respiration and inhabits slow turbid waters and swamps (Oyakawa 2003), have different ionocyte distribution in the gill epithelium when living in the same environmental water. *H. unitaeniatus* has a high ionocyte density in the primary and secondary lamella, while ionocytes are rare in the secondary lamella of *H. malabaricus* (Moron et al. 2003). Ion challenges such as exposure to deionized and hard water resulted in high proliferation of ionocytes in the secondary lamella of *H. malabaricus* without changes in the density of surface epithelium contacting water. Increased fractional surface area occurred only on the 1st and 2nd day of exposure to deionized water. In fish exposed to hard and ion-rich water, cell proliferation did not result in increased fractional surface area, although it caused Cl⁻ imbalance. These responses to ion challenges probably involved different ion transport mechanisms to ion regulation at these conditions.

The apical surface of the ionocytes represents less than 10% of the total surface area of primary lamellar epithelium contacting water, and this may change depending on environmental factors. Morphological apical surface sub-types are possible to distinguish, even in the same species, using scanning electron microscopy (Figs. 5A–D). Microvilli characterize the ionocyte surface of the erythrinids, *H. malabaricus* and *H. unitaeniatus*. Exposure to deionized water increases individual ionocyte apical surface area in *H. malabaricus* (Moron et al. 2003). A flat, slightly concave apical surface of the ionocyte with smooth or short microvilli is found in the serrasalmid fish, tambaqui (*Colossoma macropomum*) from the Amazon region; the exposure to nitrite induces a reduction in ionocyte apical surface area (Costa et al. 2017a). Ionocytes having microvilli and a sponge-like apical surface are found in *Prochilodus lineatus*; changes in water quality resulted in changes in the frequency of each ionocyte sub-type (Fig. 5B) (Camargo et al. 2009, Paulino et al. 2012a, Costa et al. 2017a). In the cichlid fish *Geophagus brasiliensis*, ionocytes have a convex apical surface (Fig. 5C) and in the loricariidae fish, *Hypostomus regani* and *Hypostomus tietensis*, ionocytes are characterized by a flat or convex apical irregular surface or short microvilli and a sponge-like apical surface (Fig. 5D).

The cellular adhesion between ionocyte and adjacent PVC in freshwater fish is characterized by a long ocludens zone, followed by a short adherens zone and one to four desmosomes linking the cells. Thus, the gill epithelial surface in freshwater fish becomes a barrier for ion diffusion, via a paracellular route. In *H. malabaricus*, the tight junctions between ionocyte and PVC reach up to 700 nm, short zonula adherens and numerous desmosome which possibly limit the passive ion loss (Moron and Fernandes 1996). Occludin protein and the claudins protein family present in the cellular junctions have been studied in teleost fish and seem to be involved in the gill permeability that determine the magnitude of paracellular solute movement (Chasiotis et al. 2010). The positive immunostaining of this protein of the PVC in the lamellae and apical junctions in ionocytes of goldfish suggests that occludin is associated with FW fish gill epithelium (Chasiotis and Kelly 2008). Acclimation of FW goldfish (*Carassius auratus*) to ion-poor water significantly elevated occludin mRNA and protein abundance in gill tissue (Chasiotis et al. 2009, 2012). Conversely, claudins, in which at least 32 isoforms are expressed in teleost, present isoform differential expression in marine and freshwater fish, resulting, more or less, in gill permeability (Chasiotis and Kelly 2011, Kelly and Chasiotis 2011).

Marine and Freshwater Elasmobranchs

Elasmobranchs are, in general, stenohaline marine species. Some species tolerate brackish water or full freshwater for long periods, migrating between both environments, and few species are freshwater stenohaline, such as the stingrays from the Amazon region in South America (Evans et al. 2004, Duncan et al. 2011, 2009). Marine elasmobranchs retain urea, having a high level of urea in plasma and extracellular fluids, to maintain osmolarity similar to sea water and those of freshwater lose plasma urea, but retain some urea in the perivisceral fluid (Duncan et al. 2009, Ballantyne and Fraser 2012). In marine elasmobranchs, the ion gains are mainly secreted by the rectal gland (Piermarine and Evans 2000, Pillans et al. 2008) and the gill ionocytes seem to be more related to acid–base regulation (Choe et al. 2007, Takabe et al. 2016). However, in freshwater elasmobranchs, the gills and kidneys are important organs for ion and water balance, as the gills are the site of ion uptake (Hazon et al. 2003, Piermarine and Evans 2000). Ion transporters have been identified in these osmoregulatory organs (Piermarini et al. 2002, Hyodo et al. 2004).

The gill primary and secondary lamellar epithelia are similar to those of teleosts and MC and ionocytes are distributed among PVC (Evans et al. 2005). The PVC architecture of the apical membrane varies, even within the same species, from microvilli to short microridges depending on their localization. In the gills of Amazon stingrays, PVC cells are called mucous PVC, as they correspond to similar cells described by Mallatt and colleagues (1995). They have numerous very small ovoid electron-dense mucous granules close to the apical membrane that are stained with Alcian blue and Schiff periodic acid (Duncan et al. 2010). MC are distributed throughout the primary lamellar's leading and trailing edges. Some large MC

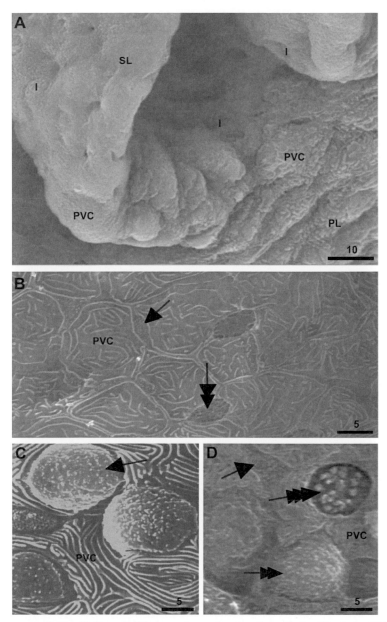

Fig. 5. SEM micrograph of gill epithelium. **A.** Primary and secondary lamellar epithelia of the loricariid fish, *Hypostomus plecostomus* showing ionocytes distributed in the interlamellar epithelium and secondary lamellar epithelium. **B.** Primary lamellar epithelium of *Prochilodus lineatus* showing two ionocyte types: type 1 with apical microvilli (arrow) and type 2 with a pical sponge-like structure (double arrow); **C.** Ionocytes in the primary lamella of the cichlid *Oreochromis niloticus* with convex apical surface (arrow); **D.** Ionocytes in the primary lamella of the loricariid *Hypostomus tietensis* showing three ionocyte types: type 1 with short apical microvilli (arrow), type 2, concex apical surface (double arrow) and type 3 with short large microvilli (triple arrow). PL, primary lamella; PVC, pavement cell; SL, secondary lamella. Scale bars in µm (Photo: S.E. Moron, M.M. Sakuragui).

arc distributed on the primary lamellar epithelium; they are stained with Alcian blue and Schiff periodic acid (Duncan et al. 2010). The ionocytes are distributed in the primary lamellar epithelium, interlamellar region and, depending on species and environment, they are found in the lamellar epithelium (Wilson et al. 2002, Duncan et al. 2010).

Ionocytes from elasmobranch marine species have a similar structure compared to sea water teleosts but have a less developed tubular system and do not have AC associated with them; ionocytes and PVC adhesion are characterized by tight junctions (Wilson et al. 2002). The ionocytes in freshwater elasmobranchs are similar to those of teleosts, but have a less developed tubular system (Figs. 3D, E). Ionocytes Na⁺-K⁺-ATPase immunoreactivity was intense at the cell periphery, suggesting basolateral membrane folds rich in Na⁺-K⁺-ATPase, and this is confirmed by transmission electron microscopy (Duncan et al. 2010). The ionocyte apical surface varied from irregular surface crypts to short microvilli (Wilson et al. 2002, Duncan et al. 2010). In the Amazon region, whose rivers are ion-poor and soft, the ionocytes are also distributed throughout the primary and secondary lamellar epithelium. Three to four ionocytes can be found grouped at the base of secondary lamella or isolated with a PVC between them. In the Negro River which has very acid water (pH 3.7–5.5), the cururu ray, *Potamotrygon wallacei* (Carvalho et al. 2016) (formely identified as *Potamotrygon* sp., and misidentified as *P.* cf. *thorsoni* or *P.* cf. *histrix*) which lives exclusively in this river presents follicular arrangement of 8–12 ionocytes distributed in the interlamellar epithelium (Fig. 6). They share a large apical pit, forming a single short channel which contacts the epithelial surface as a whole (Duncan et al. 2010). This arrangement was also described in the gill epithelium of the teleost *Tribolodon hakonensis* living in the acid water (pH 3.4–3.8) of Lake Osorezan in Japan (Hirata et al. 2003). In this species, the Na⁺-K⁺-ATPase concentration, carbonic anhydrase type II, sodium-hydrogen exchange type 3 and Na⁺-HCO₃⁻ exchange, type 1 co-transporters, as well as aquaporin 3, are upregulated.

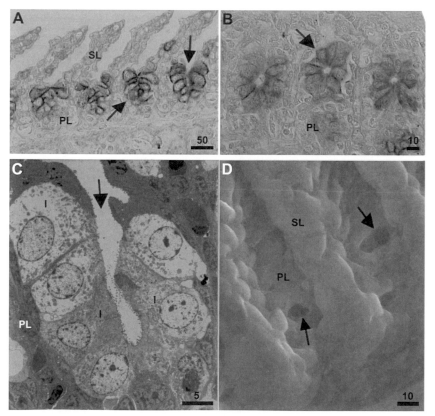

Fig. 6. *Potamotrygon wallacei.* Sagittal section **A.** and longitudinal **B.** section of primary lamellae. Na⁺/K⁺-ATPase immunostaining (basolateral membrane black stained) ionocytes showing the follicular arrangement of ionocytes (arrows) in the interlamellar epithelium. **C.** TEM micrograph of ionocytes sharing a large apical pit (arrow) in the primary lamella. **D.** SEM micrograph of the large apical pit (arrows) in the primary lamella formed by follicular arrangement of ionocytes. I, ionocyte; PL, primary lamella; SL, secondary lamella. Scale bars in μm (Photo: W.P. Duncan).

Euryhaline Fish

Euryhaline fish are those that tolerate changes in environmental salinity from ≤ 0.5 g L^{-1} (freshwater) to 30–40 g L^{-1} (seawater) by adjusting ionic and water transport to maintain osmotic and ionic homeostasis (Shultz and McCormick 2013). In general, they inhabit estuaries, living part of their life in freshwater and migrating to seawater or vice-versa or exhibiting short-term invasion of hypo- or hypersaline environments. Their tolerance to salinity changes varies with fish growth and species. Salinity tolerance is related to changes in hormones; the cortisol improves the survival in sea water and implies changes in the size and frequency of ionocytes (McCormick 2001). Other hormones are also involved in water and ionic regulation by gill, kidneys and intestinal ion fluxes, as well as drinking water, which include changes in transporter proteins and ionocyte type (Takei and McCormick 2012, McCornick 2013). The estuarine fish *Fundulus heteroclitus* (killifish) prefers salinities of 25 g L^{-1} but they are regularly exposed to rapid changes in salinity during tides. Exposure to diluted water caused changes in the membrane of ionocytes which is covered by pavement cells, resulting in a reduction of active Cl^- secretion and passive diffusive ion loss (Marshall 2003). The ionocyte density increased in hypersaline conditions and their apical membrane was exposed to the environment, increasing 100% the Cl^- secretion (Zadunaisky et al. 1995). Tilapia, *Oreochromis mossambicus*, when transferred from freshwater to seawater exhibited an increase in the ionocyte size and immunostaining intensity concomitant with an increasing Na^+/K^+-ATPase activity (Uchida et al. 2000), while other species show an increase in ionocyte size and/or number after transference from seawater to freshwater (Katoh et al. 2001, Lin et al. 2006b).

Immunohistochemistry of Na^+/K^+-ATPase in combination with cotransporters ($Na^+/K^+/2Cl^-$) and CFTR Cl^- channel provide evidence that ionocytes are functional during seawater acclimation and morphological variations in the ionocytes occur among teleost species. The development of specific isoform antibodies in freshwater and seawater allowed functional distinction of ionocytes comprising ion absorption and ion secretion in euryhaline species (Hiroi and McCormick 2012).

In euryhaline elasmobranchs, the acclimation to freshwater maintains NaCl and urea in plasma at high concentrations (Piermarini and Evans 1998, Pillans and Franklin 2004, Pillans et al. 2005, 2008). Elasmobranch gills also play important roles in body fluid regulation as a site of NaCl uptake in freshwater and as a barrier against the loss of ammonia and urea, in addition to acid-base regulation (Hazon et al. 2003, Takeda et al. 2016). Various transporters have been identified in these osmoregulatory organs (Piermarini et al. 2002), and NCC is probably one of the key molecules for hyper-osmoregulatory function in the elasmobranch gills after adaptation to low salinity (Takeda et al. 2016). The Atlantic stingray *Dasyatis sabina*, acclimated to freshwater, had greater number of Na^+/K^+-ATPase-rich cells in the gills than those acclimated to seawater and a higher activity of ouabain-sensitive Na^+/K^+-ATPase. Conversely, the Na^+/K^+-ATPase in the rectal gland was lower (Piermarini and Evans 2000). Hypersaline environmental conditions (up to 40‰) may occur in coastal lagoons having a high evaporation rate, low freshwater inputs or seawater exchange exceeding that of the open ocean (~ 34‰) (Potter et al. 2010). The juvenile bamboo shark, *Chiloscyllium punctatum*, exposed to 25, 34 and 40‰ for two weeks showed no change of Na^+/K^+-ATPase activity in the gills and rectal gland, as well as in H^+-ATPase (VHA), pendrin (Cl^-/HCO_3^- co-transporter) and the Na^+/H^+ exchanger isoform 3 (NHE3) in ionocytes of gill epithelium suggesting the limited role of gill ionocytes in maintaining Na^+ homeostasis over a high salinity range (Cramp et al. 2015).

Gill Plasticity

Gill plasticity has been intensively studied in euryhaline fish in sea water/fresh water adaptation (Kaneko et al. 2008, Bradshaw et al. 2012, Hiroi and McCormick 2012, Blair et al. 2016, 2017), after fish exposure to xenobiotics (Cerqueira and Fernandes 2002, Camargo et al. 2009, Shiogiri et al. 2012, Costa et al. 2017), as well as in response to hypoxia and changes in water temperature (Sollid et al. 2003, 2005, Sollid and Nilson 2006, Matey et al. 2008, Poleo et al. 2017). Primary and secondary lamellar epithelia proliferations, as well as ionocyte proliferation showing biochemical and morphological changes, were reported in fish acclimated to sea water and transferred to freshwater and vice-versa. In killifish (*Fundulus heteroclitus*) and tilapia (*Oreochromis niloticos*), AC developed after transfer to sea water, forming a cellular complex with the ionocytes and disappearing after transfer to freshwater. Changes in the ionocyte distribution, density, morphology of cellular apical surface, as well as in membrane co-transporters, are also reported (Shiraishi et al. 1997, Katoh et al. 2001, Hiroi et al. 2005, Kultz et al. 2013, Furukawa et al. 2015). Gill remodeling also includes fast interlamellar cell proliferation by 24 hr, persisting at 96 hr in the Arctic grayling (*Thymallus arcticus*), a strict freshwater salmonid from North America, after transfer to hypersaline water (17‰) (Blair et al. 2016). After returning to freshwater, *T. arcticus* recovered the osmotic disturbances and reverted the salinity-induced interlamellar cell proliferation to control levels following 24 hr in freshwater (Blair et al. 2017). Ionic challenge in freshwater fish such as exposure to deionized water induces ionocyte proliferation in the secondary lamellae and affects gas exchange by increasing water-blood diffusion distance and decreasing the arterial oxygen partial pressure under normoxia (Sakuragui et al. 2003).

Interlamellar cell proliferation and epithelial cell hyperthrophy are common responses of the gill epithelium to physical and chemical injuries due to xenobiotic exposure (Paulino et al. 2014, Costa et al. 2017a, b, Carmo et al. 2018b, Tavares et al. 2018) and are considered defense responses to any irritant by increasing the diffusion distance between water and blood (Mallat 1985). This response is reversible, and interlamellar cell mass tends to decrease after xenobiotic removal (Cerqueira and Fernandes 2002). Conversely, ionocyte hypertrophy and proliferation in the primary and secondary lamellae have been considered a compensatory response to ionic disturbance caused by xenobiotics (Mallat 1985, Carmo et al. 2018a).

A remarkable case was reported in carp (*Cyprinus carpio*) and goldfish (*Carassius auratus*) after exposure to hypoxia and changes in temperature in which gill remodeling are marked by intense reversible interlamellar cell proliferation, burying the secondary lamellae under normoxia and low temperature and protruding the secondary lamellae under hypoxia and high temperature, and also during exercise, suggesting a direct relationship to oxygen demand by reducing or increasing the respiratory surface area (Sollid et al. 2003, 2005, Sollid and Nilson 2006, Brauner et al. 2011, Fu et al. 2011). Ionocytes are dislocated on the outer edge of the interlamellar cell mass, which indicates a controlled process involving the migration of ionocytes during interlamellar cell proliferation (Mitrovic and Perry 2009). *C. auratus* exhibiting interlamellar proliferation have the Na^+ and Cl^- efflux and ammonia excretion decreasing in comparison to fish that did not present such cell proliferation (Mitrovic and Perry 2009, Bradshaw et al. 2012). In *Gymnocypris przewalskii*, a cold water carp species from Lake Qinghai, China, the exposure to hypoxia exhibits gill adjustments to increase O_2 uptake: reduction in the primary lamellar epithelium thickness, increasing protruded secondary lamellae and reduction of water-blood diffusion distance (Matey et al. 2008). However, such adjustments affected osmoregulation by reducing about 10–15% of the plasma Na^+ and Cl^- concentration. The ionocytes' apical surface changed from a large surface area and numerous microvilli to small cells with a flattened apical surface, suggesting that the extended gill remodeling observed in *C. carpio* and *C. auratus* may be related to fish activity and/or adaptation to a hypoxic environment (Matey et al. 2008).

Irreversible gill remodeling (Fig. 7) occurs, during development, in the obligate air-breathing fish, the pirarucu, *Arapaima gigas* which uses a modified swim bladder for O_2 uptake from the atmospheric air (Ramos et al. 2013). This species is endemic to the Amazon River Basin and one of largest freshwater fish (Val and Almeida-Val 1995). The gills of larvae of *A. gigas* (1 g, 7 to 9 days old) are similar to those of most teleost fish, and the fish is exclusively water breathing (Lüling 1964). The gills have the primary and secondary lamellae well developed up to 100 to 200 g (Costa et al. 2007, Ramos et al. 2013), but the protruded secondary lamellae became shorter in fish with 500 g (Figs. 7A, B) due to intense cell proliferation in the primary and secondary lamellae increasing epithelial thickness and decreasing interlamellar distance until it becomes almost vestigial in fish larger than 1 kg (Fig. 7C) (Ramos et al. 2013). In fish weighing from 1 to 200 g, the ionocytes are distributed in the interlamellar epithelium close to the base of the secondary lamella and, in fish larger than 600 g, the ionocytes are distributed throughout the primary lamellae occupying the outermost cell layer of primary and secondary lamellar epithelia (Figs. 8A–C). Most ionocytes exhibited an apical crypt which was almost covered by the pavement cells (Fig. 8D). They are round or elongated in shape and have a basal nucleus and the same characteristics as in other teleosts:

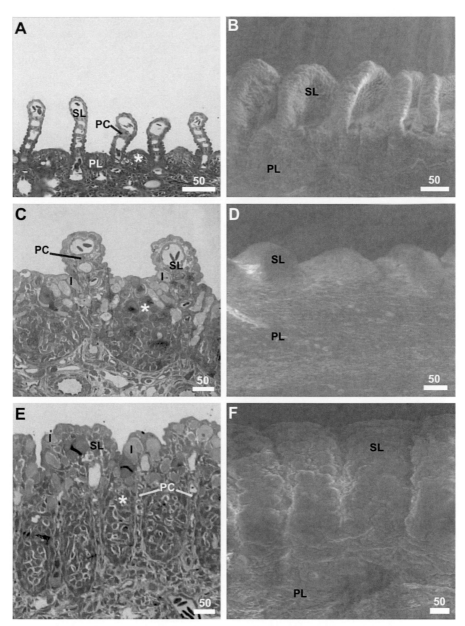

Fig. 7. Gill remodeling in *Arapaima gigas* during development. Primary and secondary lamellae of a 105 (**A** and **B**), 575 (**C** and **D**) and 1.343 (**E** and **F**) g-fish. Note the thickness of primary lamellar epithelium (*) in **A**, **C** and **E** and secondary lamellae burying into primary lamella in **B**, **D** and **F**. I, ionocyte; PC, pillar cell; PL, primary lamella; SL, secondary lamella. Scale bars in μm (Photo: C.A. Ramos).

Color version at the end of the book

numerous mitochondria, a well-developed tubular system and apical surface exhibiting short microvilli (Figs. 8E, F). During gill remodeling, extended paracellular channels appear between epithelial cells (Figs. 8E, F), which are narrow in the outermost epithelial cell layer and become large near the basement membrane. Although the ionocytes maintain contact with environmental water, the thick epithelium may affect ion uptake, as in *C. auratus* (Bradshaw et al. 2012); however, despite the interlamellar cell proliferation, the influx and efflux rates of Na$^+$ in the gills of *A. gigas* are higher in larger fish than they

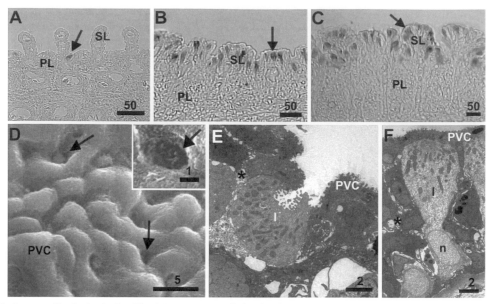

Fig. 8. Ionocyte distribution in the gills of *Arapaima gigas* during development. **A, B and C.** Na$^+$/K$^+$-ATPase immunostaining (black stained cytoplasm, arrows) ionocytes in the primary and secondary lamellae of gills of a 105 (**A**), 575 (**B**) and 1,343 (**C**) g-fish *Arapaima gigas*; **D.** SEM micrograph of primary lamellar epithelium showing ionocytes (arrows) among pavement cells. Enlarged ionocyte apical crypt in the upper corner; **E** and **F.** TEM micrographs of ionocytes with round (**E**) and elongated (**F**) shape. Note extended paracellular channels (*) among epithelial cells. I, ionocyte; n, nucleus; PL, primary lamella; SL, secondary lamella; PVC, pavement cell. Scale bars in μm (Photo: C.A. Ramos).

are in smaller fish (Gonzales et al. 2010). The short tight junctions between pavement and ionocytes and the paracellular channels contribute to ion and/or gas diffusion across the epithelium.

Acknowledgments

I acknowledge the grants from São Paulo Research Foundation (FAPESP), National Council of Scientific and Technological Development (CNPq), National Institute of Science and Technology – Aquatic Toxicology (CNPq/INCT) and Coordination of Superior Level Staff Improvement (CAPES) for supporting the original work by M.N. Fernandes and collaborators cited in this chapter.

References Cited

Alberto, A., A.F.M. Camargo, J.R. Verani, O.F.T. Costa and M.N. Fernandes. 2005. Health variables and gill morphology in the tropical fish *Astyanax fasciatus* from a sewage-contaminated river. Ecotoxicol. Environ. Safe 61: 247–255.

Araujo, N.S. and J.C.S. Borges. 2015. Rodlet cells changes in *Oreochromis niloticus* in response to organophosphate pesticide and their relevance as stress biomarker in teleost fishes. Int. J. Aquat. Biol. 3: 398–408.

Baldissetotto, B. 2003. Osmoregulatory anaptations of freshwater teleosts, pp 179–201. *In*: A.L. Val and B.G. Kapoor (eds.). Fish Adaptations. Science Publishers, Enfield, NH, USA.

Ballantyne, J.S. and D.I. Fraser. 2012. Euryhaline elasmobranchs. pp. 125–198. *In*: S.D. McCormick, A.P. Farrell and C.J. Brauner (eds.). Fish Physiology. Euryhaline Fishes. Academic Press. San Diego, USA.

Bassi, M., W. Klein, M.N. Fernandes, S.F. Perry and M.L. Glass. 2005. Pulmonary oxygen diffusing capacity of the South American lungfish *Lepidosiren paradoxa*: Physiological values by the Bohr method. Physiol. Biochem. Zool. 78: 560–569.

Bernal, D., J.K. Carlson, K.J. Goldman and C.G. Lowe. 2012. Energetics, metabolism and endothermy in sharks and rays. pp. 211–237. *In*: J.C. Carries, J.A. Musick and M.R. Heithaus (eds.). Biology of Sharks and their Relatives. 2nd Ed. CRC Press, Boca Raton, USA.

Blair, S.D., D. Matheson, Y. He and G.G. Goss. 2016. Reduced salinity tolerance in the Arctic grayling (*Thymallus arcticus*) is associated with rapid development of a gill interlamellar cell mass: implications of high-saline spills on native freshwater salmonids. Conserv. Physiol. 4. doi:10.1093/conphys/cow010.

Blair, S.D., D. Matheson and G.G. Goss. 2017. Physiological and morphological investigation of Arctic grayling (*Thymallus arcticus*) gill filaments with high salinity exposure and recovery. Conserv. Physiol. 5: cox040. doi:10.1093/conphys/cox040.

Bradshaw, J.C., Y. Kumai and S.F. Perry. 2012. The effects of gill remodeling on transepithelial sodium fluxes and the distribution of presumptive sodium-transporting ionocytes in goldfish (*Carassius auratus*). J. Comp. Physiol. B 182: 351–366.

Brauner, C.J., V. Matey, W. Zhang, J.G. Richards, R. Dhillon, Z.D. Cao and Y. Wang. 2011. Gill remodeling in crucian carp during sustained exercise and the effect on subsequent swimming performance. Physiol. Biochem. Zool. 84: 535–542.

Calabrò, C., M.P. Albanese, E.R. Lauriano, S. Martella and A. Licata. 2005. Morphological, histochemical and immunohistochemical study of the gill epithelium in the abyssal teleost fish *Coelorhynchus coelorhynchus*. Folia Histoch. Cytobiol. 43: 51–56.

Camargo, M.M.P., M.N. Fernandes and C.B.R. Martinez. 2009. How aluminium exposure promotes osmoregulatory disturbances in the neotropical freshwater fish *Prochilodus lineatus*. Aquat. Toxicol. 94: 40–46.

Carmo, T.L.L., V.C. Azevedo, P.R. Siqueira. T.D. Galvão, F.A. Santos, C.B.R. Martinez, C.R. Appoloni and M.N. Fernandes. 2018a. Mitochondria-rich cells adjustments and ionic balance in the Neotropical fish *Prochilodus lineatus* exposed to titanium dioxide nanoparticles. Aquat. Toxicol. 200: 168–177.

Carmo, T.L.L., V.C. Azevedo, P.R. Siqueira, T.D. Galvão, F.A. Santos, C.B.R. Martinez, C.R. Appoloni and M.N. Fernandes. 2018b. Reactive oxygen species and others biochemical and morphological biomarkers in the gills and kidneys of the Neotropical freshwater fish, *Prochilodus lineatus*, exposed to titanium dioxide (TiO$_2$) nanoparticles. Environ. Sci. Pollut. Res. https://doi.org/10.1007/s11356-018-2393-4.

Carvalho, M.R., R.S. Rosa and M.L.G. Araujo. 2016. A new species of Neotropical freshwater atingray (Chondrichthyes: Potamotrygonidae) from the Rio Negro, Amazonas, Brazil: the smallest species of Potamotrygon. Zootaxa 4107: 566–586.

Cerqueira, C.C.C. and M.N. Fernandes. 2002. Gill tissue recovery after copper exposure and blood parameter responses in the tropical fish *Prochilodus scrofa*. Ecotoxicol. Environ. Safe 52: 83–91.

Chasiotis, H. and S.P. Kelly. 2008. Occludin immunolocalization and protein expression in goldfish. J. Exp. Biol. 211: 1524–1594.

Chasiotis, H., J. Effendi and S.P. Kelly. 2009. Occludin expression in epithelia of goldfish acclimated to ion poor water. J. Comp. Physiol. B 179: 145–154.

Chasiotis, H. and S.P. Kelly. 2011. Effect of cortisol on permeability and tight junction protein transcript abundance in primary cultured gill epithelia from stenohaline goldfish and euryhaline trout. Gen. Comp. Endocrinol. 172: 494–504.

Chasiotis, H., D. Kolosov, P. Bui and S.P. Kelly. 2012. Tight junctions, tight junction proteins and paracellular permeability across the gill epithelium of fishes: A review. Resp. Physiol. & Neurobiol. 184: 269–281.

Choe, K.P., S.L. Edwards, J.B. Claiborne and D.H. Evans. 2007. The putative mechanism of Na$^+$ absorption in euryhaline elasmobranchs exists in the gills of a stenohaline marine elasmobranch, *Squalus acanthias*. Comp. Biochem. Physiol. A 146: 155–162.

Coolidge, E., M.S. Hedrick and M.K. Milsom. 2007. Ventilatory systems. pp. 181–211. *In*: D.J. McKenzie, A.P. Farrell and C.J. Brauner (eds.). Fish Physiology. Primitive Fishes. Academic Press. San Diego, USA.

Copeland, D.E. 1948. The cytological basis of chloride transferrin the gills of *Fundulus heteroclitus*. J. Morphol. 82: 201–228.

Costa, O.T.F., A.C. Pedreti, A. Schmitz, S.F. Perry and M.N. Fernandes. 2007. Stereological estimation of surface area and barrier thickness of fish gills in vertical sections. J. Microsc. 225: 1–9.

Costa, O.T.F., C.A. Ramos, W.P. Duncan, J.L.V. Lameiras and M.N. Fernandes. 2017a. Mitochondria-rich cells changes induced by nitrite exposure in tambaqui (*Colossoma macropomum* Cuvier, 1818). An. Acad. Bras. Ciênc. 89: 965–972.

Costa, S.T., L.T. Gressler, F.J. Sutili, L. Loebens, M.N. Fernandes, R. Lazzari and B. Baldisserotto. 2017b. Humic acid of commercial origin causes changes in gill morphology of silver catfish *Rhamdia quelen* exposed to acidic water. J. Exp. Zool. 327: 504–512.

Cozzi, R.R.F., G.N. Robertson, M. Spieker, L.N. Claus, G.M.M. Zaparilla, K.L. Garrow and W.S. Marshall. 2015. Paracellular pathway remodeling enhances sodium secretion by teleost fish in hypersaline environments. J. Exp. Biol. 218: 1259–1269.

Cramp, R.L., M.J. Hansen and C.E. Franklin. 2015. Osmoregulation by juvenile brown-banded bamboo sharks, *Chiloscyllium punctatum*, in hypo- and hyper-saline waters. Comp. Biochem. Physiol. A 185: 107–114.

Crampton, W.G.R., L.J. Chapman and J. Bell. 2008. Interspecific variation in gill size is correlated to ambient dissolved oxygen in the Amazonian electric fish *Brachyhypopomus* (Gymnotiformes: Hypopomidae). Env. Biol. Fish 83: 223–235.

Cruz, A.L., H.R. Silva, L.M. Lundstedt, A.R. Schwantes, G. Moraes, W. Klein and M.N. Fernandes. 2013. Air-breathing behavior and physiological responses to hypoxia and air exposure in the air-breathing loricariid fish, *Pterygoplichthys anisitsi*. Fish Physiol. Biochem. 39: 243–256.

Curtis, J. and C.M. Wood. 1991. The function of the urinary bladder *in vivo* in the freshwater rainbow trout. J. Exp. Biol. 155: 567–583.

Dang, Z.C., P.H.M. Balm, G. Flick and S.E. Wendelaar Bonga. 2000. Cortisol increases NA$^+$/K$^+$-ATPase density in plasma membranes of gill chloride cells in the freshwater tilapia *Oreochromis mossambicus*. J. Exp. Bio. 203: 2349–55.

Dezfuli, B.S., L. Giari, E. Simoni, D. Palazzi and M. Manera. 2003. Alteration of rodlet cells in chub caused by the herbicid Stam-M-4 (Propanil). J. Fish Biol. 63: 232–239.

Dezfuli, B.S., M. Manera, L. Giari, J.A. Pasquale and G. Bosi. 2015. Occurrence of immune cells in the intestinal wall of *Squalius cephalus* infected with *Pomphorhynchus laevis*. Fish & Shellfish Immunol. 47: 556–564.

Doyle, W.H. and D. Gorecki. 1961. The so-called chloride cell of the fish gill. Physiol. Zool. 34: 81–85.

Duncan, W.P., O.T.F. Costa, M.L.G. Araújo and M.N. Fernandes. 2009. Ionic regulation and Na$^+$/K$^+$-ATPase activity in gills and kidney of the freshwater stingray *Paratrygon aiereba* living in white and black waters in the Amazon Basin. J. Fish Biol. 74: 956–960.

Duncan, W.P., O.T.F. Costa, M.M. Sakuragui and M.N. Fernandes. 2010. Functional morphology of the gill in Amazonian freshwater stingrays (Chondrichthyes: Potamotrygonidae): Implications for adaptation to freshwater. Physiol. Biochem. Zool. 83: 19–32.

Duncan, W.P., N.F. Silva and M.N. Fernandes. 2011. Mitochondrion-rich cells distribution, Na$^+$/K$^+$-ATPase activity and gill morphometry of the Amazonian freshwater stingrays (Chondrichthyes: Potamotrygonidae). Fish Physiol. Biochem. 37: 523–531.

Duncan, W.P., M.I. Silva and M.N. Fernandes. 2015. Gill dimensions in near-term embryos of Amazonian freshwater stingrays (Elasmobranchii: Potamotrygonidae) and their relationship to the lifestyle and habitat of neonatal pups. Neotrop. Ichthyol. 13: 123–136.

Dunel-Erb, S., Y. Bailly and P. Laurent. 1982. Neuroepithelial cells in fish gill primary lamellae. J. Appl. Physiol. 53: 1342–1353.

Dymowska, A.K., P.-P. Hwang and G.G. Goss. 2012. Structure and function of ionocytes in the freshwater fish gill. Resp. Physiol. & Neurobiol. 184: 282–292.

Edwards, S.L. and W.S. Marshall. 2013. Principles and patterns of osmoregulation and euryhalinity in fishes. pp. 1–44. *In*: S.D. McCormick, A.P. Farrell and C.J. Brauner (eds.). Fish Physiology. Euryhaline Fishes. Academic Press. San Diego, USA.

Evans, D.H., P.M. Piermarine and K.P. Choe. 2004. Homeostasis: osmoregulation, pH regulation and nitrogen excretion. pp. 247–268. *In*: J.C. Carrier, J.A. Musick and M.R. Heithaus (eds.). Biology of Sharks and their Relatives. CRC Press, Boca Raton, USA.

Evans, D.H., P.M. Piermarini and K.P. Choe. 2005. The multifunctional fish gill: dominant site of gas exchange, osmoregulation, acid–base regulation, and excretion of nitrogenous waste. Physiol. Rev. 85: 97–177.

Enders, E.C. and D. Boisclair. 2016. Effects of environmental fluctuations on fish metabolism: Atlantic salmon *Salmo salar* as a case study. J. Fish Biol. 88: 344–358.

Falck, B., N.A. Hillarp, G. Thieme and A. Torp. 1962. Fluorescence of catecolamines and related compounds condensed with formaldehyde. J. Histochem. 10: 348–354.

Fernandes, M.N., F.T. Rantin, A.L. Kalinin and S.E. Moron. 1994. Comparative study of gill dimensions of three erythrinid species in relation to their respiratory function. Can. J. Zool. 72: 160–165.

Fernandes, M.N., S.A Perna and S.E. Moron. 1995a. Chloride cell apical surface changes in gill epithelia of the armoured catfish *Hypostomus plecostomus* during exposure to distilled water. J. Fish Biol. 52: 844–849.

Fernandes, M.N., S.A. Perna, C.T.C. Santos and W. Severi. 1995b. The gill filament muscles in two loricariid fish (genus *Hypostomus* and *Rhinelepis*). J. Fish Biol. 46: 1082–1085.

Fernandes, M.N. and S.A. Perna-Martins. 2001. Epithelial gill cells in the armored catfish, *Hypostomus* CF. *plecostomus* (Loricariidae). Rev. Brasil Biol. 61: 69–78.

Fernandes, M.N., A.L. Cruz, O.T.F. Costa and S.F. Perry. 2012. Morphometric partitioning of the respiratory surface area and diffusion capacity of the gills and swim bladder in juvenile Amazonian air-breathing fish, *Arapaima gigas*. Micron. 43: 961–970.

Fernandes, M.N., M.G. Paulino, M.M. Sakuragui, C.A. Ramos, C.D.S. Pereira and H. Sadauskas-Henrique. 2013. Organochlorines and metals induce changes in the mitochondria-rich cells of fish gills: An integrative field study involving chemical, biochemical and morphological analyses. Aquat. Toxicol. 126: 180–190.

Fu, S.J., C.J. Brauner, Z.D. Cao, J.G. Richards, J.L. Peng, R. Dhillon and Y.W. Wang. 2011. The effect of acclimation to hypoxia and sustained exercise on subsequent hypoxia tolerance and swimming performance in goldfish (*Carassius auratus*). J. Exp. Biol. 214: 2080–2088.

Furukawa, F., S. Watanabe, M. Inokuchi and T. Kaneko. 2011. Responses of gill mitochondria-rich cells in Mozambique tilapia exposed to acidic environments (pH 4.0) in combination with different salinities. Comp. Biochem. Physiol. A 158: 468–476.

Genz, J., A.J. Esbaugh and M. Grosell. 2011. Intestinal transport following transfer to increased salinity in an anadromous fish (*Oncorhynchus mykiss*). Comp. Biochem. Physiol. A 159: 150–158.

Gonzalez, R.J., C.J. Brauner, Y.X. Wang, J.G. Richards, M.L. Patrick, W. Xi, V. Matey and A.L. Val. 2010. Impact of ontogenetic changes in branchial morphology on gill function in *Arapaima gigas*. Physiol. Biochem. Zool. 83: 322–332.

Goss, G.G. and C.M. Wood. 1990. Na$^+$ and Cl$^-$ uptake kinetics, diffusive effluxes and acidic equivalent fluxes across the gills of rainbow trout: 1. Responses to environmental hyperoxia. J. Exp. Biol. 152: 521–547.

Goss, G.G., P. Laurent and S.F. Perry. 1992. Evidence for a morphological component in acid–base regulation during environmental hypercapnia in the brown bullhead (*Ictalurus nebulosus*). Cell Tissue Res. 268: 539–552.

Goss, G.G., P. Laurent and S.F. Perry. 1994. Gill morphology during hypercapnia in brown bullhead (*Ictalurus nebulosus*)—role of chloride cells and pavement cells in acid–base regulation. J. Fish Biol. 45: 705–718.

Goss, G.G., E.E. Orr and F. Katoh. 2005. Characterization of SLC26 anion exchanger in rainbow trout. Comp. Biochem. Physiol. A 141: S197.

Graham, J.B. 1997. Air-breathing Fishes. Evolution, Diversity and Adaptation. Academic Press, San Diego, USA.

Graham, J.B., H.J. Lee and N.C. Wegner. 2007. Transition from water to land in extant group of fishes: Air breathing and the acquisition sequence of adaptations for amphibious life in oxudercine gobies. pp. 255–288. *In*: M.N. Fernandes, M.L. Glass, F.T. Rantin and B.G. Kapoor (eds.). Fish Respiration and Environment. Science Publishers, Enfield, NH, USA.

Grosell, M. 2006. Intestinal anion exchange in marine fish osmoregulation. J. Exp. Biol. 209: 2813–2827.

Grosell, M. 2007. Intestinal transport processes in marine fish osmoregulation. pp. 333–357. *In*: B. Baldisserotto, J.M. Mancera and B.G. Kapoor (eds.). Fish Osmorregulation. Science Publishers, Enfield, USA.

Handy, R.D., F.B. Eddy and G. Romain. 1989. *In vitro* evidence for the ionoregulatory role of rainbow trout mucus in acid, acid/aluminum and zinc toxicity. J. Fish Biol. 35: 737–747.

Hawkins, W.E. 1984. Ultrastructure of rodlet cells: response to cadmium damage in the kidney of the spot *Leiostomus xanthurus* Lacépède. Gulf Res. Rep. 7: 365–372.

Hazon, N., A. Wells, R.D. Pillans, J.P. Good, W.G. Anderson and C.E. Franklin. 2003. Urea based osmoregulation and endocrine control in elasmobranch fish with special reference to euryhalinity. Comp. Biochem. Physiol. B 136: 685–700.

Hirata, T., T. Kaneko, T. Ono, T. Nakazato, N. Furukawa, S. Hasegawa, S. Wakabayashi, M. Shigekawa, M.-H. Chang, M.F. Romero and S. Hirose. 2003. Mechanism of acid adaptation of a fish living in a pH 3.5 lake. Am. J. Physiol. 284: R1199–R1212.

Hiroi, J., S.D. McCormick, R. Ohtani-Kaneko and T. Kaneko. 2005. Functional classification of mitochondrion-rich cells in euryhaline Mozambique tilapia (*Oreochromis mossambicus*) embryos, by means of triple immunofluorescence staining for Na+/K+-ATPase, Na+/K+/2Cl– cotransporter and CFTR anion channel. J. Exp. Biol. 208: 2023–2036.

Hiroi, J., S. Yasumasu, S.D. McCormick, P.P. Hwang and T. Kaneko. 2008. Evidences for an apical Na+–Cl– cotransporter involved in ion uptake in a teleost fish. J. Exp. Biol. 211: 2584–2599.

Hiroi, J. and S.D. McCormick. 2012. New insights into gill ionocyte and ion transporter junction in euryhaline and diadromous fish. Resp. Physiol. Neurobiol. 184: 257–268.

Hirose, S., T. Kaneko, N. Naito and Y. Takei. 2003. Molecular biology of major components of chloride cells. Comp. Biochem. Physiol. B. 136: 593–620.

Hughes, G.M. 1984. General anatomy of the gills. pp. 1–72. *In*: W.S. Hoar and D.J. Randall (eds.). Fish Physiology, v. XA. Academic Press, Orlando, USA.

Hwang, P.P. 1988. Multicellular complex of chloride cells in the gills of freshwater teleosts. J. Morphol. 196: 15–22.

Hwang, P.P. and T.H. Lee. 2007. New insights into fish ion regulation and mitochondria-rich cells. Comp. Biochem. Physiol. A 148: 479–497.

Hwang, P.P., T.H. Lee and L.Y. Lin. 2011. Ion regulation in fish gills: recent progress in the cellular and molecular mechanisms. Am. J. Physiol. Regul. Integr. Comp. Physiol. 301: R28–R47.

Hyodo, S., F. Katoh, T. Kaneko and Y. Takei. 2004. A facilitative urea transporter is localized in the renal collecting tubule of dogfish *Triakis scyllia*. J. Exp. Biol. 207: 347–356.

Hyodo, S., K. Kakumura, W. Takagi, K. Hasegawa and Y. Yamaguchi. 2014. Morphological and functional characteristics of the kidney of cartilaginous fishes: with special reference to urea reabsorption. Am. J. Physiol. Regul. Integr. Comp. Physiol. 307: R1381–R1395.

Inokuchi, M., J. Hiroi, S. Watanabe, P.P. Hwang and T. Kaneko. 2009. Morphological and functional classification of ion-absorbing mitochondria-rich cells in the gills of Mozambique tilapia. J. Exp. Biol. 212: 1003–1010.

Karakatsouli, N., K. Tarnaris, C. Balaskas and S.E. Papoutsoglou. 2006. Gill area and dimensions of gilthead sea bream *Sparus aurata* L. J. Fish Biol. 69: 291–299.

Katoh, F., S. Hasegawa, J. Kita, Y. Takagi and T. Kaneko. 2001. Distinct seawater and freshwater types of chloride cells in killifish, *Fundulus heteroclitus*. Can. J. Zool. 79: 822–829.

Katoh, F. and T. Kaneko. 2002. Effects of environmental Ca2+ levels on branchial chloride cell morphology in freshwater-adapted killifish *Fundulus heteroclitus*. Fish Sc. 68: 347–355.

Katoh, F and T. Kaneko. 2003. Short-term transformation and long-term replacement of branchial chloride cell in killifish transferred from seawater to freshwater, revealed by morphofunctional observations and newly established "time differential double fluorescent staining" technique. J. Exp. Biol. 206: 4113–4123.

Kelly, S.P. and H. Chasiotis. 2011. Glucocorticoid and mineralocorticoid receptors regulate paracellular permeability in a primary cultured gill epithelium. J. Exp. Biol. 214: 2308–2318.

Kenny, J.S. 1995. Views from the bridge: a memoir on the freshwater fishes of Trinidad. Julian S. Kenny, Maracas, St. Joseph, Trinidad and Tobago, 98 p.

Kaneko, T., S. Watanabe and K.M. Lee. 2008. Functional morphology of mitochondrion-rich cells in euryhaline and stenohaline teleosts. Aqua-BioSci. Monogr. (ABSM) 1: 1–62.

Keys, A. and E.N. Willmer. 1932. "Chloride secreting cells" in the gills of fishes, with special reference to the common eel. J. Physiol. 76: 368–378.

Laurent, P. 1984. Gill internal morphology. pp. 73–183. *In*: W.S. Hoar and D.J. Randall (eds.). Fish Physiology, v. XA. Academic Press, Orlando, USA.

Laurent, P., C. Chevalier and C.M. Wood. 2006. Appearance of cuboidal cells in relation to salinity in gills of *Fundulus heteroclitus*, a species exhibiting Na⁺ but not Cl⁻ uptake in freshwater. Cell Tissue Res. 325: 481–492.

Lin, L.Y., J.L. Horng, J.G. Kunkel and P.P. Hwang. 2006a. Proton pump-rich cell secretes acid in skin of zebrafish larvae. Am. J. Physiol. Cell Physiol. 290: C371–C378.

Lin, Y.M., C.N. Chen, T. Yoshinaga, S.C. Tsai, I.D. Shen and T.H. Lee. 2006b. Short-term effects of hyposmotic shock on Na⁺/K⁺-ATPase expression in gills of the euryhaline milkfish, *Chanos chanos*. Comp. Biochem. Physiol. A 143: 406–415.

Lionetto, M.G., R. Caricato, M.E. Giordan and T. Schettino. 2016. The complex relationship between metals and carbonic anhydrase: new insights and perspectives. Int. J. Mol. Sci. 17: 127.

Lüling, K. 1964. Zur biologte und ökologte von *Arapaima gigas* (Pisces, Osteoglossidae). Zoomorphol. 54: 436–530.

Mallat, J. 1985. Fish gill structural changes induced by toxicants and other irritants: a statistical review. Can. J. Fish Aquat. Sci. 42: 630–648.

Mallatt, J., J.F. Bailey, S.J. Lampa, M.A. Evans and W. Tate. 1995. Quantitative ultrastructure of gill epithelial cells in the larval lamprey *Petromyzon marinus*. Can. J. Fish. Aquat. Sci. 52: 1150–1164.

Mancera, J.M. and D. McCormick. 2007. Role of prolactin, growth hormone, insulin-like growth factor and cortisol in teleost osmoregulation. pp. 497–515. *In*: B. Baldisserotto, J.M. Mancera and B.G. Kapoor (eds.). Fish Osmorregulation. Science Publishers, Enfield, NY, USA.

Marshall, W.S. 2003. Rapid regulation of NaCl secretion by estuarine teleost fish: coping strategies for short-duration freshwater exposures. Bioch. Bioph. Acta 1618: 95–105.

Marshall, W.S. 2011. Mechanosensitive signalling in fish gill and other ion transporting epithelia. Acta Physiol. 202: 487–499.

Matisz, C.E., C.P. Goater and D. Bray. 2010. Density and maturation of rodlet cells in brain tissue of fathead minnows (*Pimephales promelas*) exposed to trematode cercariae. Int. J. Parasitol. 40: 307–312.

Matey, V., J.G. Richards, Y. Wang, C.M. Wood, J. Rogers, R. Davies, B.W. Murray, X.-Q. Chen, J. Du and C.J. Brauner. 2008. The effect of hypoxia on gill morphology and ionoregulatory status in the Lake Qinghai scaleless carp, *Gymnocypris przewalskii*. J. Exp. Biol. 211: 1063–1074.

Mayberry, L.F., A.A. Marchiondo, J.E. Ubelaker and D. Kazic. 1979. *Rhabdospora thelohani* (Laguesse, 1895) (Apicomplexa): new host and geographic records with taxonomis consideration. J. Protozool. 26: 168–178.

Mazon, A.F., M.N. Fernandes, M.A. Nolasco and W. Severi. 1998. Functional morphology of gills and respiratory area of two active rheophilic fish species, *Plagioscion squamosissimus* and *Prochilodus scrofa*. J. Fish Biol. 52: 50–61.

Mazon, A.F., M.O. Huising, A.J. Taverne-Thiele, J. Bastiaans and B.M.L. Verburg-van Kemenade. 2007. The first appearance of rodlet cells in carp (*Cyprinus carpio* L.) ontogeny and their possible roles during stress and parasite infection. Fish Shcllf. Immunol. 22: 27–37.

McCormick, S.D. 2001. Endocrine control of osmoregulation in teleost fish. Am. Zool. 41: 781–794.

McCormick, S.D. 2013. Smolt physiology and endocrinology. pp. 199–251. *In:* S.D. McCormick, A. Farrell and C. Brauner (eds.). Fish Physiology. Euryhaline Fishes, v. 32. Academic Press, New York, USA.

Mitrovic, D. and S.F. Perry. 2009. The effects of thermally induced gill remodeling on ionocyte distribution and branchial chloride fluxes in goldfish (*Carassius auratus*). J. exp. Biol. 212: 843–852.

Moraes, M.F.P.G., S. Höller, O.T.F. Costa, M.L. Glass, M.N. Fernandes and S.F. Perry. 2005. Morphometric comparison of the respiratory organs of the South American lungfish, *Lepidosiren paradoxa* (Dipnoi). Physiol. Biochem. Zool. 78: 546–559.

Moron, S.A. and M.N. Fernandes. 1996. Pavement cell ultrastructural differences on *Hoplias malabaricus* gill epithelia. J. Fish Biol. 49: 357–362.

Moron, S.E., E.T. Oba, C.A. Andrade and M.N. Fernandes. 2003. Chloride cell responses to ion chalenge in two tropical freshwater fish, the erythrinids *Hoplias malabaricus* and *Hoplerythrinus unitaeniatus*. J. Exp. Zool. 298A: 93–104.

Moron, S.E., C.A. Andrade and M.N. Fernandes. 2009. Response of mucous cells of the gills of traíra (*Hoplias malabaricus*) and jeju (*Hoplerythrinus unitaeniatus*) (Teleostei: Erythrinidae) to hypo- and hyper-osmotic ion stress. Neotrop. Ichthyol. 7: 491–498.

Nilsson, G.E., A. Dymowska and J.A. Stecyk. 2012. New insights into the plasticity of gill structure. Resp. Physiol. Neurobiol. 184: 214–222.

Oyakawa, O.T. 2003. Erythrinidae (Trahiras). pp. 238–240. *In*: R.E. Reis, S.O. Kullander and C.J. Ferraris Jr. (eds.). Checklist of the Freshwater Fishes of South and Central America. EDIPUCRS, Porto Alegre, Brasil.

Olson, K.R. 2002. Vascular anatomy of the fish gill. J. Exp. Zool. 293: 214–31.

Olson, K.R., H. Dewar, J.B. Graham and R.W. Brill. 2003. Vascular anatomy of the gills in a high energy demand teleost, the skipjack tuna (*Katsuwonus pelamis*). J. Exp. Zool. A 297: 17–31.

Paulino, M.G., M.M. Sakuragui and M.N. Fernandes. 2012a. Effects of atrazine on the gill cells and ionic balance in a Neotropical fish, *Prochilodus lineatus*. Chemosphere 86: 1–7.

Paulino, M.G., N.E.S. Souza and M.N. Fernandes. 2012b. Subchronic exposure to atrazine induces biochemical and histopathological changes in the gills of a Neotropical freshwater fish, *Prochilodus lineatus*. Ecotoxicol. Environ. Safety 80: 6–13.

Paulino, M.G., T.P. Benze, H. Sadauskas-Henrique, M.M. Sakuragui, J.B. Fernandes and M.N. Fernandes. 2014. The impact of organochlorines and metals on wild fish living in a tropical hydroelectric reservoir: bioaccumulation and histopathological biomarkcrs. Sc. Total Environ. 497-498: 293–306.

Perry, S.F. and P. Laurent. 1989. Adaptation responses of rainbow trout to lowered external NaCl concentration: contribution of the branchial chloride cell. J. Exp. Biol. 147: 147–168.

Perry, S.F. and P. Laurent. 1993. Environmental effects on fish gill structure and function. pp. 231–264. *In*: J.C. Rankin, F.B. Jensen (eds.). Fish Ecophysiology. Chapman & Hall, London, UK.

Perry, S.F. and S.G. Reid. 2002. Cardiorespiratory adjustments during hypercarbia in rainbow trout *Oncorhynchus mykiss* are initiated by external CO_2 receptors on the first gill arch. J. Exp. Biol. 205: 3357–3365.

Perry, S.F., M.G. Jonz and K.M. Gilmour. 2009. Oxygen sensing and the hypoxic ventilatory response. pp. 193–253. *In*: J.G. Richards, A.P. Farrell and C.J. Brauner (eds.). Fish Physiology, v. 27. Academic Press, New York, USA.

Piermarini, P.M. and D.H. Evans. 1998. Osmoregulation of the Atlantic stingray (*Dasyatis sabina*) from the freshwater lake Jesup of the St. Johns River, Florida. Physiol. Zool. 71: 553–560.

Piermarini, P.M. and D.H. Evans. 2000. Effects of environmental salinity on Na^+/K^+-ATPase in the gills and rectal gland of a euryhaline elasmobranch (*Dasyatis sabina*). J. Exp. Biol. 203: 2957–2966.

Piermarini, P.M., J.W. Verlander, I.E. Royaux and D.H. Evans. 2002. Pendrin immunoreactivity in the gill epithelium of a euryhaline elasmobranch. Am. J. Phys. Regul. Integr. Comp. Phys. 283: R983–R992.

Pillans, R.D. and C.E. Franklin. 2004. Plasma osmolyte concentrations and rectal gland mass of bull sharks *Carcharhinus leucas*, captured along a salinity gradient. Comp. Biochem. Physiol. A 138: 363–371.

Pillans, R.D., J.P. Good, W.G. Anderson, N. Hazon and C.E. Franklin. 2005. Freshwater to seawater acclimation of juvenile bull sharks (*Carcharhinus leucas*): plasma osmolytes and Na^+/K^+-ATPase activity in gill, rectal gland, kidney and intestine. J. Comp. Physiol. A 175: 37–44.

Pillans, R.D., J.P. Good, W.G. Anderson, N. Hazon and C.E. Franklin. 2008. Rectal gland morphology of freshwater and seawater acclimated bull sharks *Carcharhinus leucas*. J. Fish Biol. 72: 1559–1571.

Pisam, M., A. Caroff and A. Rambourg. 1987. Two types of chloride cells in the gill epithelium of a freshwater adapted euryhaline fish *Lebistes reticulatus*—their modifications during adaptation to saltwater. Am. J. Anat. 179: 40–50.

Pisam, M. and A. Rambourg. 1991. Mitochondria-rich cells in the gill epithelium of teleost fishes—an ultrastructural approach. Int. Rev. Cytol. 130: 191–232.

Pisam, M., F. Massa, C. Jammet and P. Prunet. 2000. Chronology of the appearance of beta, A, and alpha mitochondria-rich cells in the gill epithelium during ontogenesis of the brown trout (*Salmo trutta*). Anat. Rec. 259: 301–311.

Poleo, A.B.S., J. Schjolden, J. Sørensen and G.E. Nilsson. 2017. The high tolerance to aluminum in crucian carp (*Carassius carassius*) is associated with its ability to avoid hypoxia. PLoS ONE 12: e0179519. https://doi.org/10.1371/journal.pone.0179519.

Potter, I.C., B.M. Chuwen, S.D. Hoeksema and M. Elliott. 2010. The concept of an estuary: a definition that incorporates systems which can become closed to the ocean and hypersaline. Estuar. Coast. Shelf. Sci. 87: 497–500.

Ramos, C.A., M.N. Fernandes, O.T.F. Costa and W.P. Duncan. 2013. Implications for osmorespiratory compromise by anatomical remodeling in the gills of *Arapaima gigas*. Anat. Rec. 296: 1664–1675.

Regan, K.S., M.G. Jonz and P.A. Wright. 2011. Neuroepithelial cells and the hypoxia emersion response in the amphibious fish. J. Exp. Biol. 214: 2560–2568.

Sakuragui, M.M., J.R. Sanches and M.N. Fernandes. 2003. Gill chloride cell proliferation and respiratory responses to hypoxia of the Neotropical erythrinid fish *Hoplias malabaricus*. J. Comp. Physiol. B 173: 309–317.

Sardet, C., M. Pisam and J. Maetz. 1979. The surface epithelium of teleostean fish gills—cellular and junctional adaptations of the chloride cell in relation to salt adaptation. J. Cell Biol. 80: 96–117.

Schultz, E.T. and S.D. McCormick. 2013. Euryhalinity in an Evolutionary Context. EEB Articles 29. http://digitalcommons.uconn.edu/eeb_articles/29.

Schultz, A.G., P.L. Jones and T. Toop. 2014. Rodlet cells in murray cod, *Maccullochella peelii peelii* (Mitchell), affected with chronic ulcerative dermatopathy. J. Fish Dis. 37: 219–228.

Shiogiri, N.S., M.G. Paulino, S.P. Carraschi, F.G. Baraldi, C. Cruz and M.N. Fernandes. 2012. Acute exposure of a glyphosate-based herbicide affects the gills and liver of the Neotropical fish, *Piaractus mesopotamicus*. Env. Tox. Pharmacol. 34: 388–396.

Shiraishi, K., T. Kaneko, S. Hasegawa and T. Hirano. 1997. Development of multicellular complexes of chloride cells in the yolk-sac membrane of tilapia (*Oreochromis mossambicus*) embryos and larvae in seawater. Cell Tissue Res. 288: 583–590.

Siderits, D. and E. Bielek. 2009. Rodlet cells in the thymus of the zebrafish *Danio rerio* (Hamilton, 1822). Fish Shellf. Immunol. 27: 539–548.

Silva, H.R., A.L. Cruz and M.N. Fernandes. 2014. Peixes de respiração aérea. Estudos hematológicos e hemoglobínicos de Lepidosiren paradoxa (Dipnoi) e Hoplerithynus unitaeniatus (Characiformes). Novas Edições Acadêmicas, Saarbrücken.

Sollid, J., P. Angelis, K. Gundersen and G.E. Nilsson. 2003. Hypoxia induces adaptive and reversible gross morphological changes in crucian carp gills. J. Exp. Biol. 206: 3667–3673.

Sollid, J., R.E. Weber and G.E. Nilsson. 2005. Temperature alters the respiratory surface area of crucian carp *Carassius carassius* and goldfish *Carassius auratus*. J. Exp. Biol. 208: 1109–1116.

Sollid, J. and G.E. Nilsson. 2006. Plasticity of respiratory structures—Adaptive remodeling of fish gills induced by ambient oxygen and temperature. Resp. Physiol. Neurobiol. 154: 241–251.

Stiffler, D.F., J.B. Graham, K.A. Dickson and W. Stockmann. 1986. Cutaneous ion transport in the freshwater *Synbrnchus marmoratus*. Physiol. Zool. 59: 406–418.

Takabe, S., M. Inokuchi, Y. Yamaguchi and S. Hyodo. 2016. Distribution and dynamics of branchial ionocytes in hound shark reared in full-strength and diluted seawater environments. Comp. Biochem. Physiol. A 198: 22–32.

Takei, Y. and S.D. McCormick. 2013. Hormonal control of fish euryhalinity. pp. 69–123. *In*: S.D. McCormick, A. Farrell and C. Brauner (eds.). Fish Physiology. Euryhaline Fishes, v. 32. Academic Press, New York, USA.

Tavares, D., M.G. Paulino, A.P. Terezan, J.B. Fernandes, A. Giani and M.N. Fernandes. 2018. Biochemical and morphological biomarkers of the liver damage in the Neotropical fish, *Piaractus mesopotamicus*, injected with crude extract of cyanobacterium *Radiocystis fernandoi*. Environ. Sci. Pollut. Res. 25: 15349–1535.

Thorson, T.B., R.M. Wotton and T.M. Georgi. 1978. Rectal gland of freshwater stingrays, *Potamotrygon* sp. (Chondrichthyes: Potamotrygonidae). Biol. Bull. 154: 508–516.

Tresguerres, M. 2016. Novel and potential physiological roles of vacuolar-type H^+-ATPase in marine organisms. J. Exp. Biol. 219: 2088–2097.

Uchida, K., T. Kaneko, H. Miyazaki, S. Hasegawa and T. Hirano. 2000. Excellent salinity tolerance of Mozambique tilapia (*Oreochromis mossambicus*): elevated chloride cell activity in the branchial and opercular epithelia of the fish adapted to concentrated seawater. Zool. Sc. 17: 149–160.

Val, A.L. and V.M.F. Almeida-Val. 1995. Fishes of the Amazon and their Environment: Physiological and Biochemical Aspect. Springer-Verlag, New York.

Wang, Y.F., Y.C. Tseng, J.J. Yan, J. Hiroi and P.P. Hwang. 2009. Role of SLC12A10.2 a Na–Cl cotransporter-like protein, in a Cl uptake mechanism in zebrafish (*Danio rerio*). Am. J. Physiol. Reg. Integ. Comp. Physiol. 296: R1650–R1660.

Wegner, N.C. 2016. Elasmobranch gill structure. pp. 101–151. *In*: R.E. Shadwick, A.P. Farrell and C.J. Brauner (eds.). Fish Physiology, Physiology of Elasmobranch fishes: Structure and Interaction with Environment, v. 34A. Academic Press. New York, USA.

Wilson, J.M., J.D. Morgan, A.W. Vogl and D.J. Randall. 2002. Mitochondria rich-cells in the dogfish, *Squalus acanthias*. Comp. Biochem. Physiol. 132A: 365–374.

Wright, P.A. and C.M. Wood. 2015. Regulation of ions, acid–base, and nitrogenous wastes in elasmobranchs. pp. 279–345. *In*: R.E. Shadwick, A.P. Farrell and C.J. Brauner (eds.). Fish Physiology, Physiology of Elasmobranch Fishes: Internal Processes, v. 34B, Ed. Academic Press, New York, USA.

Wootton, T.P., C.A. Sepulveda and N.C. Wegner. 2015. Gill morphometrics of the thresher sharks (genus *Alopia*): correlation of gill dimensions with aerobic demand and environmental oxygen. J. Morphol. 276: 589–600.

Zadunaisky, J.A. 1984. The chloride cell: the active transport of chloride and the paracellular pathways. pp. 130–176. *In*: W.S. Hoar and D.J. Randall (eds.). Fish Physiology. Gills, Ion and Water Transfer, v. 10B. Academic Press, Orlando, USA.

Zadunaisky, J.A., S. Cardona, L. Au, D.M. Roberts, E. Fisher, B. Lowenstein, E.J. Cragoe Jr. and K.R. Spring. 1995. Chloride transport activation by plasma osmolarity during rapid adaptation to high salinity of *Fundulus heteroclitus*. J. Membr. Biol. 143: 207–217.

Chapter **16**

Sensory Organs

Guest Editor: J.F. Webb

Jacqueline F. Webb,[1,*] *Shaun P. Collin,*[2] *Michał Kuciel,*[3] *Tanja Schulz-Mirbach,*[4] *Krystyna Żuwała,*[5] *Jean-Pierre Denizot*[6] and *Frank Kirschbaum*[7]

INTRODUCTION

Sensory systems allow animals to perceive and interpret their environment and mediate key behaviors that are essential to both survival and fitness. With more than 30,000 species, fishes represent more than half of all vertebrates (see Nelson et al. 2016). They have infiltrated a remarkably diverse range of marine and freshwater habitats, so it is not surprising that their sensory systems are structurally and functionally diverse, having evolved in response to a wide range of selective pressures. The sensory organs of fishes are categorized as chemoreceptors (olfactory and gustatory systems), photoreceptors (visual system), and mechanoreceptors (auditory and lateral line systems), which are found in all species, and electroreceptors (electrosensory system), which are found only in a small subset of fish species. Here we describe the structure of the sensory organs of the major fish sensory systems (at different resolution levels, and using different methods) and its known variation among species.

The Gustatory System

Among vertebrates, only the fishes are known to have a gustatory (taste) system comprised of external taste buds on the skin in addition to internal taste buds in the epithelial lining of the oral cavity and pharynx (Hansen et al. 2002, Hanke et al. 2008, Xiong et al. 2011, Dos Santos et al. 2015). The first reports of taste buds in the skin of fishes (Weber 1827, Leydig 1851) and in the lining of their oral cavity (Schulze 1863)

[1] Department of Biological Sciences, University of Rhode Island, Kingston, RI, USA.
[2] School of Life Sciences, La Trobe University, Bundoora 3086, Victoria, Australia.
[3] Medical Collage, Poison Information Centre, Jagiellonian University in Kraków, Krakow, Poland.
[4] Department of Biology II, Ludwig-Maximilians-Universität München, Munich, Germany.
[5] Department of Comparative Anatomy, Jagiellonian University in Kraków, Krakow, Poland.
[6] Unité de Neurosciences Intégratives et Computationelles, C.N.R.S., Gif sur Yvette, France.
[7] Faculty of Life Sciences, Unit of Biology and Ecology of Fishes, Humboldt University, Berlin, Germany.
 Emails: s.collin@latrobe.edu.au; michalkuciel@gmail.com; schulz-mirbach@biologie.uni-muenchen.de; krystyna.zuwala@uj.edu.pl; denizot@unic.cnrs-gif.fr; frank.kirschbaum@staff.hu-berlin.de
* Corresponding author: jacqueline_webb@uri.edu

date to the 19th century. Rapid improvements in methods during the last 100 years allowed researchers to describe the anatomy of fish taste buds (e.g., Hirata 1966, Reutter 1978, Jakubowski and Whitear 1990, Jakubowski and Żuwała 2000), as well as their development (e.g., Hansen et al. 2002, Atkinson et al. 2016) and mechanisms of taste bud function at the molecular level (e.g., Witt 1996, Ferrando et al. 2012, Atkinson et al. 2016).

Key studies on the taste buds in fishes include the description of taste buds presented by Truillo-Cenoz (1961) and Hirata (1966), studies on taste buds carried out by Grower-Johnson and Farbman (1976) as well as results of cyto-morphological research carried out by Jakubowski and Whitear (1990) who described variations in gustatory organs in 26 bony fish species. Reutter (1971) described two types of neurotransmitters in the gustatory organs of *Ictalurus*, cholinergic and aminergic types. His work on morphology and neurophysiology was later confirmed by Nada and Hirata (1977) in other vertebrate species. The taste buds of teleost fishes are described in numerous reviews (e.g., Hirata 1966, Kapoor et al. 1975, Reutter 1986, Jakubowski and Żuwała 2000, Tagliafiero and Zaccone 2001).

An increasing number of studies on Chondrichthyes (representatives of both Holocephali and Elasmobranchii) and non-teleost bony fishes (including Dipnoi and Holostei) have compared taste buds (especially their microanatomy) among species. Reutter et al. (2000) described the cell morphology of taste buds in *Amia calva* and *Lepisosteus oculatus*. These authors compared the structure of taste buds within and among these groups and they suggested that each of these groups has its "own type" of taste buds. Differences among taste bud types are mainly found in cellular composition (in the Holostei), their ultrastructure and the position of the basal cell(s) within the taste buds (in the Chondrichthyes). Differences in the details of taste buds structure are also seen among the Teleostei. Whitear and Moate (1994a, b) described "gustatory" chemoreceptors in the Chondrichthyes (in elasmobranchs). They also demonstrated the presence of taste buds in *Scyliorhinus canicula* (Whitear and Moate 1994a). *Raja clavata* has been reported to have only solitary chemosensory cells that occur individually or in groups (Whitear and Moate 1994b). Taste buds have also been found in the mucosa of the tongue (Ferrando et al. 2012) and the palate in holocephalan (*Chimaera monstrosa*; Ferrado et al. 2016).

Distribution and Number of Fish Taste Buds

Taste buds have been described in the skin of the head, trunk, fins and the lining of the oropharyngeal cavity in representatives of Chondrichthyes, Holostei, and Teleostei. Variation in taste bud distribution has been observed among the species studied. For example, Norris (1925) described external taste buds in a sturgeon *Acipenser* sp. (Chondrostei) only on the lips and the short barbels, but they have not been found on the body and tail fin as they have been in *Amia calva* (Holostei; Allis 1889). External taste buds have not been found in the skin of *Polypterus* sp. (Holostei, Actinopterygii; Lane and Whitear 1982a). Taste buds have not been reported in *Raja clavata* (Neoselachii, Elasmobranchii), but taste reception is thought to be the function of solitary chemosensory cells in this species (Whitear and Moate 1994b).

Differences in the number of taste buds in the mucosa and skin have been described in the Teleostei (Table 1). Most frequently taste buds occur individually, but in some fish species they are located in groups close to each other, e.g., in gobiid fishes (*Bathygobius fuscus*, *Gobius bucchichi*; Fishelson and Delarea 2004), in cardinalfishes (*Ostorhinchus* [= *Apogon*] *cookii*, *Fowleria variegata*; Fishelson et al. 2004), and in oxudercine gobies (*Periophthalmus minutus*, *P. argentilineatus*; Kuciel et al. 2017) (Figs. 1A–D). In general, taste buds are most numerous in the external skin of the head and in the oral cavity lining, while they are less numerous on the trunk and fins. In many species, the highest concentration of taste buds was found on the barbels, lips, and the frontal part of the head (Żuwała and Jakubowski 1993,

Table 1. Number of taste buds (TBs) in selected fish species. TL, total length.

Species	External TBs	Internal TBs	References
Ameiurus natalis TL 25 cm (Ictaluridae)	175,000	20,000	Atema 1971
Pseudorasbora parva TL 6 cm (Cyprinidae)	1,500	6,000	Kiyohara et al. 1980

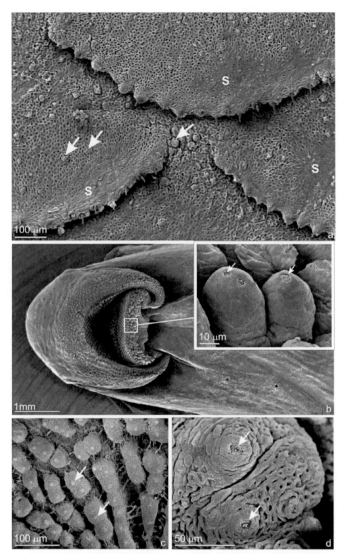

Fig. 1. Distribution of taste buds in the skin and oropharyngeal cavity lining. **A.** Scales from the body of *Serrasalmus nattereri*. **B.** Mouth of *Gyrinocheilus aymonieri*: on inset note the sensory zones of the taste buds. **C, D.** Oropharyngeal cavity lining with sensory zones of taste buds. Arrows, sensory zones of taste buds; s, scale.

Fishelson and Delarea 2004) and was most numerous on the lips in *Serrasalmus nattereri* (Raji and Norozi 2010). However, in the oral cavity lining of *Cobitis taenia* the highest density of taste buds ($307/mm^2$) was found in the epithelium of the rear part of the pharynx (Jakubowski 1983).

The distribution of taste buds varies with feeding habits among species (*Gadus morhua*, Bishop and Odense 1966; *Sparus aurata*, Cataldi et al. 1987; *Gobius paganellus*, *Istiblennius rivulatus*, Fishelson and Delarea 2004). It was shown that benthic-feeding omnivorous blenniforms have higher densities of taste buds in the oropharyngeal cavity than piscivorous species (Fishelson et al. 2012). The number of taste buds is also correlated with the environment in which a species lives (Mauri and Caprio 1992, Boudriot and Reutter 2001), its activity, the degree of development of other sensory organs, and the age of the individual. For instance, taste buds are especially numerous in the skin of less active freshwater fishes (e.g., Ostariophysi; Atema 1971, Sorensen and Caprio 1998). More taste buds are found in bottom dwelling species and those living in caves than in fish inhabiting clear waters. The number of taste buds increases with age (Atema 1971, Finger et al. 1991, Hu et al. 2018).

Gross Morphology of Taste Buds

Taste buds are onion- or pear-shaped regardless of their location in the skin or in the oral cavity mucosa (Figs. 2A, B). In both teleost and non-teleost bony fishes, taste buds are situated in the epidermis, above the basement membrane, which forms dermal papillae on which the taste buds sit (e.g., Hirata 1966, Jakubowski and Żuwała 2000, Reutter et al. 2000, Ferrando et al. 2012, 2016).

The taste buds transmit taste information to the central nervous system via fibers of the facial (VII), glossopharyngeal (IX) and vagal (X) nerves. Respective branches of the VIIth cranial nerve innervate the external taste buds present on "lips" and barbels (*N. facialis maxillaris*), on fins and body (*N. facialis recurrens*) and internal taste buds in the region of the rostral palatine bone (*N. facialis palatinus*). The Xth nerve supplies the orobranchial taste buds while the IXth supplies the remaining taste buds of the oral cavity (Atema 1971, Finger 1976). Blood vessels are present in the dermis below the taste buds and have been shown to form a capillary loop (in some freshwater species; Jakubowski 1958, 1966, 1983, Jakubowski and Żuwała 2000; Figs. 3A, B).

The size of taste buds and the number of cells in a taste bud varies among species. For instance, the height of a taste bud ranges from 25 μm (e.g., in *Gyrinocheilus aymonieri*) to ~ 100 μm (e.g., *Tinca tinca*, *Neoceratodus fosteri*). The diversity of taste bud size in selected fish species is shown in Table 2.

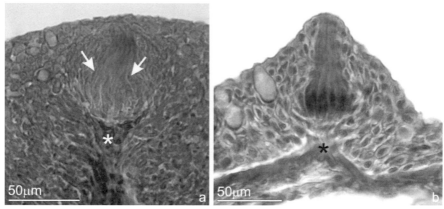

Fig. 2. Vertical section through a taste bud. **A.** *Perca fluviatilis*–paraffin section (stained with the Gomorri method), **B.** *Astyanax jordani*–paraffin section (stained with hematoxylin and eosin). Arrows, boundaries of taste buds; Asterisk, dermal papillae.

Color version at the end of the book

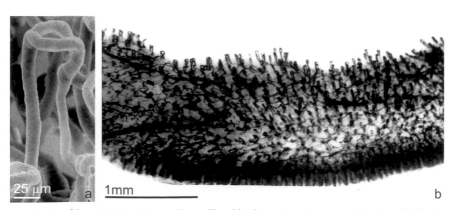

Fig. 3. Vascularization of the taste buds. **A.** Loop-like capillary blood vessels serving the papillary taste bud in the oral cavity (mercox corrosion cast). **B.** Fragment of the lip of *Tinca tinca*–India ink injection. Note many loop-like capillaries supplying the taste buds (preparation and photo made by M. Jakubowski).

In addition, there are 25 cells in the taste buds in a shark (*Scyliorhinus canicula*) but 50 to 60 cells in non-teleost bony fishes (*Lepisosteus* and *Amia*; Reutter et al. 2000). In teleost fishes, variation is more pronounced where there are ~ 23 cells in *Danio rerio* (Ohkubo et al. 2005) and 110–130 cells in *Cobitis taenia* (Jakubowski 1983).

Ultrastructure of Taste Buds

A taste bud is composed of sensory cells (gustatory receptor cells, GC and basal cells, BC) and non-sensory cells (supporting cells, SC). Within a taste bud, elongated cells (GC and SC) reaching from the apex to the basal part of the taste bud can be easily recognized even at low magnification (Figs. 2A–C). Their nuclei are located basally, in the broader part of the taste bud. One or several BCs of discoid shape are located at the bottom of the taste bud. Apical processes of gustatory and supporting cells usually form a small roundish sensory zone with a diameter from 2 μm (*Pomatoschistus minutus*) to 22 μm (*Lota lota*) (Table 2). Three types of taste buds have been distinguished in fishes (Whitear 1971, Reutter et al. 1974, Grover-Johnson and Farbman 1976, Reutter 1978, 1991, Ezeasor 1982, Jakubowski and Whitear 1990, Yashpal et al. 2006, Kuciel et al. 2017) based on the position of the sensory zone relative to the surrounding epithelial cells. The sensory zone of a taste bud may be located above the surface of the general epithelium (Type I; Fig. 4A), at the level of the epithelium (Type II; Fig. 4b), or below the level of the epithelium, forming a so-called "pore" (Type III; Fig. 4C). Such diversity is often seen among taste buds within a species and is related to their location (e.g., Boudriot and Reutter 2001).

There are different nomenclatural schemes for the cells constituting the taste buds. Some authors classify taste bud cells into "dark cells" and "light cells" based on the electron density of their cytoplasm (Welsch and Storch 1969, Storch and Welsch 1970, Reutter 1971). Reutter (1978, 1982) as well as Witt et al. (2003) point out that both "light" and "dark" cells are innervated, suggesting that they are both sensory in function.

Table 2. The size of taste buds in selected fish species.

Species (Family, Order)	Mean Diameter of the Sensory Zone (μm)	Maximal Diameter × Height (μm) of Taste Buds	References
Lota lota (Lotidae, Gadiformes)	12–22	–	Jakubowski and Whitear 1990
Pomatoschistus minutus (Gobiidae, Perciformes)	2–5	–	Jakubowski and Whitear 1990
Apteronotus albifrons (Apteronotidae, Gymnotiformes)	10–12	39 × 48	unpublished data
Gyrinocheilus aymonieri (Gyrinocheilidae, Cypriniformes)	15–19	25 × 41	unpublished data
Astyanax jordani (Characidae, Characiformes)	18	.. × 36	unpublished data
Periophthalmus barbarus *Periophthalmus gracilis* (Gobiidae, Perciformes)	4–6 3	20 × 40 –	Kuciel et al. *in press*
Amia calva (Amiidae, Amiiformes)	–	45 × 65	Baudriot and Reutter 1998, 2001
Lepisosteus oculatus (Lepistoteidae, Lepistoteiformes)	–	45 × 65	Reutter and Witt 1996
Neoceratodus forsteri (Ceratodontidae, Ceratodontiformes)	–	80 × 100	Reutter 1991
Scyliorhinus canicula (Scyliorhinidae, Carcharhiniformes)	–	30 × 50	Reutter 1994, Whitear and Moate 1994a

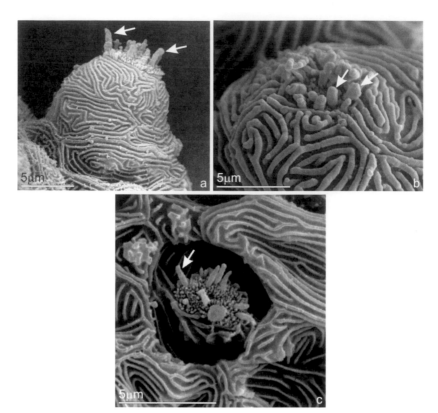

Fig. 4. Different positions of the sensory zone of the taste bud relative to the surface epithelium: **A.** Apical part of a taste bud from *Astyanax jordani* above oral epithelium; **B.** Sensory zone of *Neogobius gymnotrachelus* at the skin level; **C.** Sensory zone from a *Neogobius gymnotrachelus* gill arch below the surface epithelium. Arrows, apical finger like protrusions of receptor gustatory cells.

Gustatory Receptor Cells

The apical part of the gustatory cell forms thick finger-like process that is ~ 2.5 µm in length and 0.5 µm in width, with a core made up of tonofilaments that extend down to the apical part of cell cytoplasm (Fig. 5). In horizontal section, gustatory cells are typically round in appearance (Fig. 6) and the area between them is filled with supporting cells of irregular shape (Jakubowski and Whitear 1990 and others). Close to the apical surface, gustatory cells form tight junctions with supporting cells, and somewhat deeper desmosomal connections (Storch and Welsch 1970, Żuwała and Jakubowski 1993, Fishelson and Delarea 2004). The cytoplasm contains numerous cisterns of smooth endoplasmic reticulum, sparse microtubules and numerous mitochondria (Fig. 7A). The nucleus of the gustatory cells is located in the basal part of the cell where the cell cisterns of Golgi apparatus can also be seen. Basal projections of gustatory cells reach the basal lamina. Synaptic vesicles that are typically electron-light are found in the cytoplasm of the basal region of the cells (e.g., *Tinca tinca*, Żuwała and Jakubowski 1993; *Cobitis taenia*, Jakubowski 1983). In some species, synaptic vesicles have an electron-dense core (e.g., *Silurus glanis*, *Astyanax mexicanus*, Hansen and Reutter 2004; *Gymnocephalus cernuus*, Jakubowski unpublished data).

Basal Cells

Basal cells have a discoid shape and lie on the basal lamina, but do not form hemidesmosomes with it (Fig. 7b). Teleost taste buds can have one to several basal cells (Hirata 1966, Grover-Johnson and Farbman 1976, Jakubowski 1983, Jakubowski and Whitear 1990, Żuwała and Jakubowski 1993, Kuciel et al. 2017

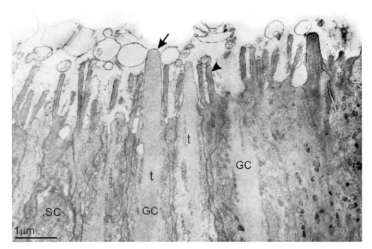

Fig. 5. Vertical section through the apical part of a taste bud of *Tinca tinca*. Apical finger-like protrusions of gustatory cells (arrow) and microvilli of supporting cells (arrowhead) are seen. GC, gustatory cell; SC, supporting cell; t, tonofilaments.

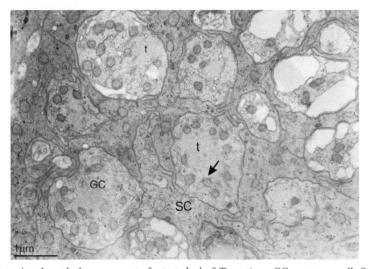

Fig. 6. Horizontal section through the upper part of a taste bud of *Tinca tinca*. GC, gustatory cell; SC, supporting cell; t, tonofilament; arrow, mitochondrium.

and others). Basal cells have not been found in the external taste buds of *Ciliata mustella* (on barbels, Crisp et al. 1975), some gadiforms, or *Anguilla anguilla*, *Poecilla reticulata* and *Colisa chuna* (Jakubowski and Whitear 1990). In the spotted dogfish, *Scyliorhinus canicula*, basal cells like those in teleosts are not found in the basal portion of the taste bud, but sit directly on top of the dermal papilla (Reutter 1994).

The nucleus of basal cells is large, usually with some indentations. Bundles of tonofilaments, cisterns of Golgi apparatus, smooth endoplasmic reticulum and glycogen granules are found in the cytoplasm. Synaptic vesicles are most frequently seen in groups close to the synapse with innervating nerve fibers. The membrane of a basal cell facing the nerve plexus often forms a few microvillar processes with a tonofilament core (Jakubowski 1983, Jakubowski and Whitear 1990). The presence of a large basal cell nucleus with an irregular membrane and finger-like processes with a tonofilament core resembles the Merkel cells found in the basal layer of the epidermis and in the mucosa lining the oral cavity in vertebrates including fish and amphibians.

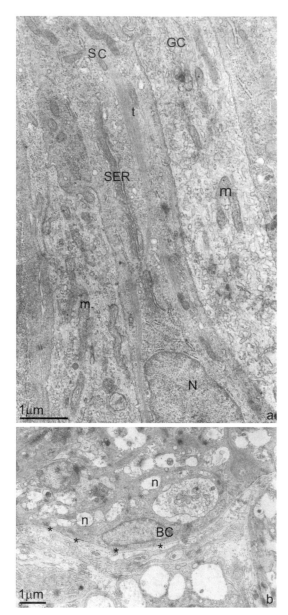

Fig. 7. Vertical section through a taste bud of *Serrasalmus nattereri*. **A.** Supranuclear part of the taste bud. Note the differences in ultrastructure between gustatory (GC) and supporting (SC) cells. **B.** Basal part of a taste bud. SER, smooth endoplasmic reticulum; t, tonofilaments; m, mitochondrium; N, nucleus; BC, basal cell; n, nerve fiber; asterisks, basal lamina.

Supporting Cells

Supporting cells surround the sensory gustatory cells (Figs. 6, 7A). The apical end of the supporting cells has few microvilli (Fig. 5), and secretory vesicles of various electron densities and size are found in the periapical cytoplasm depending on species (Jakubowski and Whitear 1990). Their cytoplasm is rich in tonofilaments, cisterns of rough endoplasmic reticulum, and polyribosomes (Fig. 7A). In *Cobitis taenia*, *Misgurnus fossilis* and *Agonus cataphractus*, the supporting cells are also rich in microtubules (Jakubowski 1983, Jakubowski and Whitear 1990). Proximal processes of supporting cells may reach the basal cell or the basal lamina with which they form hemidesmosomes.

Nerve Innervation

A nerve fiber reaches taste bud cells by coming up through the basal lamina (Fig. 8). Individual nerve fibers mingle with proximal protrusions of supporting and gustatory receptor cells and it is sometimes difficult to attribute the cell protrusions to a given cell. In some species (e.g., catfishes *Silurus glanis* and *Ameiurus* [= *Ictalurus*] *nebulosus*, Jakubowski and Whitear 1990; *Astyanax* sp., Hansen and Reutter 2004), afferent synaptic connections of neurons with gustatory cells and with basal cells (Fig. 8 inset) have been observed. Reutter (1971, 1978) also described afferent synapses between "dark cells" (supporting cells?) and nerve endings. Jakubowski and Whitear (1990) did not observe synaptic contacts between supporting cells and nerve endings in any of the 26 bony fish species studied (e.g., in Anguilliformes, Clupeiformes, Gadiformes, Atheriniformes, Perciformes). Efferent synaptic connections between nerve endings and gustatory cells and basal cells have also been described in some species of bony fishes (*Ameiurus* [= *Ictalurus*] *nebulosus*, Desgranges 1966, Royer and Kinamon 1996; *Phoxinus phoxinus*, Jakubowski and Whitear 1990; *Lepisosteus oculatus*, Reutter et al. 2000; *Astyanax mexicanus*, Boudriot and Reutter 2001).

In some teleosts (e.g., *Anguilla anguilla*, *Gasterosteus aculeatus*, *Blennius* sp., *Agonus* sp., *Triglopterus* sp., *Periophthalmus* sp.), expanded neurite profile adhere to the basal cell covering it from above (Jakubowski and Whitear 1990, Kuciel et al. 2017). In other species (e.g., *Serrasalmus nattereri*), the nerve fibers spread flatly under the basal lamina below the basal cells (Żuwała, personal observation).

Solitary Chemosensory Cells

Solitary chemosensory cells, considered to be a separate chemosensory modality, are present in the lining of the oral cavity, on the gills, and in the skin in the vicinity of taste buds (Whitear 1965, 1971, 1992, Whitear and Kotrschal 1988; Fig. 9). In *Raja clavata* (Whitear and Moate 1994b), solitary chemosensory cells seem to be the only system to perceive gustatory signals. Kotrschal (1996, 2000) suggests that these cells form a peculiar gustatory subsystem in rocklings (Gadidae, Teleostei). The apex of these cells is equipped with a solitary thick process. It had been thought that solitary chemosensory cells are present only in aquatic vertebrates, but they comprise a similar system in mammals (Sbarbati et al. 1998, Sbarbati and Osculati 2003).

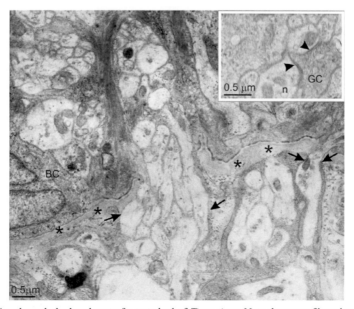

Fig. 8. Vertical section through the basal part of a taste bud of *Tinca tinca*. Note the nerve fiber plexus (arrows) entering into the taste bud. Inset: synaptic connections (arrowheads) between gustatory (GC) cell and a nerve fiber (n). BC, basal cell; asterisks, basal lamina.

Fig. 9. Solitary chemosensory cells (arrows) in the lining of the oral cavity of *Apteronotus albifrons*. Note that they are situated around a taste bud sensory zone (TB).

The Olfactory System

Like the sensory cells of taste buds and the solitary chemosensory cells, the receptor cells of the olfactory organs are stimulated when specific transmembrane protein receptor complex are bound with a specific chemical molecule (ligand). This combination causes a change in the spatial conformation of the receptor complex and a series of reactions within the cell resulting in resting potential distraction. The nerve impulse thus generated is propagated to the particular brain region responsible for the signal analysis.

Olfaction plays roles to differing degrees among species that occupy different ecological niches and this is often reflected in olfactory organ morphology. Olfaction may be a dominant sense in catfish (e.g., *Silurus glanis*, Jakubowski 1981), and a regressed sense (e.g., in Tetraodontidae, Yamamoto and Ueda 1979), and may function in combination with other senses like vision (most Cypriniformes, Burne 1909) to facilitate key behaviors. Its most common functions are identifying food and finding mates. For instance, mature salmon (*Salmo salar*) and some eels (e.g., *Anguilla vulgaris*) take long excursions to spawning areas, which are most likely found due to the sense of smell (Creutzberg 1961, Hassler 1966). Another function of smell is the detection of alarm substance released from skin after injury by predators (especially in otophysan fishes; for review see: Pfeifer 1962, 1963), which results in appropriate shelter seeking or escape behavior.

Gross Morphology of the Olfactory Organ

Variation in the structure of the olfactory organ among fishes has been reviewed by Burne (1909) and subsequent detailed reviews (Kleerekoper 1969, Doving 1977, Hara 1975, Zeiske et al. 1992, Hansen and Zielinski 2005).

The olfactory organ in the jawless fishes (hagfishes, lampreys) is unpaired and probably has the simplest structure among fishes described to date. It is found in a single recess on the dorsal surface of the head. The sensory epithelium is found at the bottom of this recess (Doving and Holmberg 1974). In the Chondrichthyes, the olfactory organ is paired and symmetrically located on the ventral side of the rostrum (Thiesen et al. 1986, Zeiske et al. 1986). Exceptions are members of the family Chlamydoselachidae in which the olfactory organ is located on the dorsal side of the rostrum (Jasiński 1982). In the Teleostei, the olfactory organ is paired and is always located on the dorsal surface of the head, rostral to the eyes. Each olfactory organ consists of an olfactory chamber containing an olfactory epithelium containing olfactory sensory neurons (OSNs) in the form of an olfactory rosette. The skin covering the olfactory chamber has paired openings (except for cichlids, where one opening is present) allowing the water to enter (incurrent nostril) and leave (excurrent nostril) the olfactory chamber. Most commonly the nares are round or oval openings, which are sometimes divided by the skin flap improving water flow (e.g., in Cyprinidae, Fig. 10A; in Salmonidae, Fig. 10B), but there are also species with simple openings without additional structures (e.g., *Gnathonemus petersii*, Fig. 10C), pipe-like nostrils as in *Pantodon buchholzi* (Fig. 10D), *Neogobius melanostomus* (Fig. 10E), and *Amia calva* (Fig. 10F) or specialized olfactory organs adapted for life out of water, as in mudskippers (Kuciel et al. 2011, 2013). Recent genetic studies by You et al. (2014) showed that mudskippers (*Boleophthalmus pectinirostris* and *Periophthalmodon schlosseri*) might be able to detect some air-borne chemicals as tetrapods do.

The most common type of olfactory rosette is composed of a longitudinal fold, the so-called middle strip or raphe (Figs. 11A, B), with lateral transverse folds (olfactory lamellae) extending from it (Figs. 11B–E). New lamellae are most often formed in the anterior end of the olfactory rosette, and increase in size as a fish grows. Observations of olfactory rosettes in *Amia calva* indicate the presence of longitudinal divisions in the central and terminal lamellae (Kuciel and Żuwala, unpublished data), which could increase the area occupied by OSN's. Different types of olfactory rosettes have been described among species, which are defined by differences in the location of the middle strip, the number and size of olfactory lamellae, the way the olfactory lamellae are attached to the chamber, and lamellar orientation in relation to the longitudinal axis of the fish's body. Yamamoto (1982) describes several different

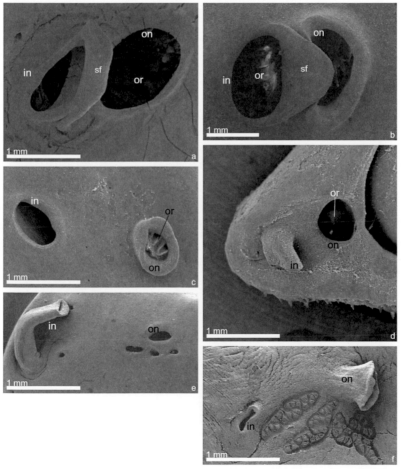

Fig. 10. Scanning electron micrographs of olfactory organ structure in different fish species. **A, B.** Simple openings with a skin flap in juvenile *Cyprinus carpio* (**A**) and in *Salmo trutta m. trutta* (**B**); **C.** Olfactory organ without a skin flap in *Gnathonemus petersii*; **D, E.** Olfactory organ with a pipe-like inlet nostril in *Pantodon buchholzi* (**D**) and *Amia calva* (**E**); **F.** Olfactory organ with a pipe-like outlet (excurrent) nostril in *Neogobius melanostomus*. In, inlet nostril; on, outlet nostril; or, olfactory rosette; sf, skin flap.

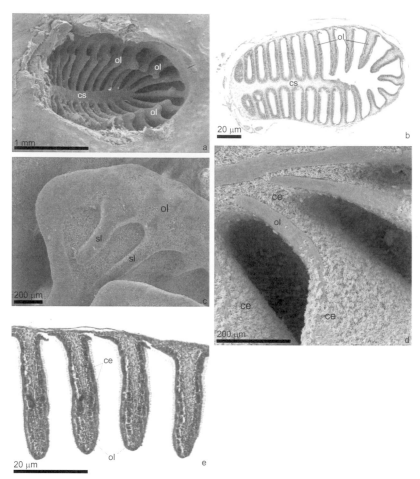

Fig. 11. Structure of the olfactory rosette. **A.** Olfactory chamber uncovered due to skin dissection of *Cyprinus carpio* with the olfactory rosette at the bottom with visible olfactory lamellas and a central strip. **B.** Cross-section of an olfactory rosette of *Amia calva*. **C.** A single olfactory lamella of *Salmo trutta m. trutta* with visible secondary lamellas; **D.** Olfactory lamella of *Salmo trutta m. trutta* at high magnification with rich ciliation of the ciliated epithelium; **E.** Olfactory lamellae of *Amia calva* at high magnification with rich ciliation of the ciliated epithelium. ol, olfactory lamella; cs, central strip; sl, secondary lamella; ce, ciliated epithelium. **A, C, D**—scanning electron micrograph; **B, E**—light micrograph, hematoxylin and eosin staining.

arrangements: (i) an olfactory chamber without rosette or olfactory plates, (ii) one plate longitudinally aligned to the long axis of the olfactory chamber, (iii) one plate aligned transversely to the long axis of the olfactory chamber, (iv) several parallel plates aligned longitudinal to the long axis of the olfactory chamber, (v) lamellae on one side in the anterior part of the olfactory chamber, (vi) lamellae growing out radially, (vii) lamellae growing radially and transversally from the median strip, (viii) lamellae growing radially and transversely to the strongly elongated median strip, and (ix) olfactory lamellae attached only to the bottom and side walls of the olfactory chamber. The number of olfactory lamellae varies among teleost taxa, and most often increases with the age (e.g., 9–18, in *Esox lucius*, Wunder, 1957; 100–150 in *Silurus glanis*, Jakubowski and Kunish 1979, Jakubowski 1981). Only a single olfactory lamella has been observed in *Tetraodon fluviatilis* (Doroshenko and Motavkin 1986), *Conidens laticephalus*, *Gadus macrocephalus*, *Cololabis saira*, and *Acanthogobius flawimanus* (Yamamoto 1982); in contrast, 230 olfactory lamellae have been observed in *Holopagrus guentheri* (Pfeifer 1964). In some gobiids (Belanger 2003, Kuciel et al. 2011, 2013), the olfactory chamber lacks an olfactory rosette and the olfactory sensory epithelium covers the walls of the olfactory chamber or olfactory canal, which is an adaptation for amphibious life in mudskippers (Kuciel et al. 2011, 2013).

Variation in the gross morphology of the olfactory organ is also defined by differences in the location of the olfactory epithelium within the olfactory chamber. The olfactory epithelium can be continuous except for borders (e.g., in *Oncorhynchus masou*, Pfeiffer 1963), zonal (e.g., in *Silurus glanis*, Jakubowski 1981), irregular (e.g., in Gasterosteiformes, Yamamoto and Ueda 1978) or in islet form (e.g., in Tetraodontiformes, Yamamoto and Ueda 1979). An olfactory rosette is lacking in gobiids. The olfactory epithelium may be located differently among species: (i) on the bottom, on lateral walls, and on the 'ceiling' of the olfactory chamber (*Neogobius melanostomus*, Belanger et al. 2003; *Neogobius gymnotrachelus*, Żuwała et al. personal observation), (ii) on the bottom and on lateral walls of the olfactory chamber (*N. fluviatilis*, Żuwała et al. personal observation), (iii) on the bottom and on lateral walls of the olfactory canals (*Boleophthalmus boddarti*, *Scartelaos histophorus*, Kuciel et al. 2013; *Parapocryptes rictuosus* and *Proterorchinus marmoratus*, Żuwała et al. unpublished data), (iv) fascicular with finger-like ramifications (*Periophthalmus argentilineatus* and *Periophthalmus minutus*, Kuciel et al. 2013) and (v) an islet form (*Periophthalmus barbarus*, Kuciel et al. 2011) (*Periophthalmus chrysospilos* and *Periophthalmus variabilis*, Kuciel et al. 2013).

In a fairly large number of teleosts, the olfactory organ is equipped with a mechanism to ventilate the olfactory epithelium by bringing in water containing odorants, using the so-called accessory nasal sacs. Doving (1977) referred to species with accessory nasal sacs as cyclosmates, and those in which the olfactory organs are flushed passively during fish movement and/or with non-sensory ciliated epithelium involvement have been called isosmates. One or two accessory nasal sacs, which are lined with a simple epithelium, have been described in several species of flatfishes (Webb 1993).

Anatomy of the Olfactory Epithelium

Olfactory sensory neurons (OSNs) are specialized neurons and are thus considered to be primary receptors, with dendrites and an axon. The central part of the OSN usually consists of the spindle-shaped pericarion and the apical part the cell forms dendritic protrusions–where the specific trans-membrane receptors are located. The axon extends from the opposite side of the cell body. The axons of all of the OSNs form the olfactory nerve, which is directed to the olfactory bulb (Gemne and Doving 1969). Depending on the species, the olfactory bulb may be located near the brain (e.g., in most teleosts, e.g., in *Anguilla anguilla*, in *Esox lucius*, in *Salmo salar*, Hara 1975), halfway between the brain and the olfactory rosette (e.g., *Raniceps raninus*, Doving 1967; *Gymnothorax kidako* and *Coryphaena hippurus*; Uchihashi 1953), or at the olfactory rosette (e.g., in all elasmobranchs, and in *Carassius*, *Ameiurus* [= *Ictalurus*], cobitids, silurids). So there are species with an elongated olfactory nerve (Fig. 12A) or an elongated olfactory tract (Fig. 12B). In the olfactory bulb, synaptic connections are formed with dendritic protrusions of the mitral cells called glomeruli. Axons of the mitral cells transfer information to other areas of the brain.

Ultrastructure

A longitudinal section through the olfactory epithelium (Fig. 13) reveals two main groups of cells: olfactory sensory neurons (OSNs) and non-sensory cells. There are three main types of OSNs: ciliated OSNs (Figs. 14A, B), microvillar OSNs (Figs. 14C, D) and crypt OSNs. Three types of non-sensory cells are present: the ciliated cells (Figs. 14B, D, G), supporting cells (Fig. 14B), and basal cells (Fig. 13).

Ciliated Olfactory Sensory Neurons

On the apical surface of the sensory neurons, dendritic protrusions (cilia) grow radially around the knob extending into the lumen of the olfactory chamber. The number of cilia on these cells varies among species, and usually ranges from 3 to 10. The cilia are 2–3 mm long in *Danio rerio* (Zeiske and Hansen 1998) and about 10 mm long in *Acipenser baeri* (Hansen and Zielinski 2005). These cilia have the typical pattern (9 + 2) of microtubule doublets when viewed in cross section. A basal body made up of two centrioles is oriented perpendicular to each other at the base of each cilium. In ciliated cells, the basal bodies are located just below the knob membrane. In some species, ciliated OSNs have additional microvillar-like

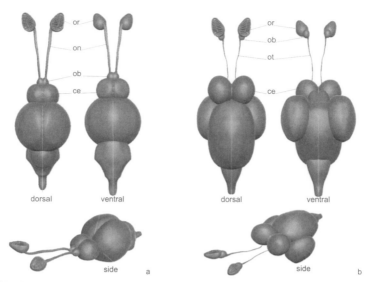

Fig. 12. Brains with elongated olfactory nerve, i.e., in *Periophthalmus barbarus* (**A**) and olfactory tract, i.e., in *Cyprinus carpio* (**B**). Or, olfactory rosette; on, olfactory nerve; ob, olfactory bulb; ce, cerebrum.

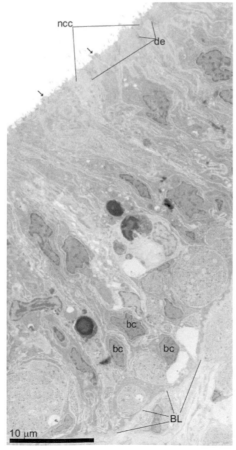

Fig. 13. A longitudinal section (transmission electron micrograph) of the olfactory epithelium with ciliated olfactory sensory neurons (arrow). Ncc, non sensory ciliated cells; de, desmosomes; bc, basal cells; BL, basal lamina.

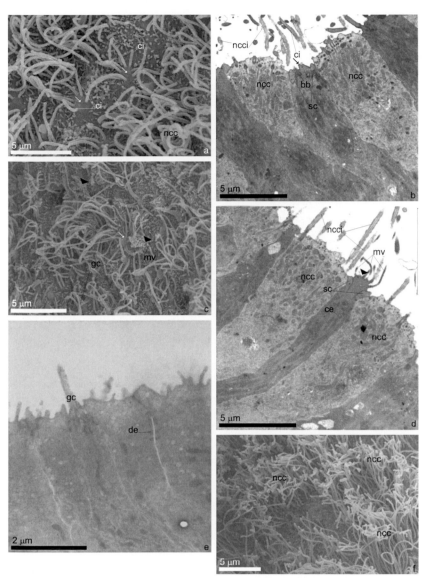

Fig. 14. Ultrastructure of the olfactory epithelium. **A, B.** ciliated olfactory sensory neurons (OSNs) and non sensory ciliated cells in *Salmo trutta m. trutta* (**A**) and in *Lepisosteus oculatus* (**B**); **C, D.** microvillar OSNs, ciliated OSNs and giant cells in *Salmo trutta m. trutta* (**C**) and *Lepisosteus oculatus* (**D**); **E.** giant cell in *Periophthalmus chrysospilos;* **F.** non sensory ciliated epithelium in *Amia calva*. Arrow, ciliated OSN; arrowhead, microvillous OSN; bb, basal body; ce, centriole; de, desmosomes; gc, giant cell; mv, microvilli; ncc, non sensory ciliated cells; ncci, cilia of non sensory ciliated cells; sc, supporting cell. **A, C, F**—scanning electron micrograph; **B, D, E**—transmission electron micrograph.

protrusions accompanying the cilia (e.g., *Acipenser baeri*, Hansen and Zielinski 2005). The cell nucleus is most often basal in position with numerous mitochondria found above the nucleus. The body of the ciliated OSN (perikarion) has an elongated shape with a spindle-ending on the cell's basal end located around two thirds of the depth of the olfactory receptor epithelium where an axon is formed.

The cells in the ciliated OSN group have one rod-like/giant OSNs (Figs. 14C, E). These cells are relatively rarely observed in fishes and are not well studied. Their structure is similar to other ciliated cells. The differences are apparent in the apical region of the cell, where all the cilia are surrounded by an outer membrane, thus forming a single thick cilium (Hansen and Zielinski 2005) that protrudes significantly above the surface of the epithelium.

Microvillar Olfactory Sensory Neurons

The body of the microvillar OSNs (Figs. 14C, D) in general is similar to that of the ciliated OSNs. It is elongated and basally, it is spindle shaped, and extends toward the axon. The cell body extends to about two thirds of the depth of the olfactory epithelium. The apical part of the cell is slightly enlarged and reaches slightly above the surface of the epithelium. The knob is generally less prominent compared to ciliated OSNs. There are small, numerous protrusions on the apical surface: 10–30 in *Dicentrarchus labrax* (Diaz et al. 2002), 30–60 in *Anguilla anguilla* (Schulte 1972), and 40–70 in *Carassius auratus* (Hansen et al. 1999). Centrioles (Fig. 14D) are always located below the contraction of the knob and may be more numerous compared to ciliated OSNs (Hansen and Zielinski 2005). The cell nucleus and the remaining organelles are located in similar locations as in ciliated OSNs.

Crypt Olfactory Sensory Neurons

These cells are characterized by the 'crypt' that occurs in the apical part of the cell and extends to about one third of the cell body. In the recess/pocket, there are a few microvilli that do not reach the surface of the sensory epithelium. The number of microvilli varies among species (usually 3–10) and is correlated with the approximate number of cilia normally observed on the ciliated OSNs. The crypts are surrounded by supporting cells (Hansen and Finger 2000). The shape of the crypt OSNs differs from that of the ciliated and microvillar OSNs. Most often they are pear-shaped, reaching up to about one third of the depth of the olfactory sensory epithelium. The cell cytoplasm contains cell organelles such as the Golgi apparatus, mitochondria, lysosomes, and the endoplasmic reticulum and the nucleus is located in the basal portion of the cell.

Non-Sensory Ciliated Cells

Non-sensory ciliated cells (Figs. 14B, D, F) are cylindrical and extend from the basal lamina to the epithelial surface. They are present in the olfactory epithelium of all teleosts and improve the circulation over the surface of the olfactory epithelium due to numerous cilia growing on the wide/extensive, flat apex of the cell (Zeiske et al. 1992). The number and length of the cilia varies among species (e.g., 40 in *Danio rerio*, Hansen and Zeiske 1998; 140 in *Anguilla anguilla*, Schulte 1972). The length of the cilia in these cells is usually greater than the length of the cilia in the OSNs (Hansen and Zielinski 2005). Numerous mitochondria are in the cytoplasm in the apical portion of the cell and a relatively small nucleus is found in the basal portion of the cell, usually with non-condensed chromatin, and is surrounded by a thick endoplasmic reticulum. The arrangement of non-sensory ciliated cells varies among species and can be of various types (Jakubowski 1981, Hansen et al. 1999) and can be: (i) arranged at the periphery of the olfactory lamella, (ii) dispersed evenly within the whole sensory area, or (iii) found on islands, strips or zones.

Supporting Cells

Supporting cells are specialized cells that surround and scaffold the OSNs. Single supporting cells are usually associated with a few OSNs. They have a cylindrical shape, often with a basal part that is clearly adjacent to the basal lamina. Bertmar (1973) distinguished supporting cells with cilia in *Salmo trutta m. trutta*, but there is no study determining if the ciliated supporting cells are identical to non-sensory ciliated cells. In the upper portion of the epithelium, the supporting cells are connected to each other by desmosomes. The most important feature of supporting cells is the presence of long tufts of tonofilaments in the cytoplasm, which are usually positioned perpendicular to the long axis of the cell (Zeiske et al. 1992).

Basal Cells

Basal cells are small cells of different shapes with a remarkably large nucleus in relation to the cell size and are located just above the basal lamina. These cells give rise to new non-sensory and sensory receptor cells (Thornhill 1967, Breipohl et al. 1976, Evans et al. 1982, Zippel et al. 1997, Hansen et al. 1999).

To date, it is unclear whether they can also produce supporting cells that are capable of mitotic division (Hansen and Zielinski 2005).

The Visual System

The eyes of fishes (Figs. 15A, B) play an important role in their survival and, in conjunction with other sensory modalities, allow them to perceive cues in a range of underwater environments, where the lighting conditions can vary enormously. Although not always the predominant sense, for most species, vision remains important for finding food, avoiding predators, social communication, finding reproductive partners, orientation, and navigation in often complex environments. The treatment of the morphology of the eye below will follow the optical pathway, as light passes through the transparent cornea, is refracted by the spherical lens, and focused on the retina, where the photoreceptors transform light energy or the "optical image" into electrical impulses or a "neural image", through the process of phototransduction. Each point in visual space is subtended by a corresponding point on the neural retina, which, in turn, is retinotopically mapped onto the visual centers in the brain. The light environment, behavioral strategies, and the complexity of each species' ecological niche have all contributed to the morphological and functional attributes of the fish retina. Such features as changes in retinal cell packing, the size of the visual field, eye size and shape, and eye mobility all appear to be under intense selective pressure.

Cornea

The cornea is a transparent window or goggle in the collagenous scleral coat of the eyecup and contributes to the structural support of the globe (Fig. 15B). The cornea also protects the inner structures of the eye from organismal invasion and unwanted environmental insults, and helps to balance the intraocular and extraocular pressures. Due to the similarity in the refractive index of the surrounding water and the aqueous humor within the eye, the fish cornea provides little, if any, refractive power compared to that provided by the spherical lens (Land 2002).

The structure of the fish cornea is similar to that of most aerial and terrestrial vertebrates. It consists of a multi-layered epithelium that forms a barrier that limits corneal hydration and maintains transparency. Microprojections (such as microridges, Fig. 15D) on the epithelial surface improve the diffusion of oxygen and nutrients and stabilize the epithelium in species that have infiltrated hostile environments (Collin and Collin 2000, 2006). Underlying the epithelium is a basement membrane, a thick stroma, and an endothelium supported by a thick basement membrane known as Desçemet's membrane. The corneal stroma consists of layers of parallel collagen fibrils (primarily collagen type I). The fibrils of adjacent lamellae are approximately perpendicular to each other to maintain transparency (Fig. 15F). In some species of fishes, two distinct stromas are present; a dermal stroma or secondary spectacle, which is continuous with the skin, and a scleral stroma, which is continuous with the scleral eyecup (Fig. 15B). The region separating the two stromas may be filled with granular material or even mucus, which allows the globe to rotate beneath a stable and clear protective goggle (Collin and Collin 2001).

Corneal iridescence is produced in some shallow water species by constructive interference of light reflected from multilayer stacks of material with different refractive indices (Lythgoe 1976, Collin and Collin 2001). The function of the iridescence includes reduction of intraocular flare, a birefringent filter and a means of camouflaging the pupil. Desçemet's membrane contributes to the mechanical strength of the cornea, while the endothelium (which is not always present in cartilaginous fishes) actively pumps bicarbonate ions and water out of the corneal stroma. Primary cilia, which protrude from the endothelial surface, are thought to monitor hydrostatic pressure and changes in the chemical composition of the anterior segment (Collin and Collin 1999). Over 100 species of diurnal teleosts are known to possess yellow corneas due to the inclusion of pigment granules (Heinermann 1984). These melanin granules (predominantly carotenoid) generally absorb short wavelength light and act as filters to protect the retina from bright light in shallow-water, reduce light scattering and chromatic aberration and enhance contrast (Kondrashev et al. 1986, Siebeck et al. 2003).

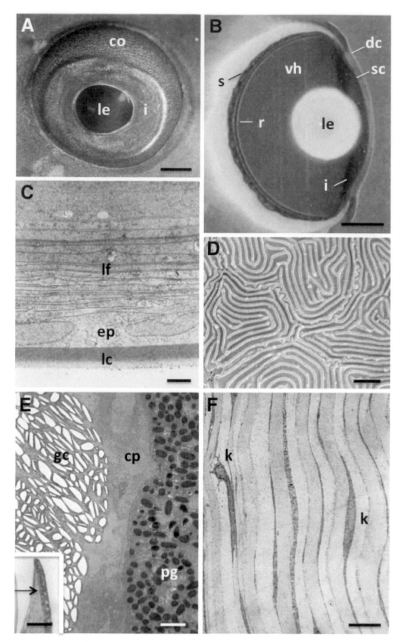

Fig. 15. The eye and cornea. **A.** Close up of the eye of the slingjaw wrasse, *Epibulus insidiator* showing the circular pupil, predominantly filled by the lens (le). Co, conjunctiva; i, iris. **B.** Eye of the Australian lungfish, *Neoceratodus forsteri* cryosectioned in the axial plane along the optical axis. Note the hemispherical shape of the globe and spherical lens (le). dc, dermal cornea; i, iris; s, sclera; sc, scleral cornea; vh, vitreous humor. **C.** Electron micrograph of the anterior region of the crystalline lens of the salamanderfish, *Lepidogalaxias salamandriodes* showing the outer lens capsule (lc), the epithelial layer (ep) and the inner lens fibers (lf). **D.** Scanning electron micrograph of the corneal surface of the blowfish *Torquigener pleurogramma* showing the complex series of microridges that extend from the outer epithelial cells. Note the interdigitations of the ridges along the cell boundaries. **E.** Transmission electron micrograph of the ultrastructure of the iris of the pipefish, *Corythoichthys paxtoni* showing the anterior collections of guanine crystals (gc) emanating from a central cell process (cp) and the posterior pigment granules (pg) or melanosomes. Inset: Low power view of the iris. Arrow depicts the direction of the incident light. **F.** Electron micrograph of the central cornea of the Florida garfish, *Lepisosteus platyrhincus* showing the alternating layers of collagenous lamellae oriented perpendicular to one another interrupted by keratocytes (k). Scale bars: 2 mm (**A**), 3 mm (**B**), 2 μm (**C, D, E**), 40 μm (**E**, inset), 4 μm (**F**).

Lens and Iris

The lens in fishes is an inelastic crystalline structure composed of protein and water. It is supported both by a suspensory ligament containing zonule fibers and the vitreous humor (Fig. 15B). The concentrically arranged lens fibers (which all lack a nucleus) are covered anteriorly by a single layer of cuboidal cells (lens epithelium) and encapsulated by a granular envelope (Fig. 15C). The lens protrudes through the pupil (defined by the circumferential edges of the pigmented iris) into the anterior chamber, often protruding beyond the surface of the body, conferring a wider field of view (Sivak 1980, Figs. 15A, B). The lenses of a large number of teleosts contain pigments, which act as filters by preventing short wavelengths of light from reaching the retina. Therefore, in addition to focusing light onto the retina, the lens may also protect the retina from the potentially harmful effects of ultraviolet light and/or may enhance visual acuity by excluding the shorter wavelengths most prone to chromatic aberration (Douglas and McGuigan 1989).

The iris controls the amount of light entering the eye through the pupil (Figs. 15A, B, E). In most teleosts, the pupil is not able to constrict or dilate in response to light (although there are exceptions, Douglas et al. 1998, 2002), while the pupil in elasmobranchs is highly mobile. An iris capable of rapid changes in pupil size (Litherland et al. 2009) provides a dynamic mechanism to regulate retinal illumination that would help maintain optimal visual performance in fluctuating lighting conditions. The iris is composed of a thick epithelium (continuous with the retinal pigment epithelium) containing pigment granules that, in many species, extends the disruptive coloration over the remainder of the head and body (Figs. 15A, E). However, the bulk of the anterior iris comprises bundles of guanine crystals by a basal lamina often separating large bundles of guanine crystals, which are highly reflective and act as a biological mirror that may help camouflage the pupil in bright light (Collin and Collin 1995; Fig. 15E). In species with mobile pupils, the iris would also include irideal sphincter or dilator pupillae muscles.

Retina

The fish retina has a basic structure similar to that in most other jawed vertebrates, but with the large number of fish species and the diversity of marine and freshwater habitats that they occupy, the range of structural, and therefore functional, variation is high. The retina of most fishes comprises three cellular layers (outer nuclear, inner nuclear and ganglion cell) and two plexiform layers (outer and inner; Figs. 16A, B). With respect to phototransduction, there are two major photoreceptor types—rods and cones. Rods are typically used in scotopic (nocturnal) light conditions, while cones are used for photopic (diurnal) vision. The presence of two or more types of cone photoreceptors with different spectral sensitivities is a prerequisite for the discrimination of color. Spectral sensitivity is governed by the absorption characteristics of the visual pigments housed in the lipid membrane of the outer segment discs and the arrangement and density of photoreceptors across the retina is one of the determining factors governing spatial resolving power and spectral sensitivity. Increased sampling by the photoreceptors in specific regions of the retina increases the resolving power in the corresponding part of the visual field.

Although the morphological criteria on which photoreceptor types are characterized is still equivocal in some groups of vertebrates (i.e., in jawless fishes, Collin et al. 1999, Collin and Trezise 2004, Collin et al. 2009), the eyes of most jawed fishes (chondrichthyans and osteichthyans) are duplex (contain both rods and cones). Photoreceptor types can be differentiated morphologically, especially if confirmed using other techniques; immunocytochemical markers, *in situ* hybrization probes to specific visual pigments and/or spectral characterisation using microspectrophotometry. The photoreceptors of most fishes are arranged into a regular mosaic, which can vary across species and in different regions of the retina. The photoreceptor patterns also vary but, in addition to optimising receptor packing, there may be functional advantages to increase photon capture and/or sensitivity at a given density.

Rod Photoreceptors

Rods are typically characterized structurally as possessing long, cylindrical outer segments containing stacks of discs, enclosed within a plasma membrane, which often contain gaps aligned along their longitudinal axes

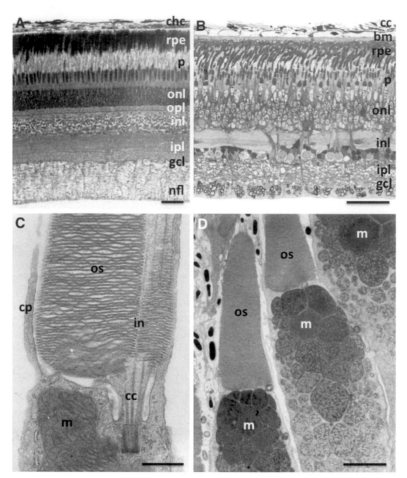

Fig. 16. Retinal and photoreceptor structure. **A.** Transverse section of the retina of the sea bream, *Girella* sp. in the dark-adapted state. chc, choriocapillaris; gcl, ganglion cell layer; inl, inner nuclear layer; ipl, inner plexiform layer; onl, outer nuclear layer; opl, outer plexiform layer; nfl, nerve fiber layer; p, photoreceptor layer; rpe, retinal pigment epithelium. **B.** Transverse section of the retina of the pink whipray, *Himantura fai*. Note the large horizontal cells within the inner nuclear layer (inl) and the bundles of myelinated axon profiles within the nerve fiber layer. bm, basement membrane. **C.** High power of the base of a rod photoreceptor in the cutlips minnow, *Exoglossum maxillingua* showing the connecting cilium (cc) linking the outer (os) and inner segment laden with mitochondria (m). Note that one of the microtubules of the connecting cilium follows the path of one of the incisures (in), where there is a longitudinal gap in the discs, which penetrates the outer segment. cp, calycal process. **D.** Long and short single cone photoreceptors in the retina of *E. maxillingua*. m, dense aggregation of mitochondria within the inner segment; os, outer segment. Scale bars: 30 μm (**A**), 50 μm (**B**), 0.5 μm (**C**) and 2 μm (**D**).

(incisures, Cohen 1963, Collin and Potter 2000, Figs. 16C, 17). The outer segment of a rod is supported by a mitochondria-rich inner segment connected via a connecting cilium (Fig. 16C). Emanating from the distal end of the inner segment are calycal processes, which project towards the back of the eye (sclerad) to form a supportive ring (Fig. 16C). The inner segment of the rod extends towards the more central parts of the eye (vitread) into a myoid process and finally to a terminal or spherule, which makes connections with both horizontal and bipolar interneurons via synaptic ribbons (typically up to three; Figs. 17A, 22B, C). The length and diameter of rods varies among species and ranges from about 20 μm in diameter in the Australian lungfish, *Neoceratodus forsteri* (Bailes et al. 2006) to < 1 μm in the deep-sea lanternfish *Symbolophorus rufinus* (De Busserrolles et al. 2014). The rods do not typically form regular mosaics (Fig. 18B), although they can be "grouped" to form macroreceptors (Locket 1971, Kunz et al. 1985, Collin et al. 1998), which are arranged into regular hexagonal arrays in some species (Figs. 18D, E). The rods (but

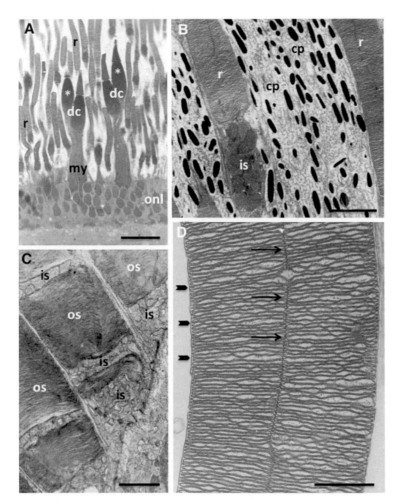

Fig. 17. Photoreceptor types. **A.** Transverse section of the retina of the cutlips minnow, *Exoglossum maxillingua* clearly showing two unequal double cones (dc) with principal (*) and accessory components supported by their long myoids (my) amidst large numbers of slender rods (r). onl, outer nuclear layer. **B.** Electron micrograph of two rod photoreceptors (r) surrounded by the processes of retinal epithelial cells (cp) containing pigment granules that migrate vitread in the light-adapted retina of the creek chub, *Semotilus atromaculatus*. **C.** Close up of the unique arrangement of aligned stacks of rod photoreceptor outer segment discs (os) and inner segments (is) in the retina of the deep-sea Sherborn's pelagic bass, *Howella sherborni*. **D.** High power view of a rod photoreceptor in the cutlips minnow, *E. maxillingua* showing an incisure (arrows) and the plasma membrane (arrowheads), which surrounds the outer segment. Scale bars: 20 μm (**A**), 3 μm (**B**), 1 μm (**C**), 0.7 μm (**D**).

also cones) of many teleosts possess a short accessory outer segment (AOS) that extends across half the diameter of the outer segment with a relatively featureless cytoplasm. Thought to be a reservoir for high energy metabolites (Fineran and Nicol 1974) or to fill the extracellular space between the photoreceptors during retinomotor movements (Wagner and Ali 1978), the AOS has been found exclusively in cones (Collin and Collin 1999) and in both rods and cones (Engström 1961, Yacob et al. 1977).

Cone Photoreceptors

The cone photoreceptor types in fishes may be classified as single, unequal double (with principal and accessory components), equal double, triple, or even quadruple cones, which are most commonly arranged into a single layer (Figs. 15A, B, D, 16D, 17A), although some deep-sea teleosts possess multiple layers or banks of receptors (rods) to increase sensitivity (Wagner et al. 1998; Fig. 17C). A cone typically possesses

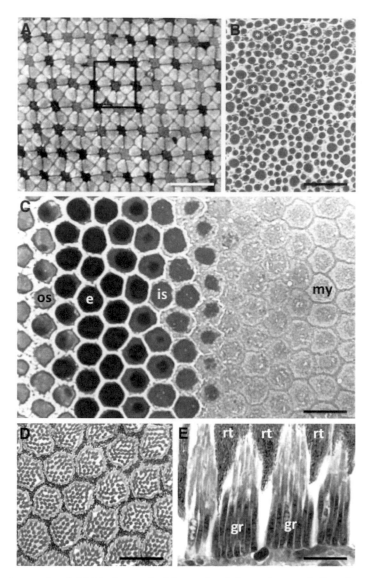

Fig. 18. Photoreceptor mosaics. **A.** Light micrograph of the regular photoreceptor array in the retina of the black bream, *Acanthopagrus butcheri* showing the repeating square mosaic of four equal double cones surrounding a central single cone (darkly stained profiles). One square unit is bordered. **B.** Photoreceptor array in the retina of the creek chub, *Semotilus atromaculatus* at the level of the outer segments. The largest profiles are the principal members of the unequal double cones (asterisks). The smallest profiles are single cones. **C.** An oblique tangential section of the outer retina of the southern hemisphere lamprey, *Mordacia mordax* showing (from vitread to sclerad) the hexagonal mosaic of rod-like photoreceptors from the level of the myoid (my) through the inner segments (is), the ellipsosomes (e) and the outer segments (os). **D.** Tangential section of the main retina of the deep-sea pearleye, *Scopelarchus michaelsarsi*. Groups of up to 35 rod photoreceptors are grouped into an array of macroreceptors separated by the processes of retinal pigment epithelial cells laden with tapetal reflecting material. **E.** Transverse section of the main retina in the tubular eye of *S. michaelsarsi* showing retinal tapetal material (rt) separating groups of rods (gr). Scale bars: 20 μm (**A**), 10 μm (**B**), 15 μm (**C**), 20 μm (**D**) and 15 μm (**E**).

an outer segment with stacks of discs that taper towards the back of the retina and are only partially enclosed in a plasma membrane (Figs. 16A, B, D, 17A). The size of the densely-packed mitochondria within the inner segment is often graded (smaller mitochondria located sclerad) and the myoid is richer in endoplasmic reticula (Fig. 16D). The cone terminal or pedicle is pyramid-shaped and contains higher numbers of synaptic ribbons (three or more) than the rod spherule (see Figs. 22B, C). Along the inner border of the opposing membranes of some cone inner segments, subsurface cisternae may exist to form what are considered double cones (equal or unequal in size). These cisternae may link large chains of photoreceptors early in retinal development, but are subsequently lost to leave regular arrangements of double cones. These may form mosaics comprising parallel rows, square patterns (four double cones surrounding a central single cone, Fig. 18A) and hexagonal patterns (six double cones surrounding a central single cone). Some of these arrangements may even occur in different retinal regions within the one eye (Reckel et al. 2001, Collin and Shand 2003). Theories regarding the function of the square mosaic include increasing both visual acuity (Engström 1963) and contrast (van der Meer 1992), providing a more uniform spectral sampling (Bowmaker 1990), allowing more detailed chromatic patterns to be resolved (Lythgoe 1979), and enhancing the detection of polarized light (Cameron and Easter 1993, Novales-Flamarique and Hawryshyn 1998). The square mosaic may also aid in the analysis of movement in all directions in contrast to a row mosaic, which may be suited to the perception of movement in two directions. The large diversity in photoreceptor mosaics among species and within different parts of the retina in an individual suggests that the environment and habitat play the most important roles in each species' sampling strategy rather than being controlled by any phylogenetic constraints (Collin and Shand 2003).

Photomechanical changes in the position of the rods and cones occur in many fishes (Douglas 1982, Douglas and Wagner 1982). In the light-adapted condition, rods migrate towards the sclera, leaving the inner cones to capture much of the available light (Fig. 17B), where the screening pigment within the retinal pigment epithelium (RPE), now located in front of the elongated rods, absorbs any extraneous light. In the dark-adapted condition, the cones migrate towards the sclera, and the rods remain more vitread to take advantage of the low illumination to which they are most sensitive. These antagonistic migrations may be rhythmic or arrhythmic.

Spectral Filters

In addition to having pigment within the cornea and lens, some cone types in fishes have intracellular inclusions such as oil droplets, ellipsoidal pigment, ellipsosomes and paraboloids, that act as spectral filters that tune the incident light path before it is absorbed by the photoreceptor visual pigments. Oil droplets are characteristic of strongly diurnal (and generally terrestrial) non-mammalian vertebrates, such as birds, turtles and lizards and are thought to improve color vision by tuning the spectral sensitivity functions of the cones. These densely pigmented spectral filters are generally absent from nocturnal or crepuscular species because they absorb light and reduce absolute sensitivity. Although common in non-teleost bony fishes, no teleosts possess oil droplets. The Australian lungfish possesses colorless (in the ultraviolet and short wavelength sensitive cones) and red (in the long wavelength sensitive cones) oil droplets that dominate the inner segment of the large photoreceptors (Bailes et al. 2006, Fig. 19A). The medium wavelength sensitive cones in *N. forsteri* and the southern hemisphere lamprey, *Geotria australis*, also possess a dense, granular, yellow pigment in both the ellipsoid and paraboloid regions of the inner segment that absorbs light strongly below 530 nm and may protect the outer segment from damage (Figs. 19A, B). In species with oil droplets and ellipsoidal pigmentation, light absorption by these filters reduces the overlap of the sensitivities of adjacent spectral cone types and improves color discrimination ability, despite the unavoidable reduction in total photon catch (Vorobyev 2003, Hart et al. 2008). Although similar in appearance to oil droplets, ellipsosomes do not act as spectral filters. Histological examination in teleosts (Nag and Battacharjee 1995) and lampreys (Collin and Trezise 2006) also reveals that they are globule-like concentrations of mitochondrial origin within the inner segment that have lost their cristae (Figs. 18C, 19B, C). These structures may play a role in trapping light and/or focusing light on the outer segments. In early ray-finned fishes, both rods and cones also possess dense concentrations of glycogen situated vitread of the inner segment ellipsoidal region. These are called paraboloids.

Retinal Pigment Epithelium, Choriocapillaris, and Choroid

The photoreceptor layer is backed by the retinal pigment epithelium (RPE), comprising a single layer of large cuboidal (or columnar) cells joined by tight junctions. These cells contain mitochondria, smooth

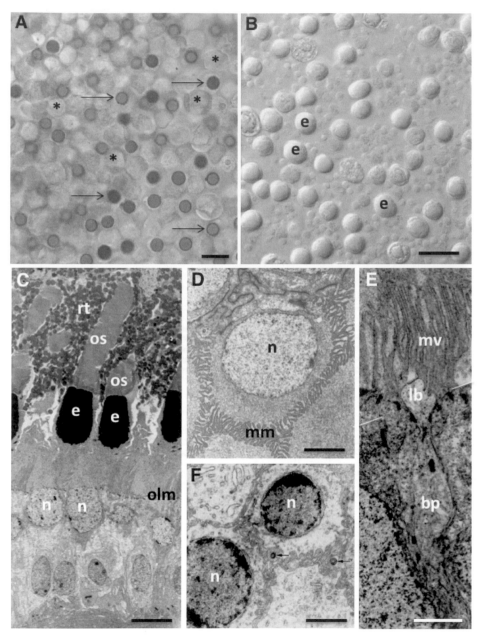

Fig. 19. Photoreceptor inclusions and the outer limiting membrane. **A.** Axial (whole mount) view of the unfixed retina of the Australian lungfish, *Neoceratodus forsteri*, showing the range of photoreceptor types containing red oil droplets (arrows), yellow myoidal pigment (faint profiles) and larger rods (asterisks). **B.** Axial view of the unfixed retina of the pouched lamprey, *Geotria australis*, viewed using differential interference contrast, showing the profiles of the five photoreceptor types at the level of the inner segments. One receptor type contains yellow pigment (small and granular), while another type contains an ellipsosome (e). **C.** Transverse section of the retina of the southern hemisphere lamprey, *Mordacia mordax*. Note the large photoreceptors containing an ellipsoidal ellipsosome (e) and long cylindrical outer segments (os). n, photoreceptor nucleus; olm, outer limiting membrane; rt, retinal tapetum. **D.** Tangential section at the level of the photoreceptor nuclei (n) showing the feathered interdigitations of the myoidal cell membranes (mm). **E.** Transverse section of the outer nuclear layer of *M. mordax* showing a bipolar process (bp) protruding through the outer limiting membrane as a Landolt's club. mv, microvillar processes of the Müller cells. **F.** Tangential section of the retina of *M. mordax* (at the level of the white lines in **E**). n, photoreceptor nuclei. Small arrows depict the profiles of the Landolt's clubs. Scale bars: 30 μm (**A**), 20 μm (**B**), 10 μm (**C**), 2 μm (**D, E**) and 4 μm (**F**).

endoplasmic reticula and Golgi bodies and are rich in melanin granules, which absorb any stray light not absorbed by the photoreceptors. In conjunction with the retinomotor movements of the photoreceptors, the melanin granules migrate in response to changes in ambient lighting conditions, where they occlude the rods in bright light conditions, allowing the incident light to freely strike the shortened cones and absorb the maximum amount of light (Fig. 17B). Under dark-adapted conditions, the pigment granules migrate towards the back of the retina with a concomitant shortening of the rods to allow the rods to maximally absorb the available light. The RPE also contains phagosomes in their apical processes, which phagocytose stacks of outer segment discs that are periodically shed (Young 1978; Fig. 20A). The RPE opposes Bruch's membrane, which comprises the basement membrane of the RPE and the endothelial cells of the blood vessels of the choriocapillaris (Figs. 20A, 21A). The choriocapillaris is a network of large and small calibre blood vessels interspersed with pigment granules, which gives rise to the choroid, a (often) pigmented layer of connective tissue and larger blood vessels and, in some species, a choroidal gland (Fig. 21). Although not a gland, this horseshoe-shaped mass of capillaries (or *rete mirabile*) surrounding the optic nerve may dampen the fluctuations in blood flow in the choroidal vessels, thereby negating any mechanical disturbance to the retina (Walls 1942).

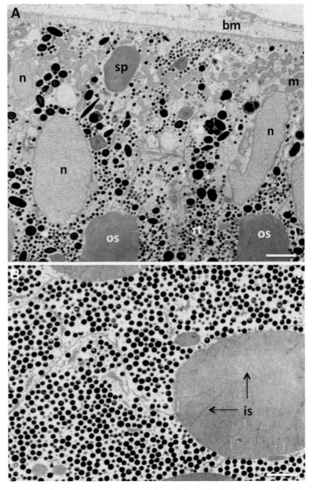

Fig. 20. The retinal pigment epithelium. **A.** Transverse section of the retinal pigment epithelium and basement membrane (bm) of the Florida garfish, *Lepisosteus platyrhinchus*. This species contains a retinal tapetum (rt) of pigment granules. m, mitochondria; n, nuclei; os, photoreceptor outer segments; sp, shedded photoreceptors discs. **B.** Tangential section through the retinal pigment epithelium of *L. platyrhinchus* showing a series of invaginations or incisures (is) of the rod outer segment. Note the density of tapetal pigment (dark, osmiophilic profiles) within the retinal pigment epithelial cells. Scale bars: 2 μm (**A, B**).

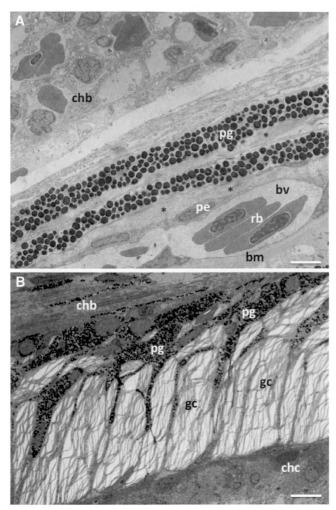

Fig. 21. The choroid and tapetum. **A.** Choriodal body (chb) and choriocapillaris in the eye of the cutlips minnow, *Exoglossum maxillingua*, showing dense aggregations of blood vessels (bv) containing red blood cells (rb), pericytes (pe) and melanosomes or pigment granules (pg) separated by collagen fibrils (asterisks). bm, basement membrane. **B.** Choroidal tapetum of the lemon shark, *Carcharhinus leucas* showing the processes of the melanocytes containing melanosomes or pigment granules (pg), which migrate vitread to occlude the reflective guanine crystals (gc) in the light-adapted condition. chb, choroidal body; chc, choriocapillaris. Scale bars: 3 μm (**A**) and 10 μm (**B**).

Tapeta

Tapeta are structural inclusions in the eyes that elicit "eyeshine" or light that is reflected out of the eye, especially under low light conditions. Tapeta can be either retinal, i.e., lying within the retinal pigment epithelium (one species of lamprey, some species of non-teleost ray-finned fishes, such as gars, as well as many species of teleosts) or choroidal, i.e., lying within the choroid (in all species of elasmobranchs examined and in some teleosts). When situated within the retina (retinal pigment epithelium), the tapetum (lucidum) is typically comprised of one or two types of reflectors: diffuse and specular. The diffuse reflectors are usually composed of small spherical or cuboidal particles (Figs. 18D, E, 20), whereas the specular reflectors consist of thin films (platelets) or needles. Both types increase sensitivity by reflecting light that has already passed through the photoreceptors back into the photoreceptor outer segments for a second time in order to improve sensitivity. Eyeshine can vary from white to blue to yellow in fishes depending on the state of light adaptation but may not fill the fundus with some retinal regions devoid of reflective material, thereby increasing sensitivity in restricted parts of the visual field. Choroidal tapeta are widespread among several groups of fishes including Chondrichthyes (sharks and rays, Denton and Nicol 1964, ratfishes, Nicol 1981, Fig. 21B), Polypteriformes (bichir), Acipenseridae (sturgeons, Nicol 1969), Dipnoi (lungfishes, Locket 1977), the coelacanth (*Latimeria chalumnae*, Locket 1973), and a few teleosts, e.g., the big-eye, *Priacanthus* sp. and the mosquitofish, *Gambusia affinis* (Lantzing and Wright 1982). Choroidal tapeta are well developed in cartilaginous fishes, in which overlapping tapetal cells are interdigitated by melanocytes containing melanosomes that migrate in response to light (Fig. 21B). This enables the tapetum to be occluded in the light-adapted state when there is no need to enhance sensitivity. The tapetal cells in elasmobranchs contain 9–20 layers of reflective crystals, typically guanine (Denton and Nicol 1964, Braekevelt 1994, Fig. 21B), where some species have the ability to vary tapetal spectral reflectance (i.e., in the sandbar shark, *Carcharhinus plumbeus*) to match ambient illumination (with respect to both intensity and spectral composition) in a variety of light environments to achieve camouflage as well as to increase retinal illumination (Best and Nicol 1967, Litherland et al. 2009).

The Outer Retina

Immediately internal to the photoreceptor layer lies the outer nuclear layer (ONL) comprised predominantly of the photoreceptor nuclei, which can often be differentiated (into rod and cone nuclei) by their size and location, characteristics following staining, and the morphology of their terminal processes (Figs. 16A, B). The ONL is bounded by the glial processes of the Müller cells, whose cell bodies are located within the inner nuclear layer and form the outer limiting membrane and the terminals of the photoreceptors. Within the ONL of some teleosts, lungfishes, elasmobranchs and lampreys, a population of bipolar cells displaced from the inner nuclear layer sends thick processes (or Landolt's clubs) towards the back of the retina (sclerad) to sit beyond the outer limiting membrane (at the base of the photoreceptors; Figs. 19E, F). The presence of glycogen and mitochondria within these projections suggests a metabolic function, where their close communication with the microvilli of the Müller cells (Fig. 19E) would provide a route for the exchange of energy stores throughout the retina (Locket 1975, Collin et al. 1999). The terminal processes (rod spherules and cone pedicles) and the sclerad processes of the bipolar and horizontal cells forms the thin outer plexiform layer (OPL; Figs. 22A–C).

The Inner Retina

The inner retina comprises the bipolar, horizontal and amacrine cell nuclei, which collectively occupy the inner nuclear layer (INL; Figs. 16A, B, 22A). The bipolar cells in fishes transmit signals directly from the photoreceptors to the ganglion cells. These neurons are diverse in their cell shape and size and receive synapses from horizontal cells, producing a receptive field organization with chromatically antagonistic center-surround response properties. Horizontal cells transfer information laterally across the retina and are capable of mediating chromatic interactions between different spectral cone types, generating the antagonistic surround of the bipolar cell receptive field and modulating spatial summation. The large diversity of amacrine cell types among teleosts suggests they have varied and complex functions but also transfer signals laterally across the retina. The processes from four types of neurons converge at the level of the inner plexiform layer (IPL). The bipolar (or input cells) make contact with the ganglion or output cell processes, in addition to the horizontal and amacrine cells, which send processes laterally throughout this layer. In some species, interplexiform cells with nuclei lying within the ONL carry information from the inner plexiform layer to the outer plexiform layer, spreading processes at both retinal layers. The innermost layers in the retina are the ganglion cell layer and the nerve fiber layer (Figs. 16A, B, 22D). Each ganglion cell possesses an axon, which carries information about the image to the visual centers of the brain via the optic nerve. The level of convergence of neural signals from the photoreceptors to the ganglion cells sets the limits on the spatial resolving power of the eye, and ultimately, visual behavior. There are many ganglion cell types with different functions (Fig. 22E). The retina is lined by the inner limiting membrane, which comprises the inner processes of the Müller cells and a basement membrane (Fig. 22D). Some species also possess vitreal vascularization, providing nutrition to the inner retina (Fig. 22D). The eyes of fishes lack an intraretinal vascular system.

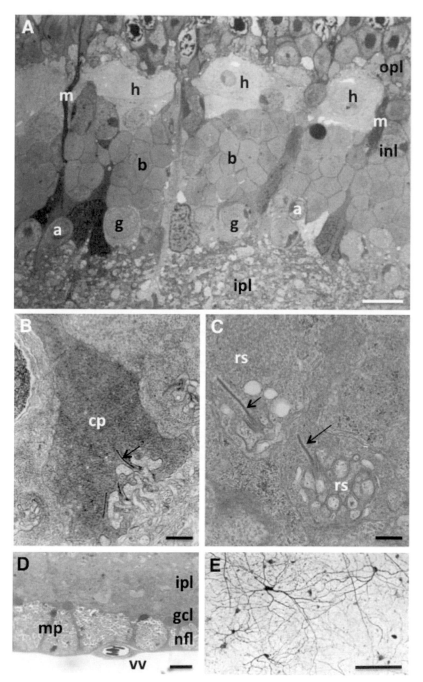

Fig. 22. The inner retina. **A.** Transverse section of the inner nuclear layer of the bamboo shark, *Chiloscyllium punctatum.* a, amacrine cells; b, bipolar cells; dg, ganglion cells displaced to the inner nuclear layer (inl); h, horizontal cells, ipl, inner plexiform layer; m, Müller cell processes; opl, outer plexiform layer. **B, C.** Transverse sections of a pyramid-shaped cone pedicle (cp, **B**) and two rod spherules (r, **C**) in the retina of the cutlips minnow, *Exoglossum maxillingua* showing their synaptic ribbons (arrows) emanating from a synaptic cleft connecting horizontal and bipolar cell processes. **D.** Transverse section of the inner retina of the creek chub, *Semotilus atramaculatus* showing bundles of ganglion cell axons separated by Müller cell processes (mp) within the nerve fiber layer (nfl). gcl, ganglion cell layer; ipl, inner plexiform layer; vv, vitreal blood vessel. **E.** Retrogradely labeled retinal ganglion cells filled with cobaltous-lysine from the optic nerve in the Florida garfish, *Lepisosteus platyrhinchus* as seen in whole mount. Note the variation in soma size and dendritic arbor. Scale bars: 10 μm (**A**), 0.5 μm (**B**), 0.3 μm (**C**), 15 μm (**D**) and 100 μm (**E**).

The Auditory System

The inner ear of fishes serves the senses of hearing (e.g., Ladich and Popper 2004, Manley 2000, Popper 1996, Popper and Fay 1997) and balance (orientation in 3-D space). The auditory and lateral line systems comprise the octavolateralis system (Popper 1996). The inner ear can function as "far-range detector" perceiving sounds in terms of particle motion and/or sound pressure (Popper 1996, Popper and Fay 2011), while the lateral line system plays an important role in detecting hydrodynamic motions within several body lengths (see above). Especially at low frequencies (up to about 200 Hz), the lateral line system and inner ear may act in concert (Braun and Coombs 2000, Higgs and Radford 2012). When detecting sound, the otolith end organs of the inner ear are primarily stimulated by particle motion (Hawkins and MacLennan 1976). This stimulation is assumed to be defined by the relative motion between the fish's body including the sensory epithelia of the ear—which have almost the same density as the surrounding water and are thus "transparent" to sound (Rogers and Cox 1988, Rogers et al. 1988)—and the denser otolith or otoconial mass (= inertial mass) overlying these sensory epithelia (de Vries 1950, Hawkins 1993, Popper and Lu 2000, Popper et al. 2005). Bony fishes, which possess a gas-filled cavity (swim bladder, lung, or any other gas-filled space), may also detect the pressure component of sound (Braun and Grande 2008, Popper and Fay 2011). In these fishes, the gas-filled bladder acts as a pressure-to-displacement transducer (Popper et al. 2003, Rogers et al. 1988).

What sensory stimuli can fishes derive from sounds and why is hearing important for them? Hearing in fishes is crucial for gaining information about abiotic sound sources like the pounding of waves on reefs, and sounds emanating from biotic sources indicating the presence of predators, prey or conspecifics (e.g., in the context of sound communication; Tavolga 1960, Hawkins and Myrberg 1983, Ladich 2014). A second essential function of the inner ear is related to the sense of balance (vestibular sense). Recent studies provide support for the "vestibular first" hypothesis introduced by van Bergeijk (1967), which suggests that the vestibular sense (= sense of balance) evolved first whereas the ability to hear (here defined as the perception and analysis of [complex] sounds at and above several hundred Hz) was developed later (see Fritzsch and Straka 2014).

The ears of jawless fishes have one (hagfishes) or two (lampreys) semicircular canals and only one otolith end organ (Popper and Hoxter 1987, Thornhill 1972). In contrast, in cartilaginous fishes (elasmobranchs and chimaeras) and bony (lobe-finned and ray-finned) fishes the inner ear has three semicircular canals and two or three otolith end organs (e.g., Evangelista et al. 2010, Platt et al. 2004, Popper 1977). Ray-finned fishes (teleost and non-teleost actinopterygians) display an amazing diversity in ear morphology (e.g., Retzius 1881, Ladich and Popper 2004, Ladich and Schulz-Mirbach 2016, Schulz-Mirbach and Ladich 2016), hearing abilities (frequency range and auditory sensitivities, Popper et al. 2004, Popper and Schilt 2008, Popper and Fay 2011), and swim bladder modifications that may enhance audition (Braun and Grande 2008, Ladich and Popper 2004, Ladich 2014, Tricas and Webb 2016).

Inner Ear Anatomy

The ear in fishes shows two types of end organs; those only covered by a gelatinous mass and those overlain by a gelatinous mass and one or more biomineralisates (otoconia, otoliths). The end organs (ampullae) of the semicircular canals are covered by a gelatinous cupula while, the otolith end organs, namely the saccule, the lagena, and the utricle are overlain by an otoconial mass or solid otoliths (Figs. 23A, B; Platt and Popper 1981, Popper 1996). In each of the three otolith end organs, the sensory epithelium is called a macula. Each macula is overlaid by numerous tiny crystals (otoconia) in cartilaginous fishes (Carlström 1963, Rosauer and Redmond 1985, Ladich and Schulz-Mirbach 2016), a solid calcareous biomineralisate (otolith, Fig. 31C) in teleost fishes (Nolf 1985, Falini et al. 2005, Popper et al. 2005), fused otoconia in lobe-finned fishes (Carlström 1963, Gauldie et al. 1986a), or by a combination of both otoconia and otoliths in non-teleost actinopterygians (Carlström 1963, Popper and Northcutt 1983, Mathiesen and Popper 1987) and some teleosts (Gauldie et al. 1986b). Otoconia or otoliths are connected to the hair cells of the sensory macula through the otolithic membrane, an extracellular pellicle (Fay and Edds-Walton 2008). The otolithic membrane is composed of a gelatinous and a fibrous region (Murayama et al. 2002,

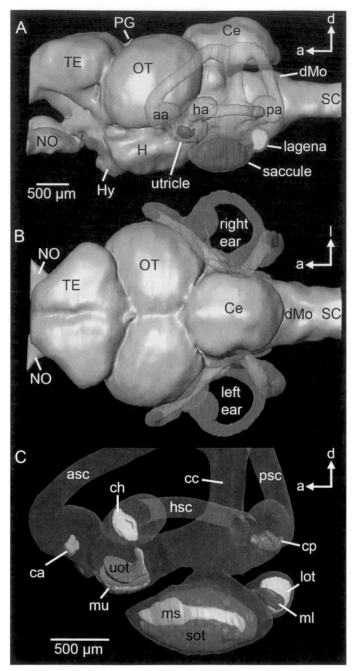

Fig. 23. Overview of the position of the left and right inner ears relative to the brain illustrating the main inner ear components. 3-D reconstructions (based on microCT imaging) of the inner ears and the brain of a teleost fish (*Steatocranus tinanti*, Cichlidae, Cichliformes) are shown in lateral (**A**), dorsal (**B**), and medial (**C**) views. Inner ear components: (1) semicircular canals: anterior (asc), horizontal (hsc), and posterior (psc) canals; ampulla of the anterior (aa), horizontal (ha), and posterior (pa) canals; cc, common canal; (2) otolith end organs: utricle, saccule, lagena; (3) sensory epithelia: cristae of the anterior (ca), horizontal (ch), posterior (cp) canals; maculae of the utricle (mu), saccule (ms), and lagena (ml); (4) otoliths: uot, utricular, sot, saccular, and lot, lagenar otoliths. a, anterior; d, dorsal; Brain parts shown in (A) and (B): Ce, cerebellum; dMo, dorsal medulla oblongata; H, hypothalamus; Hy, hypophysis; NO, optic nerve; OT, optic tectum; PG, pineal gland; SC, spinal cord; TE, telencephalic lobes. Modified from Schulz-Mirbach et al. (2013).

Color version at the end of the book

2004, 2005, Schulz-Mirbach et al. 2011; Figs. 24, 25) made up of non-collagenous as well as collagen-like glycoproteins (Hughes et al. 2006, Lundberg et al. 2006). The otolithic membrane transduces the movement of the inertial mass (either the numerous otoconia or large solid otolith) to the ciliary bundles of the sensory hair cells in the macula. The otolithic membrane also prevents the otolith from detaching from the macula, when strong forces act on the system (mechanical trauma; e.g., Benser et al. 1993). Each of the three semicircular canals ends in a dilation (ampulla) that houses an additional sensory epithelium (crista; Fig. 23C; e.g., Popper 1996). The crista has sensory hair cells with long ciliary bundles, all oriented into the same direction and it is overlaid by a cupula, a thick convex layer of a mucopolysaccharide gel (e.g., Flock and Goldstein 1978, Mathiesen 1984, Silver et al. 1998). Motion of the endolymphatic fluid in the semicircular canals, provoked by angular acceleration (i.e., rotational change in body orientation), leads to a bending of the cupula and thus to a shearing of the ciliary bundles (Flock and Goldstein 1978).

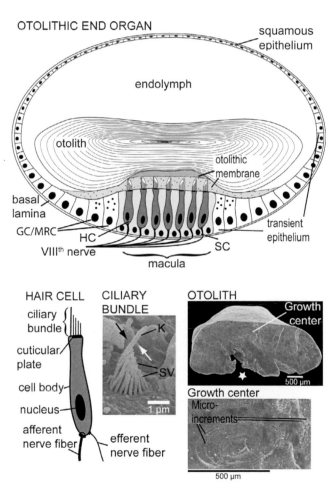

Fig. 24. Diagram of transverse section of an otolith end organ illustrating the relationship between otolith, otolithic membrane and the underlying sensory macula. Until now, it is not clear whether ciliary bundles have direct contact to the otolith (Popper et al. 2005). HC, hair cell; GC, granular cell; MRC, mitochondria-rich cell; SC, supporting cell. Modified and combined from Ibsch et al. (2001), Murayama et al. (2004), Pisam et al. (1998), Söllner and Nicolson (2004), and Takagi (1997). Hair cell and ciliary bundle: Schematic diagram of a hair cell in lateral view displaying the apical ciliary bundle (modified from Popper and Lu 2000). A scanning electron micrograph of a ciliary bundle shows the eccentrically placed kinocilium (K) and the array of stereovilli (SV) that are anchored in the cuticular plate (note that the latter is only seen in the schematic drawing). Deflection of the bundle towards the kinocilium leads to maximum depolarisation and thus maximum stimulation (white arrow), deflection in the opposite direction results in maximum hyperpolarisation (black arrow; Hudspeth 1985). Otolith: Scanning electron micrograph showing a transverse section of a lagenar otolith that clearly displays the center of otolith growth and growth rings (micro-increments). The white asterisk marks the furrow of the lagenar otolith housing the macula lagenae.

Relationship between Otolith and Macula

Tomographic sections:

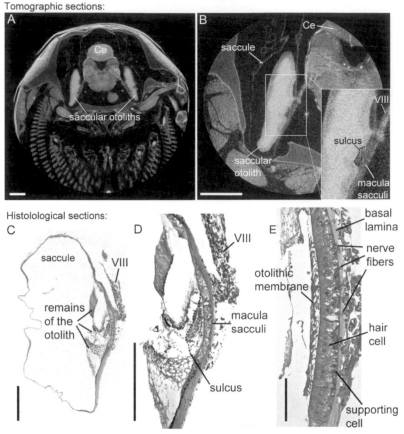

Fig. 25. Transverse µCT (**A, B**) and histological sections (**C–E**) displaying the relationship between otolith, otolithic membrane and macula in the saccule. All sections stem from the inner ears of the teleost fish species *Poecilia mexicana* or *Poecilia* sp. (Cyprinodontiformes, Poeciliidae). The higher resolution tomographic section in (**B**) shows the sulcus acusticus of the saccular otolith housing the macula sacculi. While the otoliths are well visible in the tomographic sections, the histological sections only show remains of the otolith (proteinaceous otolith matrix) due to dissolution of hard parts which is necessary to prepare samples for histological sectioning. In the inset in (**B**) and in (**D**), the sulcus acusticus is indicated by a dotted line and the macula sacculi is labeled red. The sample shown in (**A–B**) was contrast-enhanced with phosphotungstic acid prior to µCT imaging (Schulz-Mirbach et al. 2013). Histological sections were stained with Richardson's solution (Richardson et al. 1960). Ce, cerebellum; VIII, part of the VIIIth cranial nerve innervating the macula sacculi. Scale bars: 500 µm (**A–D**), 100 µm (**E**).

Color version at the end of the book

The otolith either overlies the whole macula or the otolith covers only part of the macula whereas other regions of the macula are overlain by otolithic membrane only (e.g., Platt 1977, 1993, Webb et al. 2010, Deng et al. 2011). It is hypothesized that those parts of the macula covered by the otolith and the uncovered portions are stimulated differently (Rogers and Cox 1988), but the mechanism is not yet known (Rogers and Zeddies 2008, Zeddies et al. 2011).

The macula neglecta, a fourth macula, has been identified in several species of ray-finned fishes and in all elasmobranch fishes examined (Fay et al. 1974, Corwin 1977, Saidel and McCormick 1985, Myrberg 2001, Bever and Fekete 2002, Buran et al. 2005, Casper 2011). It consists of one or two small patches of hair cells located at the base of the common canal that connects the anterior and posterior semicircular canals (see Fig. 11 in Ladich and Schulz-Mirbach 2016). The macula neglecta is covered only by a gelatinous cupula (without an overlying otoconial mass or an otolith) and hair cells are stimulated like those of the cristae (see Casper 2011). The functional role of the macula neglecta in bony fishes, such as bowfin (*Amia calva*), is still elusive, but in elasmobranch fishes there is significant evidence that it plays a major role in audition (Fay et al. 1974, Corwin 1989, Myrberg 2001, Gardiner et al. 2012).

Main Inner Ear Components

Sensory Hair Cells

The sensory hair cell is the sensory unit of the maculae and the cristae of the inner ear (Fig. 24; Hudspeth 1985, 1989) as well as the neuromasts of the lateral line system (see Section on the Lateral Line System). Each hair cell transforms mechanical force into electric signals, through mechanotransduction (Hudspeth and Corey 1977, Hudspeth 1985, 1989, Hudspeth et al. 2000, Gillespie and Müller 2009). Vestibular and auditory hair cells have a similar structure, which is like that of the neuromasts of the lateral line system (e.g., Coffin et al. 2004, Webb 2014a, b). The apical part of the hair cells is characterized by a cuticular plate supporting the ciliary bundle (Hudspeth 1985) and the hair cells are innervated by afferent (and in some cases also by efferent) neurons in their basal region. The ciliary bundle consists of one eccentrically placed kinocilium (true, microtubule-based cilium), and actin-supported stereovilli (often called stereocilia); the latter are arranged in a staircase array with the longest stereovillum being next to the kinocilium (Hudspeth 1985). The eccentric position of the kinocilium leads to a morphological and physiological polarization of the ciliary bundle, which is also referred to as ciliary bundle orientation (Hudspeth 1985, Kindt et al. 2012, Popper and Lu 2000). The kinocilium and stereovilli are connected to each other by small tip links (Gillespie and Müller 2009, Kindt et al. 2012). Deflection of the bundle towards the kinocilium leads to the opening of cation-selective transduction channels via these elastic tip links resulting in depolarization and more transmitter release at the base of the sensory hair cell (for different models see also Fig. 3 in Gillespie and Müller 2009, Hudspeth 1985, 1989, Popper and Lu 2000). Deflection of the bundle away from the kinocilium causes the closing of these channels, hyperpolarization and less transmitter release (Gillespie and Müller 2009, Hudspeth 1985, 1989, Popper and Lu 2000). Both vestibular and auditory hair cells in fishes retain the kinocilium throughout life (Kindt et al. 2012), but in mature (auditory) hair cells of amphibians (Gillespie and Müller 2009) and amniotes (auditory hair cells in birds and mammals; e.g., Kikuchi and Hilding 1965, Leibovici et al. 2005), the kinocilium is reduced during development and thus absent in adults. Based on these findings and due to contradictory experimental evidence, a potential mechanotransductive role of the kinocilium, in particular, is still controversial (see Kindt et al. 2012).

Hair Cell Heterogeneity. Like amniotes, fishes have been shown to possess different types of sensory hair cells (hair cell heterogeneity; Popper 2000): Type I-like hair cells, found in the central parts of the macula sacculi or in the striolar region of the macula utriculi, and Type II hair cells, typical of the margins of the macula sacculi or extrastriolar regions of the macula utriculi (Chang et al. 1992, Popper et al. 1993). Type I-like hair cells are characterized by: (1) afferent and efferent innervation whereas type II hair cells only display afferent innervation, (2) larger mitochondria, and (3) the presence of large subnuclear bodies of the endoplasmatic reticulum (Chang et al. 1992). The large nerve calyx typical of type I cells in amniotes (e.g., Xue and Peterson 2006) which is absent in type I-like hair cells of fishes has been identified in some hair cells of the cristae in goldfish (Chang et al. 1992, Popper et al. 1993, Popper 2000, Eatock and Songer 2011).

Ciliary Bundle Types. Within a macula, different types of ciliary bundles (Fig. 26) can be identified based on the (1) length of the kinocilium, (2) number of stereovilli, and (3) ratio of the length of the kinocilium to the length of the longest stereovillum (e.g., Deng et al. 2011, 2013, Platt and Popper 1981, 1984, Popper 1977). It is hypothesized that different hair cell types and different types of ciliary bundles have different physiological attributes (Furukawa and Ishii 1967). Taller ciliary bundles may mainly be involved in the detection of lower frequencies and smaller bundles in the detection of higher frequencies (Platt and Popper 1984, Smith et al. 2011, Deng et al. 2013).

Ciliary Bundle Orientation. The sensory maculae of the three otolith organs (the macula utriculi, macula sacculi, and macula lagenae) are either mainly oriented horizontally (macula utriculi) or vertically, i.e., along the dorso-ventral axis (macula sacculi, macula lagenae). The sensory hair cells of these maculae are generally divided into groups with different ciliary bundle orientations (Fig. 27; Popper 1976, Hudspeth and Corey 1977). The orientation pattern in the macula utriculi is highly conserved among fishes (Platt and Popper 1981, Mathiesen 1984, Ladich and Schulz-Mirbach 2016, Schulz-Mirbach and Ladich 2016), but there is variation in shape and orientation patterns in some species. Herring-like fishes (Clupeiformes) possess a tripartite macula utriculi (along with an exceptionally [tetrahedral] shape of the overlying utricular otolith; Denton and Gray 1979, Blaxter et al. 1981, Popper and Platt 1979), the catfish family Ariidae displays a thin macula running along the large utricular otolith like an equatorial band (Popper and Tavolga 1981), and the macula utriculi in the deep-sea fish family Melamphaidae is characterized by two distinct lobes in the anterior portion (Deng et al. 2013). In the cristae of the semicircular canals, ciliary bundles are all oriented in the same direction, while the ciliary bundles in the one or two patches of the macula neglecta may have the same or different directions (Popper et al. 2003, Webb et al. 2010, Gardiner et al. 2012, Ladich and Schulz-Mirbach 2016).

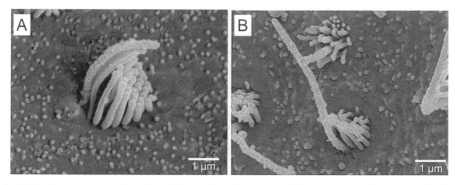

Fig. 26. Scanning electron micrographs of two different ciliary bundle types. The two types are characterized by the different absolute length of the kinocilium and the ratio between kinocilium and the longest stereovillus. **A.** Ciliary bundle with a short kinocilium and the longest stereovillus being almost as long as the kinocilium [micrograph shows a ciliary bundle on the macula sacculi in the cichlid *Etroplus maculatus*]. **B.** The kinocilium in this bundle type is at least two or three times longer than the longest stereovillus [micrograph shows a ciliary bundle on the macula sacculi in the cichlid *Steatocranus tinanti*]. The bundle type shown in (**B**) is often found at the margins of maculae.

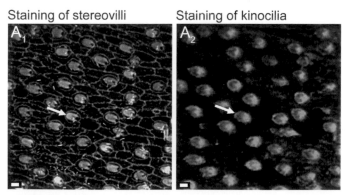

Fig. 27. Confocal images of double labeling of ciliary bundles reveals bundle orientation. In (A_1), the cuticular plate and the stereovilli of ciliary bundles are stained with TRITC-labeled phalloidin (which binds to F-actin); the kinocilium in each bundle is represented by a "black hole". (A_2), Kinocilia of the same ciliary bundles stained with anti-bovine α-tubulin mouse monoclonal antibodies and Alexa Fluor 488 conjugated anti-mouse secondary antibodies are shown as bright green dots corresponding to the positions of the "black holes" in (A_1). The white arrows mark the identical hair cell in (A_1) and (A_2) and indicate the orientation of the ciliary bundle based on the position of its kinocilium. Images stem from the macula sacculi of the teleost fish *Poecilia mexicana* (Cyprinodontiformes, Poeciliidae). Scale bars = 2 μm. Modified from Schulz-Mirbach et al. (2011).

The greatest diversity in orientation pattern of hair cells occurs in the macula sacculi (Platt and Popper 1981, Popper and Coombs 1982). In cartilaginous fishes, the macula sacculi is elongate displaying mainly two vertical groups of ciliary bundles (Lowenstein et al. 1964, Corwin 1981, Barber and Emerson 1980). In non-teleost actinopterygians (Popper and Fay 1993), the macula sacculi is hook-shaped (*Polypterus bichir*) or has a hook-shaped anterior (rostral) portion (e.g., *Amia calva*). In the anterior portion, ciliary bundle orientation follows the curvature of the closest macula margin, thereby creating horizontal groups (Popper and Fay 1993, Popper and Northcutt 1983, Lovell et al. 2005). In teleosts, the macula sacculi displays high diversity in overall shape as well as in the orientation patterns of ciliary bundles. Five main orientation patterns have been described among modern bony fishes (Popper and Coombs 1982); four have vertical and "true" horizontal orientation groups and are termed standard, dual, opposing, or alternating patterns and the fifth pattern type only has vertical orientation groups and is found in mormyrids and otophysan fishes (Popper and Platt 1983, Popper and Coombs 1982).

In cartilaginous fishes, lobe-finned fishes (sarcopterygians; Platt 1994, Platt et al. 2004), non-teleost actinopterygians (Popper 1978, Popper and Northcutt 1983, Mathiesen and Popper 1987, Lovell et al. 2005), and teleosts (for an overview see Platt and Popper 1981), the macula lagenae is crescent or half-moon shaped and contains two main orientation groups. In rays, these two orientation groups are less distinct (Barber and Emerson 1980, Lowenstein et al. 1964), while sharks show two rather clearly distinguishable groups in the macula lagenae (Lovell et al. 2007). In non-teleost actinopterygians (except *Polypterus bichir,* *Amia calva*) and some deep-sea elopomorphs (teleosts), the macula lagenae shows three orientation groups (Buran et al. 2005, Lovell et al. 2005, Mathiesen and Popper 1987, Popper and Northcutt 1983). Moreover, non-teleost actinopterygians also display a considerable diversity in the shape of the macula lagenae (Popper and Northcutt 1983, Mathiesen and Popper 1987). The macula lagenae in non-teleost actinopterygians is almost as large as or even larger than the macula sacculi (except in *Amia calva*) contrasting the condition in many teleost species (Platt and Popper 1981, Ladich and Popper 2004).

The functional roles of different orientation groups of ciliary bundles within a macula are still not completely understood. It is assumed that responses of sensory hair cells belonging to different orientation groups may play a role in directional hearing (Lu et al. 1998, Lu and Popper 2001, Lu et al. 2002, 2003, 2004, Rogers and Zeddies 2008, Zeddies et al. 2011). Moreover, correct directional adjustment of ciliary bundles on the sensory epithelium during ontogenetic development is a complex process. In chicken embryos, it was observed that the initially centrally located kinocilium moves to its eccentric position within the bundle (Sienknecht 2013, Sienknecht et al. 2011, 2014). Subsequently, ciliary bundles located near each other within a certain area are first roughly adjusted into the same direction and may undergo a final refinement step (Sienknecht 2013, Sienknecht et al. 2011, 2014). The processes of the formation of ciliary bundle orientation patterns are still poorly understood in fishes and need to be studied (Sokolowski and Popper 1987).

Other Cell Types in the Otolith End Organs

Sensory maculae are comprised of sensory hair cells (see above) and supporting cells (Fig. 24). Supporting cells are assumed to play a multifunctional role in regeneration of damaged hair cells, nutrition, transport of ions, and formation of otoliths/otoconia, among others. So far, this multifunctional role is, however, only well documented in mammals (Monzack and Cunningham 2013). The macula in fishes is flanked by large granular cells and ionocytes followed by increasingly flat cells lining the basal lamina (Pisam et al. 1998, Ibsch et al. 2001, Murayama et al. 2004). Granular cells are supposed to secret components needed for the formation of the otoliths (e.g., secretion of otolin-1 which is part of the otolith and the otolithic membrane; for an overview see Lundberg et al. 2015), whereas the mitochondria-rich cells such as large and small ionocytes may regulate the ionic composition of the endolymph (Pisam et al. 1998, Murayama et al. 2004).

Otoliths

The three otolith end organs contain the utricular otolith (lapillus), the saccular otolith (sagitta), and the lagenar otolith (asteriscus) (Figs. 23A, C), which relate to the stone-, arrow-, and star-like shape of these otoliths in otophysan fishes (such as goldfish, carp, or catfishes; Werner 1928), respectively. With the exception of the utricular otolith, which has a "stone-like" appearance in most non-otophysans as well (but see utricular otoliths in clupeiform fishes; Assis 2005), the saccular and lagenar otoliths in non-otophysan fishes possess a shape that is distinctly different from an arrow- or a star-like shape (see, e.g., Nolf 1985, Assis 2003, Tuset et al. 2008). In most non-otophysan fishes, the saccular otolith is the largest of the three otoliths, but in otophysans, the lagenar or the utricular otolith is larger than the thin and needle-shaped saccular otolith. The saccular otoliths, and to a lesser degree the lagenar and utricular otoliths, are characterized by a species- or population-specific shape (e.g., Nolf 1995, Schulz-Mirbach and Plath 2012). It is still not known how different otolith shapes affect the movement of the otolith relative to the underlying macula provoked by sound stimuli (de Vries 1950, Sand and Michelsen 1978). It is hypothesized that complex 3-D shapes of otoliths may result in a complex "rocking-like" motion rather than in a simple back and forth movement (Platt and Popper 1981, Krysl et al. 2012).

In teleost fishes, the utricular and saccular otoliths are made up of aragonite, whereas lagenar otoliths generally consist of vaterite (e.g., Oliveira and Farina 1996, Falini et al. 2005). Under stressful conditions such as overcrowding or temperature stress, the composition of the saccular otoliths may switch from aragonite to vaterite (Sweeting et al. 2004, Oxman et al. 2007). One of the exciting questions is why solid, rhythmically growing, and complex shaped otoliths evolved in teleosts while in cartilaginous fishes and most other vertebrates the macula is overlain by numerous tiny otoconia (Popper et al. 2005, Schulz-Mirbach et al. 2018).

Otoliths play an important role in fisheries management for stock discrimination (otolith shape, otolith chemistry) and determination of age structures within stocks or populations (annual rings, microincrements; Canas et al. 2012; for an overview see Campana and Thorrold 2001, Starrs et al. 2016). Otoliths are also used in paleoichthyology to reconstruct patterns of fish diversity as otoliths are often abundant in Paleogene (65 mya) and more recent sediments (references in Nolf 1985, Schulz-Mirbach and Reichenbacher 2006), including those at archaeological sites (Cook et al. 2016, Disspain et al. 2016). Otoliths are metabolically inert (Campana and Thorrold 2001) and show a daily growth pattern with daily rings (microincrements; Pannella 1971, Campana and Neilson 1985, Jolivet et al. 2008) that contain information about the age of the fish as well as features of an individual's life history and migration patterns, which are reflected in analyses of trace elements. Otoliths therefore serve as important bioarchives (Campana and Thorrold 2001, Begg et al. 2005, Kerr and Campana 2013, Starrs et al. 2016). Finally, otoliths can provide clues about the composition of the diet of piscivorous birds given that they are mostly indigestible and found in stomach contents and excrement (e.g., Martini and Reichenbacher 1997, Barrett et al. 2007).

Mechanosensory Lateral Line System

The mechanosensory lateral line system is composed of rather simple sensory receptor organs (neuromasts) that are distributed on the head, trunk, and tail of fishes (as well as larval and aquatic adult amphibians). They are composed of a population of sensory hair cells that are like those of the inner ear (see Section on the Ear) and non-sensory support and mantle cells (Fig. 23). Neuromasts respond to unidirectional and low frequency (< 200 Hz) oscillatory water flows, allowing the system to mediate key behaviors, including prey detection, predator avoidance, navigation, and communication during social interactions (McHenry et al. 2014, Montgomery et al. 2014). Neuromasts are found in pored, fluid-filled lateral line canals (canal neuromasts; Figs. 23C, D) and on the skin (superficial neuromasts, e.g., "pit organs"; Fig. 23D) and function as accelerometers and velocimeters, respectively, defining two sensory sub-modalities. Bony fishes and cartilaginous fishes have both canal and superficial neuromasts, but jawless fishes (hagfishes and lampreys) and amphibians have only superficial neuromasts. The lateral line system (lateral line canals, neuromasts) demonstrates a remarkable degree of morphological variation among species of both chondrichthyan and osteichthyan fishes (reviewed in Webb 2014a, b). The mechanosensory lateral line system was completely lost with the evolution of the amniotes and is absent in reptiles, birds and mammals.

Neuromast Structure

Neuromasts are small, multicellular sensory organs ($\sim$ 10–500 µm diameter) composed of tens to thousands of sensory hair cells as well as non-sensory cells. They are derived from cranial placodes (ectodermal thickenings on the embryonic head) and are found in the epidermis (reviewed in Webb 2014a). Each sensory hair cell has a ciliary bundle that is located on its apical surface (Fig. 28B). As in the inner ear (see section on the Ear), the hair cells have a single apical kinocilium (9 + 2 cilium), which may be quite long in neuromasts, and several shorter stereocilia (actin-based microvilli), which are graded in length and located to one side of the kinocilium defining the hair cell's axis of best physiological sensitivity to water flows and vibrations (Fig. 28B; see section on Inner Ear). Unlike the inner ear in which hair cells with four different polarities are segregated into different regions of a sensory macula, hair cells with opposing polarities (180° to one another) are distributed throughout the macula. In the small neuromasts of fish embryos and larvae, hair cells are distributed over the entire apical surface of the neuromast (Raible and Kruse 2000, Webb and Shirey 2003). In the larger neuromasts of juvenile and adult fishes, hair cells are typically restricted to a smaller, round or oval region (the "sensory zone", Jakubowski 1967b, or "sensory strip", Coombs et al. 1988), the shape of which varies among species (Figs. 28A, B; 29). The morphology of the ciliary bundle (e.g., relative length of the kinocilium, number and length of stereocilia) may vary among hair cells within a neuromast (e.g., Song and Northcutt 1991, Marshall 1996, Faucher et al. 2003), with hair cells bearing longer kinocilia found at the periphery of the sensory strip, especially in canal neuromasts. In addition, superficial neuromasts tend to have longer kinocilia than canal neuromasts. This is due to the constraints presented by canal diameter and, from a functional perspective, the need to extend beyond the boundary layer in order to function in detecting flows along the surface of a fish's body. Canal neuromast size and shape varies considerably among species, but superficial neuromasts are typically smaller than canal neuromasts (Fig. 28D) and demonstrate significantly less variation in size and shape (reviewed in Webb 2014b).

Non-sensory cells are scattered among the hair cells, and also surround the population of sensory hair cells defining the neuromast's outer perimeter and thus its shape (Figs. 28B, 29). Two types of non-sensory cells, support ("supporting") cells and mantle cells, are generally recognized based on their location (Münz 1979, Blaxter 1987) and patterns of gene expression (reviewed in Chitnis et al. 2011). Support cells, scattered among the sensory hair cells, extend from the apical surface of the neuromast to the basement membrane, below the hair cell population. In teleost fishes, they have little surface elaboration, but short microvilli are present (Fig. 28B). In non-teleost bony fishes (Webb and Northcutt 1997) and elasmobranchs (Peach and Rouse 2000, Jørgensen and Pickles 2002, Peach and Marshall 2009) they have long microvilli, which are thought to project into the cupula along with the kinocilia and stereocilia of the hair cells. Support cells give rise to hair cells during normal development and after physical or chemical damage that results in hair cell regeneration (reviewed by Coffin et al. 2014). Mantle cells (Rouse and Pickles 1991, Williams and Holder 2000, Ghysen and Dambly-Chaudière 2007), which surround the sensory hair cells and define the neuromast's perimeter, are more numerous in the larger neuromasts of juvenile and adult fishes in which the hair cells are restricted to a "sensory strip" in the center of the neuromast. Large squamous epithelial cells compose the surface of the general epithelium and the lining of the lateral line canals surrounding the neuromasts. These cells are characterized by microvillar ridges, which in teleost fishes resemble a fingerprint (Fig. 28A); they do not appear to be either a structural or functional component of the neuromast.

The ciliary bundles of all of the hair cells of a neuromast extend from the neuromast surface into a single, elongate, non-cellular, gelatinous cupula. The cupula serves as the biomechanical interface between the hair cells and the surrounding environment and its size and shape is thus critical for neuromast function (reviewed in McHenry and van Netten 2014). It is subject to damage, but grows continuously (e.g., Vischer 1989, Mukai and Kobayashi 1992) and may be many times longer than the diameter of the neuromast (e.g., McHenry and Liao 2014), especially in superficial neuromasts. Münz (1979) noted that the cupula (e.g., in Nile tilapia, *Oreochromis niloticus*) is acellular, and composed of two layers: a central layer that overlies the sensory strip and supporting cells, and an outer layer that overlies the surrounding mantle cell population. The cupula is thought to be secreted by the non-sensory cells of the neuromast (Münz 1979, Blaxter 1987, Rouse and Pickles 1991, Ghysen and Dambly-Chaudière 2007).

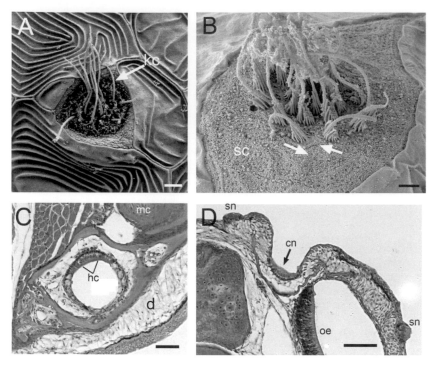

Fig. 28. Neuromast receptor organs in bony fishes. **A.** SEM of a superficial neuromast on the head of a cichlid, *Aulonocara stuartgranti* (10.5 mm SL larva). Long kinocilia (arrow) and shorter stereocilia at their base (not visible) extend from the hair cells comprising the sensory epithelium; non-sensory cells have short microvilli. The neuromast is surrounded by several squamous epithelial cells (cc) of the skin, characterized by actin-supported microridges on their surface. Scale bar = 2 µm. **B.** SEM of a superficial neuromast on the trunk of a goby, *Elacatinus lori* (9.5 mm SL, transforming juvenile) showing the sensory hair cells (each having one long kinocilia, and shorter stereocilia, which are graded in length) with opposing polarities (arrows pointing to the kinocilium of two adjacent hair cells) in a central area ("sensory strip"); non-sensory support cells surrounding the hair cells bear very short microvilli. Scale bar = 1 µm. **C.** Transverse histological section through the mandibular lateral line canal in an adult cichlid (*Labeotropheus fuelleborni*). The dentary bone (pink) is found in the dermis (d, below the basement membrane of the epidermis). The sensory hair cells (hc) have prominent nuclei and non-sensory support cells are on either side of the hair cells. The lumen of the canal is filled by water (and is in contact with the external environment via pored, not shown) and mucus-secreting cells (blue) are found in the epithelium lining the canal lumen. Meckel's cartilage (mc) is incorporated into the dentary bone, Scale bar = 50 µm. **D.** Transverse histological section through the canal neuromast in the supraorbital lateral line canal in a cichlid, *Tramitichromis* sp. (juvenile, 23 mm SL). The canal neuromast (cn) is contained within the trough-like nasal bone (pink), which has not yet enclosed the neuromast. Two superficial neuromasts (sn) as well as the ciliated olfactory epithelium (oe) in the olfactory organ are visible, Scale bar = 50 µm. **C** and **D** illustrate tissue embedded in Paraplast and stained with a modified HBQ protocol (Hall 1983).

Color version at the end of the book

The hair cells in a neuromast are innervated by a branch of one of the lateral line nerves based on its location on the head, trunk and tail (Fig. 30). Neuromasts are also innervated by efferent nerves that modulate neuromast function. Branches of the anterior lateral line nerve (ALLn) innervate neuromasts dorsal and ventral to the eye. When present, the middle lateral line nerve (MLLn) innervates a small number of neuromasts at the posterior margin of the head (Puzdrowski 1989, Gibbs 1999). The posterior lateral line nerve (PLLn) is typically divided into three branches that innervate neuromasts near the posterior margin of the head and on the trunk. A recent series of remarkably detailed papers has described the pattern of innervation of all the neuromasts on the head, trunk, and tail, in a variety of teleost fishes, including those with dramatically proliferated superficial neuromasts (Nakae and Sasaki 2005, 2006, Nakae et al. 2006, Nakae and Sasaki 2010, Asaoka et al. 2011a, b, Nakae et al. 2012a, b, c, 2013, Hirota et al. 2015, Sumi et al. 2015, Sato et al. 2017).

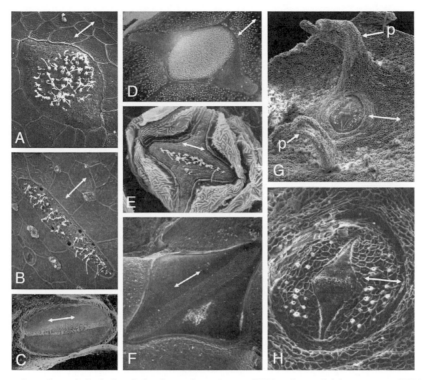

Fig. 29. A sampling of morphological variation in canal neuromasts (CNs) and superficial neuromasts (SNs) on the head of adult teleost fishes. **A.** SN on trunk of windowpane flounder, *Scophthalmus aquosus*. **B.** CN in narrow canal of zebrafish, *Danio rerio*. **C.** CN in narrow canal of mottled sculpin, *Cottus bairdi*. **D.** CN in widened canal in clown knifefish, *Notopterus chitala*. **E.** SN from the blind side of the head of California tongue sole, *Symphurus atricauda*. **F.** CN in widened canal in rex sole, *Glyptocephalus* sp. **G.** SN from the dorsal most of five horizontal lines of neuromasts on the trunk of an adult plainfin midshipman, *Porichthys notatus*, showing the pair of papillae (p) that accompany each of the trunk neuromasts. **H.** Close-up of the SN in G, showing the elongate sensory area containing the hair cells in the center of the neuromast. Arrows indicate axis of best physiological sensitivity (parallel to the long axis of the canal in all canal neuromasts), which is defined by the orientation of the hair cell bundles. From Webb 2014b, reprinted with permission of Springer-Verlag.

Canal and Superficial Neuromasts

Canal and superficial neuromasts differ morphologically and functionally (reviewed in Webb et al. 2008, Webb 2014b, McHenry and van Netten 2014). Canal neuromasts may be up to 1–2 mm in diameter and are typically larger than superficial neuromasts, which generally do not exceed 50–100 µm in diameter (Blaxter 1987, Münz 1989, Webb 1989c, Song and Northcutt 1991, Tarby and Webb 2003, Webb and Shirey 2003, Becker et al. 2016). In the cranial lateral line canals, canal neuromast size and shape appears to be correlated with the type of canal in which it is found. In narrow canals, neuromasts may be round or oval with the major axis parallel to the axis of the canal, while neuromasts in widened canals have been described as being rather large, with a prominent secondary axis perpendicular to the canal axis often resulting in a diamond shape (e.g., Garman 1899, Jakubowski 1963, 1967b, 1974). However, as more species have been studied, it has become apparent that a simple relationship between the size and shape of canal neuromasts and the type of canal (narrow, widened) in which they are found is elusive (Fig. 31; Faucher et al. 2003, Webb and Shirey 2003, Schmitz et al. 2008). Regardless of variation in neuromast shape among species, the axis of best physiological sensitivity (directional polarization or orientation) of the hair cells in canal neuromasts is always parallel with the length of the canal, which ensures the ability of neuromasts to respond to water movements within the canal.

In contrast to canal neuromasts, superficial neuromasts tend to be small and either round or diamond-shaped, occurring on the skin individually, in lines, or in clusters (e.g., Asaoka et al. 2011b, Nakae et al.

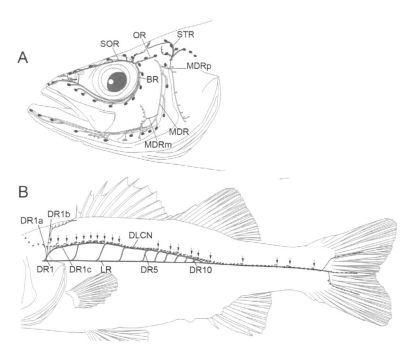

Fig. 30. Innervation of the lateral line system, in *Lateolabrax japonicus*. **A.** Lateral view of head. BR, buccal ramus (green); MDR, mandibular ramus (yellow); MDRm, medial ramule of mandibular ramus (yellow); MDRp, posterior ramule of mandibular ramus (yellow); OR, otic ramus (beige); SOR, superficial ophthalmic ramus (blue); STR, supratemporal ramus (red). Large navy and small dots indicate canal and superficial neuromasts, respectively. Based on 12 specimens, 114–157 mm SL. **B.** Lateral view of trunk and caudal fin. DLCN, dorsal longitudinal collector nerve; DR, dorsal ramule; LR, lateral ramus; STC, supratemporal canal; TRC, trunk canal. Navy and black dots represent canal and superficial neuromasts, respectively. Arrows indicate "Type II" branches. From Sato et al. 2017. Reproduced with permission by The Ichthyological Society of Japan.

Color version at the end of the book

2012a, Becker et al. 2016). They may occur in the place of canal neuromasts (replacement neuromasts), or they may accompany canals. Replacement neuromasts, which sit on the skin, or in some cases in incompletely formed canals (grooves) as the result of the evolutionary reduction of the canals (Webb 1989a, 2014b), appear to retain their ancestral orientation (parallel to the course of the canal that had been lost). In addition, "accessory neuromasts"[1] (Lekander 1949, Coombs et al. 1988) are found adjacent to, or

---

[1] The term "accessory neuromasts" has been used in two different ways in the literature and thus deserves special mention here. Studies of the morphological diversity and evolution of the lateral line system (Coombs et al. 1988, Webb 1989b) adopted the term (following Lekander's 1949 use of the term to describe lines of SNs that accompany head canals) to define those superficial neuromasts that accompany the lateral line canals on the head and on the trunk in juvenile and adult fishes. Studies of neuromast patterning in embryos and larvae in zebrafish adopted the term to define those superficial neuromasts that bud off previously differentiated superficial neuromasts as defined by Stone (1937) in *Ambystoma* embryos (discussed by Ghysen et al. 2014). The "accessory" neuromasts that accompany lateral line canals (especially the trunk canal) may indeed bud off from presumptive canal neuromasts (prior to canal formation) but data is not yet available to evaluate this hypothesis. However, some "accessory" neuromasts (those in "orthogonal pairs", Coombs et al. 1988) have hair cell orientations perpendicular to that of the CNs (Münz 1979, Webb 1989c), which would suggest that they arise from different neuromast primordia, based on work in zebrafish (Lopez-Shier et al. 2004). The "accessory" neuromasts defined in zebrafish larvae tend to share the same hair cell orientation as those from which they bud off (Ghysen et al. 2014). While historical precedent appears to favor Stone's (1937) definition of "accessory neuromasts", the two distinct morphological and developmental contexts in which this term independently evolved requires acknowledgment of both definitions pending further discussion and analysis.

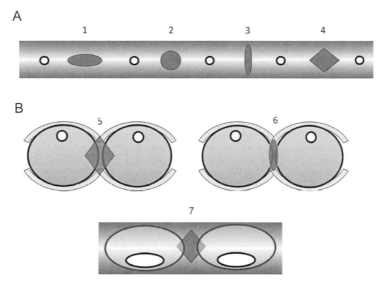

Fig. 31. Relationship of neuromast shape and canal phenotype among fishes. **A.** Narrow canals are well ossified, have small pores (open black circles) and may contain neuromasts (red) that are oval (1, mottled sculpin, *Cottus bairdi*), round (2, convict cichlid, *Amatitlania nigrofasciata*, yellow perch, *Perca flavescens*), elongate and perpendicular to the canal axis (3, zebrafish, *Danio rerio*) and diamond-shaped (4, a cichlid, *Tramitichromis*). **B.** Widened canals may contain diamond shaped (5, peacock cichlid) or elongate neuromasts (6, silverjaw minnow) in constriction between areas with membranous roofs (grey) and small pores (black) or may contain diamond-shaped neuromasts in cylindrical canals without constrictions, but with membranous roofs (grey) and larger pores (black) (7, Eurasian ruffe, *Gymnocephalus cernuus*). In all cases, hair cell orientation is parallel to the long axis of the canal, which ensures that fluid movement along the canal axis can serve as an effective stimulus.

in the epithelium overlying canals and tend to have hair cell orientations that are either parallel to and/or perpendicular to the axis of the canals they accompany (Marshall 1986, Münz 1979, Webb 1989c, Mukai et al. 2007, Schmitz et al. 2008).

Superficial neuromasts are found in pits ('pit lines') or open grooves, or they may sit flush with the skin (Webb 1989c), or on top of stalks, filaments, or papillae. Neuromasts on papillae are particularly common among deep-sea fishes (Marshall 1996, Gibbs 1999, Pietsch 2009), gobies (as 'sensory papillae'), and cave-dwelling fishes (e.g., Moore and Burris 1956). Superficial neuromasts may also be found between paired non-sensory accessory structures that appear to protect the neuromast and alter the hydrodynamic environment, thus modifying the ability of a neuromast to respond to flows (Applebaum and Schemmel 1983, Nakae and Sasaki 2010). A dramatic proliferation of hundreds to thousands of superficial neuromasts is found in several notable taxa, including goldfish (*Carissius auratus*), Mexican tetra (*Astyanax mexicanus*), gobies (Asaoka et al. 2011b, Nakae et al. 2012a), and some deep sea fishes (reviewed in Marranzino and Webb 2018).

The Lateral Line System of Bony Fishes

Coombs et al. (1988), Kasumyan (2003) and Webb (1989a, 2014a, b) provide comprehensive reviews of variation in the morphology of the lateral line system among bony fishes. The essentials are summarized here.

The Cranial Lateral Line System

Among bony fishes, the canals are contained within an evolutionarily conserved subset of dermatocranial bones regardless of variation in the shape of the head, or the size and shape of the bones in which the canals are found (see for instance, Gregory 1933). Superimposed on this conserved pattern, variation in the cranial lateral line system is characterized by variation in relative canal diameter, extent of canal development, as well as the number and distribution of superficial neuromasts (reviewed in Webb 2014a, b; Fig. 32). Five cranial lateral line canal phenotypes are found among bony fishes—three variations on narrow canals (narrow-simple, narrow-branched tubules, narrow-widened tubules), as well as reduced canals and widened canals (Webb 2014b). These phenotypes have arisen as a result of variation in the timing of lateral line canal development (e.g., Bird and Webb 2014). In addition, the location of only one neuromast between adjacent canal pores in cranial canals in teleosts, and in all other bony fishes, with the exception of the African and South American lungfishes (Webb and Northcutt 1997), is the result of the pattern of canal morphogenesis, a process that takes place at the end of the larval period (e.g., Webb 2014a, Bird and Webb 2014).

The Posterior Lateral Line System

Canal neuromasts are typically found within a single trunk canal that extends from the operculum to the caudal peduncle. The canal is contained within a horizontal series of overlapping lateral line scales, each of which contains a cylindrical, ossified canal segment containing one canal neuromast (e.g., Webb 1989c, Song and Northcutt 1991, Wonsettler and Webb 1997, Faucher et al. 2003). Lateral line scales vary within and among species with respect to size, degree of ossification, and the relative diameter of the lateral line canal segment contained within each scale (Webb 1990a, Voronina 2009, Voronina and Hughes 2013, 2017). The 3-D configuration of the lateral line scales and the lateral line canal within them has recently been reinterpreted (Webb and Ramsay 2017; Fig. 33). Eight trunk canal phenotypes are found among bony fishes and are defined by the placement and length (degree of development) of a single canal along the body axis, or by the presence of multiple trunk canals (Webb 2014b; Fig. 32). Superficial neuromasts often accompany the trunk canal and are located in the epithelium overlying the lateral line scales, with hair cell orientations parallel to or perpendicular to the canal axis, and thus to the body axis (Webb 1989c). Many teleosts also have a canal or, more typically, one or more lines of superficial neuromasts, which extend along the membranes between the fin rays of the caudal fin (e.g., Webb 1989b, Peters 1973, Asaoka et al. 2011a, b, Webb et al. in press).

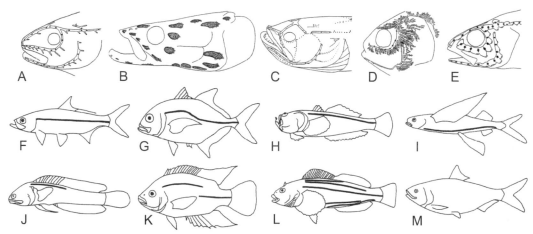

Fig. 32. Diversity of cranial **(A–E)** and trunk canal **(F–M)** morphology among teleost fishes. **A.** Narrow-simple canals (saithe, *Pollachius virens*; redrawn from Marshall 1965, reprinted with permission from Elsevier, Inc.). **B.** Narrow canals with widened tubules (*Arapaima*, from Nelson 1969, courtesy of The American Museum of Natural History). **C.** Reduced canal system with lines of superficial neuromasts (dots) in the plainfin midshipman (*Porichthys notatus*, Greene 1899). **D.** Narrow with branched canal system (Atlantic menhaden, *Brevoortia* tyrannus), from Hoss and Blaxter 1982, reprinted with permission by Wiley & Sons, Inc. **E.** Widened canal system in common percarina (*Percarina demidoffi*, from Jakubowski 1967, reprinted with permission of the author). **F.** Complete straight canal (tarpon). **G.** Complete arched canal (carangid). **H.** Dorsally placed canal (stonefish). **I.** Ventrally placed canal (flying fish). **J.** Incomplete canal (blenny). **K.** Disjunct canal (a cichlid). **L.** Multiple canal (greenling). **M.** Absent (lack of canal, herring). **F–M** Redrawn from Nelson (1984), John Wiley & Sons, Inc. Figure from Webb 2014a.

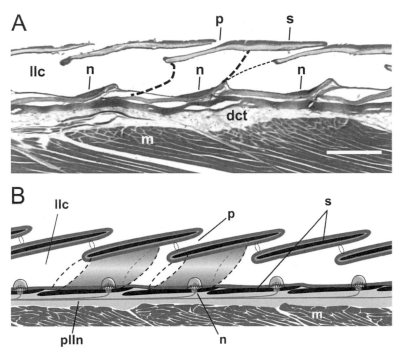

Fig. 33. Three-dimensional configuration of the lateral line scales on the trunk of teleost fishes. **A.** Sagittal section through the lateral line canal on the trunk of an embiotocid, *Embiotoca jacksoni*, showing the location of canal neuromasts (n) in the canal segments contained in each lateral line scale(s). The bony scales sit within the dermis; the lumen of the canal (llc) is lined by a general epithelium. The thin dotted line drawn between opposing tips of what appears to be one scale would suggest that the scales sit at 45° angle within the dermis. However, the thicker dashed lines indicate the location of the walls of the canal segment (out of plane of section), which connect the base and roof of each canal segment. Rostral is to left. Tissue was embedded in glycol methacrylate and stained with cresyl violet. Scale bar = 500 μm. **B.** Consensus configuration of the lateral line scales based on ten species of pomacentrids, embiotocids and pleuronectids. Overlapping lateral line scales (black) sit beneath the epidermis (gray) at a shallow angle in the dermis (light gray). Tubular canal segments (one depicted in 3-D) form a continuous, epithelium-lined canal lumen that runs parallel to the skin surface. The infrascalar and suprascalar pores (right and left, respectively, in each scale) are represented by dashed lines. Rostral is to left. Abbreviations: dct, dermal connective tissue; llc, lateral line canal; m, trunk muscle; n, canal neuromast; p, pore; s, scale; llc, lateral line canal; m, trunk muscle; n, canal neuromast; p, pore (dotted line); plln, posterior lateral line nerve; s, scale. From Webb and Ramsay 2017, reprinted with permission of the American Society of Ichthyologists and Herpetologists.

The Lateral Line System of Cartilaginous and Jawless Fishes

The lateral line system of chondrichthyan fishes (sharks, skates, rays, chimaeras) is fundamentally different from that of bony fishes with respect to the location, distribution, and morphology of the lateral line canals and the canal neuromasts within them. In elasmobranchs (sharks, skates, rays), the lateral line system is composed of both canal neuromasts and superficial neuromasts ("pit organs", Peach and Rouse 2000, Peach and Marshall 2009). The slender, cylindrical cranial lateral line canals are not associated with or embedded within the cranial cartilages, but instead, the canals are composed of epithelial cells and are located within loose dermal connective tissue. Tubules extend from the canals to the skin's surface and end in terminal pores that may appear similar to the pores of the ampullae of Lorenzini (Fig. 34). The course of the lateral line canals varies among species and with head and body shape (Chu and Wen 1979, Jordan et al. 2009, Marzullo et al. 2011, Wueringer et al. 2011a, b, Theiss et al. 2012, Collin et al. 2016). In some species, the canal neuromasts have been described as being composed of a "continuous" or "nearly continuous" sensory epithelium (Johnson 1917), or may be discrete but quite numerous within the canals. In this way they are quite distinct from the canal neuromasts of bony fishes, which are well-defined, focal populations of hair cells separated by the general epithelium that lines the canals (Webb and Northcutt 1997). However,

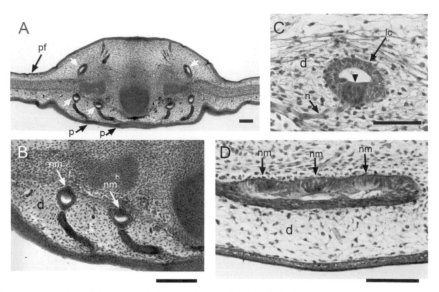

Fig. 34. Neuromasts and lateral line canals in the dermis in embryonic little skate, *Leucoraja erinacea* (65 mm TL, Stage 33). **A.** Transverse section at the level of the pectoral fins, showing lateral line canals (in transverse section, arrows) and canal tubules (in longitudinal section) extending from the canal to the pores on the ventral surface of the embryo, Scale bar = 100 µm. **B.** Enlargement of A showing small neuromasts within canals (arrows), 50 µm in diameter, sitting in the dermis (d), Scale bar = 100 µm. **C.** Transverse section through a lateral line canal on the dorsal side of the head, revealing the epithelial structure of the canal wall within the dermis (d), and showing the hair cells that comprise the neuromast (arrowhead). A nerve (n) innervates the hair cells, Scale bar = 50 µm. **D.** Longitudinal section through a ventral lateral line canal in the dermis (d) showing three closely set neuromasts (arrows) along the length of the canal, Scale bar = 100 µm. Tissue was embedded in Paraplast and stained with a modified HBQ protocol (Hall 1983).

Color version at the end of the book

recent work has shown that in some species, the neuromasts are quite distinct (Baker et al. 2013). The sensory hair cells have well-defined ciliary bundles, which are oriented 180° to one another, and parallel to the canal axis, as in bony fishes. The pores of one trunk canal and scattered superficial neuromasts are generally found on the trunk of sharks, and are surrounded by modified placoid scales (Maruska and Tricas 1998, Maruska 2001, Peach and Rouse 2000, Peach and Marshall 2009).

In the chimeras (holocephalans or ratfishes), the distribution of the lateral line canals is similar to that in elasmobranchs (sharks, skates, rays), but the cranial canals may either be narrow and tubular or they may be in the form of open grooves (with variation among holocephalan families). The tubular canals contain canal neuromasts that are elongate and almost continuous along the length of a canal, as in elasmobranchs (Ekström von Lubitz 1981). In contrast, the open grooves found on the underside of the snout of some chimaerids contain discrete neuromasts located between periodic canal dilations (Reese 1910, Ekström von Lubitz 1981, Fields et al. 1993). Superficial neuromasts have not been reported in these fishes.

The lateral line system of the jawless fishes ('Agnatha', hagfishes and lampreys) consists only of superficial neuromasts. In lampreys, neuromasts in rows on the head and body sit in pits flanked by a pair of "hillocks" (Lane and Whitear 1982b, Gelman et al. 2006). In hagfishes, a vestige of the lateral line system appears to be present only in the family Eptatretidae, where short depressions or grooves on the head contain flask-shaped sensory receptor cells (Fernholm 1985, Braun and Northcutt 1997). These hair cells have a single cilium and a corolla of shorter microvilli, but a cupula has not been reported.

The Electrosensory System

Electroreception is the ability to perceive weak electric fields via specialized receptors—electroreceptors and among fishes, is only found in lampreys, cartilaginous fishes, coelacanths, lungfishes, bichirs, sturgeon and paddlefishes, African elephantnose fishes and Aba, African knifefishes, catfishes, and South American knifefishes (Bullock and Heiligenberg 1986). The electric sense is typically found in aquatic animals, but it also occurs in some terrestrial species such as the echidna and in invertebrates such as bumblebees and the cockroach (Newland et al. 2008, Clarke et al. 2013, Collin 2017). Electroreception was discovered in the 1960s in teleost fishes, later on in amphibians (Fritzsch and Münz 1986), in monotreme mammals (Pettigrew 1999), and recently in dolphins (Czech-Damal et al. 2012). In this chapter, we restrict our description to the electrosensory organs of fishes.

Electrosensory organs were described before the discovery of electroreception (see reviews by Szabo 1974, Zupanc and Bullock 2005), so their function as electroreceptors was not immediately recognized. These include the ampullae of Lorenzini (Elasmobranchii; Dotterweich 1932); ampullae of Lorenzini (*Plotosus anguillaris*; Friedrich-Freska 1930); small pit organs (*Ameiurus*; Herrick 1901); and the ampullary organ (Stendell 1914) and tuberous organs (Franz 1920) of mormyrids. The first indication of the presence of electroreception came from behavioral experiments with the African mormyroid fish *Gymnarchus niloticus* (Lissman 1951, 1958a, Lissmann and Machin 1958).

Two distinct types of electroreceptors are recognized: ampullary and tuberous electroreceptors. Ampullary electroreceptors are sensitive to electric fields up to ca. 50 Hz. The perception of the electric fields via these receptors is termed "passive electroreception". There are three discrete sources of aquatic electric fields (reviewed by Kalmijn 1974): (1) electric fields of physical origin (e.g., water flowing through the Earth's magnetic field); (2) electric fields of electrochemical origin (e.g., contact of two chemically dissimilar media); (3) electric fields of biological origin (e.g., muscle potentials). Tuberous electroreceptors are sensitive to electric fields between ca. 50 and 2000 Hz, which are produced by the electric organs of weakly electric fish. Due to the characteristics of these receptors, the weakly electric fishes are able to detect objects ("active electrolocation"; von der Emde 1999) and to mediate both intra- and interspecific communication (Worm et al. 2018, Henninger et al. 2018, Nagel et al. 2018).

The evolution of electroreceptors from the neuromasts of the mechanosensory lateral line system was first suggested based on the innervation of the electroreceptors by the lateral line nerves, and the close proximity of the primary brain centers in the medulla oblongata that processes the information from the electroreceptors and the neuromast mechanoreceptors of the lateral line system (Bullock and Heiligenberg 1986). In non-teleost fishes, the ampullary electroreceptors are derived from embryonic lateral line placodes (reviewed by Baker et al. 2013). The evolution of ampullary electroreceptors from the mechanosensory lateral line system continues to be supported by molecular genetic data (Modrell et al. 2011, Modrell and Baker 2012, Baker et al. 2016, Modrell et al. 2017a, b).

Taxonomic Distribution of Electroreceptors

Among jawless fishes (hagfishes and lampreys), electroreceptors are found only in lampreys (end buds, in addition to mechanosensory lateral line neuromasts; Fig. 35A), but are absent in hagfishes. They occur in groups of up to 30 receptor cells on the head and the trunk, where each receptor possesses 80–90 apical microvilli (Bodznick and Northcutt 1981, Jorgensen 2005) and are open to the skin surface.

Ampullary organs are found in all chondrichthyans (sharks, skates, rays, and holocephalans) (Fig. 35B), in aquatic amphibians (e.g., the axolotl, Fig. 35C), in sarcopterygian lungfishes, in the crossopterygian *Latimeria chalumnae*, in polypterids, sturgeons, paddlefishes (Fig. 35D), and in a limited number of teleost taxa (Bullock and Heiligenberg 1986, Northcutt 1986). Thus, electroreception, via ampullary receptors, is an important vertebrate sensory system that has been conserved over hundreds of millions of years. The ampullary electroreceptors decribed so far are excited by weak cathodal stimuli, which probably open voltage-gated Ca^+ channels in the apical membrane (Teeter et al. 1980, Münz et al. 1984, Lu and Fishman 1995, Bodznick and Montgomery 2005). The neuromasts, the sensory organs of the mechanosensory lateral line system (Fig. 35E), have some anatomical similarities to the ampullary organs and possess a kinocilium and a number of microvilli of variable length (see Mechanosensory Lateral Line section). They are embedded in a gelatinous cupula, which is in contact with the water. Although the neuromast receptor cells are mechanosensory, they respond to large anodal stimuli (reviewed by Baker et al. 2013).

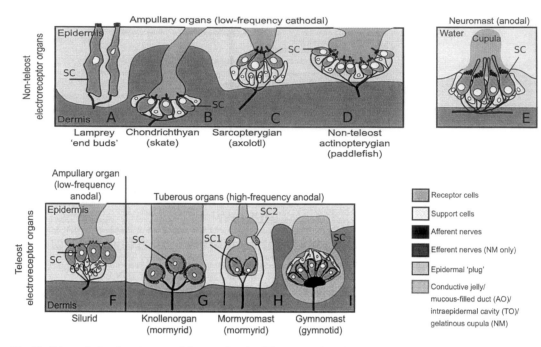

Fig. 35. Schematic drawings (not to scale) comparing the different morphologies of neuromasts and ampullary and tuberous electroreceptors in non-teleost and teleost fishes and an amphibian (axolotl). **A.** Lamprey end-buds (SC), which extend up to the surface of the epidermis apically, possess many microvilli. Ampullary organs of chondrichthyans (**B**), the axolotl (**C**), and paddlefishes (**D**) are morphologically quite similar–the sensory cells (SC) are found at the base of an intraepidermal cavity, which opens to the surface of the epidermis via an electrically-conductive canal. Each receptor cell has a single cilium and a variable number of microvilli. These ampullary electrorreceptors of non-teleost fishes all respond to low-frequency cathodal stimuli. They all receive only afferent innervation. **E.** The mechanosensory neuromasts contain many sensory cells (SC), which are equipped with a kinocilium and a number of microvilli of variable length. They are embedded in a gelatinous cupula, which is in contact with water. They receive both afferent and efferent innervation and respond to large anodal stimuli. **F.** Ampullary receptors of teleosts have a similar morphology compared to those of non-teleost receptors: The sensory cells (SC) sit at the base of an ampulla, which connects to the surface of the epidermis via an electrically-conductive canal. However, the sensory cells are lacking a cilium. They respond to low-frequency anodal stimuli. **G.** Knollenorgan (tuberous organ) of a mormyrid fish. One or several sensory cells (SC) (depending on the species) are located at the base of a canal, which is filled with modified epidermal cells. The apical ends of the sensory cells are covered by many small microvilli. **H.** Mormyromast (tuberous organ) of a mormyrid fish. Two types of sensory cells (SC1, SC2) are characteristic for this type of tuberous organ. Both sensory cell types open into a cavity, which is apically closed by a plug of modified epidermal cells. **I.** Tuberous organ of gymnotiforms. There are many sensory cells at the base of an intraepidermal cavity filled with modified epidermal cells. The different types of tuberous organs (G–I) all respond to high-frequency anodal stimuli. AO, ampullary organ; NM, neuromast; TO, tuberous organ (modified from Baker et al. 2013, with permission).

Color version at the end of the book

The ampullary organs of chondrichthyans can respond to directional information in the immediate vicinity of an electrical field. When a fish enters a local electric field, such as that produced by a potential prey item, it changes its swimming course to keep the angle between its rostro-caudal body axis and the detected electric field constant. It is suggested that this behavior is independent of the angle of approach, the polarity of the prey's electric field, changes in the strength and the direction of the field and the position of its source (Collin 2017). The sensory cells respond to low-frequency anodal stimuli. Ampullary organs respond to DC or low frequency electric fields (1–8 Hz in elasmobranchs and 6–12 Hz in silurid teleosts). Sensitivity thresholds to live prey are only detectable at short range due to the rapid dissipation of electric fields in sea water and vary across taxa, i.e., 0.1 mV/cm in lamprey (*Lampetra tridentata*), 1 nV/cm in elasmobranchs, 5 nV/cm in marine teleosts and 1 mV/cm in freshwater teleosts (Kajiura and Holland 2002, Bedore and Kajiura 2013, Kempster et al. 2015, Collin et al. 2015).

In addition to the ampullary organs, tuberous organs, which respond to high-frequency (anodal) electrical stimuli, are found in the African mormyrid elephant noses (Mormyridae), the African Aba (Gymnarchidae), and the South American knifefishes (Gymnotiformes) (Figs. 35G–H).

Anatomy of Electroreceptors

Ampullary Electroreceptors of Chondrichthyes

Passive electrorception is mediated by ampullary receptors in chondrichthyans. The electroreceptors are embedded within the dermis and connected to the surface by a canal filled with a mucopolysaccharide gel that ends in a pore (Figs. 36A, B). At the base of the ampullary canal is an ovoid capsule containing bundles of receptor cells (Fig. 36C), each of which is innervated by a sensory (afferent) axon (Fig. 36C) that projects to the medulla oblongata in the hindbrain. The canal walls are comprised of two layers of squamous epithelial cells separated from multiple layers of collagen fibers (Fig. 36C) by a basement membrane. The epithelial cells of the canal wall are joined by tight junctions and desmosomes (Fig. 36D) to ensure a high resistance so that each ampulla of Lorenzini acts as a well-insulated core conductor (Bodnick and Boord 1986). The canals that emerge from the ampullae radiate in various directions from an ampullary cluster (up to 400 ampullary tubes radiate from a single cluster), providing a mechanism of directionally sampling the surrounding electric field (Rivera-Vicente et al. 2011, Kempster et al. 2015). The length and diameter of the canals can vary substantially according to the position of the receptors over the head and, in some species, according to the osmotic conditions of the environments each species inhabits, i.e., saltwater or freshwater (Marzullo et al. 2011, Wueringer et al. 2011b, 2012). The canals may also bifurcate into multiple alveolar bulbs containing up to five alveoli, with each alveoli divided further by an alveolar septa. Generally, freshwater chondrichthyans have short canals and low number of receptor cells in comparison to marine species, which possess long canals (Wueringer et al. 2011, 2012a, b) and numerous receptor cells. There are also ultrastructural differences in the morphology of the ampullae of the bull shark, *Carcharhinus leucas* that spends appreciable periods of time as juveniles in freshwater, especially in the shape and level of convolutions of the epithelial cells lining the canals (Whitehead et al. 2015).

In elasmobranch electroreceptors, there may be up to thousands of receptor cells within a single alveolus making contact with up to 15 afferent fibers. The sensory epithelium of an ampullary organ is usually comprised of receptor cells bearing a single kinocilium and are surrounded by supportive cells that bear numerous stereocilia/microvilli (Fig. 36E). In marine elasmobranchs, only 1% of the apical surface of receptor cells is exposed to the lumen (Szabo 1974) and a single kinocilium extends from this surface (Waltman 1966). The cilium shows the unusual pattern of 8 + 1 fibers in the body and 9 + 0 in the base and does not have an apparent basal body (Boord and Campbell 1977, Wueringer et al. 2009). The basal surface of the receptor cells possesses multiple ribbon-shaped presynaptic bars (Fig. 37D) lined with synaptic vesicles (Murray 1974, Boord and Campbell 1977, Wueringer et al. 2009). The synapse makes contact with afferent nerve fibers that become unmylinated as they enter the ampulla and spread out over each alveolus (Waltman 1966). There are no efferent fibers projecting to the receptor cells. There is a wide variation in the number and distribution of electroreceptor organs among chondrichthyans (see below).

The ampullary organs are comprised of a number of grape-like structures called alveoli (Figs. 37A, B), each of which is lined with a sensory epithelium. The sensory epithelium consists of sensory cells that detect changes in electric field gradients and transmit this information to the brain via axons of the anterior lateral line nerve (ALLN). The support cells (Fig. 37C) of the sensory epithelium secrete the mucopolysaccaride gel into the ampullae of Lorenzini (Szabo 1974). According to Murray (1974), the composition of the gel is species-specific and is rich in ions and possesses electrical properties similar to those of seawater (Brown et al. 2002). At the base of each canal are ampullary bulbs (Marzullo et al. 2011, Fig. 36C), which may form distinct clusters. A cluster consists either of a loose aggregation of ampullae or a tighter collection of ampullae within a connective tissue capsule (Wueringer and Tibbetts 2008, Wueringer et al. 2011). Aggregations of ampullae and capsules ensure that different ampullae share a common internal reference potential (Kalmijn 1974). Each ampullary cluster is made up of hundreds of individual ampullary organs, which contain thousands of sensory cells, which are innervated by distinct

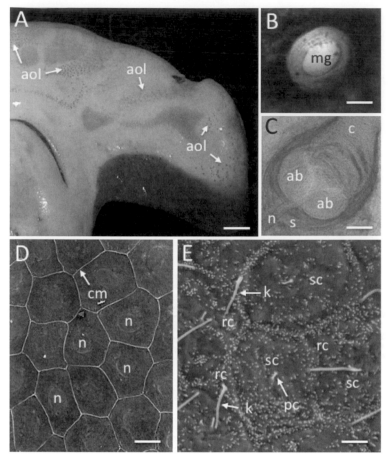

Fig. 36. Electroreceptors in chondrichthyans. **A.** Ventral side of part of the head of the hammerhead shark, *Sphyrna lewini*, showing the dense arrangement of pores of the ampullae of Lorenzini (aol). **B.** Close up of the pore of an electroreceptor on the dorsal surface of the estuary stingray, *Dasyatis fluviorum* showing the mucopolysaccaride gel (mg). **C.** Free ampulla of the freshwater whipray, *Himantura dalyensis* showing the canal ending in two ampullary bulbs (ab) surrounded by collagen sheath (s). Note the afferent nerve (n) emanating from the bulbs. **D.** Scanning electron micrograph of the epithelial surface of the lumenal canal of the wobbegong shark, *Orectolobus ornatus*. Note the prominent nuclei (n) and closely opposing cell membranes (cm). **E.** Scanning electron micrograph of the sensory epithelium lining the alveoli of the ampullary organs of wobbegong shark, *O. ornatus* showing a dense arrangement of support cells (sc) interspersed with receptor cells (rc). Protrusions from the surface include small microvilli, primary cilia (pc) and a large kinoclium (k) emanating from each receptor cell. Scale bars: 3 cm (**A**), 0.1 mm (**B**), 0.25 mm (**C**), 10 µm (**D**) and 2 µm (**E**). **B.** Kindly provided by V. Camilieri-Asch. **C** adapted from Marzullo et al. 2011. **D** and **E** adapted from Theiss et al. 2010.

branches of the ALLN (Kempster et al. 2013). Bilaterally-symmetric clusters show interspecific variation with three in carcharhiniform and lamnid sharks, four in rajid skates, and five in rhinobatid rays and pristid sawfishes (Wueringer 2012).

Ampullary and Tuberous Organs of Teleosts (Mormyridae, Gymnotiformes)

Unlike those of non-teleost fishes, the ampullary organs of teleosts (Fig. 35F) are connected via a canal to the surface of the epidermis. Their sensory cells have a small number of short microvilli, which extend into the mucous-filled duct; however, a kinocilium is absent. The anatomy of the ampullary organs among the different taxa is quite similar (Figs. 38C–E). The sensory cells and supporting cells are arranged at the base of a bulb-like ampulla (Figs. 35B–D), which is connected to the surface by a canal, filled with

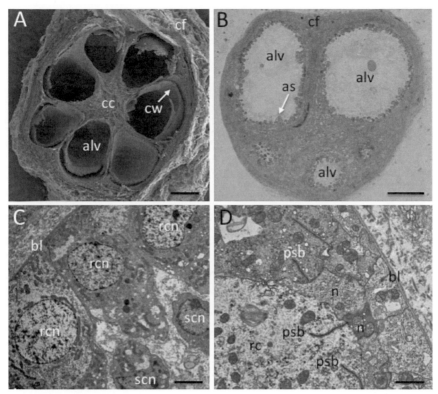

Fig. 37. The structure of ampullary receptors in elasmobranchs. **A.** Scanning electron micrograph of a central cap type ampullary organ in the wobbegong shark, *Orectolobus parvimaculatus* shown in cross section. Note the seven alveolar chambers (alv) surrounding the central cap (cc) within a sheath of collagen fibers. Cw, canal wall. **B.** Light micrograph of a cross section of an ampulla in the estuary stingray, *Dasyatis fluviorum* showing the alveolar chambers (alv) one of which contains an alveolar septum (as). cf, collagen fibers. **C, D.** Transmission electron micrographs of the ampullary epithelium of *D. fluviorum*. **C.** A row of receptor cells lining a basal lamina with support cells situated more apically lining the canal. **D.** The basal region of the epithelium where a receptor cell (rc) forms numerous connections with afferent nerves (n) via presynaptic bars (psb). bl, basal lamina; rcn, receptor cell nucleus; scn, support cell nucleus. **A** adapted from Theiss et al. (2010). **B–D.** Kindly provided by V. Camilieri-Asch. Scale bars: 100 µm (**A, B**), 5 µm (**C**) and 1 µm (**D**).

a mucopolysaccharide gel that is highly electrically-conductive. A kinocilium and a variable number of microvilli are found on the apical surface of the sensory cells (Jorgensen 2005).

In the mormyrid *Pollimyrus isidori*, the ampulla of each ampullary organ is situated in the dermis, which is below the epidermis (Figs. 38A, B). The ampullary organs of gymnotiforms possess a straight canal, which is easily seen in histological sections (Figs. 38C–E). At the base of the ampulla, a few sensory cells can be observed particularly well in the ampullary organ of the gymnotiforms, *Eigenmannia virescens* (Fig. 38C) and *Gymnotus carapo* (Fig. 38E), but is less obvious in *Rhamphichthys* sp. (Fig. 38D). The jelly-like substance in the canal consists mainly of mucopolysaccharides (Denizot 1969, 1970). The edge of the canal in gymnotiforms is equipped with long, ramified processes (see, for example, Fig. 38D) that arise from the specialized cells of the wall. In both types of ampullary organs (in mormyrids and gymnotiforms), the sensory cells are surrounded by many small supporting cells.

The tuberous organs of mormyrids and gymnotiforms differ considerably, and are a good example of convergent evolution. In adult mormyrids, the tuberous organs are comprised of Knollenorgans and mormyromasts. The Knollenorgans have large sensory cells (Fig. 35G), with their number varying according to species. The surface of each sensory cell is covered with many microvilli (Fig. 35G). In the mormyrid *Pollimyrus isidori*, the Knollenorgan consists of one large sensory cell surrounded by a sheath of thick, modified epidermal cells (Fig. 38F). The Knollenorgans perceive high frequency electrical signals used for

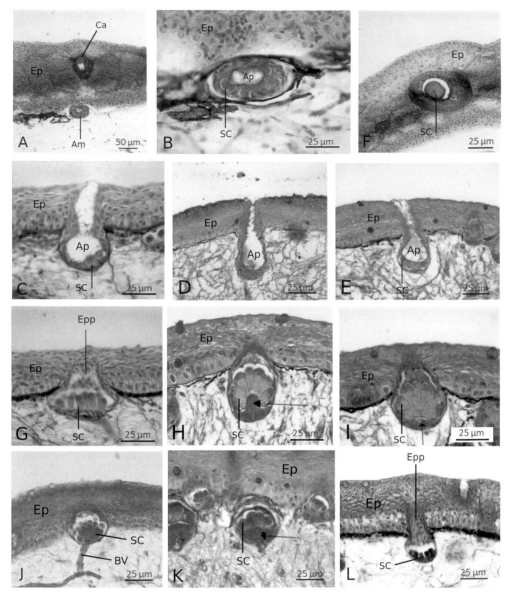

Fig. 38. Ampullary and tuberous electroreceptors in mormyrids and gymnotiforms. **A–E.** Ampullary organs, **F–L.** Tuberous organs. **A–B.** Ampullary organ of the mormyrid *Pollimyrus isidori*. **A.** The ampulla and a part of the canal (Ca) are seen. **B.** Detail of the ampulla, a sensory cell and supporting cells are shown. **C–E.** Tuberous organs of gymnotiform fishes. The ampulla and a short canal are obvious. In *Eigenmannia virescens* (**C**), note the two large sensory cells. The specific cells of the canal in gymnotiforms possess ramified processes as seen in *Rhamphichthys* sp. (**D**) and *Gymnotus carapo* (**E**). **F.** Tuberous organ (Knollenorgan) of *Pollimyrus isidori* with just one sensory cell in the skin of the head. **G.** Tuberous organ (probably type B) of *Eigenmannia virescens* comprising a few sensory cells, a base with supporting cells and the modified epidermal cells plugged on top. **H.** Tuberous organ of type A of *G. carapo*. Note the enlarged nerve ending (arrow). **I.** Tuberous organ of type B of *G. carapo* with a small number of nerve fibers at its base (arrow). **J.** Tuberous electroreceptor of *Apteronotus leptorhynchus* with a large number of sensory cells supplied by a blood vessel (BV). **K.** In *Rhamphichthys* sp. (shown here), the tuberous electroreceptor of type A resembles that of *G. carapo*, but the sensory cells are shorter. The enlarged nerve ending is indicated by the arrow. **L.** The tuberous electroreceptor in *Sternopygus macrurus* possesses a small number of sensory cells and a rather long aggregation of modified epidermal cells plugged on top. Am, ampullary organ; Ap, ampulla; Ep, epidermis; Epp, epidermal plug; Sc, sensory cell.

Color version at the end of the book

communication. Active electrolocation is performed by the mormyromasts. They comprise two different kinds of sensory cells (Fig. 35H).

Gymnotiforms have tuberous organs with many sensory cells located at the base of an intraepidermal cavity filled with modified epidermal cells (Fig. 35I). As in mormyrids, there are two types of tuberous organs, one type is used for communication, the other one for object location. However, the morphological differences are small; they differ in their innervation pattern and in their physiology (see, e.g., Szabo 1974). In type A tuberous organs, there is a pre-terminal enlargement of the nerve fiber before it contacts the sensory cells. In type B organs, a single nerve, after penetration of the basement membrane, divides into many nerve fibers before contacting the sensory cells. Each sensory cell is contacted by one nerve fiber. A tuberous organ of *Eigenmannia virescens* is shown in Fig. 38G and is probably a type B organ. Two different tuberous organs of *Gymnotus carapo* are shown in Figs. 38H (type A organ) and 38I (type B organ). *Apteronotus leptorhynchus* possesses tuberous organs with a very high number of sensory cells (Fig. 38J). The synapses of the tuberous organs are chemical synapses as determined by their morphological characteristics (Szabo 1974). In addition to chemical synapses, it appears that *Apteronotus albifrons* also has gap junctions, which are thought to increase electrical transmission (Srivastava 1972). In *Rhamphichthys* sp., the morphology of the tuberous organs (Fig. 38K) is similar to those in *Gymnotus carapo*, although the sensory cells are shorter than in *G. carapo*. In *Sternopygus macrurus*, the sensory cells of the tuberous organ are quite small, ca. 3 μm wide compared to, e.g., ca. 8 μm in *G. carapo*, and they occur in quite small numbers (Fig. 38L). Electroreception in teleosts is a new invention after the original loss of this sensory system in non-teleost actinopterygians (Bullock and Heiligenberg 1986).

In adult mormyrids, the tuberous organs are comprised of Knollenorgans and mormyromasts (see above). In contrast, larval mormyrids possess two types of larval tuberous electroreceptors (Bensouilah et al. 2002) and a promormyromast, which is lacking the second type of sensory cell (Denizot et al. 2007) of mormyromasts. Types A and B larval tuberous electroreceptors (Figs. 39A–D) possess a single electroreceptor cell inside an intraepidermal cavity, which sits on a platform of accessory cells. The cavity is filled with microvilli originating from both the sensory cells and the epidermal cells lining the intraepidermal cavity. The two types of larval tuberous electroreceptors differ in their distribution in the epidermis of the head, in the composition of their accessory cells, and with respect to their innervation. The function of these larval electroreceptors is not known. The promormyromasts (Denizot et al. 2007) are characterized by a single sensory cell and two types of accessory cells. One type of accessory cell has a dark cytoplasm, few microtubules, and contacts the sensory cell directly, whereas the second type (pyriform accessory cells) has a pale cytoplasm, many microtubules, and forms an outer layer not directly in contact with the sensory cell. The two types of tuberous organs (Types A, B) degenerate in juvenile fish whereas the promormyromasts differentiate into mormyromasts with the addition of a second type of sensory cell (Denizot et al. 2007).

Distribution and Number of Electroreceptors

Ampullary electroreceptors of adult lampreys are distributed over the entire body, but the greatest density is found on the head, often occuring in small lines or clusters (Ronan 1986). The ampullary organs of Chondrichthyes (ampullae of Lorenzini) are concentrated on the head. The distinct clustering of ampullary electroreceptors results in variation in the spatial separation of individual pores on the skin's surface. The abundance of electrosensory pores varies considerably at all taxonomic levels resulting in species-specific differences (Fig. 40). The wide variation in receptor number and distribution (Bodznick and Boord 1986, Raschi et al. 1997, Kempster et al. 2012) is thought to reflect ecological adaptations, although the specific patterns of pore distributions show a strong relationship among orders, and appear to be heavily influenced by morphology, with mouth position being a major contributing factor (Kempster et al. 2012). The electrosensory system of elasmobranchs has evolved to operate efficiently under the environmental conditions of the particular habitat in which a species lives.

The total number of ampullary pores varies greatly among elasmobranch species, from only 191 in the pluto skate, *Gurgesiella plutonia* (Raschi 1986) to 3067 in the scalloped hammerhead, *Sphyrna*

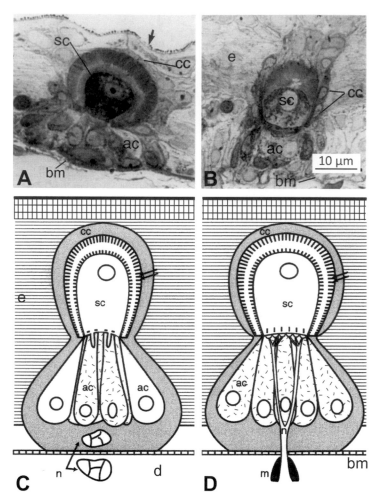

Fig. 39. Larval tuberous electroreceptors of type A (**A**) and B (**B**) shown in semithin sections in the epidermis (e) of the head of a 12-day-old larva of *Mormyrus rume proboscirostris* and the appropriate schematic drawings (**C, D**). Each type of organ consists of a single sensory cell (sc) surrounded by microvilli of the surrounding covering cells (cc). The sensory cell sits on a platform of accessory cells (ac). The outer surface of the epidermis is indicated either by a thick arrow (**A**) or a thick line (**C**). bm, basal membrane. Scale bar: 10 mm. ac, accessory cells; bm, basement membrane; c, epidermal cavity; cc, covering cells; e, epidermis; m, myelinated nerve; n, nerve (after Bensouilah et al. 2002, with permission).

lewini (Kajiura 2001). In a review of the distribution of electroreceptive pores in 117 species, Kempster et al. (2012) found that: (1) Pore number is conserved from birth throughout development and shows no variation between sexes, (2) There are significant differences in the abundance of pores in each of the two superorders of elasmobranchs (Batoidea and Selachimorpha) according to the habitat they occupy (i.e., benthic elasmobranchs possess far fewer electrosensory pores than pelagic species and possess a much greater proportion of ventrally positioned pores than pelagic species), (3) There are significant effects of phylogeny on pore distribution. For instance, batoids possess a high abundance of ventrally-located pores with most dense groupings situated around the mouth, while members of the Selachimorpha, which possess conical head shapes and are not as visually restricted upon close approach to prey, exhibit a more even dorsoventral distribution of pores, and (4) Heterodontids (e.g., the horn shark, Port Jackson shark), representing the oldest lineage of elasmobranchs, possess the lowest number of electrosensory pores, while sphyrnids (hammerhead sharks) are likely the most recent group of living elasmobranchs and possess the highest number of pores (Kempster et al. 2012). A larger number of pores, however, does

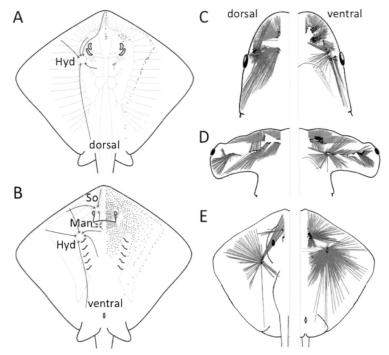

Fig. 40. Distribution of electroreceptors in elasmobranchs. **A, B.** Maps of the distribution of electrosensory ampullary pores and clusters in the bluespotted maskray, *Neotrygon kuhlii* on the dorsal (**A**) and ventral (**B**) regions of the body. On the left side of the body, the hyoid (Hyd), mandibular (Man) and supraorbital (So) ampullary clusters are denoted by large open circles with arrows indicating the direction of associated ampullary canals. On the right side of the body, the position of individual ampullary pores is denoted by small dark circles. **C–E.** Horizontal view of the electrosensory arrays of the sandbar shark, *Carcharhinus plumbeus* (**C**), scalloped hammerhead shark, *Sphyrna lewini* (**D**) and brown stingray, *Dasyatis lata* (**E**). Canals with pores on the dorsal and ventral surface are shown on the left and right side of the figure, respectively. Canals from each ampullary group are represented by different colors (buccal in blue, superficial ophthalmic anterior in green, superficial ophthalmic posterior in red and hyoid in pink). Location of ampullae is indicated by black dots at the base of the canals. **A, B.** adapted from Camilieri-Asch et al. (2013). **C–E** from Rivera-Vicente et al. (2011), reprinted with permission from PLOS.

> **Color version at the end of the book**

not necessarily translate into an increased reliance on electroreception or an increase in electrosensitivity but rather it may have an influence on the ability to resolve electric stimuli in the environment (Raschi 1986, Kempster et al. 2013).

The structure and size of electroreceptors, canal length, and pore diameter also show significant variation with ecological factors (Whitehead 2002, McGowan and Kajiura 2009, Camilieri-Asch et al. 2013). Freshwater stingrays (batoids) possess shorter canals and smaller ampullary organs when compared to marine batoids (Raschi and Mackanos 1989, Raschi et al. 1997), while species that move between freshwater and marine environments. For instance, the bull shark, *Carcharinus leucas*, possess ampullary organs with characteristics of species from both freshwater and marine environments (Whitehead et al. 2014). The euryhaline estuary stingray, *Dasyatis fluviorum*, and the marine blue-spotted maskray, *Neotrygon kuhlii* possess a higher number of ventral electrosensory pores, which correlates with a diet consisting of benthic infaunal and epifaunal prey (Camilieri-Asch et al. 2013). Both of these species also possess 'macro-ampullae' with branching canals leading to several alveoli. The size of the pores and the length of the canals in *D. fluviorum* are smaller than in *N. kuhlii*, which is likely to be an adaptation to habitats with lower conductivity.

In all species of sawfishes, ventral pores are higher in number than dorsal pores including those on the saw (Wueringer et al. 2011b). The same has been described for most species of sharks (Kajiura et al.

2010), various species of rajids (Raschi 1986), and rhinobatids (Wueringer and Tibbetts 2008). However, two types of ampullae exist in sawfishes. The ampullae of the sawfish *Pristis microdon* and *P. clavata* are of the centrum-cap type that possesses a central core of connective tissue, while the ampullae of *Anoxypristis cuspidata* are multi-alveolate, which are arranged in a grape-like formation (Andres and von Düring 1998, Wueringer et al. 2011). Ampullae with a central stage possess fewer alveoli than multi-alveolate ampullae. An increase in the number of alveoli is correlated with decreasing light availability in both rajids and sharks and increased sensitivity at lower stimulus strengths (Raschi 1984).

The distribution of ampullary electroreceptors in non-teleost fishes is well described in the review by Northcutt (1986). In *Polypterus*, the ampullary electroreceptors are restricted to the head, concentrated on the snout and around the eyes. In total, ca. 1000 receptors were counted. The ampullary organs of the coelacanth, *Latimeria chalumnae*, are not well known as they are few in number and there are limited specimens available for study. The rostral organ of *Latimeria* (Milliot and Anthony 1965) is considered to be a low resolution electrodetector with only three paired canals ending in ampullary electroreceptors (Berquist et al. 2015). In lungfishes (Dipnoi), the ampullary electroreceptors are concentrated on the head but electroreceptors are also found in lower densities over the body and the tail with ~ 2000 receptors counted on the head of lepidosirenid lungfishes (Northcutt 1986). In sturgeons, the ampullary electroreceptors are restricted to the head (Fig. 41; e.g., *Acipenser sturio*, Kirschbaum and Williot 2011), where they are found in clusters (Nachtrieb 1910, Jorgensen et al. 1972) of ~ 20 electroreceptors per cluster. Early studies reveal a very high number of ampullary electroreceptors, ~ 20,000 in the shovelnose sturgeon, *Scaphirhynchus platorynchus* and 50,000–75,000 in the paddlefish, *Polyodon spathula* (Nachtrieb 1910).

The distribution of ampullary electroreceptors in siluriform teleosts is known in just a few species. In the catfish, *Kryptopterus* (Wachtel and Szamier 1969), a concentration of receptors is found on the head, but six lines of receptors are found along the body and two lines are found along the ventral and dorsal borders of the tail (Fig. 42A). In *Plotosus anguillaris* (Bauer and Denizot 1972), long-ducted ampullary organs are found on the head and along the base of the fin rays of the dorsal and anal fins correlated with the myomere arrangement (Fig. 42B).

In mormyrid fishes, the Knollenorgans are restricted to the head region (Figs. 38F, 42C, 43B), while the tuberous mormyromasts and ampullary electroreceptors are found in the dorsal and ventral regions of the trunk (Fig. 42C) and are found in a specialized part of the epidermis (electroreceptor epidermis), which differs in thickness from the general epidermis. In larval mormyrids, a concentration of larval tuberous

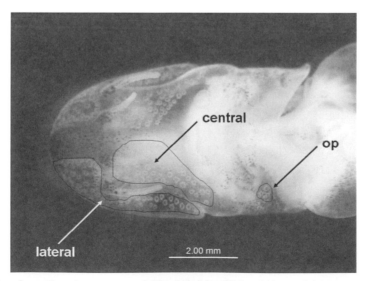

Fig. 41. Distribution of ampullary electroreceptors (white circles) in a 27 day-old larva of *Acipenser sturio* in ventral view. In the lateral sensory fields (lateral), there are about 70 receptors in each field, while in the two central fields (central) there are about 60 receptors each, with just a few being located on the ventral part of the operculum (op) (After Kirschbaum and Williot 2011, with permission).

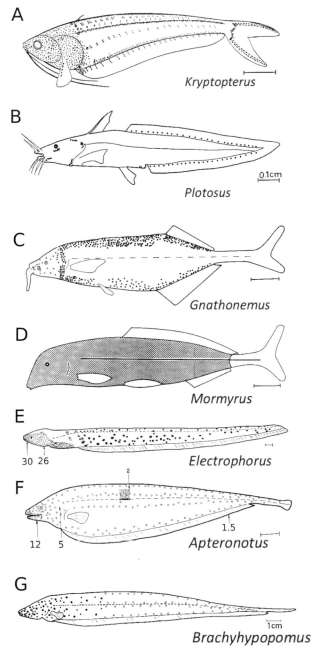

A

Kryptopterus

B

0.1cm

Plotosus

C

Gnathonemus

D

Mormyrus

E

30 26

Electrophorus

F

2

12   5

1.5

Apteronotus

G

1cm

Brachyhypopomus

Fig. 42. Distribution of ampullary and tuberous electroreceptors in teleosts: siluriform, mormyrid, and gymnotiform fishes. **A.** In the glass catfish *Kryptopterus,* the ampullary electroreceptors are concentrated on the head, in six lines along the body and also on the caudal fin. **B.** The ampullary electroreceptors in the striped eel catfish, *Plotosus anguillaris,* are concentrated on the head and metamerically at the base of the anal and dorsal fins. Scale bar also applies to *Kryptopterus*. **C.** In the elephantnose *Gnathonemus petersii*, the tuberous organs (Knollenorgans, open circles) are only found on the head, the mormyromasts (dots) are restricted to the body. **D.** *Mormyrus caballus* is an exception among the mormyrids as its electroreceptors are distributed all over the body, except the tail. **E.** In the strong electric gymnotiform, the electric eel, *Electrophorus electricus*, the distribution of the electroreceptors shows a spot-like pattern. **F.** The distribution of the ampullary receptors (diamonds) in the black ghost *Apteronotus* is similar to that found in *Kryptopterus* (**A**). Figures with numbers (in **E** and **F**) indicate tuberous organ density per mm^2. **G.** Distribution of type I (large dots), type IIA (small dots), and type IIB (open circles) tuberous organs in the knifefish *Brachyhypopomus*. Scale bar also applies to **C**–**F**. From Szabo 1965, with permission (modified).

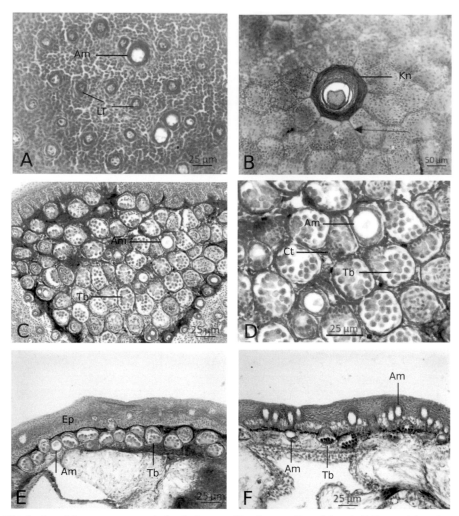

Fig. 43. Distribution of electroreceptors on the head of mormyrid and gymnotiform fishes. **A, B.** Tangential section of skin on the head of the mormyrid, *Pollimyrus isidori*, showing pores indicating ampullary, larval tuberous (LT) electroreceptors (A), and a Knollenorgan with one sensory cell (**B**). Arrow in **B** indicates boundary of epidermal cells. **C, D.** Tangential section of the skin on the head of the gymnotiform *Eigenmannia virescens*, indicating the high number of electroreceptors, in particular tuberous electroreceptors. **E, F.** Cross section of the epidermis of the head showing the high number of electroreceptors. In *E. virescens* (**E**), the tuberous organs prevail, whereas in *Sternopygus macrurus* (**F**) the ampullary receptors are more numerous.

Color version at the end of the book

electroreceptors is found on the head (Fig. 43B). The mormyrid, *Mormyrus caballus* is, compared to the distribution of receptors in *Gnathonemus petersii* (and other mormyrids), an exception because the electroreceptors are distributed over the whole body except the caudal peduncle and the fins (Fig. 42D). In the strong electric gymnotiform, the electric eel, *Electrophorus electricus*, the electroreceptors are distributed in a spot pattern (Fig. 42E). In the gymnotiform *Apteronotus*, the ampullary electroreceptors are distributed all over the body in six rows (Fig. 42F), as in *Kryptopterus* (Fig. 42A). In gymnotiforms, the different types of tuberous organs are distributed differently among species (see Szabo 1974 for more details). An extreme density of electroreceptors is seen in the head region in the gymnotiforms, *Eigenmannia virescens* (Figs. 43C–E) and *Sternopygus macrurus* (Fig. 43F).

Acknowledgements

J.F. Webb thanks all of her collaborators and the past and currrent members of her lab for their contributions to the work included here.

S.P. Collin thanks the following colleagues who have contributed to much of this work including H. Barry Collin, R. Glenn Northcutt, M.A. Ali, Lenore Litherland, Nathan S. Hart, Ian C. Potter, Charlie Braekevelt, Helena Bailes, Julia Shand, Adam Locket and Nicole Scheiber.

T. Schulz-Mirbach thanks Friedrich Ladich (University of Vienna) and Martin Heß (Ludwig-Maximilians-University Munich) for their constructive comments.

K. Żuwała thanks the co-authors for agreeing to cooperate. Thanks also to Dr. Dagmara Podkowa for technical assistance.

F. Kirschbaum thanks Rüdiger Krahe for the use of his microscope/digital camera for taking photomicrographs.

References Cited

Allis, E.P. 1889. The anatomy and development of the lateral line system in *Amia calva*. J. Morphol. 2: 463–542 + plates.

Andres, K.H. and M. von Düring. 1998. Comparative anatomy of vertebrate electroreceptors. Prog. Brain Res. 74: 113–131.

Appelbaum, S. and C. Schemmel. 1983. Dermal sense organs and their significance in the feeding behaviour of the common sole *Solea vulgaris*. Mar. Ecol. Prog. Ser. 13: 39–36.

Asaoka, R., M. Nakae and K. Sasaki. 2011a. Description and innervation of the lateral line system in two gobioids, *Odontobutis obscura* and *Pterobobius elapoides* (Teleostei: Perciformes). Ichthyol. Res. 58: 51–61.

Asaoka, R., M. Nakae and K. Sasaki. 2011b. The innervation and adaptive significance of extensively distributed neuromasts in *Glossogobius olivaceus* (Perciformes: Gobiidae). Ichthyol. Res. 59: 143–150.

Asaoka R., M. Nakae and K. Sasaki. 2012. The innervation and adaptive significance of extensively distributed neuromasts in *Glossogobius olivaceus* (Perciformes: Gobiidae). Ichthyol. Res. 59: 143–150.

Assis, C.A. 2003. The lagenar otoliths of teleosts: their morphology and its application in species identification, phylogeny and systematics. J. Fish Biol. 62: 1268–1295.

Assis, C.A. 2005. The utricular otoliths, lapilli, of teleosts: their morphology and relevance for species identification and systematics studies. Sci. Mar. 69: 259–273.

Atema, J. 1971. Structure and functions of the sense of taste in the catfish (*Ictalurus natalis*). Brain. Behav. Evolution 4: 273–294.

Atkinson, C.J.L., K.J. Martin, G.J. Fraser and S.P. Collin. 2016. Morphology and distribution of taste papillae and oral denticles in the developing oropharyngeal cavity of the bamboo shark, *Chiloscyllium punctatum*. Biol. Open 5: 1759–1769.

Bailes, H.J., S.R. Robinson, A.E.O. Trezise and S.P. Collin. 2006. Morphology, characterisation and distribution of retinal photoreceptors in the Australian lungfish *Neoceratodus forsteri* (Krefft, 1870). J. Comp. Neurol. 494: 381–397.

Baker, C.V.H., M.S. Modrell and J.A. Gillis. 2013. The evolution and development of vertebrate lateral line electroreceptors. pp. 2515–2522. *In*: R. Krahe and E. Fortune (eds.). Electric Fishes: Neural Systems, Behaviour and Evolution. J. Exp. Biol. 216. The Company of Biologists Ltd. doi:10.1242/jeb.082362.

Barber, V.C. and C.J. Emerson. 1980. Scanning electron microscopic observations on the inner ear of the skate, *Raja ocellata*. Cell Tissue Res. 205: 199–215.

Barrett, R.T., C.J. Camphuysen, T. Anker-Nilssen, J.W. Chardine, R.W. Furness, S. Garthe, O. Hüppop, M.F. Leopold, W.A. Montevecchi and R.R. Veit. 2007. Diet studies of seabirds: a review and recommendations. ICES J. Mar. Sci. 64: 1675–1691.

Baudriot, F. and K. Reutter. 1998. Ultrastructure of the taste buds in the bowfin, *Amia calva* (Holostei: Amiidae). Chem. Senses. 24: 54.

Baudriot, F. and K. Reutter. 2001. Ultrastructure of the taste buds in the blind cave fish *Astyanax jordani* ("*Anoptichthys*") and the sighted river fish *Astyanax mexicanus* (Teleostei Characidae). J. Comp. Neurol. 434: 428–444.

Bauer, R. and J.P. Denizot. 1972. Sur la presence et la repartition des organs ampullaires chez Plotosus anguillaris. Arch. Anat. Micr. Morph. Exp. 61: 85–90.

Bauman, K.I., M. Grim and Z. Halata. 2003. The Merkel Cell-Structure-Development-Function-Cancerogenesis. Springer-Verlag, Berlin.

Becker, E.A., N.C. Bird and J.F. Webb. 2016. Canal and superficial neuromast patterning and the evolution of cranial lateral line phenotype in Lake Malawi cichlid fishes. J. Morphol. 277: 1273–1291.

Bedore, C.N. and S.M. Kajiura. 2013. Bioelectric fields of marine organisms: voltage and frequency contributions to detectability by electroreceptive predators. Physiol. Biochem. Zool. 86: 298–311.

Begg, G.A., S.E. Campana, A.J. Fowler and I.M. Suthers. 2005. Otolith research and application: Current directions in innovation and implementation. Mar. Freshw. Res. 56: 477–483.

Belanger, R.M., C.M. Smith, L.D. Corkum and B.S. Zielinski. 2003. Morphology and histochemistry of the peripheral olfactory organ in the round goby, *Neogobius melanostomus* (Teleostei: Gobiidae). J. Morphol. 257: 62–71.

Benser, M.E., N.P. Issa and A.J. Hudspeth. 1993. Hair-bundle stiffness dominates the elastic reactance to otolithic-membrane shear. Hear Res. 68: 243–252.

Bensouilah, M., C. Schugardt, R. Roesler, F. Kirschbaum and J.P. Denizot. 2002. Larval electroreceptors in the epidermis of mormyrid fish: I, Tuberous organs of type A and B. J. Comp. Neurol. 447: 309–322.

Berquist, R.M., V.L. Galinsky, S.M. Kajiura and L.R. Frank. 2015. The coelacanth rostral organ is a unique low-resolution electro-detector that facilitates the feeding strike. Scientific Reports 5. DOI: 10.1038/srep08962.

Bertmar, G. 1973. Ultrastructure of the olfactory mucosa in the homing Baltic sea trout *Salmo trutta trutta*. Mar. Biol. 19: 74–88.

Best, A.C.G. and J.A.C. Nicol. 1967. Reflecting cells of the elasmobranch tapetum lucidum. Contrib. Mar. Sci. 12: 172–201.

Betancur,-R., R., R.E. Broughton, E.O. Wiley, K. Carpenter, J.A. López, C. Li, N.I. Holcroft, D. Arcila, M. Sanciangco, J.C. Cureton II, F. Zhang, T. Buser, M.A. Campbell, J.A. Ballesteros, A. Roa-Varon, S. Willis, W.C. Borden, T. Rowley, P.C. Reneau, D.J. Hough, G. Lu, T. Grande, G. Arratia and G. Ortí. 2013. The tree of life and a new classification of bony fishes. PLOS Currents Tree of Life. doi: 10.1371/currents.tol.53ba26640df0ccaee75bb165c8c26288.

Bever, M.M. and D.M. Fekete. 2002. Atlas of the developing inner ear in zebrafish. Devel. Dyn. 223: 536–543.

Bird, N.C. and J.F. Webb. 2014. Heterochrony, modularity, and the functional evolution of the mechanosensory lateral line canal system of fishes. Evo. Devo. 5: 21.

Bishop, C. and P.H. Odense. 1966. Morphology of the digestive tract of the cod, *Gadus morhua*. J. Fish. Res. Can. 23: 1607–1615.

Blaxter, J.H.S., E.J. Denton and J.A.B. Gray. 1981. Acousticolateralis system in clupeid fishes. *In*: W.N. Tavolga, A.N. Popper and R.R. Fay (eds.). Hearing and Sound Communication in Fishes. Springer, New York USA.

Blaxter, J.H.S. 1987. Structure and development of the lateral line. Biol. Rev. 62: 471–514.

Bodznick, D. and R.G. Northcutt. 1981. Electroreception in lampreys: evidence that the earliest vertebrates were electroreceptive. Science 212: 465–467.

Bodznick, D. and R.L. Boord. 1986. Electroreception in Chondrichthyes. pp. 225–256. *In*: T.H. Bullock and W. Heiligenberg (eds.). Electroreception. John Wiley & Sons, New York, USA.

Bodznick, D. and J.C. Montgomery. 2005. The physiology of low-frequency electrosensory systems. pp. 132–153. *In*: T.H. Bullock, C.D. Hopkins, A.N. Popper and R.A. Fay (eds.). Electroreception. Springer Science + Business Media, Inc., USA.

Boord, R.L. and C.B.G. Campbell. 1977. Structural and functional organization of the lateral line system of sharks. Am. Zool. 17: 431–441.

Bowmaker, J.K. 1900. Visual pigments of fishes. pp. 81–107. *In*: R.H. Douglas and M.B.A. Djamgoz (eds.). The Visual System of Fish. Chapman and Hall, New York, USA.

Braekevelt, C.R. 1994. Fine structure of the tapetum lucidum in the short-tailed stingray (*Dasyatis brevicaudata*). Histol. Histopathol. 9: 495–500.

Braun, C.B. and R.G. Northcutt. 1997. The lateral line system of hagfishes (Craniata: Myxinoidea). Acta Zool. 78: 247–268.

Braun, C.B. and S. Coombs. 2000. The overlapping roles of the inner ear and lateral line: the active space of dipole source detection. Phil. Trans. Roy. Soc., London B, Biol. Sci. 355: 1115–1119.

Braun, C.B. and T. Grande. 2008. Evolution of peripheral mechanisms for the enhancement of sound reception. *In*: J.F. Webb, R.R. Fay and A.N. Popper (eds.). Fish Bioacoustics. Springer, New York, USA.

Breipohl, W., H.P. Zippel, K. Ruckert and H. Oggolter. 1976. Morphologische und elektrophysiologische Studien zur Struktur und Funktion des olfaktorischen Systems beim Goldfisch unter normalen und experimentellen Bedingungen. Beitr Elektronenmikrosk Direkt Oberfl. 9: 561–584.

Broughton, R.E., R. Betancur-R., C. Li, G. Arratia and G. Ortí. 2013. Multi-locus phylogenetic analysis reveals the pattern and tempo of bony fish evolution. PLOS Currents Tree of Life. doi: 10.1371/currents.tol.2ca8041495ffafd0c92756e75247483e.

Brown, B.R., J.C. Hutchinson, M.E. Hughes, D.R. Kellogg and R.W. Murray. 2002. Electrical characterization of gel collected from shark electrosensors. Phys. Rev. E 65: 061903.

Budzik, K.A., K. Żuwała and R. Kerney. 2016a. Tongue and taste organ development in the ontogeny of direct-developing salamander *Plethodon cinereus* (Lissamphibia: Plethodontidae). J. Morphol. 277: 906–915.

Budzik, K.A., K. Żuwała and D.R. Buchholz. 2016b. Taste organ growth and development in the direct developing frog of *Eleutherodactylus coqui* (Lissamphibia: Eleutherodactylidae). Acta Zool. 97: 433–441.

Bullock, T.H. and W. Heiligenberg. 1986. Electroreception. John Wiley & Sons, New York, USA.

Buran, B.N., X.H. Deng and A.N. Popper. 2005. Structural variation in the inner ears of four deep-sea elopomorph fishes. J. Morphol. 265: 215–225.

Burne, R.H. 1909. The anatomy of the olfactory organ of teleostean fishes. Proc. Zool. Soc. Lond. 2: 610–663.

Cameron, D.A. and S.S. Easter Jr. 1993. The cone photoreceptor mosaic of the green sunfish (*Lepomis cyanellus*) Vis. Neurosci. 10: 375–384.

Camilieri-Asch, V., R.M. Kempster, S.P. Collin, R. Johnstone and S.M. Theiss. 2013. A comparison of the electrosensory morphology of a euryhaline and a marine stingray. Zoology 116: 270–276.

Campana, S.E. and J.D. Neilson. 1985. Microstructure of fish otoliths. Can. J. Fish Aquat. Sci. 42: 1014–1032.

Campana, S.E. and S.R. Thorrold. 2001. Otoliths, increments, and elements: keys to a comprehensive understanding of fish populations? Can. J. Fish Aquat. Sci. 58: 30–38.

Canas, L., C. Stransky, J. Schlickeisen, M. Paz Sampedro and A. Celso Farina. 2012. Use of the otolith shape analysis in stock identification of anglerfish (*Lophius piscatorius*) in the Northeast Atlantic. ICES J. Mar. Sci. 69: 250–256.

Carlström, D. 1963. A crystallographic study of vertebrate otoliths. Biol. Bull. 125: 441–463.

Casper, B.M. 2011. The ear and hearing in sharks, skates, and rays. pp. 262–269. *In*: A.P. Farrell (ed.). Encyclopedia of Fish Physiology: From Genome to Environment. Academic Press, San Diego, CA, USA.

Cataldi, E., S. Cataudella, G. Monaco, A. Rossi and L. Tancioni. 1987. A study of the histology and morphology of the digestive tract of the sea-bream, *Sparus aurata*. J. Fish Biol. 30: 135–145.

Chang, J.S.Y., A.N. Popper and W.M. Saidel. 1992. Heterogeneity of sensory hair cells in a fish ear. J. Comp. Neurol. 324: 621–640.

Chitnis, A.J., D.D. Nogare and M. Matsuda. 2011. Building the posterior lateral line system in zebrafish. Dev. Neuro. 72: 234–255.

Chu, Y.T. and M.C. Wen. 1979. Monograph of fishes of China (No. 2): A study of the lateral-line canals system and that of lorenzini ampulla and tubules of elasmobranchiate fishes of China. Shanghai: Science and Technology Press.

Clarke, D., H. Whitney, G. Sutton and D. Robert. 2013. Detection and learning of floral electric fields by bumblebees. Science 34: 66–69.

Coffin, A.B., M.W. Kelley, G.A. Manley and A.N. Popper. 2004. Evolution of sensory hair cells. pp. 55–94. *In*: G. Manley, A.N. Popper and R.R. Fay (eds.). Evolution of the Vertebrate Auditory System. Springer, New York, USA.

Coffin, A.B., H. Brignull, D.W. Raible and E.W. Rubel. 2014. Hearing loss, protection, and regeneration in the larval zebrafish lateral line. pp. 313–347. *In*: S. Coombs, H. Bleckmann, R.R. Fay, A.N. Popper (eds.). The Lateral Line System. Springer-Verlag, New York, USA.

Cohen, A.I. 1963. Vertebrate retinal cells and their organization. Biol. Rev. 38: 427–459.

Collin, H.B. and S.P. Collin. 1995. Ultrastructure and organisation of the cornea, lens and iris in the pipefish, *Corythoichthys paxtoni* (Syngnathidae, Teleostei). Histol. Histopathol. 10: 313–323.

Collin, H.B. and S.P. Collin. 2000. The corneal surface structure in aquatic vertebrates: microprojections with optical and nutritional function? Phil. Trans. Roy. Soc. B 355: 1171–1176.

Collin, S.P., R.V. Hoskins and J.C. Partridge. 1998. Seven retinal specialisations in the tubular eye of the deepsea pearleye, *Scopelarchus michaelsarsi*: A case study in visual optimisation. Brain Behav. Evol. 51: 291–314.

Collin, S.P. and H.B. Collin. 1999. A comparative study of the corneal endothelium in vertebrates. Clin. Exp. Optom. 81(6): 245–254.

Collin, S.P., I.C. Potter and C. Braekevelt. 1999. The ocular morphology of the southern hemisphere lamprey *Geotria australis* Gray, with special reference to the characterisation and phylogeny of photoreceptor types. Brain Behav. Evol. 54: 96–118.

Collin, S.P. and I.C. Potter. 2000. The ocular morphology of the southern hemisphere lamprey *Mordacia mordax* Richardson with special reference to a single class of photoreceptor and a retinal tapetum. Brain Behav. Evol. 55(3): 120–138.

Collin, S.P. and H.B. Collin. 2001. The fish cornea: adaptations for different aquatic environments. pp. 57–96. *In*: B.G. Kapoor and T.J. Hara (eds.). Sensory Biology of Jawed Fishes—New Insights. Science Publishers Inc., Enfield USA.

Collin, S.P. and J. Shand. 2003. Retinal sampling and the visual field in fishes. pp. 139–169. *In:* S.P. Collin and N.J. Marshall (eds.). Sensory Processing in Aquatic Environment. Springer-Verlag, New York.

Collin, S.P. and A.E.O. Trezise. 2004. The origins of colour vision in vertebrates. Clin. Exp. Optom. 87: 217–223.

Collin, S.P. and A.E.O. Trezise. 2006. Evolution of colour discrimination and its implications for visual communication. pp. 303–335. *In*: F. Ladich, S.P. Collin, P. Moller, and B.G. Kapoor (eds.). Communication in Fishes. Science Publishers Enfield (NH) and Plymouth (UK).

Collin, S.P. and H.B. Collin. 2006. The corneal epithelial surface in the eyes of vertebrates: environmental and evolutionary influences on structure and function. J. Morphol. 267: 273–291.

Collin, S.P., W.L. Davies, N.S. Hart and D.M. Hunt. 2009. The evolution of early vertebrate photoreceptors. Phil. Trans. Roy. Soc. Lond. B 364: 2925–2940.

Collin, S.P., R. Kempster and K.E. Yopak. 2015. How elasmobranchs sense their environment. pp. 19–99. *In*: R.E. Shadwick, A.P. Farrell and C.J. Brauner (eds.). Physiology of Elasmobranch Fishes: Structure and Interaction with Environment. Vol. 34A, 1st Edition, Elsevier, New York.

Collin, S.P., R.M. Kempster and K.E. Yopak. 2016. How elasmobranchs sense their environment. pp. 19–99. *In*: R. Shadwick, A. Farrell and C. Brauner (eds.). Physiology of Elasmobranch Fishes: Structure and Interaction with Environment. Fish Physiology, Volume 34A. Elsevier/Academic Press.

Collin, S.P. 2017. Electroreception in Vertebrates and Invertebrates. Reference Module in Life Sciences. Elsevier Press. doi:10.1016/B978-0-12-809633-8.01293-0.

Cook, P.K., E. Dufour, M.-A. Languille, C. Mocuta, S. Réguer and L. Bertrand. 2016. Strontium speciation in archaeological otoliths. J. Anal. Atom. Spectr. 31: 700–711.

Coombs, S., J. Janssen and J.F. Webb. 1988. Diversity of lateral line systems: Phylogenetic, and functional considerations. pp. 553–593. *In*: J. Atema, R.R. Fay, A.N. Popper and W.N. Tavolga (eds.). Sensory Biology of Aquatic Animals. Springer-Verlag, New York, USA.

Cordier, R. 1964. Sensory cell. pp. 313–386. *In*: J. Barchet and A.E. Mirsky (eds.). The Cell: Biochemistry, Physiology, Morphology. Academic Press, New York.

Corwin, J.T. 1977. Morphology of the macula neglecta in sharks of the genus *Carcharhinus*. J. Morphol. 152: 341–362.

Corwin, J.T. 1981. Peripheral auditory physiology in the Lemon shark: evidence of parallel otolithic and non-otolithic sound detection. J. Comp. Physiol. A 142: 379–390.

Corwin, J.T. 1989. Functional anatomy of the auditory system in sharks and rays. J. Exper. Zool. 252(S2): 62–74.

Creutzberg, F. 1961. On the orientation of migrating elvers (*Anguilla vulgaris* Turt.) in a tidal area. Neth. J. Sea Res. 1: 257–338.

Crisp, M., G.A. Lowe and M.S. Laverack. 1975. On the ultrastructure and permeability of taste buds of the marine teleost *Ciliata mustela*. Tiss. Cell 7: 191–202.

Czech-Damal, N.U., A. Liebschner, L. Miersch, G. Klauer, F.D. Hanke, C. Marschall, G. Dehnhardt and W. Hanke. 2012. Electroreception in the Guiana dolphin (*Sotalia guianensis*). Proc. Biol. Sci. 279: 663–668.

De Busserolles, F., J.L. Fitzpatrick, N.J. Marshall and S.P. Collin. 2014. The influence of photoreceptor size and distribution on optical sensitivity in the eyes of lanternfishes (Myctophidae). PLoS ONE 9(6): e99957. doi:10.1371/journal.pone.0099957.

de Vries, H. 1950. The mechanics of the labyrinth otoliths. Acta Oto-Laryng. 38: 262–273.

Deng, X.H., H.-J. Wagner and A.N. Popper. 2011. The inner ear and its coupling to the swim bladder in the deep-sea fish *Antimora rostrata* (Teleostei: Moridae). Deep-Sea Res. I. 58: 27–37.

Deng, X.H., H.-J. Wagner and A.N. Popper. 2013. Interspecific variations of inner ear structure in the deep-sea fish family Melamphaidae. Anat. Rec. 296: 1064–1082.

Denizot, J.P. 1969. Etude histichimique des mucopolysaccharides de l'epiderme et des organes recepteurs du systeme de la ligne lateral du Gymnarque, *Gymnarchus niloticus*. Z. Zellf. 98: 469–476.

Denizot, J.P. 1970. Etude histichimique comparee des mucopolysaccharides des organes recepteurs de type ampullaires de certaines poissons electrique a faible decharge: *Gnathonemus petersii* (Mormyridae), *Gymnotus carapo* (Gymnotidae) et *Gymnarchus niloticus* (Gymnarchidae). Histochemie. 23: 82–90.

Denizot, J.P., M. Bensouilah, R. Roesler, C. Schugardt and F. Kirschbaum. 2007. Larval electroreceptors in the epidermis of mormyrid Fish: II. The Promormyromast. J. Comp. Neurol. 501: 810–823.

Denton, E.J. and J.A.C. Nicol. 1964. The choroidal tapeta of some cartilaginous fishes (Chondrichthyes). J. Mar. Biol. Assoc. U.K. 44: 219–258.

Denton, E.J. and J.A.B. Gray. 1979. The analysis of sound by the sprat ear. Nature 282: 406–407.

Desgranges, J.C. 1966. Sur la double innervations des cellules sensorielles des bourgeons du gout des barbillons du Poisson chat. Comptes Rendus de l'Académie des Sciences Paris 263: 1103–1106.

Diaz, J.P., M. Priegrane, T. Noell and R. Connes. 2002. Ultrastructural study of the olfactory organ in adult and developing European sea bass, *Dicentrarchus labrax*. Can. J. Zool. 80: 1610–1622.

Disspain, M.C.F., S. Ulm and B.M. Gillanders. 2016. Otoliths in archaeology: Methods, applications and future prospects. J. Arch. Sci. Rep. 6: 623–632.

Doroshenko, M.A. and P.A. Motavkin. 1986. Olfactory epithelium of marine fishes in scanning electron microscopy. Acta Morphol. Hung. 34: 143–155.

Dos Santos, M.L., F.P. Arantes, K.B. Santiago and J.E. Dos Santos. 2015. Morphological characteristics of the digestive tract of *Schizodon knerii* (Steindachner, 1875), (Characiformes: Anostomidae): An anatomical, histological and histochemical study. Ann. Brazil Acad. Sci. 87: 867–878.

Dotterweich, H. 1932. Bau und Funktion der Lorenzinischen Ampullen. Zool. Jahrb. Physiol. 55: 347–418.

Douglas, R.H. 1982. The function of photomechanical movements in the retina of the rainbow trout (*Salmo gairdneri*). J. Exp. Biol. 96: 389–403.

Douglas, R.H. and H.J. Wagner. 1982. Endogenous patterns of photomechanical movements in teleosts and their relation to activity rythms. Cell Tiss. Res. 226: 133–144.

Douglas, R.H. and C.M. McGuigan. 1989. The spectral transmission of freshwater teleost ocular media: An interspecific comparison and a guide to potential ultraviolet sensitivity. Vision Res. 29: 871–879.

Douglas, R.H., R.D. Harper and J.F. Case. 1998. The pupil response of a teleost fish, *Porichthys notatus*: A description and comparison to other species. Vis. Res. 38: 2697–2710.

Douglas, R.H., S.P. Collin and J. Corrigan. 2002. The eyes of suckermouth armoured catfish (Loricariidae, subfamily Hypostomus): pupil response, lenticular spherical aberration and retinal topography. J. Exp. Biol. 205: 3425–3433.

Doving, K.B. 1967. Comparative electrophysiological studies on the olfactory tract of some teleost. J. Comp. Neurol. 131: 365–370.

Doving, K.B., M. Dubois-Dauphin, A. Holley and F. Jourdan. 1977. Functional anatomy of the olfactory organ of fish and the ciliary mechanism of water transport. Acta Zool. 58: 245–255.

Eatock, R.A. and J.E. Songer. 2011. Vestibular hair cells and afferents: Two channels for head motion signals. Ann. Rev. Neurosci. 34 501–534.

Ekstrom von Lubitz, D.K. 1981. Ultrastructure of the lateral-line sense organs of the ratfish, *Chimaera monstrosa*. Cell Tiss. Res. 215: 651–665.

Elsheikh, E.H., E.S. Nasr and A.M. Gamal. 2012. Ultrastructure and distribution of the taste buds in the buccal cavity in relation to the food and feeding habit of a herbivorous fish: *Oreochromis niloticus*. Tiss. Cell 44: 164–169.

Engström, K. 1961. Cone types and cone arrangement in the retina of some gadids. Acta Zool. 17: 227–243.

Engström, K. 1963. Cone types and cone arrangements in teleost retinae. Acta Zool. 44: 179–243.

Evangelista, C., M. Mills, U.E. Siebeck and S.P. Collin. 2010. A comparison of the external morphology of the membranous inner ear in elasmobranchs. J. Morphol. 271: 483–495.

Evans, R.F., B. Zielinski and T.J. Hara. 1982. Development and regeneration of the olfactory organ in rainbow trout. pp. 15–37. *In*: T.J. Hara (ed.). Chemoreception in Fishes. Elsevier, Amsterdam, Oxford, New York.

Ezeasor, D.N. 1982. Distribution and ultrastructure of taste buds in the oropharyngeal cavity of the rainbow trout, *Salmo gairdneri* Richardson. J. Fish Biol. 20: 53–68.

Falini, G., S. Fermani, S. Vanzo, M. Miletic and G. Zaffino. 2005. Influence on the formation of aragonite or vaterite by otolith macromolecules. Eur. J. Inorg. Chem. 1: 162–167.

Faucher, K., A. Aubert and J.-P. Lagardére. 2003. Spatial distribution and morphological characteristics of the trunk lateral line neuromasts of the sea bass (*Dicentrarchus labrax*, L.; Teleostei, Serranidae). Brain Behav. Evol. 62: 223–232.

Fay, R.R., J.I. Kendall, A.N. Popper and A.L. Tester. 1974. Vibration detection by macula neglecta of sharks. Comp. Biochem. Physiol. 47(4A): 1235–1240.

Fay, R.R. and P.L. Edds-Walton. 2008. Structures and functions of the auditory nervous system of fishes. pp. 49–97. *In*: J.F. Webb, R.R. Fay and A.N. Popper (eds.). Fish Bioacoustics. Springer, New York, USA.

Fernholm, B. 1985. The lateral line system of cyclostomes. pp. 113–122. *In*: R.E. Foreman, A. Gorbman, J.M. Dodd and R. Olsson (eds.). Evolution and Biology of Primitive Fishes. Springer.

Ferrando, S., L. Gallus, Ch. Gambardella, M.A. Masini, A. Cutolo and M. Vacchi. 2012. First detection of taste buds in a chimaeroid fish (Chondrichthyes: Holocephali) and their Galpha i-like immunoreactivity. Neurosci. Letters 517: 98–101.

Ferrando, S., L. Gallus, Ch. Gambardella, D. Croce, G. Damiano, Ch. Mazzarino and M. Vacch. 2016. First description of a palatal organ in *Chimaera monstrosa* (Chondrichthyes, Holocephali). Anat. Rec. 299: 118–131.

Fields, R.D., T.H. Bullock and G.D. Lange. 1993. Ampullary sense organs, peripheral, central and behavioral electroreception in chimeras (*Hydrolagus*, Holocephali, Chondrichthyes). Brain Behav. Evol. 41: 269–289.

Fineran, B.A. and J.A.C. Nicol. 1974. Studies on the eyes of New Zealand parrotfishes (Labridae). Proc. Roy. Soc. Lond. B 186: 217–247.

Finger, T.E. 1976. Gustatory pathways in the bullhead catfish. I. Connections of the anterior ganglion. J. Comp. Neurol. 165: 513–526.

Finger, T.E., S.K. Drake, K. Kotrschal, M. Womble and K.C. Dockstader. 1991. Postlarval growth of the peripheral gustatory system in the channel catfish, *Ictalurus punctatus*. J. Comp. Neurol. 314: 55–66.

Fishelson, L. and Y. Delarea. 2004. Taste buds on the lips and mouth of some blenniid and gobiid fishes: comparative distribution and morphology. J. Fish Biol. 65: 651–665.

Fishelson, L., Y. Delarea and A. Zverdling. 2004. Taste bud form and distribution on lips and in the oropharyngeal cavity of cardinal fish species (Apogonidae, Teleostei), with remarks on their dentition. J. Morphol. 259: 316–327.

Fishelson, L., C.C. Baldwin and P.A. Hastings. 2012. Comparison of the oropharyngeal cavity in the Starksiini (Teleostei: Blenniiformes: Labrisomidae): taste buds and teeth, including a comparison with closely-related genera. J. Morphol. 273: 618–628.

Flock, Å. and M.H. Goldstein. 1978. Cupular movement and nerve impulse response in the isolated semicircular canal. Brain Res. 157: 11–19.

Fox, H. and M. Whitear. 1978. Observations on Merkel cells in amphibians. Biol. Cellulaire 32: 223.

Franz, V. 1920. Zur mikroskopischen Anatomie der Mormyriden. Zool. Jb. Anat. Abt. 42: 91–148.

Friedrich-Freska, H. 1930. Lorenzinische Ampullen bei dem Siluroiden *Plotosus anguillaris* Bloch. Zool. Anz. 87: 49–66.

Fritzsch, B. and H. Münz. 1986. Electroreception in amphibians. pp. 483–496. *In*: T.H. Bullock and W. Heiligenberg (eds.). Electroreception. John Wiley & Sons, New York, USA.

Fritzsch, B. and H. Straka. 2014. Evolution of vertebrate mechanosensory hair cells and inner ears: toward identifying stimuli that select mutation driven altered morphologies. J. Comp. Physiol. A 200: 5–18.

Furukawa, T. and Y. Ishii. 1967. Neurophyiological studies on hearing in goldfish. J. Neurophys. 30: 1377–1403.

Gardiner, J.M., R.E. Hueter, K.P. Maruska, J.A. Sisneros, B.M. Casper, D.A. Mann and L.S. Demski. 2012. Sensory physiology and behavior of elasmobranchs. *In*: J.C. Carrier, J.A. Musick and M.R. Heithaus (eds.). Biology of Sharks and their Relatives, CRC Press, Boca Raton, FL USA.

Garman, S. 1899. Reports on an Exploration off the West Coasts of Mexico, Central and South America, and off the Galapogos Islands, in Charge of Alexander Agassiz, by the UW Fish Commission Steamer "Albatross" during 1891, Lieut. Commander Z.L. Tanner, USN, Commanding. XXVI—The Fishes. Mem. Mus. Comp. Zool. 24: 1–431 + 97 plates.

Gauldie, R.W., D. Dunlop and J. Tse. 1986a. The remarkable lungfish otolith. New Zealand J. Mar. Freshw. Res. 20: 81–92.

Gauldie, R.W., D. Dunlop and J. Tse. 1986b. The simultaneous occurrence of otoconia and otoliths in four teleost fish species. New Zealand J. Mar. Freshw. Res. 20: 93–99.

Gelman, S., A. Ayali, E.D. Tytell and A.H. Cohen. 2006. Larval lampreys possess a functional lateral line system. J. Comp. Physiol. A 193: 271–277.

Gemne, G. and K.B. Doving. 1969. Ultrastructural properties of primary olfactory neurons in fish (*Lota lota* L.) Amer. J. Anat. 126: 457–476.

Ghysen, A. and C. Dambly-Chaudière. 2007. The lateral line microcosmos. Genes & Devel. 21: 2118–2130.

Ghysen, A., H. Wada and C. Dambly-Chaudière. 2014. Patterning the posterior lateral line in teleosts: evolution of development. pp. 295–318. *In*: H. Bleckmann, J. Mogdans and S.L. Coombs (eds.). Flow Sensing in Air and Water—Behavioural, Neural and Engineering Principles of Operation. Springer, Berlin.

Gibbs, M.A. 1999. Lateral line morphology and cranial osteology of the rubynose brotula, *Cataetyx rubrirostris*. J. Morphol. 241: 265–274.

Gillespie, P.G. and U. Müller. 2009. Mechanotransduction by hair cells: Models, molecules, and mechanisms. Cell 139(1): 33–44.

Gregory, W.K. 1933. Fish skulls: A study of the evolution of natural mechanisms. Trans. of Amer. Phil. Soc. 23: 75–481.

Grower-Johnson, N. and A.J. Farbman. 1976. Fine structure of taste buds in the barbel of the catfish, *Ictarulus punctatus*. Cell Tiss. Res. 169: 395–403.

Hanke, B.V., A.A. Meyer and E. Oliveira. 2008. Análise histológica de estruturas sensoriais de *Centropomus parallelus* (Poey, 1896) (Centropomidae) relacionadas ao hábito alimentar. RUBS 1: 16–23.

Hansen, A. and E. Zeiske. 1998. The peripheral olfactory organ of the zebrafish, *Danio rerio*: An ultrastructural study. Chem. Sens. 23: 39–48.

Hansen, A., H.P. Zippel, P.W. Sorensen and J. Caprio. 1999. Ultrastructure of the olfactory epithelium in intact, axotomized, and bulbectomized goldfish, *Carassius auratus*. Micr. Res. Techn. 45: 325–338.

Hansen, A. and T.E. Finger. 2000. Phyletic distribution of crypt-type olfactory receptor neurons in fishes. Brain Behav. Evol. 55: 100–110.

Hansen, A., K. Reutter and E. Zeiske. 2002. Taste bud development in the zebrafish, *Danio rerio*. Devel. Dyn. 223: 483–496.

Hansen, A. and K. Reutter. 2004. Chemosensory systems in fish: structural, functional and ecological aspects. pp. 55–63. *In*: G. Von Der Emde, J. Mogdans and B.G. Kapoor (eds.). The Senses of Fish. Adaptations for the Reception of Natural Stimuli. Narosa Publishing House, New Delhi.

Hansen, A. and B.S. Zielinski. 2005. Diversity in the olfactory epithelium of bony fishes: development, lamellar arrangement, sensory neuron cell types and transduction components. J. Neurocytol. 34: 183–208.

Hara, T.J. 1975. Olfaction in fish. Prog. Neurobiol. 5(4): 271–335.

Hart, N.S., H.J. Bailes, M. Vorobyev, N.J. Marshall and S.P. Collin. 2008. Visual ecology of the Australian lungfish (*Neoceratodus forsteri*). BMC Ecology 8: 21. doi:10.1186/1472-6785-8-21.

Hasler, A.D. 1966. Underwater Guideposts—Homing of Salmon. University of Wisconsin Press, Madison.

Hawkins, A.D. and D.N. MacLennan. 1976. An acoustic tank for hearing studies on fish. pp. 149–169. *In*: A. Schuijf and A.D. Hawkins (eds.). Sound Reception in Fish. Elsevier, Amsterdam.

Hawkins, A.D. and A.J. Myrberg. 1983. Hearing and sound communication underwater. pp. 347–405. *In*: B. Lewis (ed.). Bioacoustics—A Comparative Approach. Academic Press, London.

Hawkins, A.D. 1993. Underwater sound and fish behaviour. pp. 129–169. *In*: T.J. Pitcher (ed.). Behaviour of Teleost Fishes. Chapman and Hall, London.

Heinermann, P.H. 1984. Yellow intraocular filters in fishes. Exp. Biol. 43: 127–147.

Henninger, J., R. Krahe, F. Kirschbaum, J. Grewe and J. Benda. 2018. Statistics of natural communication signals observed in the wild identify important yet neglected stimulus regimes in weakly electric fish. J. Neurosc. In press.

Herrick, C.J. 1901. The cranial nerves and cutaneous sense organs of the North American siluroid fishes. J. Comp. Neurol. 11: 177–249.

Higgs, D.M. and C.R. Radford. 2012. The contribution of the lateral line to 'hearing' in fish. J. Exp. Biol. 216: 1484–1490.

Hirata, Y. 1966. Fine structure of the terminal buds in the barbels of some fishes. Arch. Histol. Jap. 26: 507–523.

Hirota, K., R. Asaoka, M. Nakae and K. Sasaki. 2015. The lateral line system and its innervation in *Zenarchopterus dunckeri* (Beloniformes: Exocoetoidei: Zenarchopteridae): an example of adaptation to surface feeding in fishes. Ichthyol. Res. 62: 286–292.

Hudspeth, A.J. and D.P. Corey. 1977. Sensitivity, polarity, and conductance change in the response of vertebrate hair cells to controlled mechanical stimuli. US Proc. Nat. Acad. Sci. 74: 2407–2411.

Hudspeth, A.J. 1985. The cellular basis of hearing: The biophysics of hair cells. Science 230: 745–752.

Hudspeth, A.J. 1989. How the ear's works work. Nature 341: 397–404.

Hudspeth, A.J., Y. Choe, A.D. Mehta and P. Martin. 2000. Putting ion channels to work: Mechanoelectrical transduction, adaptation, and amplification by hair cells. US Proc. Nat. Acad. Sci. 97: 11765–11772.

Hughes, I., I. Thalmann, R. Thalmann and D.M. Ornitz. 2006. Mixing model systems: Using zebrafish and mouse inner ear mutants and other organ systems to unravel the mystery of otoconial development. Brain Res. 1091: 58–74.

Ibsch, M., R.H. Anken, P. Vohringer and H. Rahmann. 2001. Vesicular bodies in fish maculae are artifacts not contributing to otolith growth. Hear Res. 153: 80–90.

Jakubowski, M. 1958. The structure and vascularization of the skin in the pond-loach (*Misgurnus fossilis* L.). Acta Biol. Cracov. Ser. Zool. 1: 113–127.

Jakubowski, M. 1963. Cutaneous sense organs of fishes. I. The lateral-line organs in the stone-perch (*Acerina cernua* L.). Acta Biol. Crac. Ser. Zool. 6: 59–78.

Jakubowski, M. 1966. The lateral line organs in some Cobitidae. Acta Biol. Cracov. Ser. Zool. 71–80.

Jakubowski, M. 1967. Cutaneous sense organs of fishes. Part VII. The structure of the system of lateral-line canal organs in the Percidae. Acta Biol. Crac. Ser. Zool. 10: 69–81.

Jakubowski, M. 1974. Structure of the lateral-line canal system and related bones in the berycoid fish *Hoplostethus mediteranneus* Cuv. et Val. (Trachichthyidae, Pisces). Acta Anat. 87: 261–274.

Jakubowski, M. and E. Kunysz. 1979. Anatomy and morphometry of the olfactory organ of the wels *Silurus glanis* L. (Siluridae, Pisces). Z. mikrosk. Anat. Forsch. 93: 728–735.

Jakubowski, M. 1981. Ultrastructure (SEM, TEM) of the olfactory epithelium in the Wels, *Silurus glanis* L. (Siluridae, Pisces). Z. mikrosk. Anat. Forsch. 93: 728–735.

Jakubowski, M. 1983. New details of the ultrastructure (TEM, SEM) of taste buds in fishes. Z. mikrosk. Anat. Forsch. 97: 849–862.

Jakubowski, M. and M. Whitear. 1990. Comparative morphology and cytology of taste buds in teleosts. Z. Mikrosk. Anat. Forsch. 104: 529–560.

Jakubowski, M. and K. Żuwała. 2000. Taste organs in lower vertebrates. Taste organs in fishes. pp. 161–174. *In*: H.M. Dutta and J.S. Datta Munshi (eds.). Vertebrate Functional Morphology: Horizons of Research in the 21st Century. Science Publishers, Inc., Enfield (NH), USA.

Jasiński, A. 1982. Narządy zmysłów. pp. 428–442. *In*: H. Szarski (ed.). Anatomia Porównawcza Kręgowców. Państwowe Wydawnictwo Naukowe, Warszawa.

Johnson, S.E. 1917. Structure and development of the sense organs of the lateral line canal system of selachians (*Mustelus canis* and *Squalus acanthias*). J. Comp. Neurol. 28: 1–74.

Jolivet, A., J.-F. Bardeau, R. Fablet, Y.-M. Paulet and H. de Pontual. 2008. Understanding otolith biomineralization processes: new insights into microscale spatial distribution of organic and mineral fractions from Raman microspectrometry. Anal. Bioanal. Chem. 392: 551–560.

Jollie, M. 1984. Development of the head skeleton and pectoral girdle of salmons, with a note on the scales. Can. J. Zool. 62: 1757–1778.

Jordan, L.K., S.M. Kajiura and M.S. Gordon. 2009. Functional consequences of structural differences in stingray sensory systems. Part I: Mechanosensory lateral line canals. J. Exper. Biol. 212: 3037–3043.

Jorgensen, J.M., A. Flock and J. Wersäll. 1972. The lorenzinian ampullae of *Polyodon spathula*. Z. Zellforsch. 130: 362–377.

Jørgensen, J.M. and J.O. Pickles. 2002. The lateral line canal sensory organs of the epaulette shark (*Hemiscyllium ocellatum*). Acta Zool. (Stockholm) 83: 337–343.

Jorgensen, J.M. 2005. Morphology of electroreceptive sensory organs. pp. 47–67. *In*: T.H. Bullock, C.D. Hopkins, A.N. Popper and R.A. Fay (eds.). Electroreception. Springer Science + Business Media, Inc., USA.

Kajiura, S.M. 2001. Head morphology and electrosensory pore distribution of carcharhinid and sphyrnid sharks. Environ. Biol. Fishes 61: 125–133.

Kajiura, S.M. and K.N. Holland. 2002. Electroreception in juvenile scalloped hammerhead and sandbar sharks. J. Exp. Biol. 205: 3609.

Kajiura, S.M., A.D. Cornett and K.E. Yopak. 2010. Sensory adaptations to the environment: electroreceptors as a case study. pp. 393–434. *In*: J.C. Carrier, J.A. Musick and M.R. Heithaus (eds.). Sharks and their Relatives. 2. Biodiversity, Adaptive Physiology and Conservation. Boca Raton, CRC Press.

Kalmijn, A.J. 1974. The detection of electric fields from inanimate and animate sources other than electric organs. pp. 147–200. *In*: A. Fessard (ed.). Electroreceptors and Other Specialized Receptors in Lower Vertebrates. Springer, Berlin/Heidelberg.

Kapoor, B.C., H.E. Evans and R.A. Pevzner. 1975. The gustatory system in fish. Adv. Mar. Biol. 13: 53–108.

Kasumyan, A.O. 2003. The lateral line in fish: Structure, function and role in behavior. J. Ichthyol. 43(Suppl. 2): S175–S213.

Kempster, R.M., I.D. McCarthy and S.P. Collin. 2012. Morphological, phylogenetic and ecological factors influencing the number and distribution of electroreceptors in elasmobranchs. J. Fish Biol. 80: 2055–2088.

Kempster, R.M., E. Garza-Gisholt, C.A. Egeberg, N.S. Hart, O.R. O'Shea and S.P. Collin. 2013. Sexual dimorphism of the electrosensory system: a quantitative analysis of nerve axons in the dorsal anterior lateral line nerve of the blue spotted fantail stingray *(Taeniura lymma)*. Brain Behav. Evol. 81: 1–10.

Kempster, R.M., C.A. Egeberg, N.S. Hart and S.P. Collin. 2015. A comparison of electrosensory-driven feeding behaviour in a benthic shark (Port Jackson shark) and a benthic ray (Western shovelnose ray). Mar. Fresh. Res. http://dx.doi.org/10.1071/MF13213.

Kerr, L.A. and S.E. Campana. 2013. Chemical composition of fish hard parts as a natural marker of fish stocks. pp. 205–234. *In*: S.X. Cadrin, L.A. Kerr and S. Mariani (eds.). Stock Identification Methods: Applications in Fishery Science. Academic Press, San Diego, CA USA.

Kikuchi, K. and D. Hilding. 1965. The development of the organ of Corti in the mouse. Acta Oto-Laryng. 60: 207–221.

Kindt, K.S., G. Finch and T. Nicolson. 2012. Kinocilia mediate mechanosensitivity in developing zebrafish hair cells. Devel. Cell 23: 329–341.

Kirschbaum, F. 1977. Electric organ ontogeny: Distinct larval organ precedes the adult organ in weakly electric fish. Naturwissenschaften 64: 387–388.

Kirschbaum, F. and P. Williot. 2011. Ontogeny of the European Sturgeon, *Acipenser sturio*. pp. 65–80. *In*: P. Williot, E. Rochard, N. Desse-Berset, F. Kirschbaum and J. Gessner (eds.). Biology and Conservation of the European Sturgeon *Acipenser sturio* L. 1758. Springer, Heidelberg, Dordrecht, London, New York.

Kiyohara, S., S. Yamashita and J. Kitoh. 1980. Distribution of taste buds on the lips and inside the mouth in the minnow *Pseudorasbora parva*. Physiol. Behav. 24: 1143–1147.

Kleerkoper, H. and G.A. Van Erkel. 1960. The olfactory apparatus of *Petromyzon marinus* L. Can. J. Zool. 38: 209–223.

Kondrashev, S.L., A.G. Gamburtseva, V.P. Gnyubkina, O. Yu and P.T. My. 1986. Coloration of corneas in fishes. A list of species. Vision Res. 26: 287–290.

Kotrschal, K. 1996. Solitary chemosensory cells: why do primery aquatic vertebrates need another taste system? Trends Ecol. Evol. 11: 110–114.

Kotrschal, K. 2000. Taste(s) and olfaction(s) in fish: a review of spezialized sub-systems and central integration. Pflugers Arch. 439: R178–R180.

Krysl, P., A.D. Hawkins, C. Schilt and T.W. Cranford. 2012. Angular oscillation of solid scatterers in response to progressive planar acoustic waves: do fish otoliths rock? PLoS ONE 7 (8): e42591.

Kuciel, M., K. Żuwała and M. Jakubowski. 2011. A new type of fish olfactory organ structure in *Periophthalmus barbarus* (Gobiidae, Oxudercinae). Acta Zool. Stockholm 92: 276–280.

Kuciel, M., K. Żuwała and U. Satapoomin. 2013. Comparative morphology (SEM) of the peripheral olfactory organ in the Oxudercinae subfamily (Gobiidae, Perciformes), Zool. Anz. 252: 424–430.

Kuciel, M., K. Żuwała, E.R. Lauriano, G. Polgar, S. Malavasi and G. Zaccone. 2017. The anatomy of sensory organs. *In*: E. Murdy and Z. Jaafar (eds.). Fishes Out of Water: The Biology and Ecology of Mudskippers (in press).

Kunz, Y.W., M.N. Shuilleabhain and E. Callaghan. 1985. The eye of the venomous marine teleost *Trachinus vipera* with special reference to the structure and ultrastructure of visual cells and pigment epithelium. Exp. Biol. 43: 161–178.

Ladich, F. and A.N. Popper. 2004. Parallel evolution in fish hearing organs. pp. 95–127. *In*: G.A. Manley, R.R. Fay and A.N. Popper (eds.). Evolution of the Vertebrate Auditory System. Springer, New York, USA.

Ladich, F. 2014. Diversity in hearing in fishes: ecoacoustical, communicative, and developmental constraints. pp. 289–321. *In*: C. Köppl, G.A. Manley, A.N. Popper and R.R. Fay (eds.). Insights From Comparative Hearing Research. Springer, New York, USA.

Ladich, F. and T. Schulz-Mirbach. 2016. Diversity in fish auditory systems: one of the riddles of sensory biology. Front. Ecol. Evol. 4: 28.

Land, M.F. and D.-E. Nilsson. 2002. Animal Eyes. Oxford University Press. UK.

Lane, E.B. and M. Whitear. 1982. Sensory structures at the surface of fish skin I. Putative chemoreceptors. Zool. J. Linn. Soc. 141–151.

Lane, E.B. and M. Whitear. 1982. Sensory structures at the surface of fish skin. II. Lateralis system. Zool. J. Linn. Soc. 76: 19–28.

Lantzing, W.J.R. and R.G. Wright. 1982. The ultrastructure of the eye of the mosquitofish *Gambusia affinis*. Cell Tiss. Res. 225: 431–449.

Leibovici, M., E. Verpy, R.J. Goodyear, I. Zwaenepoel, S. Blanchard, S. Lainé, G.P. Richardson and C. Petit. 2005. Initial characterization of kinocilin, a protein of the hair cell kinocilium. Hear Res. 203: 144–153.

Lekander, B. 1949. The sensory line system and the canal bones in the head of some Ostariophysi. Acta Zool. (Stockholm) 30: 1–131.

Leydig, F. 1851. Uber die Haut einiger SuBwasserfische. Zeitschrift für wissenschaftliche Zoologie. 3: 1–12.

Lissman, H.W. 1951. Continuous electrical signals from the tail of a fish, *Gymnarchus niloticus* Cuv. Nature 167: 201–202.

Lissman, H.W. 1958a. On the function and evolution of electric organs in fish. J. Exp. Biol. 35: 156–191.

Lissman, H.W. 1958b. The mechanism of object location in *Gymnarchus niloticus* and similar fish. J. Exp. Biol. 35: 451–486.

Litherland, L., S.P. Collin and K.A. Fritsches. 2009. Visual optics and ecomorphology of the growing shark eye: a comparison between deep and shallow water species. J. Exp. Biol. 212: 3583–3594.

Locket, N.A. 1971. Retinal anatomy in some scopelarchid deepsea fishes. Proc. Roy. Soc. Lond. B 178: 161–184.

Locket, N.A. 1973. Retinal structure in *Latimeria*. Phil. Trans. Roy. Soc. Lond. B 266: 493–521.

Locket, N.A. 1975. Landolt's clubs in some primitive fishes. pp. 471–480. *In*: M.A. Ali (ed.). Vision in Fishes: New Approaches in Research. Plenum Press, New York.

Locket, N.A. 1977. Adaptations to the deep-sea environment. pp. 67–192. *In*: F. Crescitelli (ed.). The Visual System in Vertebrates. Handbook of Sensory Physiology Vol. VIII/5, Springer-Verlag, New York.

Lopez-Shier, H., C.J. Starr, F.A. Kappler, R. Kollmar and A.J. Hudspeth. 2004. Directional cell migration establishes the axes of planar polarity in the posterior lateral-line organ of the zebrafish. Dev. Cell 7: 401–412.

Lovell, J.M., M.M. Findlay, R.M. Moate, J.R. Nedwell and M.A. Pegg. 2005. The inner ear morphology and hearing abilities of the Paddlefish (Polyodon spathula) and the Lake Sturgeon (Acipenser fulvescens). Comp. Biochem. Physiol. A 142: 286–296.

Lovell, J.M., M.M. Findlay, G.M. Harper and R.M. Moate. 2007. The polarization of hair cells from the inner ear of the lesser spotted dogfish Scyliorhinus canicula. J. Fish Biol. 70: 362–373.

Lowenstein, O., M.P. Osborne and J. Wersäll. 1964. Structure and innervation of the sensory epithelia of the labyrinth in the Thornback ray (Raja clavata). Proc. R. Soc. London B Biol. Sci. 160: 1–12.

Lu, J. and H.M. Fishman. 1995. Ion channels and transporters in the electroreceptive ampullary epithelium from scates. Biophys. J. 69: 2467–2475.

Lu, Z., J. Song and A.N. Popper. 1998. Encoding of acoustic directional information by saccular afferents of the sleeper goby, *Dormitator latifrons*. J. Comp. Physiol. A—Sens. Neur. Behav. Physiol. 182: 805–815.

Lu, Z. and A.N. Popper. 2001. Neural response directionality correlates of hair cell orientation in a teleost fish. J. Comp. Physiol. A—Sens. Neur. Behav. Physiol. 187: 453–465.

Lu, Z., Z. Xu and J.H. Stadler. 2002. Roles of the saccule in directional hearing. Bioacoustics 12: 205–207.

Lu, Z., Z. Xu and W.J. Buchser. 2003. Acoustic response properties of lagenar nerve fibers in the sleeper goby, *Dormitator latifrons*. J. Comp. Physiol. A—Sens. Neur. Behav. Physiol. 189(12): 889–905.

Lu, Z., Z. Xu and W.J. Buchser. 2004. Coding of acoustic particle motion by utricular fibers in the sleeper goby, *Dormitator latifrons*. J. Comp. Physiol. A—Sens. Neur. Behav. Physiol. 190(11): 923–938.

Lundberg, Y.W., X. Zhao and E.N. Yamoah. 2006. Assembly of the otoconia complex to the macular sensory epithelium of the vestibule. Brain Res. 1091: 47–57.

Lundberg, Y.W., Y. Xu, K.D. Thiessen and K.L. Kramer. 2015. Mechanisms of otoconia and otolith development. Devel. Dyn. 244: 239–253.

Lythgoe, J.N. 1976. The ecology, function and phylogeny of iridescent multilayers in fish corneas. pp. 211–247. *In*: R. Bainbridge, G.C. Evans and O. Rackman (eds.). Light as an Ecological Factor II, Blackwell, Oxford.

Lythgoe, J.N. 1979. The Ecology of Vision. Clarendon Press, Oxford.

Ma, E.Y. and D.W. Raible. 2009. Signaling pathways regulating zebrafish lateral line development. Curr. Biol. 19: R381–R386.

Manley, G.A. 2000. Cochlear mechanisms from a phylogenetic viewpoint. US Proc. Nat. Acad. Sci. 97: 11736–11743.

Marranzino, A.N. and J.F. Webb. 2018. Flow sensing in the deep sea: The mechanosensory lateral line system of stomiiform fishes. Zool. J. Linn. Soc. 183: 945–956..

Marshall, N.J. 1986. Structure and general distribution of free neuromasts in the black goby, *Gobius niger*. J. Mar. Biol. Assoc. UK 66: 323–333.

Marshall, N.J. 1996. The lateral line systems of three deep-sea fish. J. Fish Biol. 49(Suppl.): 239–258.

Martini, E. and B. Reichenbacher. 1997. Fish remains, especially otoliths. Recent shore sediments of the Salton Sea. Courier Forschungsinstitut Senckenberg 201: 277–293.

Maruska, K.P. and T.C. Tricas. 1998. Morphology of the mechanosensory lateral line system in the Atlantic stingray, *Dasyatis sabina*: The mechanotactile hypothesis. J. Morphol. 238: 1–22.

Maruska, K.P. 2001. Morphology of the mechanosensory lateral line system in elasmobranch fishes: ecological and behavioral considerations. Env. Biol. Fish. 60: 47–75.

Marzullo, T.A., B.E. Wueringer, L. Squire Jr. and S.P. Collin. 2011. Description of the mechanoreceptive lateral line and electroreceptive ampullary system in the freshwater whipray, *Himantura dalyensis*. Mar. Freshw. Res. 62: 771–779.

Mathiesen, C. 1984. Structure and innervation of inner ear sensory epithelia in the European eel (*Anguilla anguilla* L.). Acta Zool. 65: 189–207.

Mathiesen, C. and A.N. Popper. 1987. The ultrastructure and innervation of the ear of the gar, Lepisosteus osseus. J. Morphol. 194: 129–142.

McGowan, D.W. and S.M. Kajiura. 2009. Electroreception in the euryhaline stingray, *Dasyatis sabina*. J. Exp. Biol. 212: 1544–1552.

McHenry, M.J. and J.C. Liao. 2014. The hydrodynamics of flow stimuli. pp. 93–98. *In*: S. Coombs, H. Bleckmann, R.R. Fay and A.N. Popper (eds.). The Lateral Line System. Springer-Verlag, New York, USA.

Meyer–Rochow V.B. 1981. Fish tongues—Surface fine structure and ecological considerations. Zool. J. Linn. Soc. 71: 413–426.

Millot, J. and J. Anthony. 1965. Anatomie de *Latimeria chalumnae*. Vol. II, Systeme Nerveux et Organes de Sens. Paris. Centre National de la Recherche Scientific.

Modrell, M.S., D. Buckley and C.V.H. Baker. 2011. Molecular analysis of neurogenic placode development in a basal ray-finned fish. Genesis 49: 278–294.

Modrell, M.S. and C.V.H. Baker. 2012. Evolution of electrosensory ampullary organs: conservation of *Eya4* expression during lateral line development in jawed vertebrates. Evol. Dev. 14: 277–285.

Modrell, M.S., M. Lyne, A.R. Carr, H.H. Zakon, D. Buckley, A.S. Campbell, M.C. Davis, G. Micklem and C.V.H. Baker. 2017a. Insights into electrosensory organ development, physiology and evolution from a lateral line-enriched transcriptome. eLife 2017: 1–26. DOI: 10.7554/eLife.24197.

Modrell, M.S., O.R.A. Tidswell and C.V.H. Baker. 2017b. Notch and Fgf signaling during electrosensory versus mechanosensory lateral line organ development in a non-teleost ray-finned fish. Develop. Biol. 431: 48–58.

Montgomery, J.C., H. Bleckmann and S. Coombs. 2014. Sensory ecology and neuroethology of the lateral line. pp. 121–150. *In*: S. Coombs, H. Bleckmann, R.R. Fay and A.N. Popper (eds.). The Lateral Line System. Springer-Verlag, New York, USA.

Monzack, E.L. and L.L. Cunningham. 2013. Lead roles for supporting actors: Critical functions of inner ear supporting cells. Hear Res. 303: 20–29.

Moore, G.A. and W.E. Burris. 1956. Description of the lateral-line system of the pirate perch, *Aphredoderus sayanus*. Copeia 1956: 18–20.

Mukai, Y. and H. Kobayashki. 1992. Cupular growth rate of free neuromasts in three species of cyprinid fish. Nippon Suisan Gakkaishi. 58: 1849–1853.

Mukai, Y., L.L. Chai, S.R.M. Shaleh and S. Senoo. 2007. Structure and development of free neuromasts in barramundi, *Lates calcarifer* (Block). Zool. Sci. 24: 829–835.

Münz, H. 1979. Morphology and innervation of the lateral line system in *Sarotherodon niloticus* (L.) (Cichlidae, Teleostei). Zoomorphol. 93: 73–86.

Münz, H., B. Class and B. Fritzsch. 1984. Electroreceptive and mechanoreceptive units in the lateral line of the axolotl *Ambystoma mexicanum*. J. Comp. Physiol. A 154: 33–44.

Murayama, E., Y. Takagi, T. Ohira, J.G. Davis, M.I. Greene and H. Nagasawa. 2002. Fish otolith contains a unique structural protein, otolin-1. Eur. J. Biochem. 269: 688–696.

Murayama, E., Y. Takagi and H. Nagasawa. 2004. Immunohistochemical localization of two otolith matrix proteins in the otolith and inner ear of the rainbow trout, Oncorhynchus mykiss: comparative aspects between the adult inner ear and embryonic otocysts. Histochem. Cell Biol. 121: 155–166.

Murayama, E., P. Herbomel, A. Kawakami, H. Takeda and H. Nagasawa. 2005. Otolith matrix proteins OMP-1 and Otolin-1 are necessary for normal otolith growth and their correct anchoring onto the sensory maculae. Mech. Devel. 122: 791–803.

Murray, R.W. 1974. The ampullae of Lorenzini. pp. 125–146. *In*: A. Fessard (ed.). Electroreceptors and Other Specialized Receptors in Lower Vertebrates. Springer, Berlin.

Myrberg, A.A. 2001. The acoustical biology of elasmobranchs. Envir. Biol. Fish 60: 31–45.

Nachtrieb, H.F. 1910. The primitive pores of *Polyodon spathula* (Walbaum). J. Exp. Zool. 9: 455–468.

Nada, O. and K. Hirata. 1977. The monoamine containing cell in the gustatory epithelium of some vertebrates. Arch. Histol. Japan 40: 197–206.

Nag, T.C. and J. Bhattacharjee. 1995. Retinal ellipsosomes: Morphology, development, identification, and comparison with oil droplets. Cell Tiss. Res. 279: 633–637.

Nagel, R., F. Kirschbaum, J. Engelmann, V. Hofmann, F. Pawelzik and R. Tiedemann. 2018. Male-mediated species recognition among African weakly electric fishes. R. Soc. Open Sci. 5: 170443. http://dx.doi.org/10.1098/rsos.170443.

Nakae, M. and K. Sasaki. 2005. The lateral line system and its innervation in the boxfish *Ostracion immaculatus* (Tetraodontiformes: Ostraciidae): description and comparisons with other tetraodontiform and perciform conditions. Ichthyol. Res. 52: 343–353.

Nakae, M. and K. Sasaki. 2006. Peripheral nervous system of the ocean sunfish, *Mola mola* (Tetraodontiformes: Molidae). Ichthyol. Res. 53: 233–246.

Nakae, M., S. Asai and K. Sasaki. 2006. The lateral line system and its innervation in *Champsodon snyderi* (Champsodontidae): distribution of approximately 1000 neuromasts. Ichthyol. Res. 53: 209–215.

Nakae, M. and K. Sasaki. 2010. Lateral line system and its innervation in Tetraodontiformes with outgroup comparisons: Descriptions and phylogenetic implications. J. Morphol. 271: 559–579.

Nakae, M., R. Asaoka, H. Wada and K. Sasaki. 2012a. Fluorescent dye staining of neuromasts in live fishes: an aid to systematic studies. Ichthyol. Res. 59: 286–290.

Nakae, M., E. Katayama, R. Asaoka, M. Hirota and K. Sasaki. 2012b. Lateral line system in the triplefin *Enneapterygius etheostomus* (Perciformes: Tripterygiidae): new implications for taxonomic studies. Ichthyol. Res. 59: 268–271.

Nakae, M., E. Katayama, R. Asaoka, M. Hirota and K. Sasaki. 2012c. Lateral line system in the triplefin *Enneapterygius etheostomus* (Perciformes: Tripterygiidae): new implications for taxonomic studies. Ichthyol. Res. 59: 268–271.

Nakae, M., G. Shinohara, K. Miki, M. Abe and K. Sasaki. 2013. Lateral line system in *Scomberomorus niphonius* (Teleostei, Perciformes, Scombridae): Recognition of 12 groups of superficial neuromasts in a rapidly-swimming species and a comment on function of highly branched lateral line canals. Bull. Natl. Mus. Nat. Sci. Ser. A 39: 39–49.

Nelson, J.S., T.C. Grande and M.V.H. Wilson. 2016. Fishes of the World, 5th ed. John Wiley & Sons, Inc., Hoboken, NJ.

Newland, P.L., E. Hunt and S.M. Sharkh. 2008. Static electric field detection and behavioural avoidance in cockroaches. J. Exp. Biol. 211: 3682–3690.

Nicol, J.A.C. 1969. The tapetum lucidum of the sturgeon. Contrib. Mar. Sci. 14: 5–18.

Nicol, J.A.C. 1981. Tapeta lucida of vertebrates. pp. 401–431. *In*: J.M. Enoch and F.L. Tobey Jr. (eds.). Vertebrate Photoreceptor Optics. Springer-Verlag, Berlin.

Nolf, D. 1985. Otolithi piscium. Handbook of Paleoichth. 10: 1–145.

Nolf, D. 1995. Studies on fossil otoliths—The state of the art. pp. 513–544. *In*: D.H. Secor, J.M. Dean and S.E. Campana (eds.). Recent Developments in Fish Otolith Research. University of South Carolina Press, Columbia, SC USA.

Norris, H.W. 1925. Observations upon the peripheral distribution of the cranial nerves of certain ganoid fishes (*Amia, Lepidosteus, Polyodon, Scaphirhynchus* and *Acipenser*). J. Comp. Neurol. 39: 345–432.

Northcutt, R.G. 1986. Electroreception in nonteleost bony fishes. pp. 257–285. *In*: T.H. Bullock and W. Heiligenberg (eds.). Electroreception. John Wiley & Sons, New York, USA.

Novales-Flamarique, I. and C.W. Hawryshyn. 1998. Photoreceptor types and their relation to the spectral and polarization sensitivities of clupeid fishes. J. Comp. Physiol. A 182: 793–803.

Ohkubo, Y., M. Masubuchi and K. Fujioka. 2005. Distribution and morphological features of taste buds in the zebrafish, *Danio rerio*. J. Oral Biosc. 47: 77–82.

Oliveira, A.M. and M. Farina. 1996. Vaterite, calcite, and aragonite in the otoliths of three species of piranha. Naturwiss 83: 133–135.

Oxman, D.S., R. Barnett-Johnson, M.E. Smith, A. Coffin, D.L. Miller, R. Josephson and A.N. Popper. 2007. The effect of vaterite deposition on sound reception, otolith morphology, and inner ear sensory epithelia in hatchery-reared Chinook salmon (*Oncorhynchus tshawytscha*). Can. J. Fish Aquat. Sci. 64: 1469–1478.

Pannella, G. 1971. Fish otoliths: Daily growth layers and periodical patterns. Science 173: 1124–1127.

Peach, M.B. and G.W. Rouse. 2000. The morphology of the pit organs and lateral line canal neuromasts of *Mustelus antarcticus* (Chondrichthyes: Triakidae). J. Mar. Biol. Assoc. UK 80: 155–162.

Peach, M.B. and N.J. Marshall. 2009. The comparative morphology of pit organs in elasmobranchs. J. Morphol. 270: 688–701.

Peters, H.M. 1973. Anatomie und Entwicklungsgeschichte des Laterallissystems von Tilapia (Pisces, Cichlidae). Z. fur Morphol. der Tiere. 74: 89–161.

Pettigrew, J.D. 1999. Electroreception in monotremes. J. Exp. Biol. 202: 1447–1454.

Pfeifer, W. 1962. The fright reaction of fish. Biol. Rev. 37: 495–511.

Pfeifer, W. 1963. The morphology of the olfactory organ of the Pacific salmon (*Oncorhynohus masou*). Can. J. Zool. 41: 1233–1336.

Pfeifer, W. 1964. The morphology of the olfactory organ of *Holopagrus guentheri* Gill, 1862. Can. J. Zool. 42: 235–237.

Pietsch, T.W. 2009. Oceanic Anglerfishes—Extraordinary Diversity in the Deep Sea. Berkeley: University of California Press, 557 pp.

Pisam, M., P. Payan, C. LeMoal, A. Edeyer, G. Boeuf and N. Mayer-Gostan. 1998. Ultrastructural study of the saccular epithelium of the inner ear of two teleosts, *Oncorhynchus mykiss* and *Psetta maxima*. Cell Tiss. Res. 294: 261–270.

Platt, C. 1977. Hair cell distribution and orientation in goldfish otolith organs. J. Comp. Neurol. 172: 283–297.

Platt, C. and A.N. Popper. 1981. Fine structure and function of the ear. pp. 3–37. *In*: W.N. Tavolga, A.N. Popper and R.R. Fay (eds.). Hearing and Sound Communication in Fishes. Springer, New York, USA.

Platt, C. and A.N. Popper. 1984. Variation in the lengths of ciliary bundles on the hair cells along the macula of the sacculus in two species of teleost fish. Scan. Electron Microsc. 1984: 1915–1924.

Platt, C. 1993. Zebrafish inner ear sensory surfaces are similar to those in goldfish. Hear Res. 65: 133–140.

Platt, C., J.M. Jørgensen and A.N. Popper. 2004. The inner ear of the lungfish *Protopterus*. J. Comp. Neurol. 471: 277–288.

Popper, A.N. 1976. Ultrastructure of auditory regions in inner ear of Lake whitefish. Science 192: 1020–1023.

Popper, A.N. 1977. Scanning electron microscopic study of sacculus and lagena in ears of fifteen species of teleost fishes. J. Morphol. 153: 397–417.

Popper, A.N. 1978. Scanning electron microscopic study of the otolithic organs in the bichir *Polypterus bichir* and shovel-nose sturgeon *Scaphirhynchus platorynchus*. J. Comp. Neurol. 181: 117–128.

Popper, A.N. and C. Platt. 1979. Herring has a unique receptor pattern. Nature 280(5725): 832–833.

Popper, A.N. and W.N. Tavolga. 1981. Structure and function of the ear in the marine catfish, *Arius felis*. J. Comp. Physiol. A 144: 27–34.

Popper, A.N. and S. Coombs. 1982. The morphology and evolution of the ear in actinopterygian fishes. Am. Zool. 22: 311–328.

Popper, A.N. and C. Platt. 1983. Sensory surface of the saccule and lagena in the ears of ostariophysan fishes. J. Morphol. 176: 121–129.

Popper, A.N. and R.G. Northcutt. 1983. Structure and innervation of the inner ear of the bowfin, *Amia calva*. J. Comp. Neurol. 213: 279–286.

Popper, A.N. and B. Hoxter. 1987. Sensory and nonsensory ciliated cells in the ear of the sea lamprey, *Petromyzon marinus*. Brain Behav. Evol. 30: 43–61.

Popper, A.N. and R.R. Fay. 1993. Sound detection and processing by fish—critical review and major research questions. Brain Behav. Evol. 41: 14–38.

Popper, A.N. W.M. Saidel and J.S.Y. Chang. 1993. Two types of sensory hair cell in the saccule of a teleost fish. Hear Res. 64: 211–216.

Popper, A.N. 1996. The teleost octavolateralis system: structure and function. Mar. Freshw. Behav. Physiol. 27: 95–110.

Popper, A.N. and R.R. Fay. 1997. Evolution of the ear and hearing: issues and questions. Brain Behav. Evol. 50: 213–221.

Popper, A.N. 2000. Hair cell heterogeneity and ultrasonic hearing: recent advances in understanding fish hearing. Phil. Trans. Roy. Soc. Lond. B Biol. Sci. 355: 1277–1280.

Popper, A.N. and Z.M. Lu. 2000. Structure-function relationships in fish otolith organs. Fish. Res. 46: 15–25.

Popper, A.N., R.R. Fay, C. Platt and O. Sand. 2003. Sound detection mechanisms and capabilities of teleost fishes. pp. 3–38. *In*: S.P. Collin and N.J. Marshall (eds.). Sensory Processing in Aquatic Environments. Springer, New York, USA.

Popper, A.N., D.T.T. Plachta, D.A. Mann and D.M. Higgs. 2004. Response of clupeid fish to ultrasound: a review. Ices J. Mar. Sci. 61: 1057–1061.

Popper, A.N., J. Ramcharitar and S.E. Campana. 2005. Why otoliths? Insights from inner ear physiology and fisheries biology. Mar. Freshw. Res. 56: 497–504.

Popper, A.N. and C.R. Schilt. 2008. Hearing and acoustic behavior: Basic and applied considerations. pp. 17–48. *In*: J.F. Webb, R.R. Fay and A.N. Popper (eds.). Fish Bioacoustics. Springer, New York, USA.

Popper, A.N. and R.R. Fay. 2011. Rethinking sound detection by fishes. Hear Res. 273: 25–36.

Puzdrowski, R.L. 1989. Peripheral distribution and central projections of the lateral-line nerves in goldfish, *Carassius auratus*. Brain Behav. Evol. 34: 110–131.

Quinet, P. 1971. Etude systematique des organes sensorielles de la peau des Mormyriformes. Ann. Musee Roy. Afrique Central 190: 1–97.

Raible, D.W. and G.J. Kruse. 2000. Organization of the lateral line system in embryonic zebrafish. J. Comp. Neurol. 421: 189–198.

Raji, A.R. and E. Norozi. 2010. Distribution of external taste buds in walking catfish (*Clarias batrachus*) and Piranha (*Serrasalmus nattereri*). J. Appl. Anim. Res. 37: 49–52.

Raschi, W. and L.A. Mackanos. 1989. The structure of the ampullae of Lorenzini in *Dasyatis garouaensis* and its implications on the evolution of freshwater electroreceptive systems. J. Exp. Zool. Suppl. 2: 101–111.

Raschi, W., E. Keithan and W. Rhee. 1997. Anatomy of the ampullary electroreceptor in the freshwater stingray, *Himantura signifer*. Copeia 1997: 101–107.

Raschi, W.G. 1984. Anatomical observations on the ampullae of Lorenzini from selected skates and galeoid sharks of the Western North Atlantic. Unpublished Thesis, College of William and Mary, Williamsburg, VA USA.

Raschi, W.G. 1986. A morphological analysis of the ampullae of Lorenzini in selected skates (Pisces, Rajoidei). J. Morphol. 189: 225–247.

Reckel, F., R.R. Melzer and U. Smola. 2001. Outer retinal fine structure of the garfish *Belone belone* (L.) (Belonidae, Teleostei) during light and dark adaptation: Photoreceptors, cone patterns and densities. Acta Zool. (Stockh.) 82: 89–105.

Reese, A.M. 1910. The lateral line system of *Chimaera (Hydrolagus) collei*. J. Exper. Zool. 9: 349–370.

Reno, H.W. 1966. The infraorbital canal, its lateral-line ossicles and neuromasts, in the minnows *Notropis volucellus* and *N. buchanani*. Copeia 1966: 403–413.

Retzius, G. 1881. Das Gehörorgan der Fische und Amphibien. *In*: Das Gehörorgan der Wirbelthiere. Samson & Wallin, Stockholm.

Reutter, K. 1971. Die Gesmacksknospen des Zwergwelses, *Ameiurus nebulosus* (Lesueur). Morphologische und Histochimische Untersuchungen 120: 280–308.

Reutter, K., W. Breipohl and G.J. Bijvank. 1974. Taste bud types in fishes. II. Scanning electron microscopical investigations on *Xiphophorus helleri* Heckel (Poecillidae Cyprinodontiformes Teleostei). Cell Tiss. Res. 153: 151–165.

Reutter, K. 1978. Taste organ in the bullhead (Teleostei). Adv. Anat. Embryol. Cell Biol. 55: 1–98.

Reutter, K. 1982. Taste organ in the barbell of the bullhead. pp. 77–91. *In*: T.J. Hara (ed.). Chemoreception in Fishes. Elsevier Scientific Publ. Co., Amsterdam.

Reutter, K. 1986. Chemoreceptors. pp. 586–604. *In*: J. Bereiter-Hann, A.G. Matoltsy and K.S. Richards (eds.). Biology of the Integument 2-Vertebrates. Springer-Verlag, Berlin.

Reutter, K. 1991. Ultrastructure of taste buds in the Australian lungfish *Neoceratodus forsteri* (Dipnoi). Chem. Sens. 16: 404.

Reutter, K. 1994. Ultrastructure of taste buds in the spotted dogfish *Scyliorhinus caniculus* (Selachii). *In*: K. Kurihara, N. Suzuki and H. Ogawa (eds.). Olfaction and Taste XI. Springer, Tokyo.

Reutter, K. and M. Witt. 1996. Ultrastructure of the taste buds in the spotted gar, *Lepisosteus oculatus* (Holostei). Chem. Senses 21: 663–664.

Reutter, K., F. Baudriot and M. Witt. 2000. Heterogeneity of fish taste bud ultrastructure as demonstrated in the holosteans *Amia calva* and *Lepisosteus oculatus*. Phil. Trans. Roy. Soc. Lond. B 355: 1225–1228.

Richardson, K.C., L. Jarett and E.H. Finke. 1960. Embedding in epoxy resins for ultrathin sectioning in electron microscopy. Stain Technol. 35: 313–323.

Rivera-Vicente, A.C., J. Sewell and T.C. Tricas. 2011. Electrosensitive spatial vectors in elasmobranch fishes: Implications for source localization. PLOS ONE 6(1): e16008. doi:10.1371/journal.pone.0016008.

Rogers, P.H. and M. Cox. 1988. Underwater sound as a biological stimulus. pp. 131–149. *In*: J. Atema, R.R. Fay, A.N. Popper and W.N. Tavolga (eds.). Sensory Biology of Aquatic Animals. Springer, New York, USA.

Rogers, P.H., A.N. Popper, M.C. Hastings and W.M. Saidel. 1988. Processing of acoustic signals in the auditory system of bony fish. J. Acoust. Soc. Amer. 83: 338–349.

Rogers, P.H. and D.G. Zeddies. 2008. Multipole mechanisms for directional hearing in fish. pp. 233–252. *In*: J.F. Webb, R.R. Fay and A.N. Popper (eds.). Fish Bioacoustics. Springer, New York, USA.

Ronan, M. 1986. Electroreception in cyclostomes. pp. 209–224. *In*: T.H. Bullock and W. Heiligenberg (eds.). Electroreception. John Wiley & Sons, New York, USA.

Rosauer, E.A. and J.R. Redmond. 1985. Comparative crystallography of vertebrate otoconia. J. Laryng. Otol. 99: 21–28.

Rouse, G.W. and J.O. Pickles. 1991. Ultrastructure of free neuromasts of *Bathygobius fuscus* (Gobiidae) and canal neuromasts of *Apogon cyanosoma* (Apogonidae). J. Morphol. 209: 111–120.

Royer, S. and J.C. Kinamon. 1996. Comparison of high-pressure freezing/freeze substitution and chemical fixation of catfish barbel taste buds. Microsc. Res. Techn. 35: 385–412.

Saidel, W.M. and C.A. McCormick. 1985. Morphology of the macula neglecta in the bowfin, *Amia calva*. Soc. Neurosci. Abstr. 11: 1212.

Sand, O. and A. Michelsen. 1978. Vibration measurements of perch saccular otolith. J. Comp. Physiol. A 123: 85–89.

Sarrazin, A.F., V.A. Nuñez, D. Sapede, V. Tassin, C. Dambly-Chaudiere and A. Ghysen. 2010. Origin and early development of the posterior lateral line system of zebrafish. J. Neurosci. 30: 8234–44.

Sato, M., R. Asaoka, M. Nakae and K. Sasaki. 2017. The lateral line system and its innervation in *Lateolabrax japonicus* (Percoidei *incertae sedis*) and two apogonids (Apogonidae), with special reference to superficial neuromasts (Teleostei: Percomorpha). Ichthyol. Res. 64: 308–330.

Sbarbati, A., C. Crescimanno, D. Benati and F. Osculati. 1998. Solitary chemosensory cells in the developing chemoreceptorial epithelium of the vallate papilla. J. Neurocytol. 27: 631–635.

Sbarbati, A. and F. Osculati. 2003. Solitary chemosensory cells in mammals? Cell Tiss. Organs 175: 51–55.

Schmitz, A., H. Bleckmann and J. Mogdans. 2008. Organization of the superficial neuromast system in goldfish, *Carrasius auratus*. J. Morphol. 269: 751–761.

Schulte, E. 1972. Untersuchungen an der Regio olfactoria des Aals, *Anguilla anguilla* L. Zeit. Zellforsch. Mikrosk. Anat. 125: 210–228.

Schulz-Mirbach, T. and B. Reichenbacher. 2006. Reconstruction of Oligocene and Neogene freshwater fish faunas—an actualistic study on cypriniform otoliths. Acta Palaeontol. Polon 51: 283–304.

Schulz-Mirbach, T., M. Heß and M. Plath. 2011. Inner ear morphology in the Atlantic molly *Poecilia mexicana*—First detailed microanatomical study of the inner ear of a cyprinodontiform species. PLoS ONE 6: e27734.

Schulz-Mirbach, T. and M. Plath. 2012. All good things come in threes—species delimitation through shape analysis of saccular, lagenar and utricular otoliths. Mar. Freshw. Res. 63: 934–940.

Schulz-Mirbach, T., M. Heß and B.D. Metscher. 2013. Sensory epithelia of the fish inner ear in 3D: studied with high-resolution contrast enhanced microCT. Front. Zool. 10: 63.

Schulz-Mirbach, T. and F. Ladich. 2016. Diversity of inner ears in fishes: Possible contribution towards hearing improvements and evolutionary considerations. pp. 341–391. *In*: J.A. Sisneros (ed.). Fish Hearing and Bioacoustics—An Anthology in Honour of Arthur N. Popper and Richard R. Fay. Springer, New York, USA.

Schulz-Mirbach, T., F. Ladich, M. Plath and M. Heß. 2018. Enigmatic ear stones: what we know about the funcitonal role and evolution of fish otoliths. Biol. Rev. (in press).

Schulze, F.E. 1863. Uber die becherf Ormigen Orange der Fische. Zeitschrift für wissenschaftliche Zoologie. 12: 218–222.

Siebeck, U.E., S.P. Collin, M. Ghoddusi and N.J. Marshall. 2003. Occlusable corneas in toadfishes: Light transmission, movement and ultrastructure of pigment during light and dark adaptation. J. Exp. Biol. 206: 2177–2190.

Sienknecht, U.J., B.K. Anderson, R.M. Parodi, K.N. Fantetti and D.M. Fekete. 2011. Non-cell-autonomous planar cell polarity propagation in the auditory sensory epithelium of vertebrates. Devel. Biol. 352: 27–39.

Sienknecht, U.J. 2013. Origin and development of hair cell orientation in the inner ear. pp. 69–109. *In*: C. Köppl, G.A. Manley, A.N. Popper and R.R. Fay (eds.). Insights from Comparative Hearing Research. Springer, New York, USA.

Sienknecht, U.J., C. Köppl and B. Fritzsch. 2014. Evolution and development of hair cell polarity and efferent function in the inner ear. Brain Behav. Evol. 83: 150–161.

Silver, R.B., A.P. Reeves, A. Steinacker and S.M. Highstein. 1998. Examination of the cupula and stereocilia of the horizontal semicircular canal in the toadfish *Opsanus tau*. J. Comp. Neurol. 402: 48–61.

Sivak, J.G. 1980. Accommodation in vertebrates: a contemporary survey. Curr. Topics Eye Res. 3: 281–330.

Smith, M., J. Schuck, R. Gilley and B. Rogers. 2011. Structural and functional effects of acoustic exposure in goldfish: evidence for tonotopy in the teleost saccule. BMC Neurosci. 12: 19.

Sokolowski, B.H.A. and A.N. Popper. 1987. Gross and ultrastructural development of the saccule of the toadfish Opsanus tau. J. Morphol. 194(3): 323–348.

Söllner, C. and T. Nicolson. 2004. The zebrafish as a genetic model to study otolith formation. pp. 229–242. *In*: E. Bäuerlein (ed.). Biomineralization—Progress in Biology, Molecular Biology and Application. Wiley-VHC, Weinheim.

Song, J. and R.G. Northcutt. 1991. Morphology, distribution and innervation of the lateral-line receptors of the Florida gar, *Lepisosteus platyrhincus*. Brain Behav. Evol. 37: 10–37.

Sorensen, P.W and J. Caprio. 1998. Chemoreception. pp. 375–405. *In*: D.H. Evans (ed.). The Physiology of Fishes. CRC Press, Boca Raton.

Srivastava, C.B.L. 1972. Morphological evidence for electrical synapse of "gap" junction type in another vertebrate receptor. Experientia 28: 1029–1030.

Starrs, D., B.C. Ebner and C.J. Fulton. 2016. All in the ears: unlocking the early life history biology and spatial ecology of fishes. Biol. Rev. 91: 86–105.

Stendell, W. 1914. Morphologische Studien an Mormyriden. Verh. Dtsch. Zool. Ges. 24: 254–261.

Stone, LS. 1937. Further experimental studies of the development of lateral-line sense organs in amphibians observed in living preparations. J. Comp. Neurol. 68: 83–115.

Storch, V.N. and U.N. Welsch. 1970. Electron microscopic observations of the taste buds of some bony fishes. Arch. Histol. Japan 32: 145–153.

Sumi, K., R. Asaoka, M. Nakae and K. Sasaki. 2015. Innervation of the lateral line system in the blind cavfish *Astyanax mexicanus* (Characidae) and comparisons with the eyed surface-dwelling form. Ichthyol. Res. 62: 420–430.

Sweeting, R.M., R.J. Beamish and C.M. Neville. 2004. Crystalline otoliths in teleosts: Comparisons between hatchery and wild coho salmon (*Oncorhynchus kisutch*) in the Strait of Georgia. Rev. Fish Biol. Fish 14: 361–369.

Szabo, T. 1965. Sense organs of the lateral line system in some electric fish of the Gymnotidae, Gymnarchidae and Mormyridae. J. Morph. 117: 229–250.

Szabo, T. 1974. Anatomy of the specialized lateral line organs of electroreception. pp. 13–58. *In*: A. Fessard (ed.). Electroreceptors and other Specialized Receptors in Lower Vertebrates. Springer Verlag, Berlin, Heidelberg, New York.

Tagliafiero, G. and G. Zaccone. 2001. Morphology and immunohistochemistry of taste buds in bony fishes. pp. 335–345. *In*: B.G. Kapoor and T.J. Hara (eds.). Sensory Biology of Jawed Fishes. Science Publishers Inc., Enfield, NH.

Takagi, Y. 1997. Meshwork arrangement of mitochondria-rich, Na$^+$, K$^+$-ATPase-rich cells in the saccular epithelium of rainbow trout (*Oncorhynchus mykiss*) inner ear. Anat. Rec. 248: 483–489.

Tarby, M.L. and J.F. Webb. 2003. Development of the supraorbital and mandibular lateral line canals in the cichlid, *Archocentrus nigrofasciatus*. J. Morphol. 254: 44–57.

Tavolga, W.N. 1960. Sound production and underwater communication in fishes. pp. 93–136. *In*: W.E. Lanyon and W.N. Tavolga (eds.). Animal Sounds and Communication. American Institute of Biological Sciences, Washington DC USA.

Teeter, J.H., R.B. Szamier and M.V.L. Bennett. 1980. Ampullary electroreceptors in the sturgeon *Scaphirhynchus platorynchus* (Rafinesque). J. Comp. Physiol. A 138: 213–223.

Tekye, T. 1990. Morphological differences in neuromasts of the blind cave fish Astyanax hubbsi and the sighted river fish *Astyanax mexicanus*. Brain Behav. Evol. 35: 23–30.

Theiss, S.M., S.P. Collin and N.S. Hart. 2012. The mechanosensory lateral line system in two species of wobbegong shark (Orectolobidae). Zoomorphology 131: 339–348.

Thiesen, B., E. Zeiske and H. Breucker. 1986. Functional morphology of the olfactory organs in the spiny dogfish (*Squalus acanthius* L.) and the small-spotted catshark (*Scyliorhinus canicula* (L.)). Acta Zool. 76: 73–86.

Thornhill, R. 1972. The development of the labyrinth of the lamprey (*Lampetra fluviatilis* Linn. 1758). Proc. Roy. Soc. Lond. B Biol. Sci. 181: 175–198.

Thornhill, R.A. 1967. The ultrastructure of the olfactory epithelium of the lamprey *Lampetra fluviatilis*. J. Cell Sci. 2: 591–602.

Toyoshima, K. 1994. Role of Merkel cells in the taste organ morphogenesis of frog. pp. 13–15. *In*: K. Kurihara, N. Suzuki and H. Ogawa (eds.). Olfaction and Taste XI. Springer-Verlag, Tokyo.

Toyoshima, K., Y. Seta, H. Harada and T. Toyono. 2000. Morphology and distribution of Merkel cells in some vertebrate. pp. 83–87. *In*: H. Suzuki and T. Ono (eds.). Merkel Cells, Merkel Cell Carcinoma and Neurobiology in the Skin. Elsevier, Amsterdam.

Tricas, T.C. and J.F. Webb. 2016. Acoustic communication in butterflyfishes: Anatomical novelties, physiology, evolution, and behavioral ecology. pp. 57–92. *In*: J.A. Sisneros (ed.). Fish Hearing and Bioacoustics: An Anthology in Honor of Arthur N. Popper and Richard R. Fay. Advances in Experimental Medicine and Biology, Vol. 877. Springer, New York, USA.

Truillo-Cenoz, O. 1961. Electron microscope observations on chemo-and mechano-receptor cells of fishes. Zeitschrift fur Zellforschung und Mikroskopische Anatomie. 54: 654–676.

Tuset, V.M., A. Lombarte and C.A. Assis. 2008. Otolith atlas for the western Mediterranean, north and central eastern Atlantic. Sci. Mar. 72: 7–198.

Uchihashi, K. 1953. Ecological study of the Japanese teleost in relation to the brain morphology. Bull. Japan Reg. of Fisheries Research Lab. 2: 1–66.

van Bergeijk, W.A. 1967. The evolution of vertebrate hearing. pp. 1–49. *In*: W.D. Neff (ed.). Contributions to Sensory Physiology. Academic Press, New York, USA.

Van der Meer, H.-J. 1992. Constructional morphology of photoreceptor patterns in percomorph fish. Acta Biotheor. 40: 51–85.

Vischer, H.A. 1989. The development of lateral-line receptors in *Eigenmannia* (Teleostei, Gymnotiformes). I. The mechanoreceptive lateral-line system. Brain Behav. Evol. 33: 205–222.

von der Emde, G. 1999. Active electrolocation of objects in weakly electric fish. J. Exp. Biol. 202: 1205–1215.

Vorobyev, M. 2003. Coloured oil droplets enhance colour discrimination. Proc. Roy. Soc. Lond. B 270: 1255–1261.

Voronina, E.P. 2009. Structure of lateral-line scales in representatives of families of the order Pleuronectiformes. J. Ichthyol. 49: 940–961.

Voronina, E.P. and D.R. Hughes. 2013. Types and developmental pathways of lateral line scales in some teleost species. Acta Zool. 94: 154–166.

Voronina, E.P. and D.R. Hughes. 2017. Lateral line scale types and review of their taxonomic distribution. Acta Zoologica. DOI: 10.1111/azo.12193.

Wachtel, A.W. and R.B. Szamier. 1969. Special cutaneous receptor organs of fish: Ampullary organs of the non electric catfish *Kryptopterus*. J. Morphol. 128: 291–308.

Wagner, H.-J. and M.A. Ali. 1978. Retinal organisation in goldeye and mooneye (Teleostei: Hiodontidae). Rev. Can. Biol. 37: 65–84.

Wagner, H.-J., E. Fröhlich, K. Negishi and S.P. Collin. 1998. The eyes of deep-sea fishes. II. Functional morphology of the retina. Prog. Retinal Eye Res. 17: 637–685.

Walls, G.L. 1942. The Vertebrate Eye and its Adaptive Radiation. Cranbrook Institute of Science Bulletin, London.

Waltman, B. 1966. Electrical ties and fine structure of the ampullary canals of Lorenzini. Acta Physiol. Scand. 66(suppl. 264): 1–60.

Webb, J.F. 1989a. Developmental constraints and evolution of the lateral line system in teleost fishes. pp. 79–98. *In*: S. Coombs, P. Görner and H. Münz (eds.). The Mechanosensory Lateral Line: Neurobiology and Evolution. Springer-Verlag, New York, USA.

Webb, J.F. 1989b. Gross morphology and evolution of the mechanosensory lateral line system in teleost fishes. Brain, Behav. Evol. 33: 34–53.

Webb, J.F. 1989c. Neuromast morphology and lateral line trunk ontogeny in two species of cichlids: an SEM study. J. Morphol. 202: 53–68.

Webb, J.F. 1990. Comparative morphology and evolution of the lateral line system in the Labridae (Perciformes: Labroidei). Copeia 1990: 137–146.

Webb, J.F. 1993. Accessory nasal sacs of flatfishes: Systematic significance and functional implications. Bull. Mar. Sci. 52: 541–553.

Webb, J.F. and R.G. Northcutt. 1997. Morphology and distribution of pit organs and canal neuromasts in non-teleost bony fishes. Brain, Behav. Evol. 50: 139–151.

Webb, J.F. and J.E. Shirey. 2003. Post-embryonic development of the lateral line canals and neuromasts in the zebrafish. Dev. Dyn. 228: 370–385.

Webb, J.F., J. Montgomery and J. Mogdans. 2008. Mechanosensory lateral line and fish bioacoustics. pp. 145–182. *In*: J.F. Webb, R.R. Fay and A.N. Popper (eds.). Fish Bioacoustics Springer-Verlag, New York, USA.

Webb, J.F., J.L. Herman, C.F. Woods and D.R. Ketten. 2010. The ears of butterflyfishes: "Hearing generalists" on noisy coral reefs? J. Fish Biol. 77: 1434–1451.

Webb, J.F. 2011. Lateral line structure. pp. 336–346. *In*: A.P. Farrell (ed.). Encyclopedia of Fish Physiology: From Genome to Environment. Academic Press, San Diego, USA.

Webb, J.F. and J. Ramsay. 2017. A new interpretation of the 3-D configuration of lateral line scales and the lateral line canal contained within them. Copeia 105: 339–347.

Webb, J.F., K.P. Maruska, J.M. Butler and M.A.B. Schwalbe. The lateral line system of cichlid fishes: Anatomy to Behavior. *In*: M.E. Abate and D.L.G. Noakes (eds.). The Behavior, Ecology and Evolution of Cichlid Fishes: A Contemporary Modern Synthesis. Springer Academic, New York, USA, in press.

Webb, J.F. 2014a. Lateral line morphology and development and implications for the functional ontogeny of flow sensing of fishes. pp. 247–270. *In*: H. Bleckmann, J. Mogdans and S.L. Coombs (eds.). Flow Sensing in Air and Water—Behavioural, Neural and Engineering Principles of Operation. Springer, Berlin.

Webb, J.F. 2014b. Morphological diversity, development, and evolution of the lateral line system. pp. 17–72. *In*: S. Coombs, H. Bleckmann, R.R. Fay and A.N. Popper (eds.). The Lateral Line System. Springer Handbook of Auditory Research, Vol. 48. Springer-Verlag, New York, USA.

Webb, J.F., N.C. Bird, L. Carter and J. Dickson. 2014. Comparative development and evolution of two lateral line phenotypes in Lake Malawi cichlids. J. Morphol. 275: 678–692.

Weber, H. 1827. Über das Geschmacksorgan der Karpfen und den Ursprung seiner Nerven. Arch. Anat. Physiol. 1827: 309–315.

Welsch, U. and V. Storch. 1969. Die Feinstruktur der Geschmacksknospen von Welsen (*Clarias batrachus* (L.) und *Kryptopterus bicirrhis* (Cuvier et Velenciennes)). Z. Zellforsch. Mikrosk. Anat. 100: 552–559.

Werner, C.F. 1928. Studien über die Otolithen der Knochenfische. Zeitschrift für wissenschaftliche. Zoologie. 131: 501–587.

Whitear, M. 1965. Presumed sensory cells in fish epidermis. Nature 208: 703–704.

Whitear, M. 1971. Cell specialization and sensory function on fish epidermis. J. Zool. Lond. 163: 552–559.

Whitear, M. and K. Kotrschal. 1988. The chemosensory anterior dorsal fin in rocklings *Gaidropsarus* and *Ciliata* (Teleostei, Gadidae): activity, fine structure and innervations. J. Zool. 216: 339–366.

Whitear, M. and R.M. Moate. 1994a. Microanatomy of taste buds in the dogfish, *Scyliorhinus canicula*. J. Submicrosc. Cytol. Pathol. 26: 357–367.

Whitear, M. and R.M. Moate. 1994b. Chemosensory cells in the oral epithelium of *Raja clavata* (Chondrichthyes). J. Zool. 232: 295–312.

Whitehead, D.L. 2002. Ampullary organs and electroreception in freshwater *Carcharhinus leucas*. J. Physiol. Paris 96: 391–395.

Whitehead, D.L., A.R.G. Gauthier, E.W.H. Mu, M.B. Bennett and I.R. Tibbetts. 2014. Morphology of the ampullae of Lorenzini in jubvenile *Carcharhinus leucas*. J. Morphol. http://dx.doi.org/10.1002/jmor.20355.

Williams, J.A. and N. Holder. 2000. Cell turnover in neuromasts of zebrafish larvae. Hear Res. 143: 171–181.

Withear, M. 1992. Solitary chemoreceptor cells. pp. 103–125. *In*: T.J. Hara (ed.). Chemoreception in Fishes. Chapman and Hall, London.

Witt, M. 1996. Carbohydrate histochemistry of vertebrate taste organs. Progr. Histochem. Cytochem. 30: 1–172.

Witt, M., K. Reutter and I.J. Miller. 2003. Morphology of the peripheral taste system. pp. 651–677. *In:* R.L. Doty (ed.). Handbook of Olfaction and Gustation. Marcel Dekker, New York.

Wonsettler, A.L. and J.F. Webb. 1997. Morphology and development of the multiple lateral line canals on the trunk in two species of *Hexagrammos* (Scorpaeniformes: Hexagrammidae). J. Morph. 233: 195–214.

Worm, M., F. Kirschbaum and G. von der Emde. 2017. Social interactions between live and artificial weakly electric fish: Electrocommunication and locomotor behavior of *Mormyrus rume proboscirostris* towards a mobile dummy fish. PLOS ONE 12(9): e0184622.

Wueringer, B.E. and I.R. Tibbetts. 2008. Comparison of the lateral line and ampullary system of two species of shovelnose ray. Rev. Fish Biol. Fish 18: 47–64.

Wueringer, B.E., I.R. Tibbetts and D.L. Whitehead. 2009. Ultrastructure of the ampullae of Lorenzini of *Aptychotrema rostrata* (Rhinobatidae). Zoomorphol. (Berlin) 128: 45–52.

Wueringer, B.E., S.C. Peverell, J. Seymour, L. Squire Jr. and S.P. Collin. 2011a. Sensory systems in sawfishes. 2. The lateral line. Brain, Behav. Evol. 78: 150–161.

Wueringer, B.E., S.C. Peverell, J. Seymour, L. Squire Jr., S.M. Kajiura and S.P. Collin. 2011b. Sensory systems in sawfishes. 1. The ampullae of Lorenzini. Brain Behav. Evol. 78: 139–149.

Wueringer, B.E. 2012. Electroreception in elasmobranchs: sawfish as a case study. Brain Behav. Evol. 80: 97–107.

Wueringer, B.E., L. Squire Jr., S.N. Kajiura, I.R. Tibbetts, N.S. Hart and S.P. Collin. 2012. Electric field detection in sawfish and shovelnose rays. PLoS ONE 7(7): e41605. DOI:10.1371/journal.pone.0041605.

Wueringer, B.E., L. Squire Jr., S.M. Kajiura, N.S. Hart and S.P. Collin. 2012. The function of the sawfish's saw. Curr. Biol. 22: R150–R151.

Wunder, W. 1957. Die Sinnesorange der Fische. Allgem Fichereizeit. 82: 171–173.

Xiong, D., L. Zhang, H. Yu, C. Xie, Y. Kong, Y. Zeng and Z. Huo Band Liu. 2011. A study of morphology and histology of the alimentary tract of *Glyptosternum maculatum* (Sisoridae, Siluriformes). Acta Zool. 92: 161–169.

Xue, J. and E.H. Peterson. 2006. Hair bundle heights in the utricle: Differences between macular locations and hair cell types. J. Neurophys. 95: 171–186.

Yacob, A., C. Wise and Y.W. Kunz. 1977. The accessory outer segment of rods and cones in the retina of the gruppy *Poecilia reticulata* P. (Teleostei). An electron microscopical study. Cell Tiss. Res. 177: 181–193.

Yamamoto, M. and K. Ueda. 1978. Comparative morphology of fish olfactory epithelium. V. Gasterosteiformes, Channiformes and Synbranchiformes. Bull. Japan Soc. Sci. Fish 44: 1309–1314.

Yamamoto, M. and K. Ueda. 1979. Comparative morphology of fish olfactory epithelium. IX. Tetraodontiformes. Zoological Magazine (Tokyo) 88: 210–218.

Yamamoto, M. 1982. Comparative morphology of the peripheral olfactory organ in teleosts. pp. 39–59. *In:* T.J. Hara (ed.). Chemoreception in Fishes. Elsevier, Amsterdam.

Yashpal, M., U. Kumari, S. Mittal and A.K. Mittal. 2006. Surface architecture of the mouth cavity of a carnivorous fish *Rita rita* (Hamilton 1822) (Siluriformes, Bagridae). Belgian J. Zool. 136: 155–162.

You, X., C. Bian, Q. Zan, X. Xu, X. Liu, J. Chen, J. Wang, Y. Qiu, W. Li, X. Zhang, Y. Sun, S. Chen, W. Hong, Y. Li, S. Cheng, G. Fan, C. Shi, J. Liang, Y. Tom Tang, C. Yang, Z. Ruan, B. Jie, C. Peng, Q. Mu, J. Lu, M. Fan, S. Yang, Z. Huang, X. Jiang, X. Fang, G. Zhang, Y. Zhang, G. Polgar, H. Yu, J. Li, Z. Liu, G. Zhang, V. Ravi, S.L. Coon, J. Wang, H. Yang, B. Venkatesh, J. Wang and Q. Shi. 2014. Mudskippers genome provide insight into the terrestial adaptation of amphibious fishes. Nat. Comm. DOI: 10.1038/ncomms6594.

Young, R.W. 1978. Visual cells, daily rhythms and vision research. Vis. Res. 18: 573–578.

Zeddies, D.G., R.R. Fay and J.A. Sisneros. 2011. Sound source localization and directional hearing in fishes. pp. 298–303. *In:* A.P. Farrell (ed.). Encyclopedia of Fish Physiology: From Genome to Environment. Elsevier, San Diego, USA.

Zeiske, E., J. Caprio and S.H. Gruber. 1986. Morphological and electrophysiological studies on the olfactory organ of the lemon shark, *Negaprion brevirostris*. pp. 381–391. *In:* T. Uyeno, R. Arai, T. Taniuchi and K. Matsuura (eds.). Indo-Pacific Fish Biology: Proceedings of the Second International Conference on Indo-Pacific Fishes. Ichthyological Society of Japan, Tokyo.

Zeiske, E., B. Thiesen and H. Breucker. 1992. Structure, development, and evolutionary aspects of the peripheral olfactory system. pp. 13–39. *In:* T.J. Hara (ed.). Fish Chemoreception. Chapman and Hall, London.

Zippel, H.P., A. Hansen and J. Caprio. 1997. Renewing olfactory receptor neurons do not require contact with the olfactory bulb to develop normal responsiveness. J. Comp. Physiol. A-Sens. Neur. Behav. Physiol. 181: 425–437.

Zupanc, G.K.H. and T.H. Bullock. 2005. From electrogenesis to electroreception: An overview. pp. 5–46. *In:* T.H. Bullock, C.D. Hopkins, A.N. Popper and R.A. Fay (eds.). Electroreception. Springer Science + Business Media, Inc., USA.

Żuwała, K. and M. Jakubowski. 1993. Light and electron microscopy (SEM, TEM) of taste buds in the tench, *Tinca tinca* L. Acta Zool. 74: 277–282.

Chapter **17**

Morphology and Ecomorphology of the Fish Brain

The Rhombencephalon of Actinopterygians

Anastasia S. Kharlamova* and Sergei V. Saveliev

INTRODUCTION

The brain of the anamnia represents the basal organization type of all vertebrates including amniotes (Nieuwenhuys et al. 1998). The diversity of fish explains brain features characterizing high rank taxa. At the same time, the anamnia brain consists of less associative centres and superstructures in comparison with higher vertebrates. In the absence of the complicated associative brain centres, which cover primary sensory and motor centres, the brain external morphology correlates with species biology and the spectacular diversity in the ecology of fishes provides a unique occasion for demonstrating vertebrate richness in gross brain morphology (Nikonorov 1985, Saveliev 2005, White and Brown 2014) (Fig. 1). Thus, the anamnia brain is a perfect object for eco- and paleomorphology. Futhermore, structural features of the brain associated with the species' ecology and behaviour are often more prominent in the case of extreme specializations than features associated with the phylogenetic position (Kurepina 1981, Kotschal et al. 1998, Devitsina 2004).

The brainstem is a very conservative structure in vertebrates (Northcutt 2002); thus it is suitable for comparison of taxa of high rank. At the same time, the brainstem structure also presents a great diversity at the morphological and histological level due to the specific ecology of species (Figs. 1J–L). The rhombencephalic primary somatic and visceral centres are the most variable brain structures of the anamnia (Kotrschal et al. 1998, Kotrschal and Palzenberger 1992).

In this chapter, we discuss classical academic studies with special reference to an ecomorphological approach illustrated by rhombencephalon atlases (from the caudal border of the rhombencephalon to the isthmus region (not encluded)) of five actinopterygian species: non-specialized and extremely specialized species belonging to the major taxa were chosen as model objects for the brainstem description: (1) The senegal bichir (*Polypterus senegalus* Cuvier, 1829, Polypteridae) (Figs. 3–5). (2) The siberian sturgeon

Laboratory of Nervous System Development, Research Institute of Human Morphology, 117418, Russia, Moscow, Tsurupi st., 3.
Email: embrains@hotmail.com
* Corresponding author: grossulyar@gmail.com

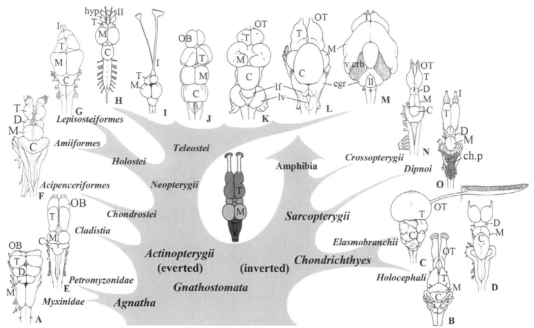

Fig. 1. Phylogenetic tree of the anamnia with reference to brain morphology: hypothetical common brain type of anamnia is inserted. Dorsal views of the brain of *Myxine glutinosa* (Myxinidae, Myxiniformes, Agnatha) (**A**); and Gnathostomata: Chondrichtyes: *Chimaera monstrosa* (Chimaeridae, Chimaeriformes, Holocephali) (**B**); *Squalus acanthias* (Squalidae, Squaliformes, Elasmobranchii) (**C**); *Sphyrna zygaena* (Sphyrnidae, Carcharhiniformes, Elasmobranchii) (**D**); Teleostomi: Osteichthyes: Actinopterygii: *Polypterus senegalus* (Polypteridae, Polypteriformes, Cladistia), *Acipenser baerii* (Acipenseridae, Acipenseriformes, Chondrostei) (**E**); Actinopterygii: Neopterygii: *Lepisosteus oculatus* (Lepisosteidae, Lepisosteiformes, Holostei) (**G**); *Mola mola* (Molidae, Tetraodontiformes, Teleostei) (**H**); *Pseudopleuronectes americanus* (Pleuronectidae, Pleuronectiformes, Teleostei) (**I**); *Anguilla anguilla* (Anguillidae, Anguilliformes, Elopomorpha, Teleostei) (**J**); *Cyprinus carpio* (Cyprinidae, Cypriniformes, Teleostei) (**K**); *Pseudopimelodus bufonius* (Pseudopimelodidae, Siluriformes, Teleostei) (**L**); *Brienomyrus brachyistius* (Mormyridae, Osteoglossiformes, Teleostei) (**M**); *Latimeria chalumnae* (Latimeriidae, Coelacanthiformes (Crossopterygii), Teleostomi) (**N**) and Sarcopterygii: *Lepidosiren paradoxa* (Lepidosirenidae, Ceratodontiformes, Dipnoi) (**O**) (adapted from Saveliev 2001 (**A, B, D, H, N, O**)). C, *corpus cerebelli*; ch.p., choroid plexus; D, diencephalon; egr, *eminentia granularis*; hyp, hypophysis; lf, *lobus facialis*; ll, *lobus lateralis*; lv, *lobus vagus*; M, *tectum mesencephalic*; OB, *bulbus olfactorius*; OT, *tractus olfactorius*; T, *lobus hemisphericus telencephali*; v.crb, *valvula cerebelli*; I, *nervus olfactorius*; II, *nervus opticus*.

(*Acipenser baerii* Brandt 1869, Acipenseridae) (Figs. 4–9). (3) The spotted gar (*Lepisosteus oculatus* Winchell 1864, Lepisosteidae) (Figs. 10–12). (4) The European eel (*Anguilla anguilla* Linnaeus 1758, Anguillidae) (Figs. 13–16). (5) The common carp (*Cyprinus carpio*) (Figs. 17–22). We have also briefly mentioned the brain of cartilaginous fish (Chondrichthyes), coelacanths (Sarcopterygii, Coelacanthiformes) and lungfishes (Sarcopterygii, Dipnoi) for comparative purposes.

The Fish Brain: Common Features and Ecomorphology

The common brain type of the anamnia can be referred to as basal for vertebrates and the brain of jawless fishes (Agnatha) is the most archetypical type in that sense. Evenly developed five brain divisions are well seen in the anamnia brain (Fig. 1A): the forebrain with olfactory bulbs, diencephalon, mesencephalon and metencephalon—the cerebellum and rhombencephalon, developing from the corresponding embryonic anlagen.

   The organization of the anamnia telencephalon in relation to the taxonomic position is shown in Fig. 2. The telencephalon is inverted in jawless and cartilaginous fishes, i.e., it has a structure type which was further developed in the evolutionary line leading to terrestrial vertebrates–amniotes: the telencephalic

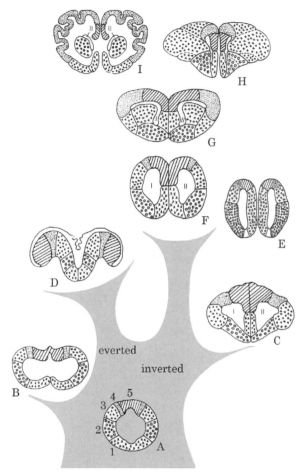

Fig. 2. Phylogenetic tree of vertebrates according to forebrain organisation: there are evolutional lines of the Actinopterygii with the everted type of the forebrain and the line of the Sarcopterygii, including terrestrial vertebrates, with an inverted forebrain. The primitive condition of the inverted type is represented by Agnatha, Chondrichtyes and Dipnoi. Hypothetical monospheric ancestor type (**A**), schemes of the forebrain of Agnatha (**B**), Chondrichthyes (**C**), Actinopterygii (Teleostei) (**D**), Dipnoi (**E**), Amphibia (**F**), Reptilia (**G**), Aves (**H**), Mammalia (**I**). 1, *Septum*; 2, *Striatum*; 3, *Pallium lateralis*; 4, *Pallium dorsalis*; 5, *Pallium medialis* (adapted from Saveliev 2005).

hemispheres represent closed "bubbles" with the ventricular cavity and the branching vascular plexus turned inside. The telencephalon is everted in actinopterygian fishes—the hemispheres of the telencephalon turned outwards. In bichirs (Polypteriformes), the dorso-caudal part of the everted telencephalon is bent back and forms "pseudoventricles" (Nieuwenhuys et al. 1998). The dorsolateral hemispheres are in extreme lateral position and are connected to each other by a thin ependymal layer that covers the upper hemisphere in bony fish (Osteichthyes). The vascular plexus may sometimes cover the forebrain of bony fish superiorly as the asymmetry of the plexus of the third ventricle continues. Such a forebrain structure is an evolutionary dead-end variant, as it has no further development in the terrestrial radiation of vertebrates, but is widely presented among species of the aquatic habitat (actinopterygians). In lower sarcopterygians (i.e., coelocanths, lungfishes and amphibians), the telencaphalon comprises the olfactory bulb, paired hemispheres and small impaired telencephalon. Coelacanths (Sarcopterygii, Actinistia) show an intermediate forebrain type. The telencephalic hemisphere of the African coelochanth (*Latimeria chalumnae*) is formed on the base of the inverted type with an opened medial wall (lateral ventricles are open, so that the telencephalic cavity becomes covered by a thin ependymal layer). The telencephalon of lungfishes (Dipnoi) is inverted (Fig. 6A) (Nieuwenhues et al. 1998). The size of the telencephalon in both

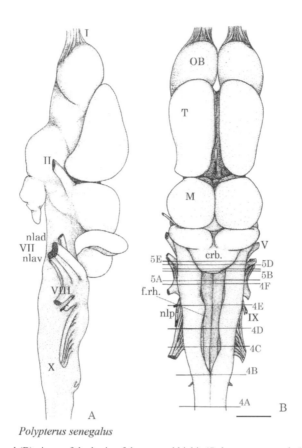

Polypterus senegalus

Fig. 3. Lateral (**A**) and dorsal (**B**) views of the brain of the senegal bichir (*Polypterus senegalus*) (scale bar = 1 mm). Level of cross sections shown in Figs. 4 and 5 are indicated. For abbreviations see page 368.

inverted and everted evolutionary lines varies from large hemispheres in species with well-developed olfaction to an almost reduced state, for example, in the ocean sunfish (*Mola mola*) (Fig. 1H).

Cartilaginous fishes are a good example for the ecomorphological approach. The metencephalic (cerebellum and rhombencephalon) centres of motor coordination are well developed in relation to other parts of the brain in non-specialized species of chondrichthyans. The rhombencephalon is well developed in ratfish (*Chimaera monstrosa*) (Fig. 1B). The sense of smell and the telencephalon are well developed in active predators. The telencephalon of sharks and rays demonstrates a great variability depending on the ecological habitats the species are inhabiting. Pelagic active predators, hunting in the upper water layers, can detect food distantly by sense of smell. They capture water by the input nostrils during the forward movement; it provides the ability to capture air bubbles. These bubbles are trapped and retained by the lamellae of olfactory sac if the speed of the shark is higher than 1.2 m/s and its rostrum is raised 1–2 cm above the water surface. The substances from the air bubbles dissolve in water and the shark can thus sense odours from outside the water. By increasing the speed, the shark pushes air out, and the cycle can be repeated (Saveliev and Chernikov 1994). Thus such pelagic sharks have a highly developed telencephalon, often looking like a mono-hemispheric formation (Fig. 1C). Benthic predators have two hemispheres of the telencephalon. The bilateral olfactory centres of the telencephalon in rays are formed around a mono-hemispheric formation, thus it looks like a three-hemispheric structure (Fig. 1D) (Saveliev 2001).

The majority of fishes have a well-developed visual system, and correspondingly, the tectum of the mesencephalon. Motor centres of fish are mainly localized in the spinal cord (except for a few descending pathways and the system of the Mauthner cell) (Nortcutt and Devis 1983, Nieuwenhues 1998). Metencephalic centres (cerebellum and rhombencephalon) of motor coordination are well developed in

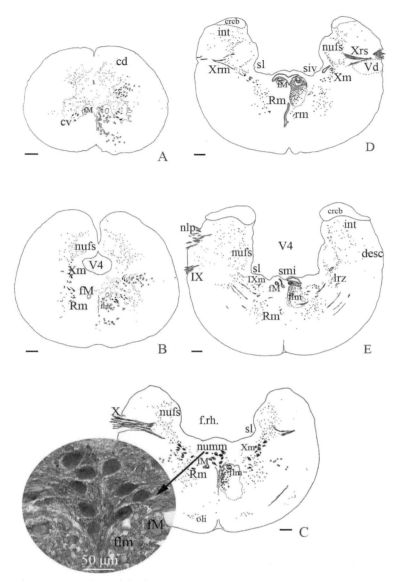

Fig. 4. Drawings of transverse sections of the rhombencephalon of *Polypterus senegalus* at different levels (**A–E**) (scale bar = 100 μm). Microphotograph of the histology of the *nucleus medianus magnocellularis* of the median reticular zone is inserted (**D**) (scale bar = 50 μm).

relation to other parts of brain in non-specialized species of chondrichthyans. The rhombencephalon is well developed in ratfish (Fig. 1B). The cerebellum varies widely from a small rise in the rhombencephalon (for example, in *Amia calva*, Holostei) up to the folded structure developed in cartilaginous fishes (for details see Yopak et al. 2015) and to the extremely large folded structure covering both the rhombencephalon and mesencephalon in electroreceptive actinopterygian species (see below).

Coelacanths present a non-specialized brain: it has a relatively poorly developed cerebellum, telencephalon and mesencephalon, but disproportionally well-developed motor centres of the rhombencephalon (Fig. 1N). The brain is uniformly developed in lungfish, which are characterized by a shift in respiration type, seasonal sleep, and nest building. The large vascular plexus is a distinguishing feature of the South American lungfish (*Lepidosiren paradoxa*) brain; it covers both the cavity of the *fossa rhomboidea* with the rhombencephalon, the diencephalon, and the telencephalon (Fig. 10).

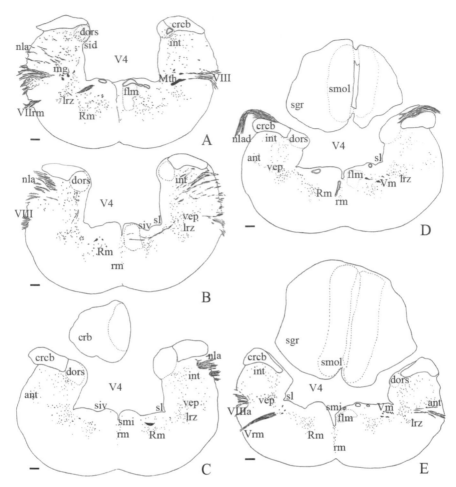

Fig. 5. Drawings of transverse sections of the rhombencephalon of *Polypterus senegalus* at different levels (**A–E**) (scale bar = 100 µm).

The telencephalon is comparatively large in the brain of bichirs and the *tectum mesencephali* is moderately developed (Nieuwenhuys 1983, Nieuwenhuys and Oey 1983) (Fig. 3). Sturgeons present a highly developed rhombencephalon and large *auriculi cerebelli* (Sbikin 1973, see also Fig. 6 for *Acipencer baerii*) and the cerebellar velum enlarges under the hemispheres of the *tectum mesencephali* (Nieuwenhuys et al. 1998).

The mesencephalon, cerebellum and the coordination centres for movement of the brainstem are typically well developed in bony fishes (Nieuwenhuys et al. 1998). However, the telencephalon with the olfactory bulbs is much smaller in the brain of *Amia calva* (Heijdra and Nieuwenhuys 1994). The cerebellum and the tectum of the mesencephalon of the bowfin and the spotted gar are not very large (Nieuwenhuys et al. 1998). The *tectum mesencephali* prevails in the brain of young *Lepisosteus oculatus*. However, the telencephalon and cerebellum are of significant size in adults (Platel et al. 1977) (Fig. 10). The process of individual development, including the shift of the feeding strategy, could explain these differences in the relative sizes of the telencephalon between young and adult spotted gars. The brain of the spotted gar presented in Fig. 9 displays a fish 1–1.5 years old, while adult males reach sexual maturity at the age of 2–3 years, and females in the third or fourth year (Parker and McKee 1984).

The brain of the European eel is close to a hypothetical non-specialized anamnia brain type (Fig. 13). It has a fairly uniformly developed telencephalon, mesencephalon and cerebellum covering the rostral part of the *fossa rhomboidea* (Okamura et al. 2002 (Japanese eel (*Anguilla japonica*)), see also Fig. 13 for the

European eel). The rhombencephalon presents a typical structure with longitudinal grooves formed by the bilateral *sulcus limitans* and *sulcus intermedius dorsalis*, and the *sulcus medianus inferior* and slight prominences are formed by the somatosensory, viscerosensory and motor columns of the rhombencephalon and the *crista cerebelli* (see below) (Figs. 14–16).

Flatfishes (Pleuronectiformes) provide an interesting example of morphological rearrangement during ontogenesis. Flatfishes become one-side oriented during individual development. The telencephalon of flatfishes turns to asymmetry, due to partial degeneration of one side of the olfactory tract resulting from larval metamorphosis (Prasada Rao and Finger 1984). The asymmetry of the tectum of the mesencephalon is visible during metamorphosis and eye shifting is smoothed after metamorphosis, thus the *tectum mesencephalli* is almost symmetrical in adults (Brinon et al. 1993) (Fig. 1I).

In some cases, the unique biology of a species results in the *a priori* development of one brain division and poor development of other compartments. The biology of adult sunfish (*Mola mola*) is very specific ("simple"): it is a zooplankton feeding species with a breeding strategy based on the reproduction of a large number of eggs (r-strategy). Adult sunfish are very large and possess a very small brain. The sunfish has a shortened spinal cord due to the absence of its tail, a virtually invisible forebrain owing to its reduced sense of smell, relatively large pituitary that can be seen rostral to the telencephalon with the olfactory bulbs and well developed mesencephalon. Neurohumoral and visual centres present the main part of the sunfish's brain, this situation completely corresponds with the biology of the sunfish (Saveliev 2001) (Fig. 1G).

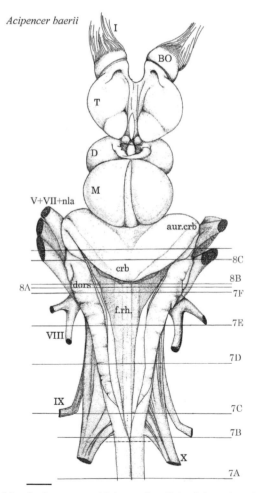

Fig. 6. Dorsal view of the brain of the siberian sturgeon (*Acipenser baerii*) (scale bar = 1 mm). Level of cross sections shown in Figs. 7 and 8 are indicated. For abbreviations see page 368.

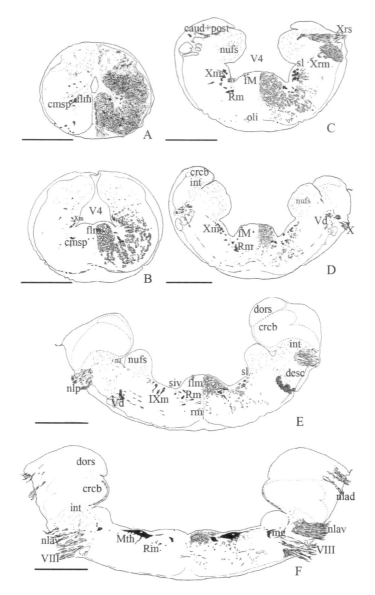

Fig. 7. Drawings of transverse sections of the rhombencephalon of *Acipenser baerii* at different levels (see Fig. 6) (bar = 1 mm). For abbreviations see page 368.

Adaptive radiation of anamnia, and especially actinopterygians, results in a wide range of brain structures. The same specialization results in similar nervous system features even in different evolutionary lines. The primary centres of the lateral line and hearing are generally well presented in meso- and epipelagic species. Vision is also very important in the shallow marine and fresh water species. The significance of vision and taste decreases with depth. Chemoreception is well developed in benthic fishes and inhabitants of turbid water (Motta and Kotrschal 1992, Kotrschal and Palzenberger 1992, Motta et al. 1995). Rhombencephalic features of specialized bony fishes are clearly visible by gross preparation and are usually associated with the hypertrophy of its dorsal sensory columns—the primary viscera- and/ or somatosensory centres (Nieuwenhuys 2011) (Figs. 1L–M, 23). Special "electrical hemispheres" of the cerebellum are well visible in the brain of electroreceptive fishes. Electroreceptive centres—the lateral

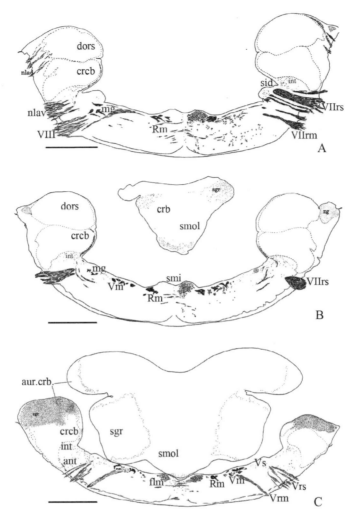

Fig. 8. Drawings of transverse sections of the rhombencephalon of *Acipenser baerii* at different levels (see Fig. 6) (bar = 1 mm). For abbreviations see page 368.

lobes of the rhombencephalon—are also well developed. "Electrical" centres of some fishes extremely expand covering the rhombencephalon, tectum mesencephali and even the forebrain (*Gnatonemus petersii*, Mormyridae) (Fig. 1M). The *torus semicularis* of the mesencephalon—a secondary centre of mechanical-, audio- and electroreception—is also highly developed. Viscerosensory and visceromotor centres achieve a significant development in the brain of highly gustatory sensitive cyprinids (Cyprinidae), such as the common carp, the crucian carp (*Carassius carassius*) and the bream (*Abramis brama*). Visceral centres form large vagal lobes (*lobi vagi*) of the rhombencephalon and the facial lobe (*lobus facialis* in common carp). However, cyprinids have also a well developed mesencephalon and cerebellum (Fig. 17).

   Thus, the gross morphology of the fish brain compared to the ideal hypothetical brain of non-specialized anamnia provides evidence of its habitat and biology. Codfish (Gadidae) are well studied by ecomorphologists, as are two families of perciform fish—the polar Nototheniidae and African freshwater cichlids. Closely related cyprinid species with different biology are the classic object of ecomorphology (Evans 1931, Kotrschal et al. 1998). Evans' studying a large sample of carps and codfishes also finds that brain centre development correlates with the biology of the species. Species feeding on invertebrates (e.g., *Merluccius merluccius*), also hunters (Gadus morua) feeding on active moving prey, have an enlarged

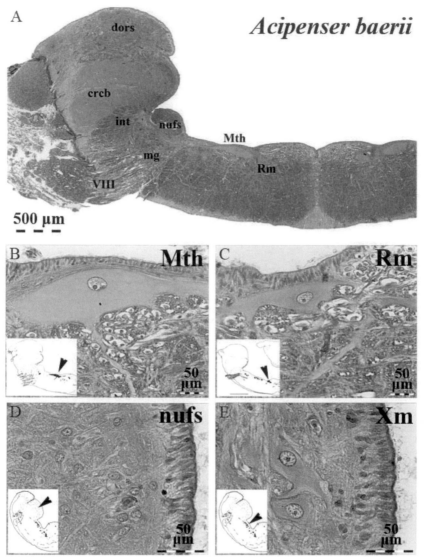

Fig. 9. Cross sections of the rhombencephalon of *Acipenser baerii*. Mallory trichrome preparations at the level of the Mauthner cell (Mth) (**A**). **B–E**. High-magnification at the levels indicated in the inlets by arrowheads representing Figs. **7F** (**B**), **8C** (**C**), **8D** (**D**, **E**). crcb, *crista cerebelli*; dors, *nucleus dorsalis areae octavolateralis;* int, *nucleus intermedius areae octavolateralis;* mg, *nucleus magnocellularis areae octavolateralis;* Mth, Mauthner cell; nufs, *nucleus fasciculi solitarii (nucleus tractus solitarius);* Rm, *area reticularis medialis;* Xm–motor centre of the *nervus vagus;* VIII, *nervus octavus.*

Color version at the end of the book

mesencephalic tectum and octavo-lateral primary and secondary centres (especially the crista cerebelli and eminentia granularis) and reduced gustatory centres (lobus vagus). Based on the peculiarities of the external brain morphology (and especially the rhombencephalon), three ecological subgroups of Cyprinids can be distinguished: from the ecological type of benthophages (crucian carp, common carp, bream) to the roach-like type up to active predators (rudd, chub, dace) (Evans 1931).

Accurate measurements and correct evaluation of such results without overstating its significance may mark ecological groups among closely related species, as well as identify some morphological tendencies. Up to five ecological subgroups of cyprinids can be distinguished according to the detailed quantitative study of the morphology and histology of cyprinids brain (brain centre volumes were calculated in relation to its total volume) (Kotrschal and Palzenberger 1992): (1) Species hunting for fast moving prey with dominant visual sensation, and therefore, well-developed tectum mesencephali and moderately represented octavo-lateral and gustatory centres (so-called basal brain type of cyprinids as presented in roach, rudd, dace (the majority of cyprinids belong to this group)). (2) Plankton-feeding species with the bream (*Abramis ballerus*) type brain characterized by relatively well developed visual, octavo-lateral and gustatory centres. (3) Species feeding on detritus and benthic invertebrates specialized on chemosensory perception (including extra-oral gustatory reception) with highly developed vision and highly developed gustatory centres (common carp, tench (*Tinca tinca*)). (4) Species with well-developed visual and lateral line centres, but comparatively low-developed gustatory centres (sabrefish, *Pelecus cultratus*). (5) Species feeding on invertebrates with well-presented visual and gustatory brain centres and a moderately developed octavo-lateral system (common minnow, *Phoxinus phoxinus*).

Searching for the morpho-functional correlations between biology and brain of fishes was especially popular in the Soviet Union (Svetovidov 1955, Nikitenko 1964, Sbikin 1973, 1980, Nikonorov 1985). Correlations between brain structure and characteristics such as habitat (Pavlovsky and Kurepina 1953, Kurepina 1981), orientation (Sbikin 1973), and feeding (Braginskaya 1948a, b) were discovered. The relative size of different brain departments determines feeding groups. It was postulated that fishes with primary development of the mesencephalon are often predators hunting for large mobile prey. Species with relatively developed brain divisions and a well-presented telencephalon are considered to be benthic nocturnal predators (for example, the burbot (*Lota lota*) and the cusk (*Brosme brosme*)) (Braginskaya 1948a). Based on the hypothesis that the mass of brain compartments is directly proportional to its volume, the special sensory index was proposed. According to the evaluation of relative sizes of appropriate brain centres, the micro- (*Esox lucius*), medio- (*Cyprinus carpio*) and macrosmatic (European eel, Siberian sturgeon) groups were distinguished (Devitsina 2004).

To-date, there are three main methods to quantify the volume of the fish brain: the conventional techniques of histology (1), approximating brain volume to an idealized ellipsoid (or half ellipsoid) and measurements from the magnetic resonance imaging (2) or X-ray tomography (3) (Ullman 2010). For an example of the recent volumetric analysis provided by ecomorphologists, see original paper by White and Brown (2015). The authors have described different brain pattern organisations for gobiid fishes depending on the environmental complexity: complex rock pool habitat leads to a comparatively large telencephalic brain pattern; a simple sandy habitat is correlated with a larger optic tectum. They have also discussed several volumetric methods, noted shrinkage artifacts during histological preparation and postulated that X-ray microtomography produces the most reliable results for volumetric calculations (White and Brown 2015). In both of the above studies, the greatest histological artefacts were found concerning the telencephalon and optic tectum. Unfortunately, even recent X-ray microtomography does not give a sufficient resolution for a detailed quantitative analysis of separate brain centres at the histological level.

The evaluation of rhombencephalic centres based on gross preparation is not always directly associated with the quantitative and cytoarchitectural organization of the nuclei at the histological level. However, the immediately visible rhombencephalic anatomy correlates with physiological functions in the case of extreme specialization of species. For example, the relative volume of the rhombencephalic nuclei of five actinopterygians is presented in Table 1. Corresponding values were calculated in relation to the whole volume of the rhombencephalon, including tracts and fascicles and in relation to the volume of all nuclei measured. The quantitative analysis allows the recognition of developed centres that are not obvious from the gross anatomy, e.g., gustatory centres in *Acipenser baerii*. The dominant development of viscerosensory brain centres of sturgeon corresponds to the data on the highly developed gustatory sensitivity of intra- and extra-oral perception (see below).

The method of topological analysis of the brainstem proposed by Nieuwenhaus is very useful for interspecific comparison of anamnia (Nieuwenhaus 1974, Kremers and Nieuwenhaus 1979, Oey and Nieuwenhaus 1993, Heijdra and Nieuwenhaus 1994). 16 species of anamnia were carefully examined, among which there were representatives of all major anamnia taxa, including three amphibian species. This method provides information on both the qualitative and quantitative levels.

Table 1. Relative volume of rhombencephalic nuclei of five actinopterygians.

Nucleus $(V_{nucleus}/V_{rhomben})/(V_{nucleus}/V_{nuclei})100$ (%) Specimen	*Acipencer baerii*		*Polypterus senegalus*		*Lepisosteus oculatus*		*Anguilla anguilla*		*Cyprinus carpio*	
motor centre of the VII, IX, X nerves	0,4	3	0,3	3,5	0,2	2	0,8	6,5	2,4	8,4
caudal part of the motor centre of the vagal nerve									0,4	0,2
motor zone of the lobus vagus									1,9	7,7
motor centre of the facial nerve									0,1	0,5
motor centre of the VI nerve			0,3	0,3						
motor centre of the trigeminal nerve	0,1	1	0,1	1	0,1	1	0,25	2	0,1	0,4
medial reticular zone	0,6	5	1,7	19	1,6	11,5	1,65	14	1,3	5
nucleus reticularis medialis							0,4	4		
oliva superior									0,1	0,6
oliva inferior	0,08	0,6	0,05	0,6	0,07	0,5			0,15	0,4
median reticular zone							0,16	0,7		
nucleus medianus magnocellularis of polipterids			0,1	0,8						
lateral reticular zone			0,4	5	0,3	2				
nucleus fasciculi solitarii	2,7	20	1	11	0,1	8	1,2	10		
sensory zone of the vagal and glossopharingeal lobes									10	41
sensory zone of the facial nerve (*lobus facialis*)									10	21
nuclei caudalis + posterior areae octavolateralis	0,3	2					0,15	1	0,1	0,4
nucleus octavolateralis medius	1,1	8	2,7	30	5,5	39	2,5	21	0,8	3
nucleus descendens areae octavolateralis	0,38	3	0,41	5	0,63	4	0,70	6	1,44	6
nuclei magnocellularis + tangentialis areae octavolateralis							0,2	2	0,2	1
nuclei magnocellularis + *vestibularis parvocellularis* parvocellularis *areae octavolateralis*			0,3	3						

Table 1 contd. ...

...Table 1 contd.

Nucleus $(V_{nucleus}/V_{rhomben})/(V_{nucleus}/V_{nuclei})100$ (%) — Specimen	*Acipencer baerii*	*Polypterus senegalus*	*Lepisosteus oculatus*	*Anguilla anguilla*	*Cyprinus carpio*
nucleus magnocellularis *areae octavolateralis*	0,05 / 0,4		0,2 / 1		
nucleus anterius (octavius) areae octavolateralis		0,2 / 2,5	0,2 / 1	0,3 / 2	0,5 / 2
nucleus dorsalis areae octavolateralis	2,6 / 19	0,4 / 4			
crista cerebellum	4,9 / 37	1,6 / 18	3,7 / 29	3,7 / 32	1,2 / 5
nucleus medialis nervi trigemini	0,1 / 1		0,1 / 1	0,2 / 1,5	0,3 / 1
secondary gustatory nucleus					0,8 / 3
$V_{nuclei}/V_{rhomben}$	26,9	18,8	25,6	23	49

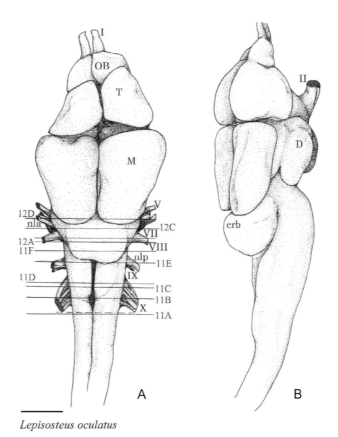

Lepisosteus oculatus

Fig. 10. Dorsal (**A**) and lateral (**B**) views of the brain of the spotted gar (*Lepisosteus oculatus*) (scale bar = 1 mm). Level of cross sections shown in Figs. 11 and 12 are indicated. For abbreviations see page 368.

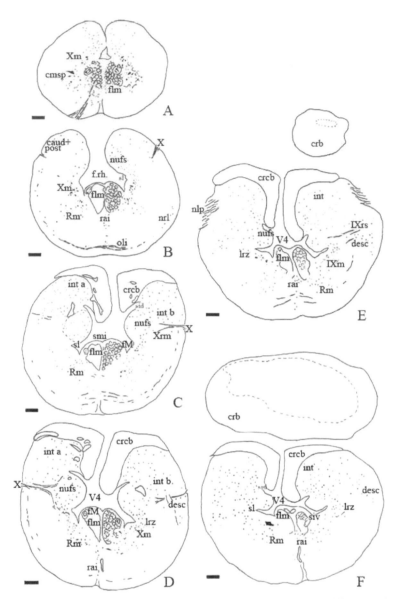

Fig. 11. Drawings of transverse sections of the rhombencephalon of *Lepisosteus oculatus* at different levels (A–F) shown in Fig. 10A (bar = 100 μm). For abbreviations see page 368.

Organization of the Rhombencephalon: Motor and Sensory Centres

The rhombencephalon is the caudal part of vertebrate brain developing from one of the primary brain vesicles–metencephalon (or hindbrain, rhombencephalon)—during the embryogenesis of vertebrates (Johnston 1907, Nieuwenhuys et al. 1998). It rostrally borders the mesencephalon, and merges with the spinal cord caudally (Saveliev 2001, Nieuwenhuys 1982). It is generally accepted to consider the brainstem as a whole, including rhombencephalic (derivatives of the metencephalon) and mesencephalic (derivatives of the primary vesicle of the midbrain) structures (Thors and Nieuwenhuys 1979, Nicundiwe and Nieuwenhuys 1983, Nieuwenhuys et al. 1998, Nieuwenhuys 2011, McCormick 2001).

Since the times of W. His, two longitudinal zones of the brainstem are distinguished: the dorsal (sensory) plate (*planum alare*) and ventral or basal (motor) plate (*planum basale*) separated by the *sulcus*

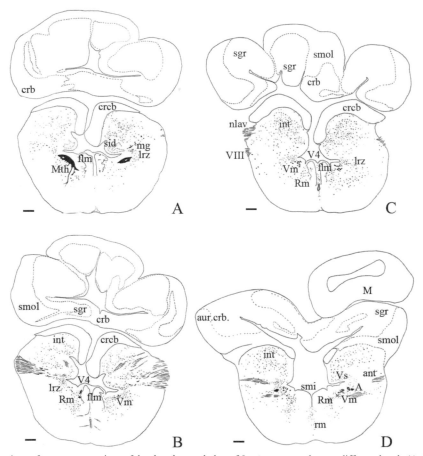

Fig. 12. Drawings of transverse sections of the rhombencephalon of *Lepisosteus oculatus* at different levels (A–D) shown in Fig. 10A (bar = 100 μm). For abbreviations see page 368.

limitans of His (Nieuwenhuys et al. 1998, Nieuwenhuys 2011). Four functional zones (somatosensory, viscerosensory, visceromotor and somatomotor columns) are also described by Herrick-Johnston. According to the classic description of Herrick, the somatosensory (dorsal) zone takes a lateral position (dorsolateral), the viscerosensory zone a—medial position (dorsomedial or intermediodorsal), and the visceromotor (intermedioventral) and somatomotor (ventral or medial) zones have a medial and lower position (Herrick 1906). Four longitudinal columns comprise of the principal pattern of the brainstem organization of vertebrates (Johnston 1907, Nieuwenhuys 1998, Saveliev 2001) which is visible early in embryogenesis in four areas of mitotic activity. In general, the names of the longitudinal morphological zones correspond to the functional content with certain exceptions: the somatomotor nucleus of nerve IV of *Lampetra fluviatilis* actually lies in the dorsal "sensory" zone, the "visceromotor" or intermedioventral zone which also includes the Mauthner cells, efferent neurons of nerve VIII and some others (Nieuwenhuys 2011).

Visceral Sensory and Motor Zones of the Rhombencephalon: General and Special Visceral Centres

Primary rhombencephalic centres of visceral sensitivity are divided into general and special (gustatory) centres. General visceral information—afferentiation of internal organs—projects into the rhombencephalon with the coelomic root of *nervus vagus* (X). Information from gustatory buds is projected into brain via sensory roots of the *nervus facialis* (VII), *nervus glossopharyngeus* (IX) and *nervus vagus* (X) (Johnston 1907, Nieuwenhuys et al. 1998).

In the rhombencephalon of sturgeons and bichirs, the general and special viscerosensory centres (*nucleus fasciculi solitarii* or *nucleus tractus solitarius*) form a single column that is limited by the *sulcus limitans* and *sulcus intermedius dorsalis*. This column is situated from the posterior edge of the rhomboid fossa caudally up to the area of the sensory roots of the facial nerve rostrally (Nieuwenhuys 1983, Nieuwenhuys and Oey 1983) (Figs. 4B–E, 7C–F). In the spotted gar, the viscerosensory centre takes a similar position (Figs. 11C–E).

In Teleostei, the general viscerosensory centre (general visceral nucleus (Kanwal and Caprio 1987, Finger and Kanwal 1992)) or the nucleus commissurialis of Cajal (nC) (Morita and Finger 1985, 1987) is located in the caudal rhombencephalon just rostrally to the obex (Johnston 1907) (Fig. 17B). In cyprinids, medial (nCm) and lateral (nCl) subdivisions of the nucleus of Cajal can be distinguished immunohistochemically: the medial nucleus consists of a large number of tyrosine hydroxylase and substance P immunoreactive fibres in contrast to the lateral nucleus. The afferents of the coelomic root of nervus vagus project exclusively to the nucleus of Cajal, and some fibres end contralaterally in the medial subnucleus (Morita and Finger 1987). The coelomic root of *nervus vagus* is considered to be divided into an upper root (supra) (*ramus pharyngeus superior* and *ramus intestinalis*) and a lower root (*ramus pharyngeus inferior* и *ramus cardiacus*). According to the terminology of pharyngeal muscles, the coelomic root is divided into the supradiaphragmatic division (*ramus pharyngeus superior*, *ramus cardiacus* and *ramus pharyngeus inferior*) and the subdiaphragmatic division (*ramus intestinalis*). Supradiaphragmatic afferents project to the lateral sub-nucleus of Cajal; in addition, afferents of the upper part of the pharynx go into the dorsal part of the lateral nucleus, and subdiaphragmatic afferents are found in the ventral subnucleus. Afferents from the area between the palatal organ and the upper pharynx end dorsolaterally. These afferents are ipsilateral. Afferents of the internal organs (gastrointestinal tract) project bilaterally into the medial subnuclei.

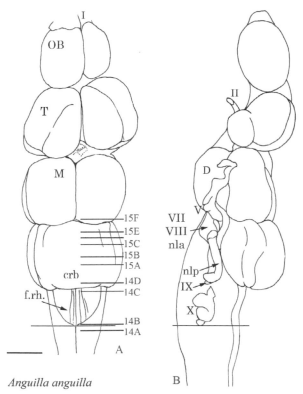

Anguilla anguilla

Fig. 13. Dorsal (**A**) and lateral (**B**) views of the brain of the European eel (*Anguilla anguilla*) (scale bar = 1 mm). Level of cross sections shown in Figs. 14 and 15 are indicated. For abbreviations see page 368.

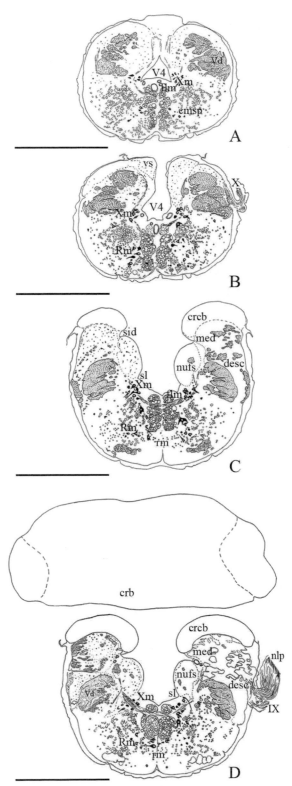

Fig. 14. Drawings of transverse sections of the rhombencephalon of *Anguilla anguilla* at different levels (see Fig. 13) (scale bar = 1 mm). For abbreviations see page 368.

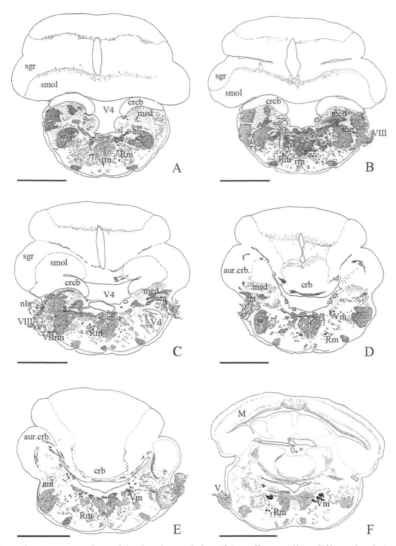

Fig. 15. Drawings of transverse sections of the rhombencephalon of *Anguilla anguilla* at different levels (see Fig. 13) (scale bar = 1 mm). For abbreviations see page 368.

The area postrema is an additional centre of the coelomic root of the nervus vagus of the Neopterygii (Morita and Finger 1987). The ventral edge of the area postrema is located in the area of the cavity of the fourth ventricle opening according to a dopaminergic system study of the european eel (Roberts et al. 1989).

Afferents of vagal, glossopharyngeal and facial nerves project to special viscerosensory centres which form the common nucleus the *fasciculi solitarii* in the majority of bony fishes (Nieuwenhuys et al. 1998). The *nucleus fasciculi solitarii* is a primary gustatory centre (Devitsina 2004, Nieuwenhuys et al. 1998), facing into the cavity of the fourth ventricle between the *sulcus limitans* and *sulcus intermedius dorsalis*. The caudal part of the roller is occupied by the intra-oral gustatory sensitivity centre and thus receives primary projections of nerves IX and X (Devitsina 2004, Kuperina 1981, Nieuwenhuys et al. 1998). Rostrally, it passes into the extra-oral gustatory sensitive centre of nerve VII, which is highly developed in carp, catfish and eel forming an independent *lobus facialis* in carps (Devitsina 2004). Extra- and intra-oral gustatory reception systems are two separate systems of chemical reception (Finger and Morita 1985), having different peripheral sources, primary centres (Morita and Finger 1985, 1987) and pathways into the brainstem (Morita et al. 1980).

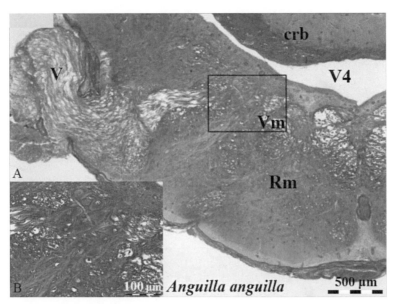

Fig. 16. Rhombencephalon of *Anguilla Anguilla.* Mallory tichrome preparation at the level of the Vth nerve. High-magnification of the motor nucleus of Vth nerve is inserted and indicated by frame (scale bar = 100 μm). crb, *corpus cerebelli*; Rm, *area reticularis medialis*; V, *nervus trigeminus*; Vm, *nucleus motorius nervi trigemini*; V4, *ventriculus quartus.*

In highly specialized gustatory sensitive ciprinids, the caudal part of the special viscerosensory zone forms a well-marked formation in the posterior part of the brain—the *lobus vagus* (Figs. 17, 18C–E, 19A–B, 22). The vagal lobes of *Cyprinus carpio* are formed both as viscersensory and visceromotor centres of nerve X and IX that are organized as laminar and somatotopical structures in certain Cyprinidae (Morita and Finger 1985, 1987, Nieuwenhuys et al. 1998). Up to 16 layers are described in the goldfish (*Carassius auratus*) (Morita and Finger 1985, Morita et al. 1980). Special sensory fibres of nervus vagus are divided into two parts entering the rhombencephalon: the first rises dorsally along the lateral wall of the *lobus vagus* (*radix sensorious nervi vagi, ramus superfacialis*), the second one (*radix sinensorius nervi vagi, ramus profundus*) is localized more medially and is involved in the formation of the fibrous layer of the *lobi vagi* (layer XII). The glossopharyngeal lobe as a separate component is described only in cyprinids (Nieuwenhuys et al. 1998). The sensory component of nerve IX is an independent centre in the caudal edge of *lobus facialis*, located slightly lateral to the sensory centre of *nervus facialis*, which fuses with the rostral part of the *lobi vagi*. The viscerosensory centre is formed as an unpaired *lobus facialis* receiving afferents from the *nervus facialis* (external gustatory buds afferentation) in the brain of *Cyprinus carpio* (Kuperina 1981, Devitsina 2004, Puzdrowski 1987, Nieuwenhuys et al. 1998) (Figs. 18E, 19A–B). Responses to tactile stimulations are also registered in the facial lobe of some Teleostei (including carp) by means of physiological methods (Nieuwenhuys et al. 1998). It is considered that some responses can be promoted by both facial fibres itself and trigeminal collaterals (Nieuwenhuys et al. 1998, Puzdrowski 1988, 1987).

Secondary Gustatory Nucleus

The secondary gustatory nucleus is well defined in the rhombencephalon of common carp. It is directly localized under the descending tract of the trigeminal nerve (McCormick 1994). Secondary gustatory tract fibres (Figs. 19B–C, 20A–C) are projected to the large secondary gustatory centre that is localized in the most rostral part of the rhombencephalon and is limited dorsally by cerebellar structures. Secondary gustatory centres are bilaterally connected by a large fascicle crossing the midline over the fourth ventricle (Fig. 20D).

The visceral motor column of the rhombencephalon is a continuation of the same area of the spinal cord that is localized in the ventral intermediate area (*area intermedius ventralis*) (Johnston 1907). Neurons of the visceromotor column are situated in the periventricular zone between the *sulcus limitans* and the *sulcus intermedius ventralis*. Motor cells of cranial nerves VII, IX and X in the majority of actinopterygians form a single motor column from the caudal edge of the rhomboid fossa up to the level of entrance of the roots of the VIIIth nerve. Motor cells of cranial nerves VII, IX, X form a single cluster in holosteans (van der Horst 1916, Kappers et al. 1936): there is a single visceromotor column in *Lepisosteus oculatus*, while the motor centre of nervus facialis is separated from a single motor centre of nerves X, IX in *Amia calva* (Heijdra and Nieuwenhuys 1994). As a rule, motoneurons of cranial nerves VII, IX, X also form a single visceromotor column in Teleostei (Nieuwenhuys et al. 1998). The caudal border of motor nucleus X is located in the caudal area of the medulla oblongata near the *sulcus limitants* of His, almost behind the level of the opening of ventricle IV. The X, IX, VII nerve motor column can be crossed and distinguished by the plexus of transverse fibres of the bilateral pathways of the rhombencephalon (dorsal arcuate fibres, Black 1917).

The common visceromotor centre is located in the caudal part of the visceromotor column and considered as a motor complex of *nervus vagus* (Black 1917, Kapers et al. 1936). Special motoneurons of nervus vagus are correlated with the *nucleus ambiguus* of the Amniota (Morita and Finger 1987, Nieuwenhuys et al. 1998). The motor complex of *nervus vagus* in fish, as a rule, does not directly lie at the level of the special and visceral components, as seen in amniotes (special visceral (*neevus ambiguus*) and general visceral populations (dorsal motor nucleus of the vagus)). Motoneurons' operating organs of the abdominal cavity are the most caudal motoneurons of the complex, the foremost of them are localized parallel to the special visceromotor column (the level of separation of the vagal motor area in species specialized in gustatory sensitivity). Neurons operating the heart are localized medially, overlapping a special motor column of the rhombencephalon. Cells operating the musculature of pneumatocyst are diffusely distributed in the caudal part of the medulla oblongata and the rostral part of the spinal cord under

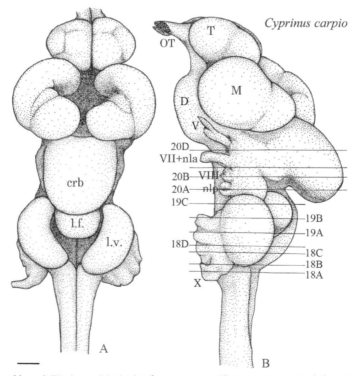

Fig. 17. Dorsal (**A**) and lateral (**B**) views of the brain of common carp (*Cyprinus carpio*) (scale bar = 1 mm). Level of cross sections shown in Figs. 18 and 19 are indicated. crb, *corpus cerebelli*; D, *diencephalon*; l.f., *lobus facialis*; l.v., *lobus vagus*; M, *tectum mesencephalic*; nla, *nervus lateralis anterior*; nlp, *nervus lateralis posterior*; OT, *tractus olfactorius*; T, *lobus hemisphericus telencephali*; V, *nervus trigeminus*; VII–*nervus facialis*; VIII, *nervus octavus*.

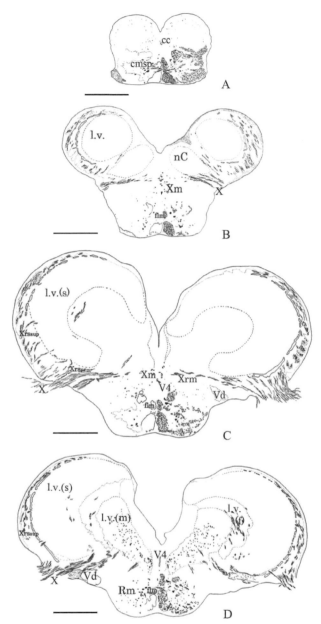

Fig. 18. Drawings of transverse sections of the rhombencephalon of *Cyprinus carpio* at different levels (A–D) shown in Fig. 17B (scale bar = 20 μm). For abbreviations see page 368.

the spinal canal (Nieuwenhuys et al. 1998). Cells of *nervus glossopharyngeus* are localized rostrally; the foremost neurons of the visceromotor column are the cells of *nervus facialis*. Goldfish, highly specialized in gustatory sensitivity, is an exception. The glossopharyngeal vagal motor zone is divided into the following subdivisions (subpopulations) in goldfish: the motor zone of the vagal lobe (it is separated from the special motor zone of the vagal lobes, which innervates the skeletal musculature of the palatal organ) (1); lateral (innervating *arcus branchialis* and "supradiaphragmatic" organs) (2) and medial ("subdiaphragmatic" organs) columns of the vagal motor complex (3); paramedian motor neurons, innervating the heart (are diffusely localized in the intermitant zone between the ependymal and the main substance of the *medulla oblongata*) (4). The special visceromotor zone partially forms vagal lobes in the goldfish, taking

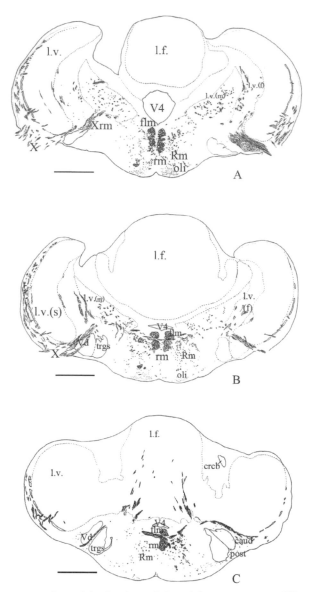

Fig. 19. Drawings of transverse sections of the rhombencephalon of *Cyprinus carpio* at different levels (A–C) shown in Fig. 17B (scale bar = 20 μm). For abbreviations see page 368.

two profound layers in its composition. Vagal motoneurons of the goldfish innervate the musculature of *arcus bronchialis* and the palatal organ (Morita and Finger 1985, Goehler and Finger 1992).

The visceromotor centre of nervus facialis of cyprinids is separated from the visceromotor centres of the glossopharyngeal and vagal nerves (Morita and Finger 1985) (Fig. 20A). It is found at the level of the ventral area of the vagal lobes up to the level of the entrance of the posterior root of the VIII nerve in the common carp. Motoneurons of the centre are predominantly radially oriented and can be distinguished as larger, located medially, and smaller and numerous motoneurons, located laterally. Facial nerve motoneurons operate the adductor (adductor operculi) and levator (levator operculi) of the operculum in carp (Luiten 1976).

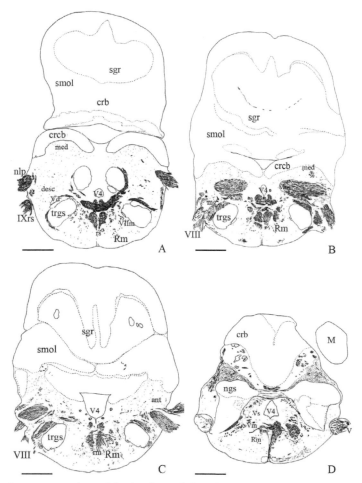

Fig. 20. Drawings of transverse sections of the rhombencephalon of *Cyprinus carpio* at different levels (A–D) shown in Fig. 17B (scale bar = 20 μm). For abbreviations see page 368.

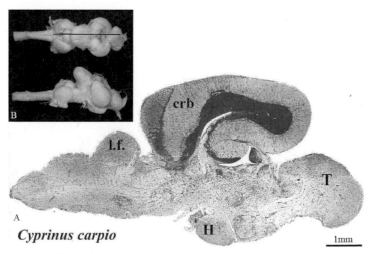

Fig. 21. Sagittal section of the brain of *Cyprinus carpio* (Nissl preparation) (**A**). Macrophotos (**B**) of the dorsal and lateral views of the brain are inserted. crb, *corpus cerebelli*; l.f., *lobus facialis*; H, hypothalamus; T, *lobus hemisphericus telencephali*.

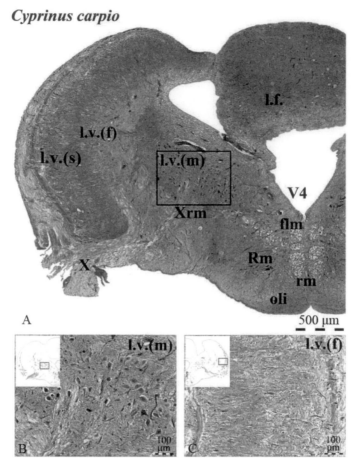

Fig. 22. Cross section of the rhombencephalon of *Cyprinus carpio* (Mallory trichrome) (**A**). High-magnification (**B, C**) of the vagal lobe. **B.** Motor zone of the *lobus vagus* (see insertion). **C.** Fibrous zone of the *lobus vagus.* flm, *fasciculus longitudinalis medialis*; l.f., *lobus facialis*; l.v., *lobus vagus* (s, sensory zone) (m, motor zone) (f, fibrous zone); oli, *oliva inferior*; Rm, *area reticularis medialis*; rm, *area reticularis medianus*; V4, *ventriculus quartus*; X, *nervus vagus*; Xrm, motor root of *nervus vagus*.

Color version at the end of the book

Special Somatic Sensitivity Area of the Rhombencephalon: Nuclei of the Octavolateral System

The rhombencephalon area of special somatic sensitivity receives primary projections from the octavolateralis system. This area is divided into a dorsal (dorsal and intermediate), a lateral (mechano- and electroreceptive) and a ventro-octaval (inner ear) part in actinopterygians (Table 2). In general, the dorsal part is electrosensitive, while the intermediate part is mechanoreceptive (McCormick 1982).

The *nucleus octavolateralis medius* (McCormick 1982) or *nucleus octavolateralis intermedius* (Nieuwenhuys et al. 1998) is located in the lateral part of the somatosensitive column in actinopterygians (mechanoreception of lateral line organs) (McCormick 1981 (*Amia calva*), 1982, Puzdrowski 1989 (*Carassius auratus*), Song and Northcutt 1991a, b (*Lepisosteus platyrhincus*), New and Northcutt 1984a (*Scaphirhynchus platorynchus*)). In addition, the *nucleus caudalis* is found in the lateral zone of certain holosteans and teleosts (the *nucleus caudalis* is not always present) (McCormick 1982, Song and Northcutt 1991) and is situated caudally in the dorsal part of the rhombencephalon with the most caudal cells occurring at the level of the posterior edge of the rhomboid fossa. These nuclei receive the majority of the lateral line mechanoreceptor projections, some part of the lateral fibres also project to the *nucleus magnocellularis*

Table 2. Nuclei of the area octavolateralis of actinopterygians.

Taxa	Dorsal Part	Intermediate Part	Ventral Part
Brachiopterygii (polypterids)	nucleus dorsalis	nucleus intermedius (medialis)	nucleus ventralis (magnocellularis + parvocellularis) (Nieuwenhuys et al. 1998); nuclei anterior, magnocellularis, descending and posterior (McCormick 1982)
Chondrostei	nucleus dorsalis	nuclei intermedius, caudalis (Nieuwenhuys et al. 1998)	nuclei anterior, magnocellularis, descendens and posterior (New and Northcutt 1984a, McCormick 1982)
Holostei	–	nuclei intermedius, caudalis	nuclei anterior, magnocellularis, descendens
Teleostei (not electroreceptive)	–	intermediate lateral zone	nuclei anterior, magnocellularis, tangencialis, descendens and posterior
Dipnoi	nucleus dorsalis	nucleus intermedius	nuclei magnocellularis, parvocellularis
Crossopterygii	nucleus dorsalis	nucleus intermedius	nuclei anterior, magnocellularis, descending (Northcutt 1980)

of the octave area in *Amia calva* (see below) (McCormick 1981). The cerebellar crest—a neuropil layer consisting of the fibers originating in the caudal cerebellar granilar eminence—is also referred to as molecular part of mechanoreceptive lateral region.

The nucleus dorsalis of the area *octavolateralis* receives information from ampullar receptors (Ampullae of Lorentzini) related to the electric field reception in bichirs and sturgeons (Bullock 1982). It is located mediodorsally to the mechanoreceptive column of the lateral line (McCormick and Braford 1993, McCormick 1982, 1981, New and Northcutt 1984a). The *nucleus dorsalis* of Polypteriformes is located medial to the mechanoreceptive lateral nuclei and below the *crista cerebellum* (Figs. 5A–D). The *nucleus dorsalis* is located above the *crista cerebelli* in sturgeons (Figs. 7E–F, 8A–B). Thus, in sturgeons the mechanoreceptive part clearly is divided in the large *nucleus medialis* (*intermedius*) and the small *nucleus caudalis*, and the electrosensitive area, the *nucleus dorsalis*. Besides, the anterior lateral line nerve is divided into the superior and inferior roots (that is not a characteristic of Neopterygii).

The electroreceptive organs in Teleostei are new inventions: the Holostei, and the progenitors of the present-day Teleostei, have lost this type of sensitivity during evolution and the dorsal part of the *area octavolateralis* (Bullock 1982). The Ampullae of Lorentzini are low-frequency electroreceptors (see chapter sensory organs. It is assumed that the Ampullae of Lorentzini were present among the whole group of protoaquatic vertebrates, including the jawless vertebrates. Among the extant species, they are found only in Chondrostei and Elasmobranchii, they disappeared in the Neopterigii. Ampullary organs are also found in various Osteichthyes (see chapter Sensory Organs); they perceive, as in the whole teleostean line, low-frequencies up to 50 Hz. They differ from the Ampullae of Lorentzini of the *Chondrostei* and they are not homologous to them (Bullock 1982). The *nucleus dorsalis* is also absent. In addition, tuberous organs (sensitive to frequencies between 50 and 2000 Hz) developed in the Teleostei (Bullock 1982).

The ability to orient in the geomagnetic field is widely spread among fishes, some species also possess the high-developed electroreception apparatus and have special differentiated centres of the lateral system (Bullock 1982). Among the extant Teleostei, the Mormyroidei, Gymnotiformes and Siluriformes possess well developed electroreceptive structures (Bass 1982, Bullock 1999). Siluriformes possess only low-frequency "ampullary" receptors and the so-called ability for "passive" electroreception. Mormyroidei and Gymnotiformes possess also high-frequency receptive organs (tuberous organs) (Bullock 1982) (see chapter Sensory Organs). These species are characterized by the so-called "active" electroreception due to the production of self-generated weakly electric fields (see chapter Structure and Function of Electric Organs). This ability is used for object location and intra- and interspecific communication (Bell 1979). The lateral zone of the area octavolateralis of these species consists of two columns—a medial mechanoreceptive part and a lateral electroreceptive part. The lateral area in Siluriformes consists of a mechanoreceptive part consisting of medial and caudal nuclei, which are homologous to the single mechanoreceptive lateral

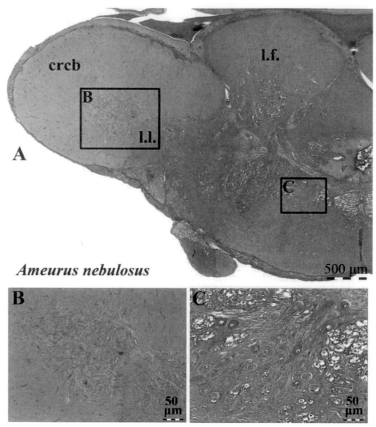

Fig. 23. Cross section of the rhombencephalon of the brown bullhead (*Ameurus nebulosus*) (Mallory trichrome) (**A**). High magnifications of a fragment of the lateral line lobe (**B**) and motoneurons of the visceral motor rhombencephalon column (**C**) are inserted and indicated by frames. crcb, *crista cerebelli*; l.f., *lobus facialis*; l.l., *lobus lateralis.*

Color version at the end of the book

area of the Neopterygii, not possessing electrosensitivity, and the lateral cell column (lateral line lobe). Electroreceptive information goes to special *lobi lateralis* (Fig. 23). The lateral line lobes are located in the anterior part of the octavo-lateral area and are covered by the crista cerebellum (McCormick 1982). The *lobus lateralis* forms multilayer structures in Mormyroidei (up to 7 layers) (Bullock 1982). Therefore, the electroreceptive apparatus of the Teleostei is constructed differently than that in Chondrostei; the *lobus lateralis* of the rhombencephalon of Teleostei is not homologous to the *nucleus dorsalis* of bichirs and sturgeons (McCormick 1982, Northcutt 1991). In addition, all electroreceptors of the Teleostei are presumably developing via some common anlagen of the lateral line (Bullock 1982) (see chapter Sensory Organs).

Thus, three types of organization of the rhombencephalic lateral line area can be distinguished among Actinopterygii (McCormick 1982, Bell 1979): I—Dorsal electroreceptive, lateral (intermediate) and ventral parts found in bichirs and sturgeons (also seen in coelacanths and lungfishes); II—The dorsal electroreceptive part is lacking in Holostei and the majority of Teleostei; III—Teleostei with "active" electroreception possess well-developed lateral line lobes.

In general, the following nuclei are found in the ventral part of the *area octavolateralis* of the rhombencephalon of actinopterygians: the *nucleus octavolateralis anterior*, the *nucleus magnocellularis*, the *nucleus octavolateralis tangentialis* (only found in Teleostei), the descending octaval nucleus (*nucleus descendens areae octavolateralis*), and the *nucleus octavolateralis posterior* (McCormick 1982, Meredith and Butler 1983, McCormick and Braford 1994) (Table 2).

The nucleus ventralis is described in the ventral part of the area octavolateralis in bichirs. This nucleus comprises two subdivisions, which are referred to as *nucleus vestibularis magnocellularis* and *parvocellularis* in lungfishes (Nieuwenhuys et al. 1998). In teleosts, the *nucleus magnocellularis* and the *nucleus tangentialis* receive octaval projections. The dorsal region of the *nucleus magnocellularis* receives also mechanoreceptive input (Bullock et al. 1982, McCormick 1983, Meredith 1984, Puzdrowski 1984). The *nucleus magnocellularis* is also referred to as the *nucleus vestibularis lateralis* (Korn et al. 1977) and even as homologue to the mammalian nucleus Deiters (Kappers et al. 1936).

The projections from the inner ear are examined in detail in various species of Teleostei (Meredith and Butler (*Astronotus ocellatus*) 1983, Meredith et al. 1987 (*Anguilla anguilla*), McCormick and Braford 1993, 1994 (*Ictalurus punctatus* and *Carassius auratus*, respectively), McCormick 2001 (*Arius felis*), Bass et al. 2001 (*Opsanus beta*)). It is found that practically all nuclei of the ventral octavo-lateral area process both vestibular and acoustic information, the exception is the nucleus *octavo-lateralis tangentialis* of Teleostei, receiving only vestibular inputs (*Astronotus ocellatus*) (Meredith and Butler 1983).

Efferent cells of the area octavolateralis are found within the rhombencephalon of Neopterygii (Roberts and Meredith 1986 (*Anguilla anguilla*), Puzdrowski 1989 (*Carassius auratus*), Song and Nortcutt 1991 (*Lepisosteus platyrhincus*)). Efferent cells of the area octavo-lateralis in the brain of the eel, studied with HPR-method, are closely associated with the olive superior (Roberts and Meredith 1986) (see below).

Rhombencephalic Centres of *Nervus Trigeminus*: Motor Nucleus of the Trigeminal Nerve and General Somatosensory Area

Sensory centres of the *nervus trigeminus* (V) belong to the general somatic sensory system (Nieuwenhuys et al. 1998). Trigeminal nerve fibres also participate in the innervation of the fish's olfactory sac (Devitsina and Chervova 1983). Input projects into the centre with four sensory nerve rami: ophthalmic, mandibular, maxillar (maxillomandibular) ramus, and the fourth so-called *ramus communicans* from the complex of the hyomandibular tract (Puzdrowski 1988).

The main sensory nucleus of nerve V, also named medial sensory nucleus of *nervus trigeminus* (New and Northcutt 1984b), and the nucleus of the descending bundle of the trigeminal nerve (*radix descendens nervi trigemini*) receive the primary sensory inputs of nerve V in actinopterygians (Puzdrowski 1988, Luiten 1975, 1979, Nieuwenhuys et al. 1998) (Figs. 7–8, 11–12, 14–15, 20). The mesencephalic sensory nucleus of *nervus trigeminus* is also recognized as a primary centre (Nieuwenhuys et al. 1998), also named as the nucleus of the ascending mesencephalic ramus of *nervus trigeminus* (Woodburne 1935). Trigeminal nerve centres lack anatomical structures which could be identified in gross preparation. Nevertheless, there are some differences at the histological level which have systematic significance and characterize taxa. The main sensory nucleus of *nervus trigeminus* is absent in the bichir rhombencephalon: the sensory component of *nervus trigeminus* is represented exclusively by the *nucleus descendens nervi trigemini* in bichirs and in Agnatha (Woodburne 1935, Nieuwenhuys 1993, Nieuwenhuys and Oey 1983) (Figs. 4–5). The main sensory nucleus of nerve V (Fig. 8C), the nucleus *descendens nervi trigemini pars ventralis et dorsalis* and the mesencephalic sensory nucleus of *nervus trigeminus* are found in sturgeons (*Scaphirhynchus platorynchus*, *Acipenser oxyrinchus* (New and Northcutt 1984)). The maxillar and mandibular rami of *nervus trigeminus* do not fuse but enter, separately, the rhombencephalon only in sturgeons (New and Northcutt 1984). The *nucleus descendens nervi trigemini* of Holostei has two subdivisions (Heijdra and Nieuwenhuys 1994 (*Amia calva*)). The mesencephalic nucleus of *nervus trigeminus* is also described in holosteans (*Amia calva*: Heijdra and Nieuwenhuys 1994, Woodburne 1935). In cyprinids, small neurons of the main sensory nucleus of *nervus trigeminus* are located in the periventricular area dorsally and medially limited by the centre of the secondary gustatory tract (Fig. 20D). Sensory fibres of *nervus trigeminus* are divided into two parts after their entrance into the brain: a rostromedial bundle and a caudally oriented descending bundle. The rostromedial bundle ends up in the main sensory nucleus of nerve V and also projects to the mesencephalic nucleus of *nervus trigeminus*. Sensory fibres of the caudally oriented fascicle form a descending tract of nerve V (*radix descendens nervi trigemini*). Cells of the *nucleus descendens nervi trigemini* are diffusely arranged. Various subdivisions of the nucleus are found in *Cyprinus carpio* (Luiten 1975): among them are described the medial funicular nucleus (MFn) (Luiten 1975, Puzdrowski

1988) and the *nucleus spinalis* of *nervus trigeminus* (Spv) (spinal trigeminal nucleus). The last one is considered to be an independent nucleus (Luiten and van der Pers 1977, *Carassius auratus*, Puzdrowski 1988). The trigeminal projections to the main nucleus of *nervus trigeminus* and the medial funicular nucleus are topographically organized: afferents of ramus ophtalmicus of the nerve V project into the ventral part of the nuclei, and the maxilla- and hyomandibular rami into dorsal part (Puzdrowski 1988).

The motor nucleus of *nervus trigeminus* is located in the rostral part of the rhombencephalon within the intermedial ventral zone. It is referred to as visceral motor centre (Nieuwenhuys et al. 1998). Trigeminal motoneurons innervate the mandibular musculature: muscles of the mandibular itself, the hyoid arch and partially the operculum (Luiten 1976). In brain of the Senegal bichir the caudal border of the Vth nerve motor nucleus lies rostrally to the Mauthner cells. The anterior border of the nucleus is located between the levels of the motor and sensory roots of the nerve V entrance (Figs. 5D–E). Motoneurons of *nervus trigeminus* form a compact column in the rhombencephalon of sturgeons (Figs. 8B–C). Its drop-shaped and oval motoneurons of the nucleus are radially oriented.

The Nucleus A—the rostral part of the motor column of *nervus tregiminus*—is described in bowfin as a small group of cells found just rostrally to the main motor nucleus of *nervus tregiminus* (Heijdra and Nieuwenhuys 1994). In spotted gar, there are cells situated just rostrally to the main motoneurons of the nerve V. These motoneurons are larger than motoneurons of the main motor nucleus of *nervus trigeminus* and can be considered as a nucleus A (Fig. 12D).

In the European eel, the anterior border of the motor nucleus of nerve V is located in the region of the entrance of its roots (Figs. 15D–F). In the common carp, large motor cells ($28–39 \times 81–85$ µm) lie at the level of entrance of the sensory fibres of *nervus trigeminus* parallel to the area of the main motoneurons of *nervus trigeminus* (Fig. 20D). These cells are considered to be the large cellular part of motor nucleus of *nervus trigeminus*.

The Reticular Formation of the Rhombencephalon of Actinopterygians

The reticular formation of the rhombencephalon is a very conservative structure (Northcutt 2002). The actinopterigian reticular nuclei form dense accumulations. The medial (*area reticularis medialis*), median (*area reticularis medianus*), and lateral (*area reticularis lateralis*) zones are described in reticular formation of actinopterygians. The reticular formation does not demonstrate any special development depending on the ecology of fish, but there are still a number of differences characterizing taxa.

The medial zone of the reticular formation forms the rhombencephalic continuation of the somatosensory column of the spinal cord. Neurons of the medial and median zones are located in the brainstem including both the rhombencephalic and mesencephalic parts. The lateral zone of the reticular formation lies only within the rhombencephalic area. The *oliva inferior*, *oliva superior* and *nucleus reticularis lateralis* (or *nucleus funiculi lateralis*) are located in the caudal part of the rhombencephalon and are also referred to as reticular formation (Nieuwenhuys et al. 1998).

The midline of the rombencephalon is a positional criterion for the median reticular zone—no more than 1/20 of rhombencephalon's width lateral to the midline. According to immunohistochemical studies, serotoninergic neurons of the raphe area are referred to as the median zone of the reticular formation (Nieuwenhuys et al. 1998). The lower (posterior) and upper (anterior) raphe nuclei (*nucleus raphes inferior et superior*) are recognized in the median reticular formation of actinopterygians (Cohen and Kriebel 1989, Meek and Joosten 1989, Johnston et al. 1990). The *nucleus raphes inferior* can also be divided into the *nucleus raphes inferior* (or *nucleus raphes magnus*) and the *nucleus raphes intermedius* (Meek et al. 1989, Meek and Joosten 1989 (*Gnathonemus petersii*) (or *nucleus raphes pallidus/obscurus*, according to Ekstrom and Ebbesson 1989 (*Oncorhinchus nerka*))).

In reedfish, the border between the superior and inferior raphe nuclei passes at the level of the Mauthner cell (Nieuwenhuys 1983, Nieuwenhuys and Oey 1983). In the Senegal bichir, raphe cells are localized more dorsally and form a slightly denser accumulation in the area of nucleus raphe superior—in front of the Mauthner cells. Furthermore, bichirs are the only living fish possessing a *nucleus medianus magnocellularis*. It is described in reedfish (Nieuwenhuys 1983, Nieuwenhuys and Oey 1983, Nieuwenhuys et al. 1998) and in the Senegal bichir (Fig. 4C). In the brain of the spotted gar, the border of the *nuclei raphes inferior* and *superior* lies also at the level of the Mauthner cells. In the brain of the European eel, caudally located neurons of the *nucleus raphes inferior* are larger than the rostral ones.

Neurons of the medial zone (*area reticularis medialis*) together with motor cells of the VIth nerve lie in the ventral motor zone representing a continuation of the somatic motor column of the spinal cord (*columna motorius spinalis*) (Johnston 1907, Nieuwenhuys et al. 1998). The medial reticular zone is located under the *sulcus intermedius ventralis*, alongside with the *fasciculus longitudinalis medialis* in the basal plate of the rhombencephalon.

Three nuclei in the medial zone of the reticular formation are distinguished in all *Osteichthyes*: the *nucleus reticucularis inferior*, the *nucleus reticucularis medius* and *nucleus reticularis superior* (Roberts 1992 (*Anguilla anguilla*), Prasada Rao et al. 1987 (*Carassius auratus*)). Caudal cells of the *nucleus reticularis inferior* are situated from the border of the spinal cord caudally to the level of the Mauthner cells rostrally (Nieuwenhuys et al. 1998). The Mauthner cells are considered to be the largest neurons of the medial reticular formation (Nieuwenhuys et al. 1998). The pair of the extremely large Mauthner cells is situated in the upper part of the basal plate laterally to the medial reticular zone at the level of the caudal edge of *nucleus magnocellularis octavolateralis* (Figs. 7F, 9). Very large cells (70 × 130 µm) also appear at this level in the medial reticular formation of siberian sturgeon (Figs. 7F, 8A–B). In the brain of the spotted gar, large cells (26 × 31 µm) lie in the rostral part of the *nucleus reticularis inferior*. Cells of the *nervus abducens* are located in the region of the medial reticular formation, dorsally and laterally to the main number of neurons of *nucleus reticucularis medius* (*Salmo gaerdneri* (Nieuwenhuys 1998), *Solea vulgaris* (Quensel 1806, Black 1917)). The border of the *nucleus reticularis medius* and *nucleus reticularis superior* is described at the level of the rostral motoneurons of *nervus trigeminus* (Nieuwenhuys et al. 1998, 1983) and in front of the anterior nucleus of the octavo-lateral system according to a HPR-study (Prasada Rao et al. 1987).

The lateral zone of the reticular formation (*area reticularis lateralis*) is found in bichirs (Nieuwenhuys and Oey 1983) and holosteans (Heijdra and Nieuwenhuys 1994). The caudal border of the lateral zone is located in the area of the entrance of the rostral roots of *nervus vagus*. In the brain of the reedfish, the *nucleus reticularis lateralis* is absent. It is described in Acipenseridae and Teleostomi (Nieuwenhuys 1998). Small triangle and fusiform cells of the *nucleus reticularis lateralis* in *Acipencer baerii* are diffusely arranged and do not form dense aggregations. The lateral area of the reticular formation in the brain of the spotted gar extends from the level rostrally to the caudal edge of the rhomboid fossa up to the level of the exit of the caudal roots of *nervus trigeminus*. The *nucleus reticularis lateralis*, described in bowfin (Heijdra and Nieuwenhuys 1994), can be identified in the posterior part of rombencephalon of the spotted gar (Fig. 11B).

The oliva inferior are also found in all actinopterygian species in the caudal part of the rhombencephalon (Nieuwenhuys et al. 1998). The oliva superior is well presented in cyprinids (Echteler 1984). Cyprinids have a well developed *oliva inferior* and *oliva superior* (McCormick and Braford 1994) (Figs. 9A, 18E, 22). It is also described in the European eel (Meredith and Roberts 1986). It is assumed that these structures take part in the relay of the inputs of the primary octavo-lateral centres (Echteler 1984, Nieuwenhuys et al. 1998). Anterior and posterior efferent nuclei of the octavo-lateral system are identified in cyprinids (Puzdrowski 1989): according to HPR-studies, the *oliva superior* is associated with efferent cells of the octaval system and it is located in the region near the lateral bundle (*funiculus lateralis*) (Meredith and Roberts 1986, McCormick and Braford 1994).

Summary

The brain of anamnia is characterized by an archaic organization and a wide range of adaptive features unknown for brains of other vertebrates. The structural brain adaptations lead to an evolutionary deadlock for many deeply specialized taxa, but at the same time, it allows us to see archaic features of non-specialized subdivisions in hyperspecialized species. The study of anamnia in a comparative way with a hypothetical brain archetype provides evidence for the species' biology for both ecomorphological and evolutionary approaches.

Abbreviations for Figures 3–23

A	nucleus A (of holosteans)
aur.crb	auriculi cerebelli
ant	*nucleus anterius (octavius) areae octavolateralis*
caud	*nucleus caudalis areae octavolateralis*
cc	*canalis centralis*
cd	*cornu dorsalis*
cmsp	*columna motorius spinalis*
crb	*corpus cerebelli*
crcb	*crista cerebelli*
cv	*cornu ventralis*
D	Diencephalon
desc	*nucleus descendens areae octavolateralis*
dors	*nucleus dorsalis areae octavolateralis*
egr	*eminentia granularis*
fi	*foramen intraventriculare* (Monroi)
flm	*fasciculus longitudinalis medialis*
f.rh.	*fossa rhomboidea*
hyp.	*hyphotalamus*
int	*nucleus intermedius areae octavolateralis*
l.el.	lobus electricus
l.g.	*lobus glossopharyngeus*
l.ll.	lateral line lobe
l.f.	*lobus facialis*
l.v.	*lobus vagus*
M	*tectum mesencephali*
Mth	Mauthner cell
fM	fiber of the Mauthner cell
Mz	motor zone *lobus vagus*
nC	nucleus commissurialis of Cajal
ngs	*nucleus gustatorius secundarius*
nla	*nervus lateralis anterior*
nlad	*nervus lateralis anterior pars dorsalis*
nlav	*nervus lateralis anterior pars ventralis*
nufs	*nucleus fasciculi solitarii (nucleus tractus solitarius)*
OB	*bulbus olfactorius*
oli	*oliva inferior*
oli(s)	*oliva superior*
OT	*tractus olfactorius*
P	*pallium* (P1-3 – pallial field)
PO	*pedunculus olfactorius*
pv	pseudoventricle of polipterids
rai	*nucleus raphes inferior*
ras	*nucleus raphes superior*
Rm	*area reticularis medialis*
rm	*area reticularis medianus*
Rl (lrz)	*area reticularis lateralis*
sep	septum
sgr	*stratum granulare cerebelli*
sid	*sulcus intermedius dorsalis*
siv	*sulcus intermedius ventralis*

sl	*sulcus limitans*
smi	*sulcus medianus inferior*
smol	*stratum moleculare cerebelli*
T	*lobus hemisphericus telencephali*
tt	*tela lelencephali*
V4	*ventriculus quartus*
v.crb.	*valvula cerebelli*
vep	*nucleus vestibularis parvocellularis*
vimp	*ventriculus impar telencephali*
vs	viscerosensory zone
zgml	*zona granularis marginalis (pars lateralis)*
I	*nervus olfactorius*
II	*nervus opticus*
V	*nervus trigeminus*
Vd	*fasciculus descendens* (of the *nervus trigeminus*)
Vm	*nucleus motorius nervi trigemini*
Vs	*nucleus medialis nervi trigemini* (main sensory nucleus of trigeminal nerve)
V(VII/IX/X)rm	motor root of the *nervus trigeminus* (*nervus facialis/nervus glossopharyngeus/ nervus vagus*)
V(VII/IX/X)rs	sensory roots of the *nervus trigeminus* (*nervus facialis/nervus glossopharyngeus/nervus vagus*)
Vt	*ventriculus telencephali*
VI	*nervus abducens*
VII	*nervus facialis*
VIII	*nervus octavus*
IX	*nervus glossopharyngeus*
VIIm	motor centre of the *nervus vagus*
IXm	motor centre of the *nervus glossopharyngeus*
X	*nervus vagus*
Xm	motor centre of the *nervus vagus*
Xrm	motor root of the *nervus vagus*
Xrssup/pr	superior/posterior sensory roots of the *nervus vagus*

References Cited

Bass, A.H. 1982. Evolution of the vestibulolateral lobe of the cerebellum in electroreceptive and nonelectroreceptive teleosts. J. Morphol. 174: 335–348.

Bass, A.H., D.A. Bodnar and M.A. Marchaterre. 2001. Acoustic nuclei in the medulla and midbrain of the vocalizing gulf toadfish (Opsanus beta). Brain Behav. Evol. 57: 63–79.

Bell, C.C. 1979. Central nervous system physiology of electroreception, a review. J. Physiol. Paris.

Bell, C.C., H. Zakon and T.E. Finger. 1989. Mormyromast electroreceptor organs and their afferent fibers in mormyrid fish: I. Morphology. J. Comp. Neurol. 286(3): 391–407.

Black, D. 1917. The motor nuclei of the cerebral nerves in phylogeny. J. Comp. Neurol. 27: 467–599.

Braginskaya, R.Y. 1948a. [Fish brain morphology in connection with its feeding mode]. Dokl Akad Nauk SSSR 59(6): 1213–1216. Russian.

Braginskaya, R.Y. 1948b. [Stages of the brain development of sazan]. Dokl Akad Nauk SSSR 60(3): 505–507. Russian.

Briñón, J.G., M. Médina, R. Arévalo, J.R. Alonso, J.M. Lara and J. Aijón. 1993. Volumetric analysis of the telencephalon and tectum during metamorphosis in a flatfish, the turbot Scophthalmus maximus. Brain Behav. Evol. 41: 1–5.

Bullock, T.H. 1982. Electroreception. Annu. Rev. Neurosci. 5: 121–170.

Bullock, T.H., R.G. Northcutt and D.A. Bodznick. 1982. Evolution of electroreception. Trends Neurosci. 5: 50–53.

Bullock, T.H. 1999. The future of reseach on electroreception and electrocommunication. J. Exp. Biol. 202: 1455–1458.

Cohen, S.L. and R.M. Kriebel. 1989. Brainstem location of serotonin neurons projecting to the caudal neurosecretory complex. Brain Res. Bull. 22(3): 481–487.

Devitsina, G.V. and L.S. Chervova. 1983. Structural characteristics of the trigeminal ganglion of the cod (*Gadus morhua*). Zh. Ev. Biochim. Phisiol. 19(3): 293–299. In Russian.

Devitsina, G.V. 2004. [Chemosensory system of fishes: structural and functional organization and interconnection]. Avtoref. dis.... dok. boil. nauk. MSU, Moscow. Russia. In Russian.

Echteler, S.M. 1984. Connections of the auditory midbrain in a teleost fish, *Cyprinus carpio*. J. Comp. Neurol. 230: 536–551.

Ekström, P. and S.O. Ebbesson. 1988. Distribution of serotonin-immunoreactive neurons in the brain of sockeye salmon fry. J. Chem. Neuroanat. 2(4): 201–213.

Evans, H.M. 1931. A comparative study of the brains in british cyprinoids in relation to their habits of feeding, with special reference to the anatomy of the medulla oblongata. Proceed. Royal Soc. London B: Biol. Sciences 108: 233–157.

Evans, H.M. 1935. The brain of gadus, with special reference to the medulla oblongata and its variations according to the feeding habits of different gadida. Proceed. Royal Soc. London. B: Biol. Sciences 117: 367–399.

Finger, T. and J. Kanwal. 1992. Ascending general visceral pathways within the brainstems of two teleost fishes: *Ictalurus punctatus* and *Carassius auratus*. J. Comp. Neurol. 4: 509–520.

Goehler, L.E. and T.E. Finger. 1992. Functional organization of vagal reflex systems in the brain stem of the goldfish, *Carassius auratus*. J. Comp. Neurol. 319: 463–478.

Greenwood, P.H. 1984. Polypterus and erpetoichtys: Anachronistic osteichtyans. pp. 143–147. *In*: N. Eldridge and S. Stanley (eds.). Living Fossils. New York, Springer-Verlag, USA.

Heijdra, Y.F. and R. Nieuwenhuys. 1994. Topological Analisis of the Brainstem of the Bowfin, Amia calva. J. Comp. Neurol. 339: 12–26.

Herrick, C.J. 1906. On the centres for taste and touch in the medulla oblongata of fishes. J. Comp. Neurol. 16: 403–437.

Holmgren, N. and C.J. van der Horst. 1925. Contribution to the morphology of the brain of Ceratodus. Acta Zool. 6: 59–165.

Johnston, J.B. 1901. The brain of Acipenser. A contribution to the morphology of the vertebrate brain. Zool. Jahrb. Abt. Anat. Ontog. 15: 59–260.

Johnston, J.B. 1907. The Nervous System of Vertebraes. John Murray, London. UK.

Kanwal, J.S. and J. Caprio. 1987. Central projections of the glossopharyngeal and vagal nerves in the channel catfish, Ictalurus punctatus: clues to differential processing of visceral inputs. J. Comp. Neurol. 264(2): 216–30.

Kikugawa, K., K. Katoh, S. Kuraku, H. Sakurai, O. Ishida, N. Iwabe and T. Miyata. 2004. Basal jawed vertebrate phylogeny inferred from multiple nuclear DNA-coded genes. BMC Biol. 11; 2: 3.

Kotrschal, K. and M. Palzenberger. 1992. Neuroecology of cyprinids: comparative, quantitative histology reveals diverse brain patterns. Environ. Biol. Fish 33: 135–152.

Kotrschal, K., M.J. Van Staaden and R. Huber. 1998. Fish brains: evolution and environmental relationships. Rev. Fish. Biol. Fish 8: 373–408.

Kremers Jan-Willem, P.M. and R. Nieuwenhuys. 1979. Topological analysis of the brain stem of the crossopterygian fish Latimeria chalumnae. J. Comp. Neurol. 187: 613–638.

Kurepina, M. 1981. Brain of animals. Nauka, Moscow. Russia. Russian.

Lauder, G.V. and K.F. Liem. 1983. The evolution and interrelationships of the actinopterygian fishes. Bull. Mus. Comp. Zool. 150: 95–197.

Luiten, P.G.M. 1975. The central projections of the trigeminal, facial and anterior lateral line nerves in the carp (*Cyprinus carpio* L.). J. Comp. Neurol. 160: 399–417.

Luiten, P.G.M. 1976. A somatotopic and functional representation of the respiratory muscles in the trigeminal and facial motor nuclei of the carp (*Cyprinus carpio* L.). J. Comp. Neur. 166: 191–200.

Luiten, P.G.M. and J.N.C. van der Pers. 1977. The connections of the trigeminal and facial motor nuclei in the brain of the carp (*Cyprinus carpio* L.) as revealed by anterograde and retrograde transport of HRP. J. Comp. Neurol. 174: 575–590.

Luiten, P.G.M. 1979. Proprioceptive reflex connections of head musculature and the mesencephalic trigeminal nucleus in the carp. J. Comp. Neurol. 183(4): 903–12.

McCormick, C.A. 1981. Central projections of the lateral line and eight nerves in the bowfin, Amia calva. J. Comp. Neurol. 197: 1–15.

McCormick, C.A. 1982. The organization of the octavolateralis area in actinopterygian fishes: A new interpritation. J. Morphol. 171: 159–181.

McCormick, C.A. and M.R. Braford. 1993. The primary octaval nuclei and inner ear afferent projections in the otophysan Ictalurus punctatus. Brain Behav. Evol. 42: 48–68.

McCormick, C.A. and M.R. Braford. 1994. Organisation of the inner iar endorgan projections in the goldfish, *Carassius auratus*. Brain Behav. Evol. 43: 189–205.

McCormick, C.A. 2001. Brainstem acustic areas in the marine catfish, Arius felis. Brain Behav. Evol. 57: 134–149.

Meek, J. and H.W.J. Joosten. 1989. The distribution of serotoninin the brain of the mormyrid teleost Gnathonemus petersii. J. Comp. Neurol. 281: 206–222.

Meek, J., H.W.J. Joosten and H.W.M. Steinbusch. 1989. Distribution of dopamine immunoreactivity in the brain of the mormyrid teleost Gnathonernus petersii. J. Comp. Neurol. 281: 362–38.

Meredith, G.E. and A.B. Butler. 1983. Organization of eighth nerve afferent projections from individual endorgans of the inner ear in the teleost, Astronotus ocellatus. J. Comp. Neurol. 220: 44–62.

Meredith, G.E. and B.L. Roberts. 1986. The relationship of saccular efferent neurons to the superior olive in the eel. J. Comp. Neurol. 265: 494–506.

Meredith, G.E. and B.L. Roberts. 1987. Distribution and morphological characteristics of efferent neurons innervated end organs in the ear and lateral line of the european eel. J. Comp. Neurol. 265: 494–506.

Meredith, G.E., B.L. Roberts and S. Maslam. 1987. Distribution of afferent fibers in the brainstem from endorgans in the ear and lateral line in the european eel. J. Comp. Neurol. 265: 507–520.

Morita, Y., H. Ito and H. Masai. 1980. Central gastatory paths in the crucian carp, *Carassius carassius*. J. Comp. Neurol. 191: 119–132.

Morita, Y., T. Murakami and H. Ito. 1983. Cytoarchitecture and topographic projections of the gastatory centres in a teleost, *Carassius carassius*. J. Comp. Neurol. 218: 378–394.

Morita, Y. and T.E. Finger. 1985. Topographic and laminar organisation of the vagal gastatory system in the goldfish, *Carassius auratus*. J. Comp. Neurol. 238: 187–201.

Morita, Y. and T.E. Finger. 1987. Topographic representation of the sensory and motor roots of the medulla of goldfish, Carassius auratus. J. Comp. Neurol. 264: 231–249.

Motta, P.J. and K. Kotrschal. 1992. Correlative, experimental, and comparative evolutionary approaches in ecomorphology. Netherlands J. Zool. 42(2-3): 400–415.

Motta, P.J., S.F. Norton and J.J. Luczkovich. 1995. Perspectives on the ecomorphology of bony fishes. Envir. Biol. Fishes 44: 11–20.

New, J.G. and R.G. Northcutt. 1984a. Central projections of the lateral line nerves in the shovelnose sturgeon. J. Comp. Neurol. 225: 129–140.

New, J.G. and R.G. Northcutt. 1984b. Primary projections of the trigeminal nerve in two species of sturgeon: Acipencer oxyrhynchus and Scaphirhynchus platorynchus. J. Morphol. 182: 125–136.

Nicundiwe, A.M. and R. Nieuwenhuys. 1983. The cell masses in the brainstem of the south African clawed frog xenopus laevus: A topographical and topological analysis. J. Comp. Neurol. 213: 199–219.

Nieuwenhuys, R. 1982. An overview of the organisation of the brain of actinopterygian fishes. Amer. Zool. 22: 287–310.

Nieuwenhuys R. 1983. The central nervous system of the brachiopterygian fish Erpetoichthys calabaricus. J. Hirnforsch. 24: 501–533.

Nieuwenhuys, R. and L. Oey. 1983. Topological analysis of the brain stem of the reedfish, Erpetoichthys calabaricus. J. Comp. Neurol. 213: 220–232.

Nieuwenhuys, R., H. Donkelaar and C. Nicholson. 1998. The Central Nervous System of Vertebrates. Springer-Verlag, Berl. Germany.

Nieuwenhuys, R. 2011. The structural, functional, and molecular organization of the brainstem. Front. Neuroanat. 5: 33. 22.

Nikitenko, MF. 1964. [On the size and structure of brains of a fishes in connection with its habitats]. Vopr. Ichtyol. 4: 34–44.

Nikonorov, S.I. 1985. [Neuromorphological features of brain in relation to ecology of chum salmon Oncorhynchus keta (Walb.) Oncorhynchus gorbuscha (Wakb.) (Salmonidae)]. Vopr. Ichtyol. 25: 313–319. In Russian.

Noack, K., R. Zardoya and A. Meyer. 1996. The complete mitochondrial DNA sequence of the bichir (Polypterus ornutipinnis), a basal ray-finned fish: ancient establishment of the consensus vertebrate gene order. Genetics 144: 1165–1180.

Northcutt, R.G. 1980. Anatomical evidence of electroreception in the coelacanth (Latimeria chalumnae). Zbl Vet. Med. A Anat. Histol. Embriol. 9: 289–295.

Northcutt, RG. 1991. Morphology, distribution and innervation of the lateral-line receptors of the Florida gar, Lepisosteus platyrhincus. Brain Behav. Evol. 37: 10–37.

Northcutt, R.G. 2002. Understanding vertebrate brain evolution. Integ. Comp. Biol. 42: 743–756.

Northcutt, R.G. and R.E. Davis. 1985. Fish Neurobiology: Brain stem and sense organs. University of Michigan Press. Ann Arbor. Vol. 1.

Parker, B. and P. McKee. 1984. Status of the spotted gar, *Lepisosteus oculatus*, in Canada. Can. Field Natur. 98(1): 80–86.

Patterson, C. 1982. Morphology and interrelationships of primitive actinopterygian fishes. Am. Zool. 22: 241–259.

Pavlovskii, E.N. and M. Kurepina. 1953. Fish brain organization in connection with its habitats. pp. 134–182. *In*: E.N. Pavlovskii (ed.). Studies on the General Problems of Ichtyology. Izd. Ak. Nauk SSSR., M.-L, Russia. In Russian.

Platel, R., J.-M. Ridet, R. Bauchot and M. Diagne, 1977. L'organisation encephalgue chez Amia, Lepisosteus et Polypterus: Morphologie et analyse guantative compares. J. Hirnforsch. 18: 69–73.

Prasada Rao, P.D. and T. Finger. 1984. Asymmetry of the olfactory system in the brain of the winter flounder, Pseudopleuronectes americanus. J. Comp. Neurol. 225: 492–510.

Prasada, R.P.D., A.G. Jadhao and S.C. Sharma. 1987. Descending projection neurons to the spinal cord of the goldfish, *Carassius auratus*. J. Comp. Neurol. 265(1): 96–108.

Puzdrowski, R.L. 1987. The peripheral distribution and central projections of the sensory rami of the facial nerve in goldfish, *Carassius auratus*. J. Comp. Neurol. 259: 382–392.

Puzdrowski, R.L. 1988. Afferent projections of the trigeminal nerve in the goldfish, *Carassius auratus*. J. Morphol. 192: 131–147.

Puzdrowski, R.L. 1989. Peripheral distribution and central projections of the lateral-line nerves in goldfish, *Carassius auratus*. Brain Behav. Evol. 34: 110–131.

Roberts, B.L. and G.E. Meredith. 1986. The relationship of succular efferent neurons to the superior olive in the eel, *Anguilla anguilla*. Neurosci. Lett. 68: 69–72.

Roberts, B.L., G.E. Meredith and S. Maslam. 1989. Immunocytochemical analysis of the dophamine system in the brain and spinal cord of the European eel, *Anguilla anguilla*. Anat. Embryol. (Berl.) 180: 401–412.

Roberts, BL. 1992. Neural mechanisms underlying escape behavior in fishes. Rev. Fish. Biol. Fish. 2: 243–266.

Romer, A.S. 1962. The vertebrate body. Saunders, Philadelphia, USA.

Saveliev, S.V. and V.P. Chernikov. 1994. Oceanic whitetip shark (Carcharhinus longimanus) can use air smell for food searching. Vopr. Ichtyol. 34(2): 219–225. In Russian.

Saveliev, S.V. 2001. Comparative anatomy of the nervous system of vertebrates. GEOTAR-MED, Moscow. Russia. In Russian.

Saveliev, S.V. 2005. The brain origin. Vedi, Moscow. Russia. In Russian.

Sbikin, Y.N. 1973. On external morphology of the brain of sturgeons (fem. Acipenseridae). Vopr. Ichtyol. 13(5): 945–948. In Russian.

Sbikin, Y.N. 1980. Age changes of vision of fishes in connection with its behavior. Nauka, Moscow, Russia. In Russian.

Song, J. and R.G. Northcutt. 1991a. Morphology, distribution and innervation of the lateral-line receptors of the florida gar, Lepisosteus platyrhincus. Brain Behav. Evol. 37: 10–37.

Song, J. and R.G. Northcutt. 1991b. The primary projections of the lateral-line nerves of the florida gar, *Lepisosteus platyrhinchus*. Brain Behav. Evol. 37: 38–63.

Thors, F. and R. Nieuwenhuys. 1979. Topological analysis of the brain stem of the lungfish *Lepidosiren paradoxa*. J. Comp. Neurol. 187: 589–612.

Ullmann, J.F., G. Cowin and S.P. Collin. 2010. Quantitative assessment of brain volumes in fish: comparison of methodologies. Brain Behav. Evol. 76(3-4): 261–70.

White, G.E. and C. Brown. 2015. Variation in brain morphology of intertidal gobies: A comparison of methodologies used to quantitatively assess brain volumes in fish. Brain Behav. Evol. 85: 245–256.

Yopak, K.E., T.J. Lisney and S.P. Collin. 2015. Not all sharks are "swimming noses": variation in olfactory bulb size in cartilaginous fishes. Brain Struct. Funct. 220: 1127–1143.

Yopak, K.E., V.L. Galinsky, R.M. Berquist and L.R. Frank. 2016. Quantitative classification of cerebellar foliation in cartilaginous fishes (Class: Chondrichthyes) using three-dimensional shape analysis and its implications for evolutionary biology. Brain Behav. Evol. 87(4): 252–64.

Chapter **18**

Endocrine System

Werner Kloas[1] and *Frank Kirschbaum*[2],*

INTRODUCTION

Endocrinology developed as a subdiscipline of animal physiology at the end of the 19th century using the concept of "internal secretion" for the regulation of physiological functions (Oliver and Schaefer 1895). Historically, early endocrinological research has been performed mainly with mammals in order to increase the medicinal knowledge about physiology of humans. The endocrine system has been ascribed to be an additional regulator to the nervous system (Bayliss and Starling 1902) and Starling defined already in 1905 the classical concept for endocrinology including a hormone producing gland secreting hormones into the blood stream (internal secretion) that have, in turn, physiological effects on target organs in a distant organ. This classical concept which, in general, still holds true for vertebrates became further defined as "endocrinology" by Pende (1909) and introduced as a scientific subdiscipline of physiology by Crookshank (1914). Since the concept for endocrinology has been established in mammals, emerging interest of zoologists arose to investigate whether this basic physiological principle is also present in the various groups of the animal kingdom including fishes and invertebrates. Especially, endocrine research in fishes led to a rise of comparative endocrinology to increase our understanding of evolution and of functions for physiological regulation. In general, the evolution from single cell to multicellular organisms caused differentiation of physiological actions and communication within severe parts of the organismic body concerning metabolism and development creating three communication networks: (1) The immune system for recognition of self and non-self (e.g., pathogens), (2) the nervous system for fast uptake and response of external stimuli/information, and (3) the endocrine system for regulation of long lasting body homeostasis by hormones affecting reproduction, metabolism, osmomineral regulation, stress, development (metamorphosis), color change, and behavior. What all three communication networks have in common is that they communicate via chemical messengers (immunomodulators, neurotransmitters, hormones) and thus they often interact with each other. From an evolutionary point of view, the immune system is the most ancient one and the endocrine system evolved in parallel to the nervous system already in the phylum *Coelenterata* (Kloas and Lutz 2006). Because comparative endocrinology also includes invertebrates,

[1] Leibniz-Institute of Freshwater Ecology and Inland Fisheries, Department of Ecophysiology and Aquaculture, Müggelseedamm 310, 12587 Berlin, Germany & Humboldt University, Faculty of Life Sciences, Department of Endocrinology, Institute of Biology and Albrecht-Daniel-Thaer-Institute.
[2] Humboldt University, Faculty of Life Sciences, Unit of Biology and Ecology of Fishes, Philippstr. 13, Haus 16, D-10115 Berlin, Germany.
Email: werner.kloas@igb-berlin.de
* Corresponding author: frank.kirschbaum@staff.hu-berlin.de

with most of them having no closed blood circulation system, the classical definition of hormones had to be modified. Hormones are chemical messengers, synthesized in specific cells or glands and released internally in order to cause physiological reactions in distant cells or organs at very low concentrations. The biological actions are committed according to the principle of key and key hole by specific receptors of target cells recognizing and binding the hormone to elicit a cellular response. Thus, not the hormone itself but the specificity of the hormone receptor of a target cell is responsible for any endocrine action causing physiological changes. The binding of a hormone to its receptor, in turn, leads to specific cellular responses via a receptor mediated activation or inactivation of signal transduction pathways ranging from short lasting metabolic to long lasting genomic responses. The biochemical nature of hormones refers also to their mode of action. Lipophilic hormones such as steroids and thyroid hormones act mainly via genomic effects whereas hydrophilic ones such as catecholamines and peptide hormones affect target cells via membrane bound receptors affecting physiological cellular activities. Recently, the number of hormones detected and identified is exponentially increasing because the application of modern molecular genetic techniques reveals, especially concerning peptide hormones, additional ones in a fast manner but their functions are mostly not known so far (Kloas and Lutz 2006, Takei and Loretz 2006, Norris and Carr 2013).

The most advanced development of endocrine systems seems to be evolved in vertebrates already in the various classes of fishes by closed interactions of nervous and endocrine systems, leading to the neuroendocrine regulation via a strict, hierarchy regulating endocrine effects of hormones produced in the classical peripheral endocrine glands. In the central nervous system (CNS), the highest hierarchic level is the hypothalamus being a kind of main computer to get all inputs from the nervous system and in turn triggering the pituitary by releasing or inhibiting hormones via the hypophyseal portal vein system. The pituitary releases tropic hormones into the blood circulation to stimulate the classical endocrine glands: thyroid gland, gonads (ovary and testis), and adrenal gland, or the liver being a source of hormones (insulin like growth factor-I) as well as a target organ. The classical endocrine glands in turn secrete their hormones into the blood circulation to affect their target organs via binding to their specific receptors to induce physiological effects regulating body homeostasis. At each hierarchic level like the hypothalamus, the pituitary, the endorine gland and finally the target organ further endogenous and exogenous factors can influence the regulation at these different levels. However, the most known regulatory inputs on these classical endocrine systems are the so-called negative feedback mechanisms meaning that the hormones of the peripheral endocrine glands themselves can decrease the stimulatory actions of the hypothalamus and the pituitary. Beside that strong neuroendocrine regulation of peripheral glands, there are independent endocrine systems such as: pineal gland, parathyroid gland, ultimobranchial bodies, endocrine pancreas, gastrointestinal tract, blood vessels, corpuscles of Stannius, thymus, kidney, and heart demonstrating, at least in part, the occurrence of a dispersed endocrine system by hormone synthesizing cells or cell groups being complementary to the classical glandular endocrine systems. Despite the classical endocrine systems characterized by peripheral endocrine glands which are present throughout all classes of vertebrates in a homologous way also having similar neuroendocrine regulation mechanisms, the appearance of endocrine systems varies in a wide range of morphological structures differing drastically even within the same class of vertebrates, for instance in teleosts (for review Norris and Carr 2013).

In this chapter, we restrict our description only on the endocrine systems of teleosts representing the vertebrate class Osteichthyes. In teleosts, additional special endocrine organs have been exclusively identified such as corpuscles of Stannius, the urophysis or caudal neurosecretory organ, and the pseudobranch whereas any morphologically structured parathyroid organ is missing. A multitude of hormones regulate the complex interactions of the many piscine organ systems and their interaction with the environment (reviewed by Takei and Loretz 2006). This diversity is reflected in the morphological appearance and distribution of the various piscine endocrine systems (see Fig. 1).

Neuroendocrine Control and Classical Endocrine Glands in Teleosts

The task of the endocrine systems is to maintain the homeostasis within an individual to cope with the environment including adaptations and to fulfill physiological requests for survival and reproduction

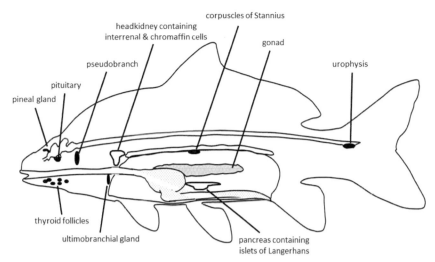

Fig. 1. Schematic presentation of endocrine systems in teleosts. The histological appearance and distribution of endocrine organs in teleosts can vary greatly but has common features as ascribed here.

of the species. In teleosts, the classical endocrine systems with peripheral endocrine glands (Fig. 1) are mainly under neuroendocrine control (Fig. 2) where external and internal stimuli affect the CNS to influence the highest hierarchic level of endocrine systems, the hypothalamus, which in turn regulates the adenohypophysis of the pituitary as next level to secrete tropic hormones that in turn stimulate the peripheral endocrine glands to produce and secrete their hormones to cause physiological effects for homeostasis maintenance in target organs.

Hypothalamus

In teleosts, as in all vertebrate classes, the hypothalamus as a part of the diencephalon of the CNS serves as a kind of main computer to assess all inputs by external and internal stimuli in order to regulate homeostasis at the highest responding level by secreting releasing or inhibiting hormones to regulate the secretion of tropic hormones by the *pars distalis* and *pars intermedia* of the anterior pituitary (adenohypophysis).

Pituitary

The pituitary is the central regulatory endocrine system and is ontogenetically derived from two different parts. The neurohypophysis or posterior pituitary, also named *pars nervosa* (Figs. 3A, B, D), is derived from neuroectoderm and represents a part of the diencephalic hypothalamus, whereas the anterior pituitary or adenohypophysis (Figs. 3A–D) originates from the ectodermal dorsal nasopharhyngeal epithelium, the so-called Rathke's pouch, and contacts the hypothalamus during early ontogeny and evolves into the adenohypophysis with the *pars distalis* and the *pars intermedia* connecting to the neurohypophyseal *pars nervosa*.

The neurosecretory cells of the hypothalamus send their hormones via axons to the neurohypophysis (*pars nervosa*) or via the hypophyseal portal vein system to the adenohypophysis. These hormones either stimulate hormone producing cells in the adenohypophysis or act as release-inhibiting factors. The most important hormones of the hypothalamus are: (1) Growth hormone releasing hormone (GRH) to stimulate in *pars distalis* the secretion of growth hormone (GH) and the counteracting release inhibiting hormone somatostatin (SS). GH has, as main targets, the liver to produce insulin like growth factor-I but also the gills to affect osmoregulation. (2) Thyrothropin-releasing hormone (TRH) or thyroliberin acts also on cells of the *pars distalis* of the adenohypophysis to secrete thyrotropin or thyroid stimulating hormone (TSH) to regulate the activity of the thyroid gland. (3) Gonadotropin releasing hormone (GnRH) to increase

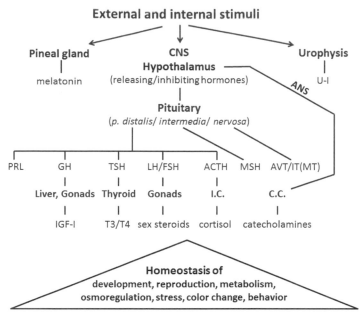

Fig. 2. Endocrine regulation in teleosts. Schematic representation of the classical endocrine organs and their secretagogues together with the hierarchic neuroendocrine control. A more detailed description of the endocrine regulation in teleosts, especially concerning hypothalamic releasing and inhibiting hormones, is given in the text of this chapter. Abbreviations: ACTH—adrenocorticotropes hormone; ANS—autonomous nervous system; AVT—arginine vasotocin; C.C.—chromaffin cells; CNS—central nervous system; FSH—follicle stimulating hormone; GH—growth hormone; I.C.—interrenal cells; IGF-I—insulin like growth factor-I; IT—isotocin; LH—luteinizing hormone; MSH—melanocyte stimulating hormone; MT— mesotocin; T3—triiodothyronine; T4—tetraiodothyronine or thyroxine; TSH—thyroid stimulating hormone; PRL— prolactin; U-I—urotensin-I.

secretion of the gonadotropins, luteinizing hormone (LH) and follicle stimulating hormone (FSH), by *pars distalis* cells to affect in gonads gametogenesis and steroidogenesis of the sex steroids, gestagens, androgens, and estrogens. (4) Corticotropin-releasing-hormone (CRH) and urotensin I (U-I) act on *pars distalis* cells to secrete the adrenocorticotropic-hormone (ACTH) to stimulate in turn the interrenal tissue to produce cortisol. (5) Secretion of prolactin (PRL) by *pars distalis* seems to be regulated antagonistically by hypothalamically derived prolactin releasing hormone (PRH) and prolactin release inhibiting hormone (PRIH). However, in teleosts a clear identification of specific PRH and PRIH is still pending and PRL release is due to multifactorial regulation because PRL also affects manifold functions concerning osmoregulation and reproduction varying greatly among various species. (6) The second part of the adenohypophysis, *pars intermedia*, is known to release melanocyte stimulating hormone (MSH) to affect in many fishes color change and further physiological effects related to photoperiod. MSH secretion is thought to be mainly under inhibitory hypothalamic control of a melanotropin release inhibiting hormone (MRIH) such as dopamine but also a melanotropine releasing hormone (MRH) is discussed as well as neural stimulation. (7) In the hypothalamus two nonapeptides, arginine vasotocin (AVT) and one oxytocin-like hormone, isotocin or mesotocin, is produced in teleosts. The axons of the nonapeptide synthesizing hypothalamic cells project to the neurohypophysis or *pars nervosa* from where the nonapeptides are released into blood circulation to affect mainly osmoregulation but also providing further physiological effects.

Pineal Gland

The pinealocytes of the pineal organ or epiphysis produce the hormone melatonin. The hormone production is regulated by illumination of the pineal gland. The pineal gland is a further neuroendocrine organ derived from the dorsal part of the diecephalon as an outgrowth and extends in general over the forebrain

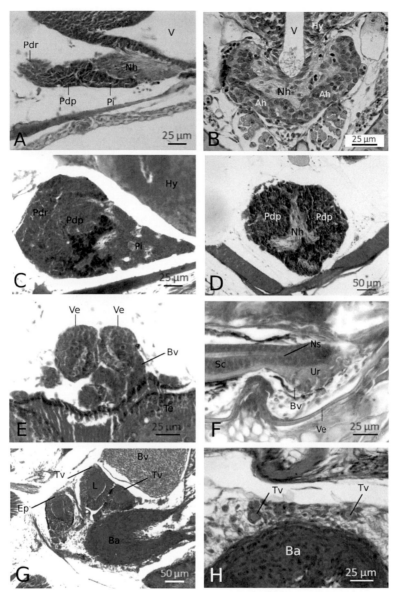

Fig. 3. Endocrine systems connected to the central nervous system (**A–F**) and follicles of the thyroid gland (**G–H**). **A.** Sagittal section through the hypophysis of the mormyrid fish *Pollimyrus isidori* depicting the neurohypophysis (Nh) and the *pars intermedia* (Pi) and the proximal (Pdp) and the rostral part (Pdr) of the *pars distalis* of the adenohypophysis. **B.** Cross section through the hypothalamus-hypophyseal system of the Congo tetra *Phenacogrammus interruptus* showing the neurohypophysis (Nh) and its connection to the hypothalamus (Hy) and the surrounding part of the adenohypophysis (Ah). **C, D.** Sagittal (**C**) and cross section (**D**) through the hypophysis of the glass knifefish *Eigenmannia virescens*. **C.** The three sections of the adenohypophysis, *pars intermedia* (Pi) and the proximal (Pdp) and the rostral part (Pdr) of the *pars distalis* are seen. **D.** A central part of the neurohypophysis (Nh) surrounded by the proximal (Pdp) part of the *pars distalis* of the adenohypophysis. **E.** Cross section through the pineal organ of *Eigenmannia virescens*. Two end vesicles (Ve) are seen to be comprised of peripheral red stained cells and more centrally by darkly stained cells. **F.** Sagittal section through the urophysis (Ur) of the rainbow fish *Melanotaenia herbertaxelrodi*. Note the many blood vessels (Bv) in the urophysis and red stained cells in the spinal cord, probably neurosecretory cells (Ns). **G, H.** Cross sections of the thyroid follicles (Tv) of *Eigenmannia virescens* (**G**) and the banded knifefish *Gymnotus carapo* (**H**). **G.** The lumen (L) of the follicle is surrounded by a thick epithelium (Ep) indicative for active epitheliar cells. **H.** Follicle surrounded by a thin epithelium indicative for non-active epitheliar cells. Ba, bulbus arteriosus; Bv, blood vessel; Hy, hypothalamus; Sc, spinal cord; V, ventricle; Ve, vertebra.

Color version at the end of the book

(Harder 1975). It is composed of a pineal stalk connected with its lumen to the third ventricle and a distal vesicle (Fig. 3E). In some species, e.g., the trout, there may be many vesicles and the pineal gland is quite voluminous (Hibiya 1982). The pineal vesicles are composed of an outer layer of cells with nerve cell character, supporting cells and cells with photoreceptor characteristics (Ekström and Meissl 1997) which extend into the lumen of the vesicle (Omura and Oguri 1969, Fenwick 1970, Oguri and Omura 1973). Its main secretagogue is melatonin derived from serotonin as a precursor and the release of melatonin is strictly depending on light perception because the pineal gland in fishes is extremely light sensitive and even lower light intensities such as moon light are perceived through the scull. Illumination inhibits melatonin secretion whereas darkness stimulates melatonin secretion. Melatonin is extremely important for triggering circadian but also seasonal rhythmicity and acts on circadian locomotor activity, diel thermal preference, immune system and growth and furthermore on change in coloration influenced by photoperiod and temperature (reviewed by Takei and Loretz 2006).

Urophysis

It has to be noted that fishes exclusively possess, in comparison to other vertebrates, an accessory neurosecretory organ that is located caudally in the tail region, the urophysis (Fig. 3F). The structure of the caudal urophysis is similar to the structure of the neurohypophysis. The organ is found at the caudal end of the spinal cord. In the spinal cord, there is a population of neurosecretory cells (Dahlgren cells) which send their unmyelinated axons ventral into a neurohemal organ, the urophysis (Fig. 3F) comprised of many capillary blood vessels and nerve fibers (Fridberg and Bern 1968, Bern 1969). There is a considerable variation in the structure of the caudal neurosecretory system in teleosts (Kobayashi et al. 1986). The caudal neurosecretory system produces two kinds of urotensins (U-I and U-II) (reviewed by Takei and Loretz 2006) which affect smooth muscle activity, hypophyseal secretion of ACTH and ion transport processes related to osmoregulation (Bern et al. 1985, Rivas et al. 1986, Lederis et al. 1994, Winter et al. 2000). In addition, the organ shows a high concentration of acetylcholine (Bern et al. 1985, Conlon et al. 1996).

Thyroid Gland

The thyroid gland follicles are comprised of a simple epithelium and a central lumen containing a colloid. These follicles ontogenetically derive endodermally from the epithelium of the gill clefts and the bottom of the oral cavity. They aggregate along the *aorta ventralis* and are sometimes also found in the head kidney, the pericard and even in the eyes (Spannhof 1995) or in the liver (Norris and Carr 2013). A thick epithelium is indicative of active epithelial cells (Fig. 3G), whereas a thin epithelium indicates inactive cells (Fig. 3H). The thyroid is the only endocrine gland in vertebrates where its secretory products, the thyroid hormones (TH), are stored extracellularly within the follicular lumen. The epithelial cells produce the glycoprotein thyroglobulin (Tg) incorporating, several times, the amino acid tyrosine. The enzyme thyroid peroxidase at the basolateral site of the follicular epithelium iodinates the tyrosine of Tg by introducing 3 or 4 iodine residues leading to triiodothyronine (T3) and tetraiodothyronine/thyroxine (T4), respectively. Tg is stored in the central lumen of the follicles. The pituitary hormone TSH induces the enzymatical transformation of Tg leading to the release of T4 and T3 that are subsequently secreted into the circulation. Circulating T4 is furthermore converted in the peripheral tissues into T3 which is the biologically more active form of TH (reviewed by Norris and Carr 2013). TH in particular promotes development, e.g., growth and differentiation, together with other hormones and plays a pivotal role in any metamorphic processes such as smoltification of salmon or metamorphosis of flatfish (Power et al. 2001). In addition, TH is known to have permissive actions in cooperation with other hormones important for reproduction, growth, osmoregulation, differentiation and metabolism.

Ultimobranchial Body

The ultimobranchial body is comprised of a few follicles together with a loose arrangement of cells (Fig. 4A). It is situated below the esophagus and above the pericard. It is histologically difficult to identify

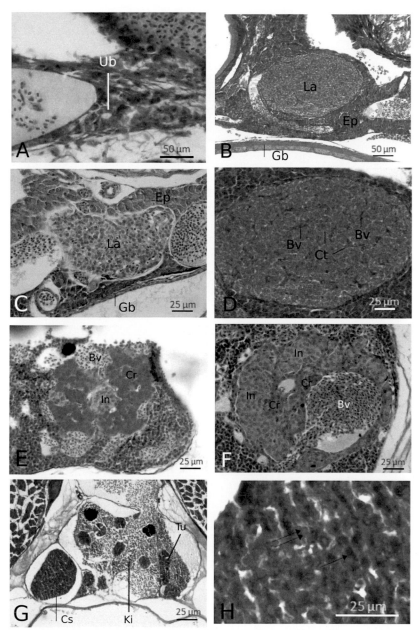

Fig. 4. Ultimobranchial body (**A**), islets of Langerhans (**B–D**), interrenal and chromaffin tissues (**E, F**), and corpuscule of Stannius (**G, H**). **A.** Cross section through a follicle of the ultimobranchial body (Ub) of the glass knifefish *Eigenmannia virescens*. Sagittal sections of the islets of Langerhans of *Eigenmannia virescens* (**B, D**) and the Cichlid *Apistogramma borelli* (**C**). **B.** islet of Langerhans embedded in the exocrine tissue (Ep) of the pancreas. **C.** Detail of the islet of Langerhans. Many blood vessels (Bv) are seen, connective tissue strings (Ct) and many red stained nuclei. **D.** Islet of Langerhans (La) embedded in the exocrine tissue (Ep) of the pancreas. **E, F.** Cross section through the head kidney of *Eigenmannia virescens* (**E**) and a side branch of the cardinal vein in the banded knifefish *Gymnotus carapo* (**F**) showing interrenal and chromaffin cells. **E.** Interrenal (In) and chromaffin cells (Cr) in the head kidney. **F.** Interrenal cells (In) encircling a blood vessel (Bv). **G, H.** Cross sections through the corpuscule of Stannius of *Eigenmannia virescens*. **G.** One of the two corpuscules of Stannius (Cs) in the caudal part of the body kidney (Ki). **H.** Detail of the corpuscule of Stannius. Cells with dark blue stained cytoplasm (arrow) and others with red stained cytoplasm (double arrow) are seen. Bv, blood vessel; Gb, gas bladder epithelium; M, striated muscle of the esophagus; Tu, tubule.

because of its small size. Ontogenetically, it is derived from the last pair of pharyngeal arches and then migrate backwards (Genten et al. 2009). The organ produces calcitonin (CT). Histologically recognisable are granulated cells producing CT and nongranulated supporting cells (U-cells) (Takei and Loretz 2006), probably having a holocrine role in addition to the CT function (Robertson 1986, Sasayama 1999). The physiological role of CT is not quite clear. It affects plasma Ca levels as a hypercalcemic factor, but there are also studies showing a hypocalcemic action (Sasayama et al. 2002). Sexual maturation and spawning is also influenced by CT (Takei and Loretz 2006, Norris and Carr 2013).

Corpuscules of Stannius

The corpuscules of Stannius were discovered by Stannius in 1839 (after Hibiya 1982). In general, it is a paired organ, but in some fish there can be ten or even more discrete glands (Genten et al. 2009). They are found on the lateroventral (Fig. 4G) or laterodorsal border of the kidneys. They differentiate ontogenetically from the pronephric duct or mesonephric tubules (after Hibiya 1982). The corpuscules are divided by connective tissue septa to form lobules or strings. In the corpuscule of Stannius of the glass knifefish *Eigenmannia virescens,* histologically, two cell types can be distinguished: cells with red stained plasma and others with dark blue stained plasma (Fig. 4H), probably two stages in the activity of the glandular cells. The corpuscules of Stannius produce a glycoprotein hormone, named stanniocalcin or hypocalcin, which is involved in Ca homeostasis and counteracts the hypercalcemic action of calcitonin derived from ultimobranchial body (Norris and Carr 2013).

Islets of Langerhans

The pancreas is composed of both, exocrine and endocrine tissues (Figs. 4B and C). The endocrine part is represented by one or several islets (islets of Langerhans); in some fishes, some elements of the endocrine part are found outside the exocrine gland, for instance in most cyprinids as Brockmann bodies located on the gall bladder (Epple and Brinn 1986). The endocrine pancreas apparently has evolved from diffusely arranged endocrine cells of the epithelium of the intestine, which afterwards migrated out of the gastrointestinal system (Epple and Brinn 1986, Youson 2000). Light microscopical investigations reveal three cell types in the Islets of Langerhans (Hibiya 1982): A, B, and D cells. However, in the histological sections of the islets of Langerhans of the banded knifefish *Gymnotus carapo* (Fig. 4C) and the glass knifefish *Eigenmannia virescens* (Fig. 4D), these different cell types cannot be distinguished from each other.

The A cells produce and secrete glucagon, which induces elevation of blood glucose levels. The B cells synthesize insulin which decreases blood glucose levels. Somatostatin (SST) is produced and released by the D cells. This hormone stimulates lipid break down and glycogen mobilization and is an important regulator of metabolism and growth (Sheridan et al. 2000). A fourth cell type, F cells, has also been identified. F cells produce the pancreatic polypeptide (PPP) of yet unknown physiological functions (Norris and Carr 2013).

Interrenal and Chromaffin Tissues

The interrenal and chromaffin tissues do not represent a compact organ in fish; instead, they are intermingled with other tissues (Henderson and Kime 1987). In teleosts, interrenal and chromaffin cells occur in the head kidney (Fig. 4E) and are located around the epithelium of the venous vessels merging into the cardinal veins (Fig. 4F). There is a great variability in the size, shape and appearance of interrenal and chromaffin cells in teleosts (Genten et al. 2009). The interrenal cells, which are of mesodermal origin, are stimulated mainly by hypophyseal ACTH and produce corticosteroids, of which cortisol is the most important hormone. This hormone regulates energy metabolism and controls water and electrolyte balance. The chromaffin cells are of neuroectodermal origin and synthesize catecholamines, of which adrenaline and noradrenaline are the most important hormones. These hormones are released via stimulation of the autonomous nervous system in stress situations and influence metabolism, blood circulation and gill function. In teleosts, as in higher vertebrates, all stress is mediated via the coordinated actions of chromaffin and interrenal cells known as fight-and flight response, being under multifactorial regulation (Kloas et al. 1994, Kloas 1999) and causing a multitude of physiological effects concerning primary, secondary and tertiary responses (Wendelaar Bonga 1997).

Pseudobranch

We have added to this chapter Endocrine Systems a description of the pseudobranch. This organ was first described by Broussonet in 1758. The pseudobranch represents a modified hemibranch, which is derived from the first gill arch and represents, in most teleosts, a bilateral paired organ (Fig. 5A). In some species, it is either located inside the operculum (Hibiya 1982) or it can be completely lacking as in the eel (Wittenberg

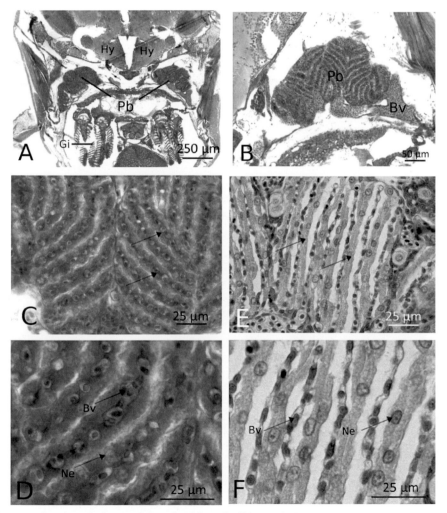

Fig. 5. Cross section through the pseudobranch of the zebrafish *Brachydanio rerio* (**A–C, E**) and the Congo tetra *Phenacogrammus interruptus* (**D, F**). **A.** The paired pseudobranch (Pb) outside the branchial cavity. **B.** Anatomy of one pseudobranch (Pb) with the fused secondary lamellae and the efferent blood supply (Bv). **C.** Many fused, in parallel arranged secondary lamellae (arrows). **D.** Several secondary lamellae and their central capillary blood vessels (Bv) and the two peripheral epithelia with their nuclei (Ne) surrounding the central capillary blood vessels. **E.** Several secondary lamellae (arrows) arranged in parallel. **F.** Many capillary blood vessels (Bv) and the peripheral epithelium with their nuclei (Ne) attached to one side of the string of blood vessels. Gi, gills; Hy, hypothalamus.

and Haedrick 1974). Since its discovery, different functions have been attributed to this organ (Gwyneth and Holliday 1960): (1) It is a salt regulatory organ; (2) It has respiratory function as a supplementary gill; (3) It is an ocular regulator by controlling blood pressure in the eye; (4) It is an endocrine organ. More recent data indicate (reviewed by Bridges et al. 1998) that the pseudobranch is connected to the choroid rete and acts on the supply of oxygen to the retina of the fish eye. Figure 5B shows one of the two pseudobranch organs of the zebra fish with the many fused secondary lamellae and the large supplying blood vessels. Details of the secondary lamellae with the central capillary blood vessels and the surrounding epithelia are shown in Figs. 5C and E. In the Congo tetra *Phenacogrammus interruptus*, the capillary blood vessels are only on one face interconnected with a strip of gill epithelium (Figs. 5D and F). Thus, despite the earlier hypothesis that the pseudobranch might be an endocrine organ in teleosts, nowadays more evidence exists that it is restricted to a physiological rather than an endocrine organ (Mölich et al. 2009).

Final Remark

The chapter about Endocrine Organs is restricted to tissues and organs creating a morphologically distinct appearance to be able to be identified histologically as specific endocrine organs. It is noteworthy that we neglected on purpose the great number of diffusely disseminated endocrine cells of the gastrointestinal tract (see chapter Digestive System), the kidney and the peripheric nervous system, producing a large variety of hormones with various physiological functions as well as the heart as an endocrine organ secreting natriuretic peptides affecting osmoregulation (see chapter Cardiovascular System and Blood) (for review see Takei and Lorentz 2006).

Acknowledgements

We thank Dr. Andrea Ziková for providing Fig. 1 and Prof. Rüdiger Krahe for his offer to use his microscope/digital camera device for taking the photomicrographs.

References Cited

Bayliss, W.M. and E.H. Starling. 1902. The mechanism of pancreatic secretion. J. Physiol. 28(5): 325–353.

Bern, H.A. 1969. Urophysis and caudal neurosecretory system. pp. 399–418. *In*: W.S. Hoar and D.J. Randall (eds.). Fish Physiology II. Academic Press, New York and London.

Bern, H.A., D. Pearson, B.A. Larson and R.S. Nishioka. 1985. Neurohormones from fish tails: the caudal neurosecretory system I. "Urophysiology" and the caudal neurosecretory system of fishes. Horm. Res. 41: 533–552.

Bridges, C.R., M. Berenbrink, R. Müller and W. Waser. 1998. Physiology and biochemistry of the pseudobranch: an unanswered question? Comp. Biochem. Physiol. 119A: 67–77.

Broussonet, P.M.A. 1785. Mémoire pour servir à l'histoire de la respiration des poissons. Mem. Acad. R. Sci. 174–196.

Conlon, J.M. and R.J. Balment. 1996. Synthesis and release of acetylcholin by the isolated perfused trout caudal neurosecretory system. Gen. Comp. Endocrinol. 103: 46–50.

Crookshank, F.G. 1914. Case of (?) insufficiency of endocrinic glands. Proc. Royal Soc. Medicine 7: 69–70.

Ekström, P. and H. Meissl. 1997. The pineal organ of teleost fishes. Rev. Fish. Biol. Fish 7: 199–284.

Epple, A. and J.E. Brinn. 1986. Pacratic islets. pp. 279–317. *In*: P.K.T. Pang and M.P. Schreibmann (eds.). Vertebrate Endocrinology: Fundamental and Biomedical Implications, Vol. 1. Morphological Considerations. Academic Press, San Diego.

Fenwick, J.C. 1970. The pineal organ. pp. 91–108. *In*: W.S. Hoar and D.J. Randall (eds.). Fish Physiology IV. Academic Press, New York and London.

Fridberg, G. and H.A. Bern. 1968. The urophysis and the caudal neurosecretory system of fishes. Biol. Rev. 43: 175–199.

Genten, F., E. Terwinghe and A. Danguy. 2009. Atlas of Fish Histology. Science Publishers, Enfield, Jersey, Plymouth.

Gwyneth, P. and F.G.T. Holliday. 1960. An experimental analysis of the function of the pseudobranch in teleosts. J. Exp. Biol. 37: 344–354.

Harder, W. 1975. Anatomy of Fishes. E. Schweizerbartsche Verlagsbuchhandlung. Nägele und Obermiller, Stuttgart.

Henderson, I.W. and D.E. Kime. 1987. The adrenal cortical steroids. pp. 121–142. *In*: P.K.T. Pang and M.P. Schreibmann (eds.). Vertebrate Endocrinology: Fundamental and Biomedical Implications, Vol. 2, Morphological Considerations. Academic Press, San Diego.

Hibiya, T. (ed.). 1982. An Atlas of Fish Histology. Normal and Pathological Features. Kodansha Ltd., Tokyo. Gustav Fischer Verlag, Stuttgart, New York.

Kloas, W., M. Reinecke and W. Hanke. 1994. Role of the atrial natriuretic peptide (ANP) for adrenal regulation in the teleost fish *Cyprinus carpio*. Am. J. Physiol. 267: R1034–R1042.

Kloas, W. 1999. Stress physiology in fish. pp. 157–160. *In*: E. Roubos, S. Wendelaar Bonga, H. Vaudry and A. De Loof (eds.). Rec. Developm. Comp. Endocrinol. Neurobiol., Shaker Publishing.

Kloas, W. and I. Lutz. 2006. Amphibians as model to study endocrine disrupters. J. Chromat. A 1130: 16–27.

Kobayashi, H., K. Owada, C. Yamada and Y. Okawara. 1986. The caudal neurosecretory system in fishes. pp. 147–174. *In*: P.K.T. Pang and M.P. Schreibmann (eds.). Vertebrate Endocrinology: Fundamental and Biomedical Implications, Vol. 1 Morphological Considerations. Academic Press, San Diego.

Lederis, K., J.N. Fryer, Y. Okawara, C. Schonrock and D. Richter. 1994. Corticotropin-releasing-factors acting on the fish pituitary: experimental and molecular analysis. pp. 67–100. *In*: N.M. Sherwood and C. Hew (eds.). Fish Physiology. Academic Press, San Diego.

Mölich, A., W. Waser and N. Heisler. 2009. The teleost pseudobranch: a role for preconditioning of ocular blood supply? Fish Physiol. Biochem. 35(2): 273–86.

Norris, D. and J. Carr. 2013. Vertebrate Endocrinology: Fifth Edition, Academic Press.

Oguri, M. and Y. Omura. 1973. Ultrastructure and functional significance of the pineal organ of teleosts. pp. 412–434. *In*: C.C. Thomas (ed.). Responses of Fish to Environmental Changes. Springfield, Illinois.

Oliver, G. and E.A. Schaefer. 1895. Recording of blood pressure of anaesthetized dog and response to suprarenal gland extract. J. Physiol. 18: 246.

Omura, Y. and M. Oguri. 1969. Histological studies on the pineal organ of 15 species of teleosts. Bull. Japan Soc. Sci. Fish. 35: 991–1000.

Pende, N. 1909. Sistema nervoso sympatico e glandole a secrezione interna, distrofie endocrino-simpatiche. Napoli, ed. G. Civelli, 23 p.

Power, D.M., L. Llewellyn, M. Faustino, M.A. Nowell, B.T. Björnsson, I.E. Einarsdottir, A.V.-M. Canario and G.E. Sweeney. 2001. Thyroid hormones in growth and development of fish. Comp. Biochem. Physiol. C 130: 447–459.

Rivas, R.J., R.S. Nishioka and H.A. Bern. 1986. *In vitro* effects of somatostatin and urotensin II on prolactin and growth hormone secretion in Tilapia, *Oreochromis mossambicus*. Gen. Comp. Endocrinol. 63: 245–251.

Robertson, D.R. 1986. The ultimobranchial body. pp. 235–259. *In*: P.K.T. Pang and M.P. Schreibmann (eds.). Vertebrate Endocrinology: Fundamental and Biomedical Implications, Vol. 1 Morphological Considerations. Academic Press, San Diego.

Sasayama, Y. 1999. Hormonal control of Ca homeostasis in lower vertebrates: considering the evolution. Zool. Sci. 16: 857–869.

Sasayama, Y., Y. Takei, S. Hasegawa and D. Zuzuki. 2002. Direct raises in blood Ca levels by infusing a high-calcium solution in the blood stream accelerate the secretion of calcitonin from the ultimobranchial gland in eels. Zool. Sci. 19: 1039–1049.

Sheridan, M.A., J.D. Kittilson and B.J. Slagter. 2000. Structure-function relationships of the signalling system for the somatostatin peptide hormone family. Am. Zool. 40: 269–286.

Spannhof, L. 1995. Einführung in die Fischphysiologie. Verlag Dr. Kovac, Hamburg.

Starling, E.H. 1905. The Croonian Lectures. I. On the chemical correlation of the functions of the body. Lancet. 166: 339–341.

Takei, Y. and C.A. Loretz. 2006. Endocrinology. pp. 271–318. *In*: D.H. Evans and J.B. Claiborne (eds.). The Physiology of Fishes. CRC Taylor & Francis, Boca Raton, London, New York.

Wendelaar Bonga, S.E. 1997. The stress response in fish. Physiol. Rev. 77: 591–625.

Winter, M.J., A. Ashworth, H. Bond, M.J. Brierley, C.A. McCrohan and R.J. Balment. 2000. The caudal neurosecretory system: control and function of a novel neuroendocrine system in fish. Biochem. Cell Biol. 78: 193–2003.

Wittenberg, J.B. and R.L. Haedrick. 1974. The choroid rete mirabile oft he fish eye—II. Distribution and relation to the pseudobranch and to the swim-bladder rete mirabile. Biol. Bull. 146: 137–156.

Youson, J.H. 2000. The agnathan enteropancreatic endocrine system: phylogenetic and ontogenetic histories, structure, and function. Am. Zool. 40: 179–199.

Index

Family Names Index

English Names Index

Species Index

Color Plate Section

Chapter 2

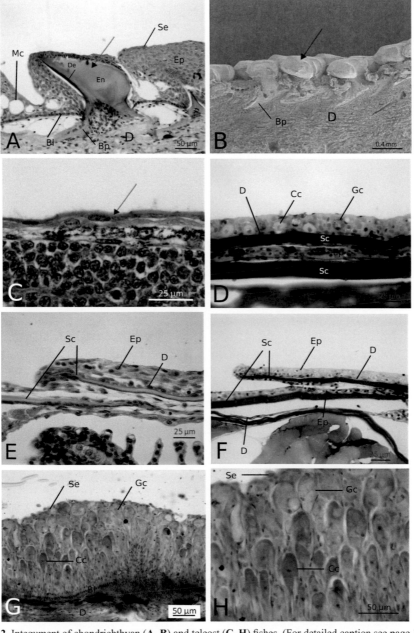

Fig. 2. Integument of chondrichthyan (**A, B**) and teleost (**C–H**) fishes. (For detailed caption see page 17.)

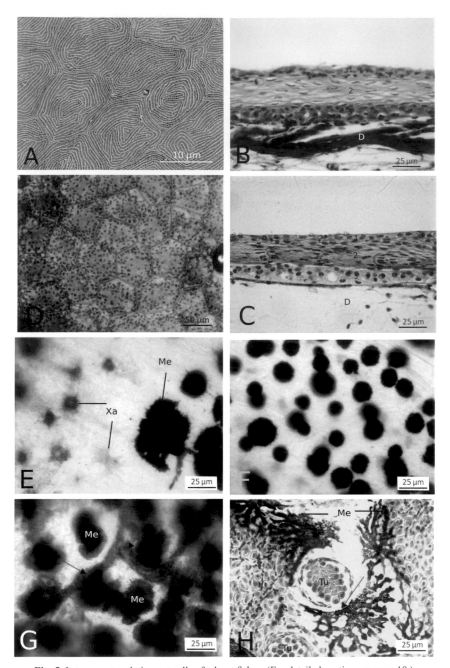

Fig. 5. Integument and pigment cells of teleost fishes. (For detailed caption see page 19.)

Chapter 4

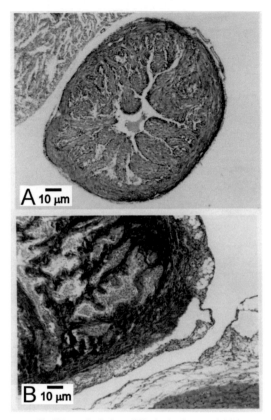

Fig. 4. Comparison of the two types of bulbi arteriosi; they differ in the form and thickness of their trabeculae. (For detailed caption see page 57.)

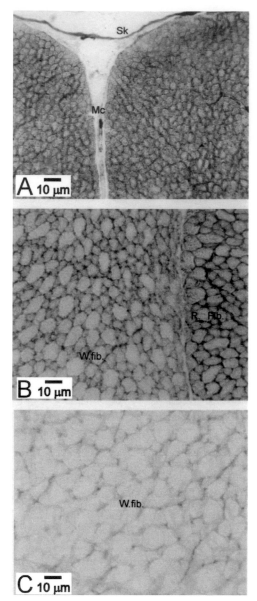

Fig. 9. Cryostat sections of the trunk muscle of the crucian carp (*Carassius aureatus gibelio* Bloch) displaying a high activity of succinate dehydrogenase (SDH). (For detailed caption see page 69.)

Chapter 5

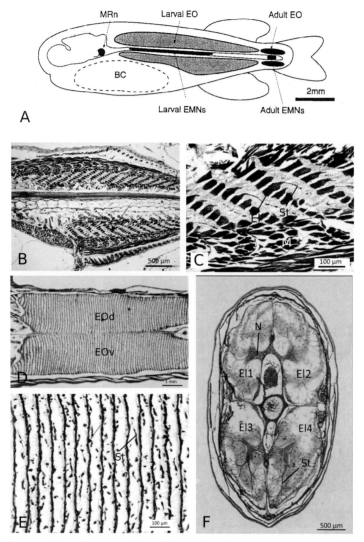

Fig. 4. Topographical, anatomical and histological data of the larval and the adult electric organs of *Pollimyrus isidori*. (For detailed caption see page 80.)

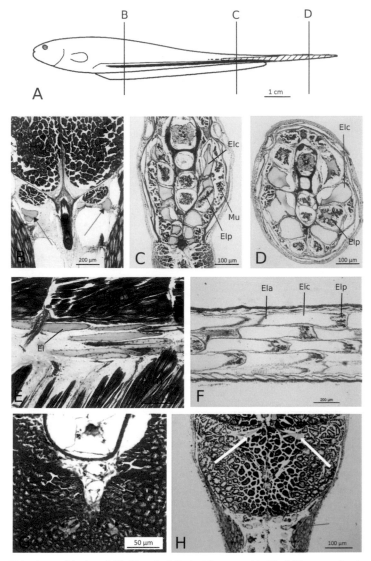

Fig. 6. Anatomy and histology of the larval (**G, H**) and adult electric organs (**A–F**) of *Eigenmannia virescens*. (For detailed caption see page 82.)

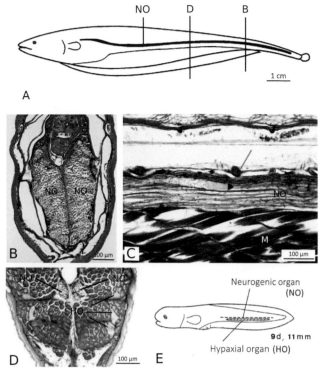

Fig. 7. Anatomy and histology of the neurogenic and myogenic electric organs in *Apteronotus leptorhynchus*. (For detailed caption see page 83.)

Chapter 6

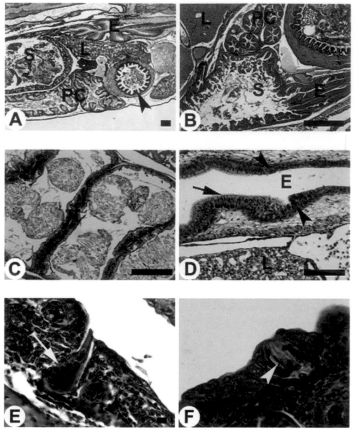

Fig. 1. A–D. Longitudinal sections. (For detailed caption see page 89.)

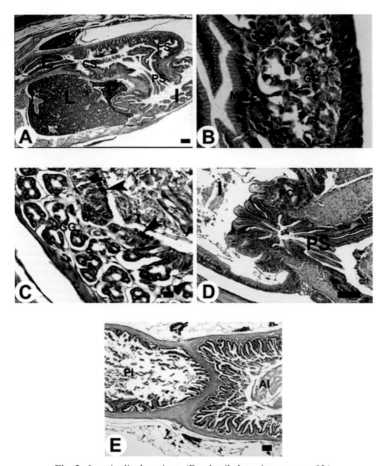

Fig. 2. Longitudinal sections. (For detailed caption see page 92.)

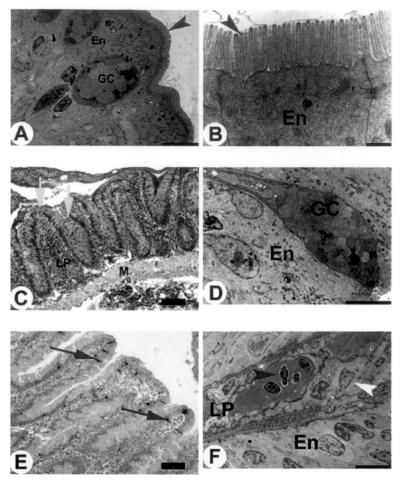

Fig. 3. Cross (**E**) and longitudinal sections (**A–D, F**) of the anterior intestine. (For detailed caption see page 95.)

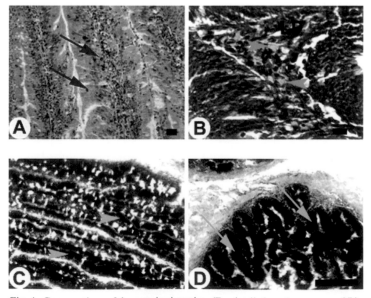

Fig. 4. Cross sections of the anterior intestine. (For detailed caption see page 97.)

Chapter 8

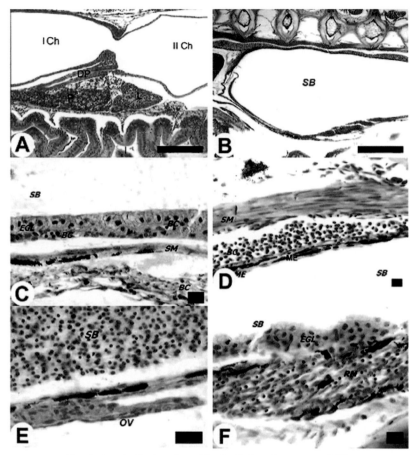

Fig. 1. Longitudinal sections. (For detailed caption see page 118.)

Chapter 9

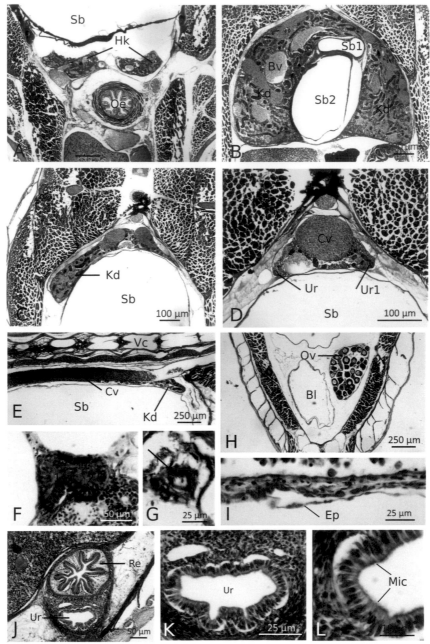

Fig. 1. Anatomy and histology of the kidney and the appropriate ducts in *Eigenmannia virescens* (Sternopygidae, Gymnotiformes) shown in cross sections and a sagittal section (E). Azan stain. (For detailed caption see page 122.)

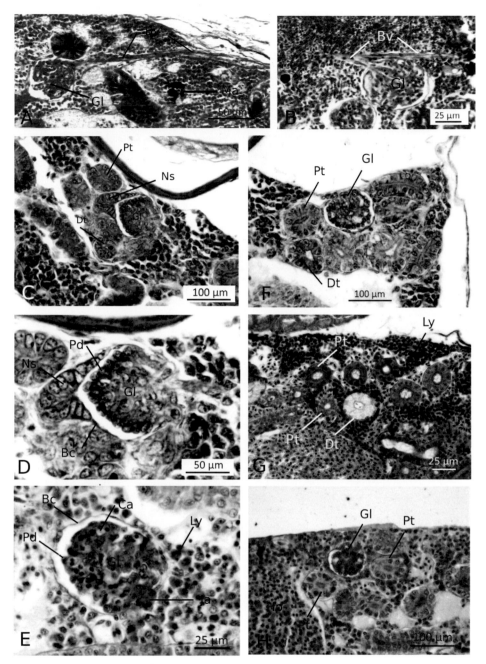

Fig. 2. Details of the nephrons of various freshwater teleosts. Azan stain. (For detailed caption see page 124.)

Chapter 10

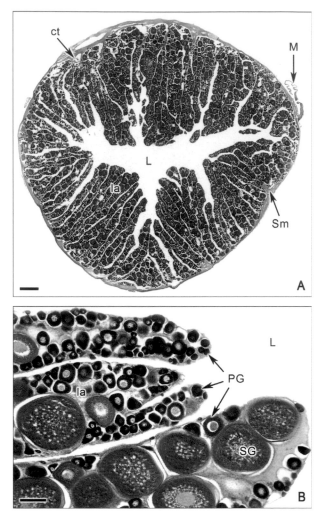

Fig. 1. Structure of the ovary. Ovary of Common Snook *Centropomus undecimalis*. (For detailed caption see page 128.)

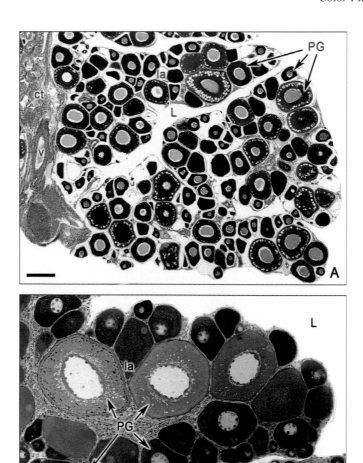

Fig. 2. Structure of the ovarian lamellae. Ovaries of Red Drum *Sciaenops ocellatus*. (For detailed caption see page 129.)

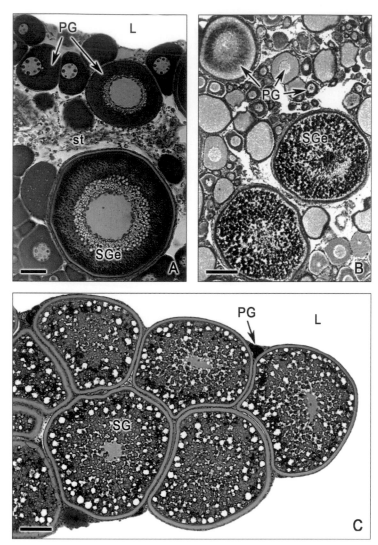

Fig. 3. Structure of the ovarian lamellae. The lumen (L) surrounds the lamellae. (For detailed caption see page 130.)

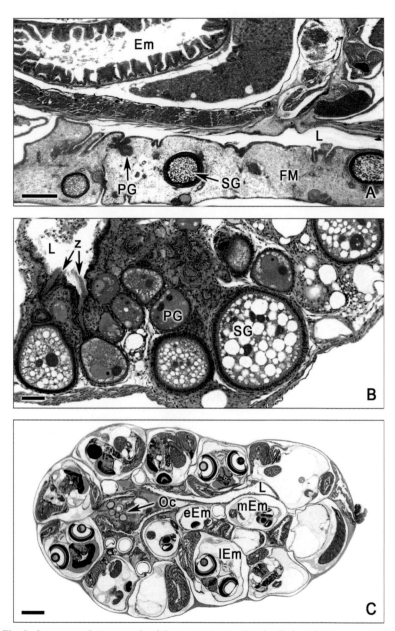

Fig. 5. Structure of the ovary in viviparous teleosts. (For detailed caption see page 132.)

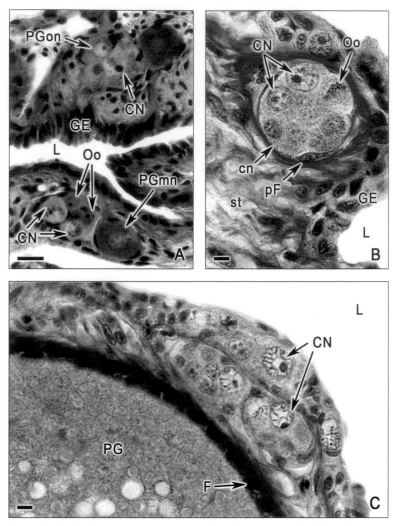

Fig. 7. Oogenesis. Oogonia and oocytes in Primary Growth stage. (For detailed caption see page 135.)

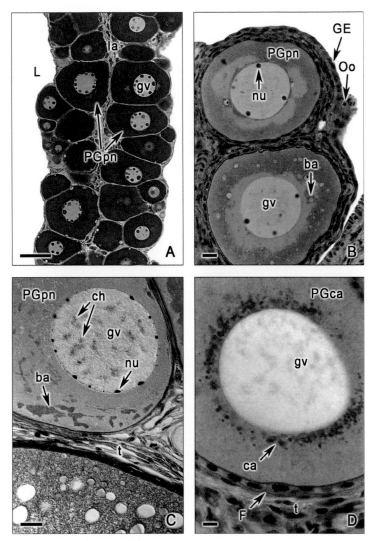

Fig. 8. Oogenesis. Oocytes in Primary Growth stage. (For detailed caption see page 137.)

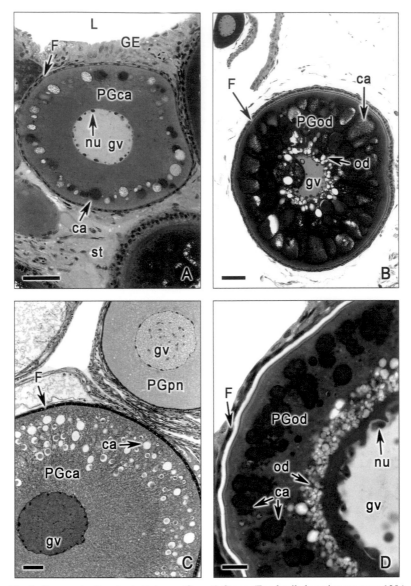

Fig. 9. Oogenesis. Oocytes in Primary Growth Stage. (For detailed caption see page 139.)

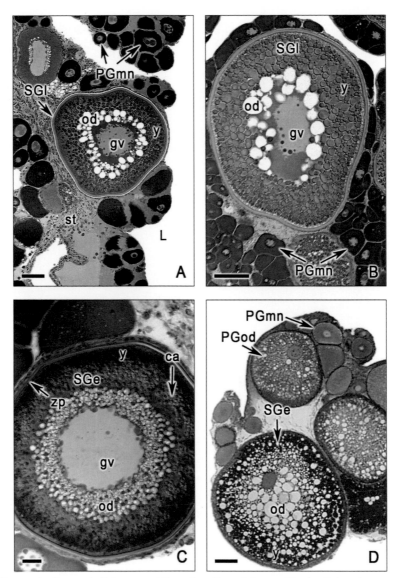

Fig. 10. Oogenesis. Oocytes in Primary and Secondary Growth Stages. (For detailed caption see page 140.)

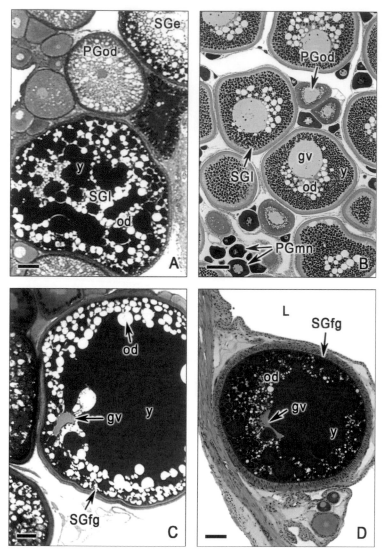

Fig. 11. Oogenesis. Oocytes in Secondary Growth Stage. (For detailed caption see page 141.)

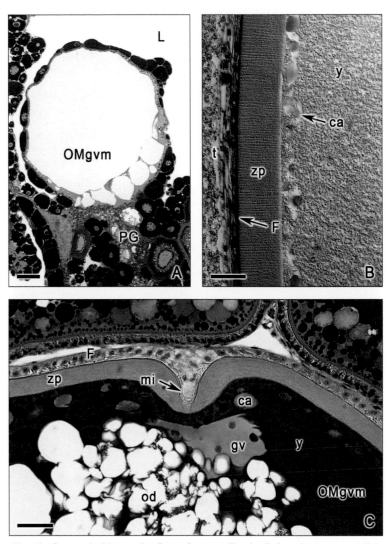

Fig. 12. Oogenesis. Maturation Stage Oocytes. (For detailed caption see page 142.)

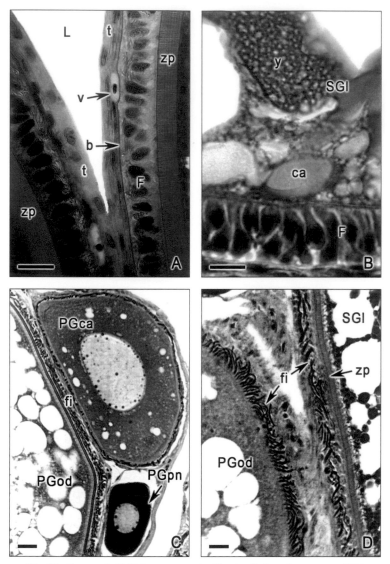

Fig. 13. Oogenesis. Follicle components. (For detailed caption see page 143.)

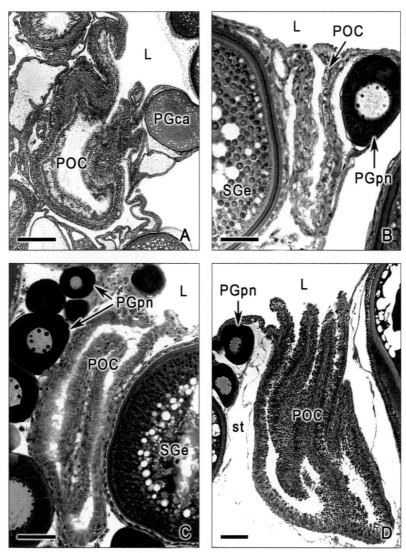

Fig. 14. Postovulatory follicle complex (POC). (For detailed caption see page 145.)

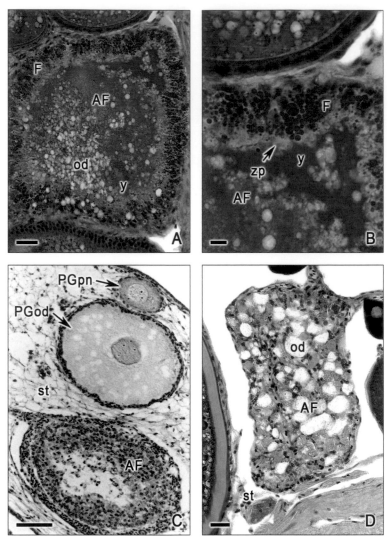

Fig. 15. Atretic follicles. (For detailed caption see page 146.)

Chapter 12

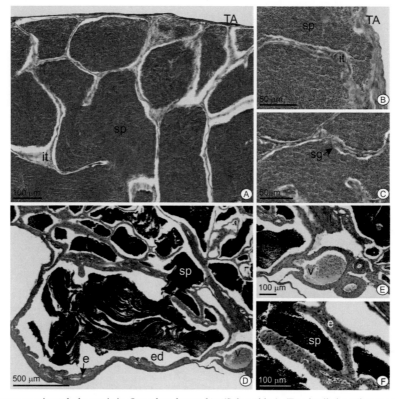

Fig. 1. Anastomosing tubular testis in *Oncorhynchus mykiss* (Salmonidae). (For detailed caption see page 179.)

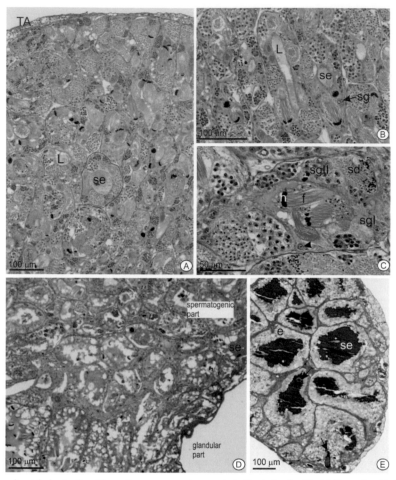

Fig. 2. Anastomosing tubular testis with continuous spermatogenesis in *Pantodon buchholzi* (Pantodontidae). (For detailed caption see page 180.)

Chapter 13

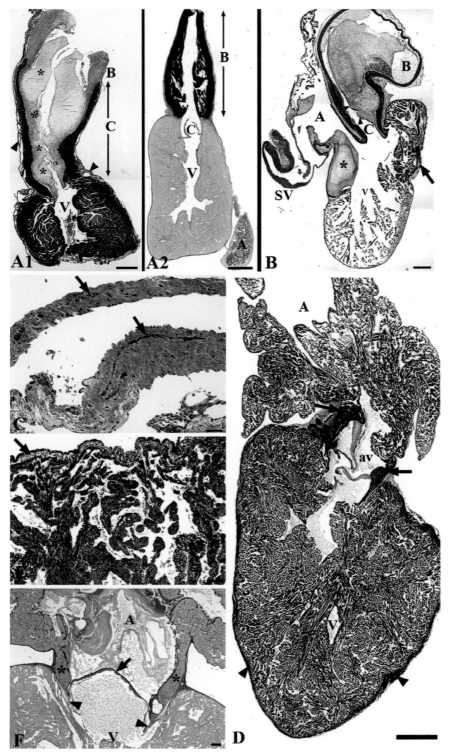

Fig. 2. Anatomy and histology of the heart of various fish groups. (For detailed caption see page 211.)

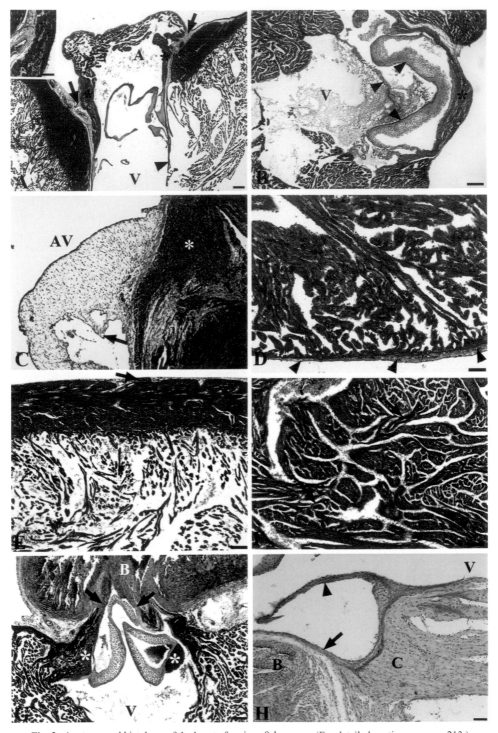

Fig. 3. Anatomy and histology of the heart of various fish groups. (For detailed caption see page 213.)

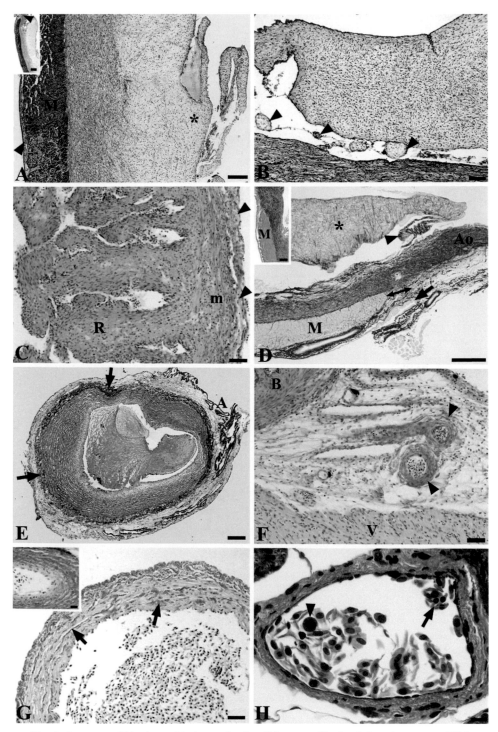

Fig. 6. Anatomy and histology of the heart of various fish groups. (For detailed caption see page 221.)

Chapter 14

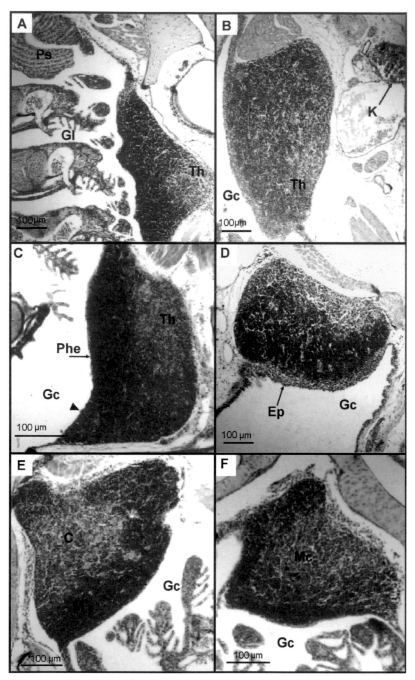

Fig. 1. Anatomy and histology of the thymus of various fish species. (For detailed caption see page 233.)

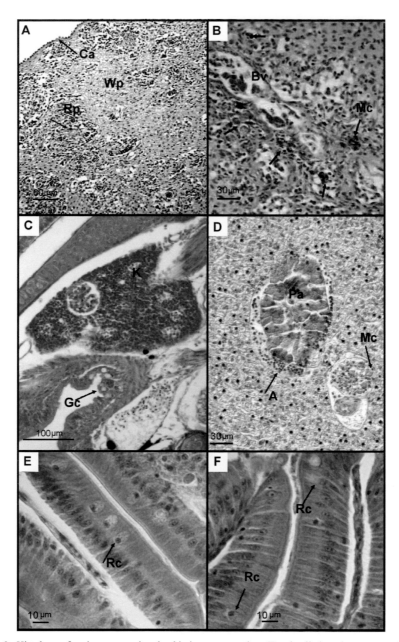

Fig. 2. Histology of various organs involved in immune reaction. (For detailed caption see page 236.)

Chapter 15

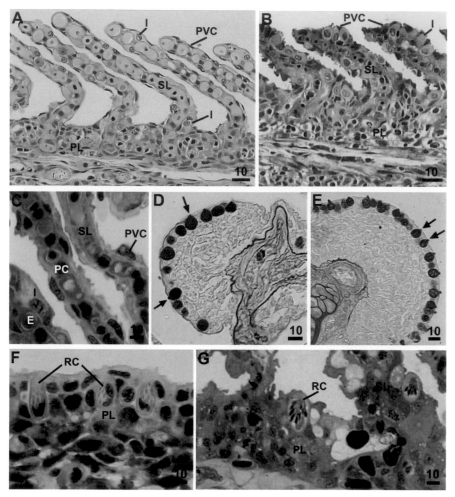

Fig. 2. Primary and secondary lamellae of curimbatá, *Prochilodus lineatus*, a water-breathing fish. (For detailed caption see page 250.)

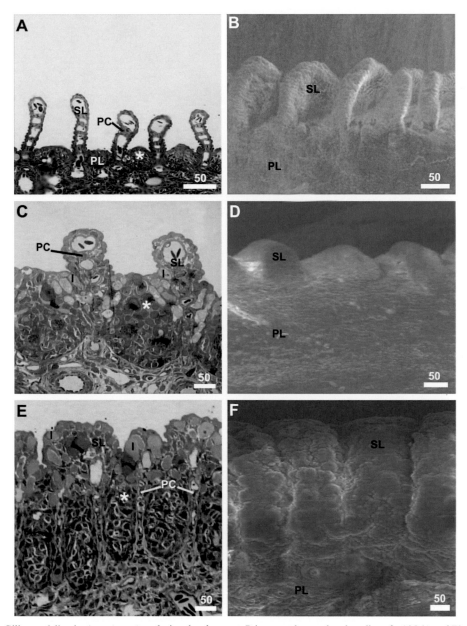

Fig. 7. Gill remodeling in *Arapaima gigas* during development. Primary and secondary lamellae of a 105 (**A** and **B**), 575 (**C** and **D**) and 1.343 (**E** and **F**) g-fish. (For detailed caption see page 259.)

Chapter 16

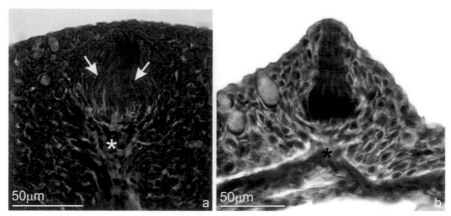

Fig. 2. Vertical section through a taste bud. (For detailed caption see page 270.)

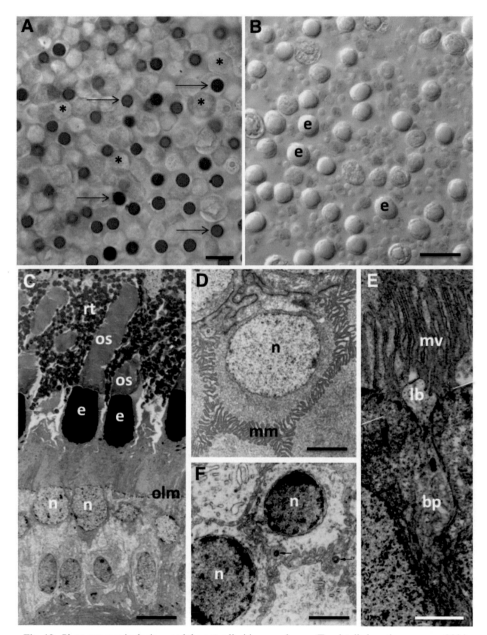

Fig. 19. Photoreceptor inclusions and the outer limiting membrane. (For detailed caption see page 290.)

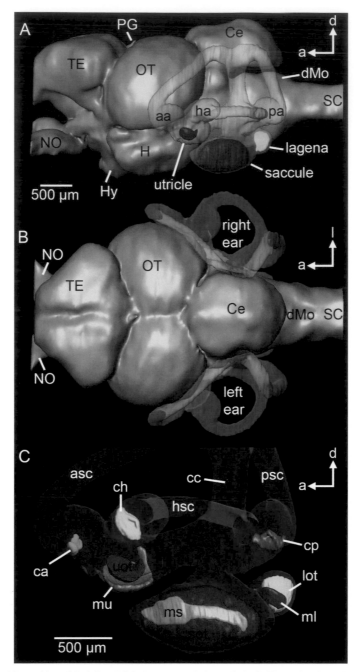

Fig. 23. Overview of the position of the left and right inner ears relative to the brain illustrating the main inner ear components. 3-D reconstructions (based on microCT imaging) of the inner ears and the brain of a teleost fish (*Steatocranus tinanti*, Cichlidae, Cichliformes) are shown in lateral (**A**), dorsal (**B**), and medial (**C**) views. (For detailed caption see page 296.)

Relationship between Otolith and Macula

Tomographic sections:

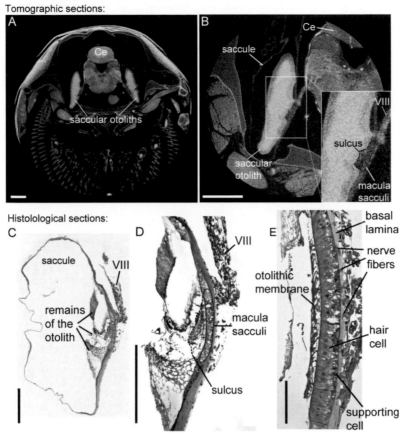

Fig. 25. Transverse μCT (**A, B**) and histological sections (**C–E**) displaying the relationship between otolith, otolithic membrane and macula in the saccule. (For detailed caption see page 298.)

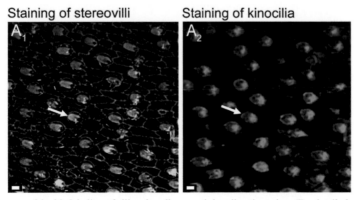

Fig. 27. Confocal images of double labeling of ciliary bundles reveals bundle orientation. (For detailed caption see page 300.)

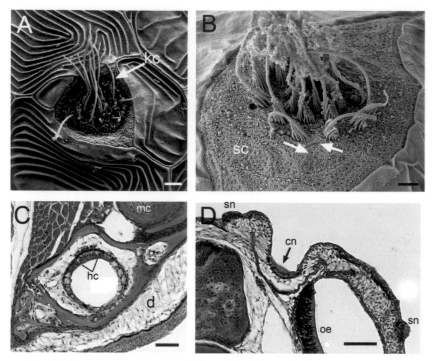

Fig. 28. Neuromast receptor organs in bony fishes. (For detailed caption see page 304.)

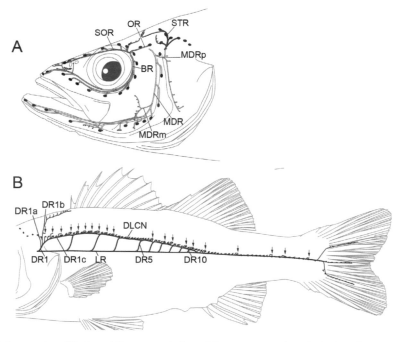

Fig. 30. Innervation of the lateral line system, in *Lateolabrax japonicus.* (For detailed caption see page 306.)

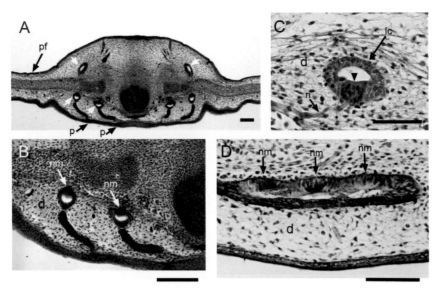

Fig. 34. Neuromasts and lateral line canals in the dermis in embryonic little skate, *Leucoraja erinacea* (65 mm TL, Stage 33). (For detailed caption see page 310.)

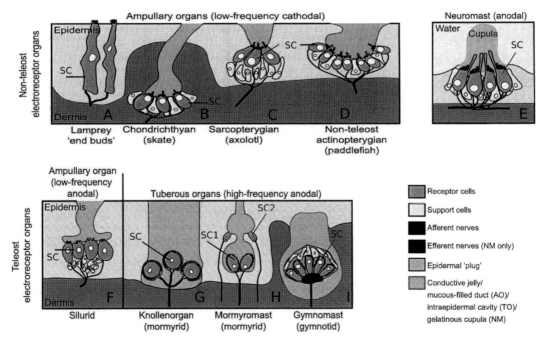

Fig. 35. Schematic drawings (not to scale) comparing the different morphologies of neuromasts and ampullary and tuberous electroreceptors in non-teleost and teleost fishes and an amphibian (axolotl). (For detailed caption see page 312.)

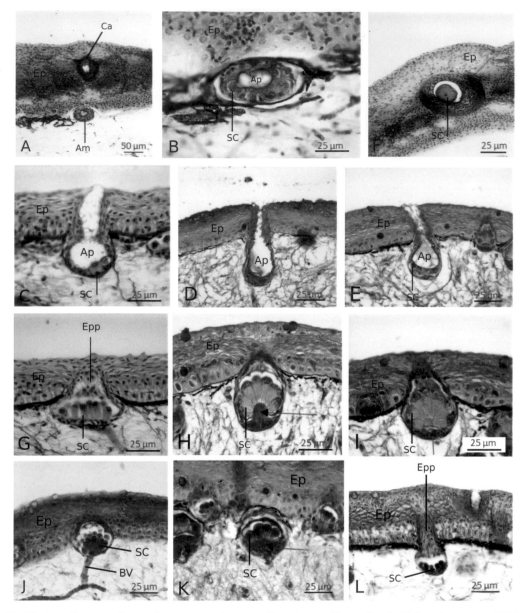

Fig. 38. Ampullary and tuberous electroreceptors in mormyrids and gymnotiforms. (For detailed caption see page 316.)

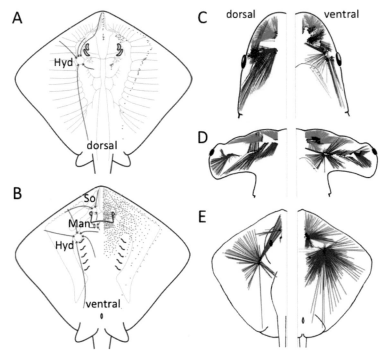

Fig. 40. Distribution of electroreceptors in elasmobranchs. (For detailed caption see page 319.)

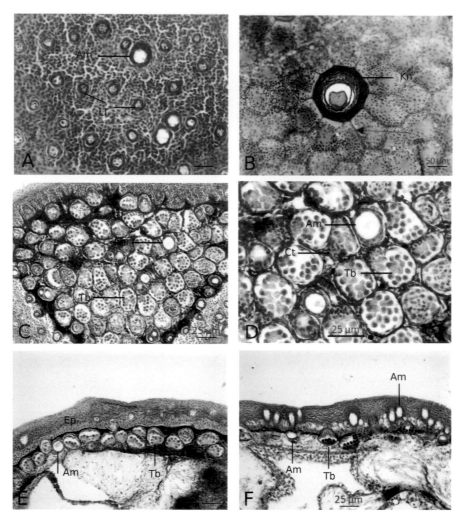

Fig. 43. Distribution of electroreceptors on the head of mormyrid and gymnotiform fishes. (For detailed caption see page 322.)

Chapter 17

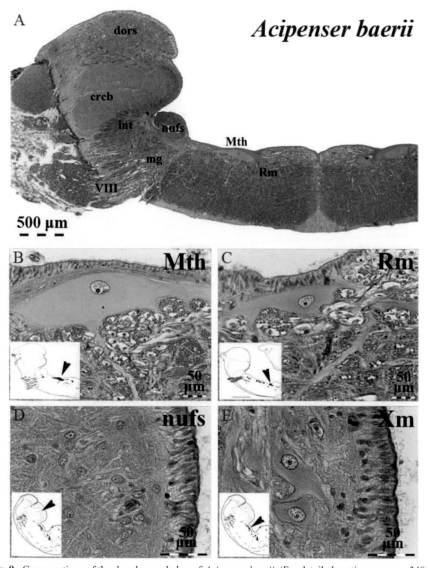

Fig. 9. Cross sections of the rhombencephalon of *Acipenser baerii*. (For detailed caption see page 348.)

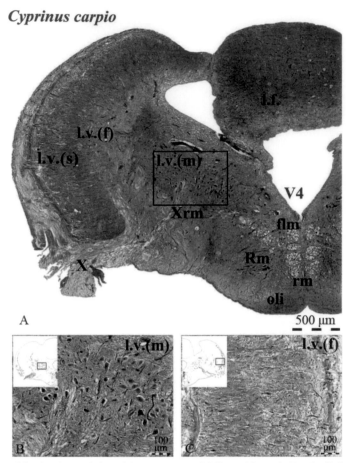

Fig. 22. Cross section of the rhombencephalon of *Cyprinus carpio* (Mallory trichrome). (For detailed caption see page 362.)

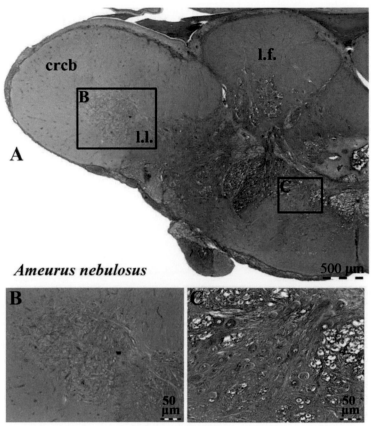

Fig. 23. Cross section of the rhombencephalon of the brown bullhead (*Ameurus nebulosus*) (Mallory trichrome). (For detailed caption see page 364.)

Chapter 18

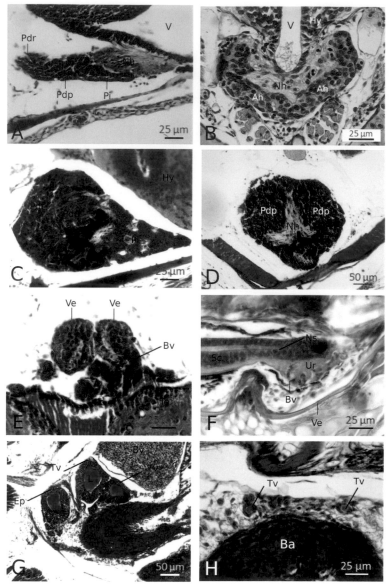

Fig. 3. Endocrine Systems connected to the central nervous system (**A–F**) and follicles of the thyroid gland (**G–H**). (For detailed caption see page 377.)

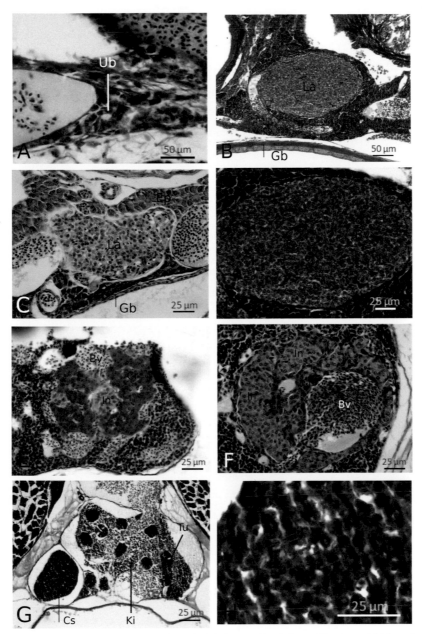

Fig. 4. Ultimobranchial body (**A**), islets of Langerhans (**B–D**), interrenal and chromaffin tissues (**E, F**), and corpuscule of Stannius (**G, H**). (For detailed caption see page 379.)

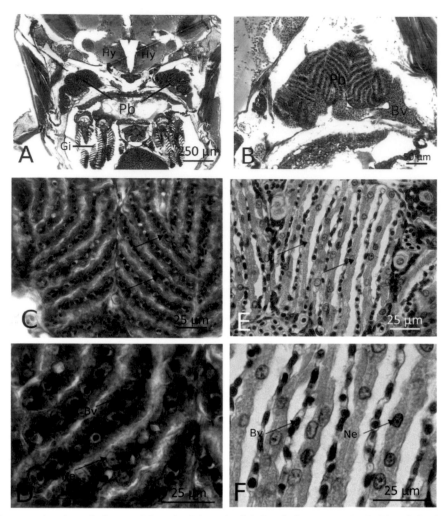

Fig. 5. Cross section through the pseudobranch of the zebrafish *Brachydanio rerio* (**A–C, E**) and the Congo tetra *Phenacogrammus interruptus* (**D, F**). (For detailed caption see page 381.)

Printed and bound by CPI Group (UK) Ltd, Croydon, CR0 4YY

25/10/2024

01779362-0001